Chemical Engineering Fluid Mechanics

Chemical Engineering Fluid Mechanics

Second Edition, Revised and Expanded

Ron Darby

Texas A&M University
College Station, Texas

CRC Press is an imprint of the
Taylor & Francis Group, an **informa** business

Published in 2001 by
CRC Press
Taylor & Francis Group
6000 Broken Sound Parkway NW, Suite 300
Boca Raton, FL 33487-2742

Printed in the United States of America on acid-free paper
15 14 13 12 11 10 9 8 7

International Standard Book Number-10: 0-8247-0444-4 (Hardcover)
International Standard Book Number-13: 978-0-8247-0444-5 (Hardcover)

Library of Congress Cataloging-in-Publication Data

Catalog record is available from the Library of Congress

Visit the Taylor & Francis Web site at
http://www.taylorandfrancis.com

and the CRC Press Web site at
http://www.crcpress.com

Preface

The objectives of this book are twofold: (1) for the student, to show how the fundamental principles underlying the behavior of fluids (with emphasis on one-dimensional macroscopic balances) can be applied in an organized and systematic manner to the solution of practical engineering problems, and (2) for the practicing engineer, to provide a ready reference of current information and basic methods for the analysis of a variety of problems encountered in practical engineering situations.

The scope of coverage includes internal flows of Newtonian and non-Newtonian incompressible fluids, adiabatic and isothermal compressible flows (up to sonic or choking conditions), two-phase (gas–liquid, solid–liquid, and gas–solid) flows, external flows (e.g., drag), and flow in porous media. Applications include dimensional analysis and scale-up, piping systems with fittings for Newtonian and non-Newtonian fluids (for unknown driving force, unknown flow rate, unknown diameter, or most economical diameter), compressible pipe flows up to choked flow, flow measurement and control, pumps, compressors, fluid-particle separation methods (e.g., centrifugal, sedimentation, filtration), packed columns, fluidized beds, sedimentation, solids transport in slurry and pneumatic flow, and frozen and flashing two-phase gas–liquid flows. The treatment is from the viewpoint of the process engineer, who is concerned with equipment operation, performance, sizing, and selection, as opposed to the details of mechanical design or the details of flow patterns in such situations.

For the student, this is a basic text for a first-level course in process engineering fluid mechanics, which emphasizes the systematic application of fundamental principles (e.g., macroscopic mass, energy, and momentum balances and economics) to the analysis of a variety of fluid problems of a practical nature. Methods of analysis of many of these operations have been taken from the recent technical literature, and have not previously been available in textbooks. This book includes numerous problems that illustrate these applications at the end of each chapter.

For the practicing engineer, this book serves as a useful reference for the working equations that govern many applications of practical interest, as well as a source for basic principles needed to analyze other fluid systems not covered explicitly in the book. The objective here is not to provide a mindless set of recipes for rote application, however, but to demonstrate an organized approach to problem analysis beginning with basic principles and ending with results of very practical applicability.

Chemical Engineering Fluid Mechanics is based on notes that I have complied and continually revised while teaching the junior-level fluid mechanics course for chemical engineering students at Texas A&M University over the last 30 years. It has been my experience that, when being introduced to a new subject, students learn best by starting with simple special cases that they can easily relate to physically, and then progressing to more generalized formulations and more complex problems. That is the philosophy adopted in this book. It will certainly be criticized by some, since it is contrary to the usual procedure followed by most textbooks, in which the basic principles are presented first in the most general and mathematical form (e.g., the divergence theorem, Reynolds transport theorem, Navier Stokes equations, etc.), and the special cases are then derived from these. Esoterically, it is very appealing to progress from the general to the specific, rather than vice versa. However, having taught from both perspectives, it is my observation that most beginning students do not gain an appreciation or understanding from the very general, mathematically complex, theoretical vector expressions until they have gained a certain physical feel for how fluids behave, and the laws governing their behavior, in special situations to which they can easily relate. They also understand and appreciate the principles much better if they see how they can be applied to the analysis of practical and useful situations, with results that actually work in practice. That is why the multi-dimensional vector generalizations of the basic conservations laws have been eschewed in favor of the simpler component and one-dimensional form of these laws.

It is also important to maintain a balanced perspective between fundamental, or theoretical, and empirical information, for the practicing engineer must use both to be effective. It has been said that all the tools of mathematics and physics in the world are not sufficient to calculate how much water will flow in a given time from a kitchen tap when it is opened. However, by proper formulation and utilization of certain experimental observations, this is a routine problem for the engineer. The engineer must be able to solve certain problems by direct application of theoretical principles only (e.g., laminar flow in uniform conduits), others by utilizing hypothetical models that account for a limited understanding of the basic flow phenomena by incorporation of empirical parameters (e.g., :turbulent flow in conduits and fittings), and still other problems in which important information is purely empirical (e.g., pump efficiencies, two-phase flow in packed columns). In many of these problems (of all types), application of dimensional analysis (or the principle of "conservation of dimensions") for generalizing the results of specific analysis, guiding experimental design, and scaling up both theoretical and experimental results can be a very powerful tool.

This second edition of the book includes a new chapter on two-phase flow, which deals with solid–liquid, solid–gas, and frozen and flashing liquid–gas systems, as well as revised, updated, and extended material throughout each chapter. For example, the method for selecting the proper control valve trim to use with a given piping configuration is presented and illustrated by example in Chapter 10. The section on cyclone separators has been completely revised and updated, and new material has been incorporated in a revision of the material on particles in non-Newtonian fluids. Changes have made throughout the book in an attempt to improve the clarity and utility of the presentation wherever possible. For example, the equations for compressible flow in pipes have been reformulated in terms of variables that are easier to evaluate and represent in dimensionless form.

It is the aim of this book to provide a useful introduction to the simplified form of basic governing equations and an illustration of a consistent method of applying these to the analysis of a variety of practical flow problems. Hopefully, the reader will use this as a starting point to delve more deeply into the limitless expanse of the world of fluid mechanics.

Ron Darby

Contents

Unit Conversion Factors

Dimension	*Equivalent Units*
Mass	1 kg = 1000 g = 0.001 metric ton (tonne) = 2.20461 lb_m = 35.27392 oz
	1 lb_m = 453.593 g = 0.453593 kg = 5 × 10^{-4} ton = 16 oz
Force	1 N = 1 kg m/s^2 = 10^5 dyn = 10^5 g cm/s^2 = 0.22418 lb_f
	1 lb_f = 32.174 lb_m ft/s^2 = 4.4482 N = 4.4482 × 10^5 dyn
Length	1 m = 100 cm = 10^6 μm = 10^{10} Å = 39.37 in. = 3.2808 ft = 1.0936 yd = 0.0006214 mi
	1 ft = 12 in. = 1/3 yd = 0.3048 m = 30.48 cm
Volume	1 m^3 = 1000 liters = 10^6 cm^3 = 35.3145 ft^3 = 264.17 gal
	1 ft^3 = 1728 in.^{3} = 7.4805 gal = 0.028317 m^3 = 28.317 liters = 28,317 cm^3
Pressure	1 atm = 1.01325 × 10^5 N/m^2 (Pa) = 1.01325 bar = 1.01325 × 10^6 dyn/cm^2 = 760 mm Hg @ 0°C (torr) = 10.333 m H_2O @ 4°C = 33.9 ft H_2O @ 4°C = 29.921 in. Hg @ 0°C = 14.696 lb_f/in.^{2} (psi)
Energy	1 J = 1 N m = 10^7 erg = 10^7 dyn cm = 2.667 × 10^{-7} kWh = 0.23901 cal = 0.7376 ft lb_f = 9.486 × 10^{-4} Btu [550 ft lb_f/(hp s)]
Power	1 W = 1 J/s = 0.23901 cal/s = 0.7376 ft lb_f/s = 9.486 × 10^{-4} Btu/s = 1 × 10^{-3} kW = 1.341 × 10^{-3} hp
Flow Rate	1 m^3/s = 35.3145 ft^3/s = 264.17 gal/s = 1.585 × 10^4 gal/min = 10^6 cm^3/s
	1 gpm = 6.309 × 10^{-5} m^3/s = 2.228 × 10^{-3} ft^3/s = 63.09 cm^3/s

Example: The factor to convert Pa to psi is 14.696 psi/(1.01325 × 10^5 Pa)

Some values of the gas constant: R = 8.314 × 10^3 kg m^2/(s^2 kg mol K)
= 8.314 × 10^7 g cm^2/(s^2 g mol K)
= 82.05 cm^3 atm/(g mol K)
= 1.987 cal/(g mol K) or Btu/(lb mol °R)
= 1545 ft lb_f/(lb mol °R)
= 10.73 ft^3 psi/(lb mol °R)
= 0.730 ft^3 atm/(lb mol °R)

Chemical Engineering Fluid Mechanics

1

Basic Concepts

I. FUNDAMENTALS

A. Basic Laws

The fundamental principles that apply to the analysis of fluid flows are few and can be described by the "conservation laws":

1. Conservation of mass
2. Conservation of energy (first law of thermodynamics)
3. Conservation of momentum (Newton's second law)

To these may also be added:

4. The second law of thermodynamics
5. Conservation of dimensions ("fruit salad" law)
6. Conservation of dollars (economics)

These *conservation laws* are basic and, along with appropriate rate or *transport models* (discussed below), are the starting point for the solution of every problem.

Although the second law of thermodynamics is not a "conservation law," it states that a process can occur spontaneously only if it goes from a state of higher energy to one of lower energy. In practical terms, this means that energy is dissipated (i.e., transformed from useful mechanical energy to low-level thermal energy) by any system that is in a dynamic (nonequilibrium) state. In other words, useful (mechanical) energy associated with resistance to motion, or "friction," is always "lost" or transformed to a less useful form of (thermal) energy. In more mundane terms, this law tells us that, for example, water will run downhill spontaneously but cannot run uphill unless it is "pushed" (i.e., unless mechanical energy is supplied to the fluid from an exterior source).

B. Experience

Engineering is much more than just applied science and math. Although science and math are important tools of the trade, it is the engineer's ability to use these tools (and others) along with considerable judgment and experiment to "make things work"—i.e., make it possible to get reasoable answers to real problems with (sometimes) limited or incomplete information. A key aspect of "judgment and experience" is the ability to organize and utilize information obtained from one system and apply it to analyze or design similar systems on a different scale. The conservation of dimensions (or "fruit salad") law enables us to design experiments and to acquire and organize data (i.e., experience) obtained in a lab test or model ssytem in the most efficient and general form and apply it to the solution of problems in similar systems that may involve different properties on a different scale. Because the vast majority of problems in fluid mechanics cannot be solved without resort to experience (i.e., empirical knowledge), this is a very important principle, and it will be used extensively.

II. OBJECTIVE

It is the intent of this book to show how these basic laws can be applied, along with pertinent knowledge of system properties, operating conditions, and suitable assumptions (e.g., judgment), to the analysis of a wide variety of practical problems involving the flow of fluids. It is the author's belief that engineers are much more versatile, valuable, and capable if they approach the problem-solving process from a basic perspective, starting from first principles to develop a solution rather than looking for a "similar" problem (that may or may not be applicable) as an example to follow. It is this philosophy along with the objective of arriving at workable solutions to practical problems upon which this work is based.

III. PHENOMENOLOGICAL RATE OR TRANSPORT LAWS

In addition to the conservation laws for mass, energy, momentum, etc., there are additional laws that govern the *rate* at which these quantities are transported from one region to another in a continuous medium. These are called *phenomenological* laws because they are based upon observable phenomena and logic but they cannot be derived from more fundamental principles. These rate or "transport"models can be written for all conserved quantities (mass, energy, momentum, electric charge, etc.) and can be expressed in the general form as

$$\text{Rate of transport} = \frac{\text{Driving force}}{\text{Resistance}} = \text{Conductance} \times \text{Driving force} \tag{1-1}$$

This expression applies to the transport of any conserved quantity Q, e.g., mass, energy, momentum, or charge. The rate of transport of Q per unit area normal to the direction of transport is called the *flux* of Q. This transport equation can be applied on a microscopic or molecular scale to a stationary medium or a fluid in laminar flow, in which the mechanism for the transport of Q is the intermolecular forces of attraction between molecules or groups of molecules. It also applies to fluids in turbulent flow, on a "turbulent convective" scale, in which the mechanism for transport is the result of the motion of turbulent eddies in the fluid that move in three directions and carry Q with them.

On the microscopic or molecular level (e.g., stationary media or laminar flow), the "driving force" for the transport is the negative of the gradient (with respect to the direction of transport) of the concentration of Q. That is, Q flows "downhill," from a region of high concentration to a region of low concentration, at a rate proportional to the magnitude of the change in concentration divided by the distance over which it changes. This can be expressed in the form

$$\text{Flux of } Q \text{ in the } y \text{ direction} = K_{\mathrm{T}}\left[-\frac{d(\text{Conc. of } Q)}{dy}\right] \tag{1-2}$$

where K_{T} is the *transport coefficient* for the quantity Q. For microscopic (molecular) transport, K_{T} is a property only of the medium (i.e., the material). It is assumed that the medium is a continuum, i.e., all relevant physical properties can be defined at any point within the medium. This means that the smallest region of practical interest is very large relative to the size of the molecules (or distance between them) or any substructure of the medium (such as suspended particles, drops, or bubbles). It is further assumed that these properties are homogeneous and isotropic. For macroscopic systems involving turbulent convective transport, the driving force is a representative difference in the concentration of Q. In this case, the transport coefficient includes the effective distance over which this difference occurs and consequently is a function of flow conditions as well as the properties of the medium (this will be discussed later).

Example 1-1: What are the dimensions of the transport coefficient, K_T?

Solution. If we denote the dimensions of a quantity by brackets, i.e., [x] represents "the dimensions of x," a dimensional equation corresponding to Eq. (1-2) can be written as follows:

$$[\text{Flux of } Q] = [K_T]\frac{[Q]}{[\text{volume}][y]}$$

Since [flux of Q] = $[Q]/\mathrm{L}^2\mathrm{t}$, [volume] = L^3, and $[y] = \mathrm{L}$, where L and t are the dimensifons of length and time, respectively, we see that $[Q]$ cancels out from the equation, so that

$$[K_T] = \frac{\mathrm{L}^2}{\mathrm{t}}$$

That is, the dimensions of the transport coefficient are independent of the specific quantity that is being transported.

A. Fourier's Law of Heat Conduction

As an example, Fig. 1-1 illustrates two horizontal parallel plates with a "medium" (either solid or fluid) between them. If the top plate is kept at a temperature T_1 that is higher than the temperature T_0 of the bottom plate, there will be a transport of thermal energy (heat) from the upper plate to the lower plate through the medium, in the $-y$ direction. If the flux of heat in the y direction is denoted by q_y, then our transport law can be written

$$q_y = -\alpha_T \frac{d(\rho c_v T)}{dy} \tag{1-3}$$

where α_T is called the *thermal diffusion coefficient* and $(\rho c_v T)$ is the "concentration of heat." Because the density (ρ) and heat capacity (c_v) are assumed to be independent of position, this equation can be written in the simpler form

$$q_y = -k\frac{dT}{dy} \tag{1-4}$$

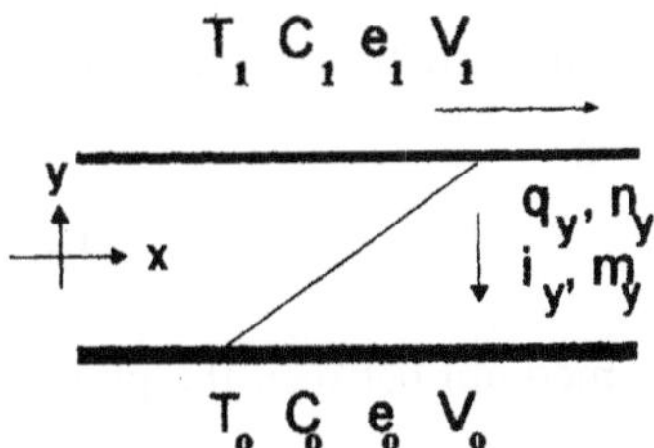

FIGURE 1-1 Transport of energy, mass, charge, and momentum from upper to lower surface.

where $k = \alpha_T \rho c_v$ is the thermal conductivity of the medium. This law was formalized by Fourier in 1822 and is known as *Fourier's law* of heat conduction. This law applies to stationary solids or fluids and to fluids moving in the x direction with straight streamlines (e.g., laminar flow).

B. Fick's Law of Diffusion

An analogous situation can be envisioned if the medium is stationary (or a fluid in laminar flow in the x direction) and the temperature difference $(T_1 - T_0)$ is replaced by the concentration difference $(C_1 - C_0)$ of some species that is soluble in the fluid (e.g., a top plate of pure salt in contact with water). If the soluble species (e.g., the salt) is A, it will diffuse through the medium (B) from high concentration (C_1) to low concentration (C_0). If the flux of A in the y direction is denoted by n_{Ay}, then the transport law is given by

$$n_{Ay} = -D_{AB} \frac{dC_A}{dy} \tag{1-5}$$

where D_{AB} is the molecular diffusivity of the species A in the medium B. Here n_{Ay} is negative, because species A is diffusing in the $-y$ direction. Equation (1-5) is known as *Fick's law of diffusion* (even though it is the same as Fourier's law, with the symbols changed) and was formulated in 1855.

C. Ohm's Law of Electrical Conductivity

The same transport law can be written for electric charge (which is another conserved quantity). In this case, the top plate is at a potential e_1 and the bottom plate is at potential e_0 (electric potential is the "concentration of charge"). The resulting "charge flux" (i.e., current density) from the top plate to the bottom is i_y (which is negative, because transport is in the $-y$

direction). The corresponding expression for this situation is known as *Ohm's law* (1827) and is given by

$$i_y = -k_e \frac{de}{dy} \tag{1-6}$$

where k_e is the "electrical conductivity" of the medium between the plates.

D. Newton's Law of Viscosity

Momentum is also a conserved quantity, and we can write an equivalent expression for the transport of momentum. We must be careful here, however, because velocity and momentum are vectors, in contrast to mass, energy, and charge, which are scalars. Hence, even though we may draw some analogies between the one-dimensional transport of these quantities, these analogies do not generally hold in multidimensional systems or for complex geometries. Here we consider the top plate to be subject to a force in the x direction that causes it to move with a velocity V_1, and the lower plate is stationary ($V_0 = 0$). Since "x-momentum" at any point where the local velocity is v_x is mv_x, the concentration of momentum must be ρv_x. If we denote the flux of x-momentum in the y direction by $(\tau_{yx})_{mf}$, the transport equation is

$$(\tau_{yx})_{mf} = -\nu \frac{d(\rho v_x)}{dy} \tag{1-7}$$

where ν is called the *kinematic viscosity*. It should be evident that $(\tau_{yx})_{mf}$ is negative, because the faster fluid (at the top) drags the slower fluid (below) along with it, so that "x-momentum" is being transported in the $-y$ direction by virtue of this drag. Because the density is assumed to be independent of position, this can also be written

$$(\tau_{yx})_{mf} = -\mu \frac{dv_x}{dy} \tag{1-8}$$

where $\mu = \rho\nu$ is the *viscosity* (or sometimes the *dynamic viscosity*). Equation (1-8) applies for laminar flow in the x direction and is known as *Newton's law of viscosity*. Newton formulated this law in 1687! It applies directly to a class of (common) fluids called *Newtonian fluids*, which we shall discuss in detail subsequently.

1. Momentum Flux and Shear Stress

Newton's law of viscosity and the conservation of momentum are also related to Newton's second law of motion, which is commonly written $F_x = ma_x = d(mv_x)/dt$. For a steady-flow system, this is equivalent to

$F_x = \dot{m}v_x$, where $\dot{m} = dm/dt$ is the mass flow rate. If F_x is the force acting in the x direction on the top plate in Fig. 1. to make it move, it is also the "driving force" for the rate of transport of x-momentum ($\dot{m}v_x$) which flows from the faster to the slower fluid (in the $-y$ direction). Thus the force F_x acting on a unit area of surface A_y is equivalent to a "flux of x-momentum" in the $-y$ direction [e.g., $-(\tau_{yx})_{\text{mf}}$]. [Note that $+A_y$ is the area of the surface bounding the fluid volume of interest (the "system"), which has an *outward* normal vector in the $+y$ direction.] F_x/A_y is also the "shear stress," τ_{yx}, which acts on the fluid—that is, the force $+F_x$ (in the $+x$ direction) that acts on the area A_y of the $+y$ surface. It follows that a *positive* shear stress is equivalent to a *negative* momentum flux, i.e., $\tau_{yx} = -(\tau_{yx})_{\text{mf}}$. [In Chapter 3, we define the rheological (mechanical) properties of materials in terms that are common to the field of mechanics, i.e., by relationships between the *stresses* that act upon the material and the resulting material *deformation*.] It follows that an equivalent form of Newton's law of viscosity can be written in terms of the shear stress instead of the momentum flux:

$$\tau_{yx} = \mu \frac{dv_x}{dy} \tag{1-9}$$

It is important to distinguish between the momentum flux and the shear stress because of the difference in sign. Some references define viscosity (i.e., Newton's law of viscosity) by Eq. (1-8), whereas others use Eq. (1-9) (which we shall follow). It should be evident that these definitions are equvialent, because $\tau_{yx} = -(\tau_{yx})_{\text{mf}}$.

2. Vectors Versus Dyads

All of the preceding transport laws are described by the same equation (in one dimension), with different symbols (i.e., the same game, with different colored jerseys on the players). However, there are some unique features to Newton's law of viscosity that distinguish it from the other laws and are very important when it is being applied. First of all, as pointed out earlier, momentum is fundamentally different from the other conserved quantities. This is because mass, energy, and electric charge are all *scalar* quantities with no directional properties, whereas momentum is a *vector* with directional character. Since the gradient (i.e., the "directional derivative" dq/dy or, more generally, ∇q) is a vector, it follows that the gradient of a scalar (e.g., concentration of heat, mass, charge) is a vector. Likewise, the flux of mass, energy, and charge are vectors. However, Newton's law of viscosity involves the gradient of a vector (e.g., velocity or momentum), which implies *two* directions: the direction of the vector quantity (momentum or velocity) and the direction in which it varies (the gradient direction). Such quantities are called *dyads* or *second-order tensors*. Hence, momentum flux is a dyad, with the direction of the momentum (e.g., x) as well as the direction in which this momentum is transported (e.g., $-y$). It is also evident that the equivalent shear stress (τ_{yx}) has two directions, corresponding to the direction in which the force acts (x) and the direction (i.e., "orientation") of the surface upon which it acts (y). [Note that all "surfaces" are vectors because of their orientation, the direction of the surface being defined by the (outward) vector that is normal to the surface that bounds the fluid volume of interest.] This is very significant when it comes to generalizing these one-dimensional laws to two or three dimensions, in which case much of the analogy between Newton's law and the other transport laws is lost.

3. Newtonian Versus Non-Newtonian Fluids

It is also evident that this "phenomenological" approach to transport processes leads to the conclusion that fluids should behave in the fashion that we have called Newtonian, which does not account for the occurrence of "non-Newtonian" behavior, which is quite common. This is because the phenomenological laws inherently assume that the molecular "transport coefficients" depend only upon the thermodyamic state of the material (i.e., temperature, pressure, and density) but not upon its "dynamic state," i.e., the state of stress or deformation. This assumption is not valid for fluids of complex structure, e.g., non-Newtonian fluids, as we shall illustrate in subsequent chapters.

The flow and deformation properties of various materials are discussed in Chapter 3, although a completely general description of the flow and deformation (e.g., rheological) properties of both Newtonian and non-Newtonian fluids is beyond the scope of this book, and the reader is referred to the more advanced literature for details. However, quite a bit can be learned, and many problems of a practical nature solved, by considering relatively simple models for the fluid viscosity, even for fluids with complex properties, provided the complexities of elastic behavior can be avoided. These properties can be measured in the laboratory, with proper attention to data interpretation, and can be represented by any of several relatively simple mathematical expressions. We will not attempt to delve in detail into the molecular or structural origins of complex fluid properties but will make use of information that can be readily obtained through routine measurements and simple modeling. Hence, we will consider non-Newtonian fluids along with, and in parallel with, Newtonian fluids in many of the flow situations that we analyze.

IV. THE "SYSTEM"

The basic conservation laws, as well as the transport models, are applied to a "system" (sometimes called a "control volume"). The *system* is not actually the volume itself but the *material within a defined region*. For flow problems, there may be one or more streams entering and/or leaving the system, each of which carries the conserved quantity (e.g., Q) into and out of the system at a defined rate (Fig. 1-2). Q may also be transported into or out of the system through the system boundaries by other means in addition to being carried by the in and out streams. Thus, the conservation law for a flow problem with respect to any conserved quantity Q can be written as follows:

$$\begin{matrix}\text{Rate of } Q \\ \text{into the system}\end{matrix} - \begin{matrix}\text{Rate of } Q \\ \text{out of the system}\end{matrix} = \begin{matrix}\text{Rate of accumulation of } Q \\ \text{within the system}\end{matrix} \tag{1-10}$$

If Q can be produced or consumed within the system (e.g., through chemical or nuclear reaction, speeds approaching the speed of light, etc.), then a "rate of generation" term may be included on the left of Eq. (1-10). However, these effects will not be present in the systems with which we are concerned. For example, the system in Fig. 1-1 is the material contained between the two plates. There are no streams entering or leaving this system, but the conserved quantity is transported into the system by microscopic (molecular) interactions through the upper boundary of the system (these and related concepts will be expanded upon in Chapter 5 and succeeding chapters).

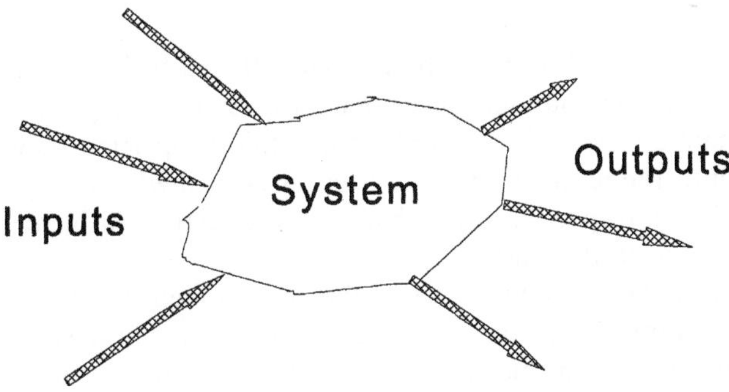

FIGURE 1-2 The "system."

V. TURBULENT MACROSCOPIC (CONVECTIVE) TRANSPORT MODELS

The preceding transport laws describe the rate of transfer of heat, mass, charge, or momentum from one region of a continuum to another by virtue of molecular interactions only. That is, there is no actual bulk motion of material in the transport direction (y), which means that the medium must be stationary or moving only in the direction (x) normal to the transport direction. This means that the flow (if any) must be "laminar"; i.e., all fluid elements move in straight, smooth streamlines in the x direction. This occurs if the velocity is sufficiently low and is dominated by stabilizing viscous forces. However, as the velocity increases, destabilizing inertial forces eventually overcome the viscous forces and the flow becomes turbulent. Under turbulent conditions, a three-dimensional fluctuating flow field develops that results in a high degree of mixing or "convection" due to the bulk motion of the turbulent eddies. As a result, the flow is highly mixed, except for a region near solid boundaries that is called the *boundary layer* (δ). The fluid velocity approaches zero at a stationary boundary, and thus there is a region in the immediate vicinity of the wall that is laminar. Consequently, the major resistance to transport in turbulent (convective) flow is within this boundary layer, the size of which depends upon the dynamic state of the flow field as well as fluid properties but in turbulent flows is typically quite small relative to the dimensions of the total flow area (see Chapter 6, Sec. III.B.).

The general transport models for the turbulent convective transport of heat and mass can be expressed as follows:

$$\text{Heat flux:} \quad q_y = k_e \frac{\Delta T}{\delta} = h\,\Delta T \tag{1-11}$$

$$\text{Mass flux:} \quad n_{Ay} = D_e \frac{\Delta C_A}{\delta} = K_m\,\Delta C_A \tag{1-12}$$

where k_e is a turbulent or "eddy" thermal conductivity, D_e is a turbulent or "eddy" diffusivity, and δ is the boundary layer thickness. Since k_e, D_e, and all depend on the dynamic state of low as well as the fluid properties, they are combined with δ into the terms h, the *heat transfer coefficient*, and K_m, the *mass transfer coefficient*, respectively, which are the convective (turblent) transport coefficients for heat and mass.

The situation with regard to convective (turbulent) momentum transport is somewhat more complex because of the tensor (dyadic) character of momentum flux. As we have seen, Newton's second law provides a correspondence between a force in the x direction, F_x, and the rate of transport of x-momentum. For continuous steady flow in the x direction at a bulk velocity V_x in a conduit of cross-sectional area A_x, there is a transport of x-momentum in the x direction given by

$$F_x = d(mV_x)/dt = \dot{m}V_x = (\rho V_x A_x)V_x = \rho V_x^2 A_x \tag{1-13}$$

The corresponding flux of x-momentum in the x direction is $F_x/A_x = \rho V_x^2$. This x-momentum is also the driving force for convective transport of x-momentum in the $-y$ direction (toward the wall), i.e., $\tau_{yx} = F_x/A_y$. Therefore, the convective flux of x-momentum from the fluid to the wall (or the stress exerted by the fluid on the wall) can be expressed as

$$\text{Momentum flux:} \quad (\tau_{yx})_{\text{wall}} = \tau_{\text{w}} = \frac{f}{2}\rho V_x^2 \tag{1-14}$$

where f is called the *Fanning friction factor* (other definitions of the friction factor are also used, which differ by a factor of 2 or 4 from the Fanning friction factor). Although Eq. (1-14) is the counterpart of the turbulent flux expressions for heat and mass, the form of this equation appears somewhat different because of the correspondence between force and rate of momentum and the dyadic nature of the momentum flux and stress. Like the heat and mass transfer coefficients, the friction factor depends upon dynamic flow conditions as well as upon fluid properties. It should be evident from Eq. (1-9) that laminar flows are dominated by the fluid viscosity (which is stabilizing), whereas Eq. (1-14) indicates that turbulent flows are dominated by the fluid density (i.e., inertial forces), which is destabilizing. The proper definition of f and its dependence on flow conditions and fluid properties, is consistent for either laminar or turbulent flow (as explained in Chapters 5 and 6).

PROBLEMS

1. Write equations that define each of the following laws: Fick's, Fourier's, Newton's, and Ohm's. What is the conserved quantity in each of these laws? Can you represent all of these laws by one general expression? If so, does this mean that all of the processes represented by these laws are always analogous? If they aren't, why not?
2. The general conservation law for any conserved quantity Q can be written in the form of Eq. (1-10). We have said that this law can also be applied to "dollars" as the conserved quantity Q. If the "system" is your bank account,
   (a) Identify specific "rate in," "rate out," and "rate of accumulation" terms in this equation relative to the system (i.e., each term corresponds to the rate at which dollars are moving into or out of your account).
   (b) Identify one or more "driving force" effects that are responsible for the magnitude of each of these rate terms, i.e., things that influence how fast the dollars go in or out. Use this to define corresponding "transport con-

stants" for each "in" and "out" term relative to the appropriate "driving force" for each term.

3. A dimensionless group called the Reynolds number is defined for flow in a pipe or tube

$$N_{Re} = \frac{DV\rho}{\mu} = \frac{\rho V^2}{\mu V/D}$$

where V is the average velocity in the pipe, ρ is the fluid density, μ is the fluid viscosity, D is the tube diameter. The second form of the group indicates that it is a ratio of the convective (turbulent) momentum flux to the molecular (viscous) momentum flux, or the ratio of inertial forces (which are destabilizing) to viscous forces (which are stabilizing). When viscous forces dominate over inertial forces, the flow is laminar and fluid elements flow in smooth, straight streamlines, whereas when inertial forces dominate, the flow is unstable and the flow pattern break up into random fluctuating eddies. It is found that laminar flow in a pipe occurs as long as the value of the Reynolds number is less than 2000.

Calculate the maximum velocity and the corresponding flow rate (in cm^3/s) at which laminar flow of water is possible in tubes with the following diameters:

$$D = 0.25,\ 0.5,\ 1.0,\ 2.0,\ 4.0,\ 6.0,\ 10.0\,\text{in.}$$

4. A layer of water is flowing down a flat plate that is inclined at an angle of 20° to the vertical. If the depth of the layer is 1/4 in., what is the shear stress exerted *by* the plate *on* the water? (Remember: Stress is a dyad.)
5. A slider bearing consists of a sleeve surrounding a cylindrical shaft that is free to move axially within the sleeve. A lubricant (e.g., grease) is in the gap between the sleeve and the shaft to isolate the metal surfaces and support the stress resulting from the shaft motion. The diameter of the shaft is 1 in., and the sleeve has an inside diameter of 1.02 in. and a length of 2 in.
   (a) If you want to limit the total force on the sleeve to less than 0.5 lb_f when the shaft is moving at a velocity of 20 ft/s, what should the viscosity of the grease be? What is the magnitude of the flux of momentum in the gap, and which direction is the momentum being transported?
   (b) If the lubricant is a grease with a viscosity of 400 cP (centipoise), what is the force exerted on the sleeve when the shaft is moving at 20 ft/s?
   (c) The sleeve is cooled to a temperature of 150°F, and it is desired to keep the shaft temperature below 200°F. What is the cooling rate (i.e., the rate at which heat must be removed by the coolant), in Btu/hr, to achieve this? Properties of the grease may be assumed to be: specific heat = 0.5 Btu/(lb_m °F); SG (specific gravity) = 0.85; thermal conductivity = 0.06 Btu/(hr ft °F).
   (d) If the grease becomes contaminated, it could be corrosive to the shaft metal. Assume that this occurs and the surface of the shaft starts to corrode at a rate of 0.1 μm/yr. If this corrosion rate is constant, determine the maximum

concentration of metal ions in the grease when the ions from the shaft just reach the sleeve. Properties of the shaft metal may be assumed to be MW = 65; SG = 8.5; diffusivity of metal ions in grease = $8.5 \times 10^{-5}\ \text{cm}^2/\text{s}$.

6. By making use of the analogies between the molecular transport of the various conserved quantities, describe how you would set up an experiment to solve each of the following problems by making electrical measurements (e.g., describe the design of the experiment, how and where you would measure voltage and current, and how the measured quantities are related to the desired quantities).
   (a) Determine the rate of heat transfer from a long cylinder to a fluid flowing normal to the cylinder axis if the surface of the cylinder is at temperature T_0 and the fluid far away from the cylinder is at temperature T_1. Also determine the temprature distribution within the fluid and the cylinder.
   (b) Determine the rate at which a (spherical) mothball evaporates when it is immersed in stagnant air, and also the concentration distribution of the evaporating compound in the air.
   (c) Determine the local stress as a function of position on the surface of a wedge-shaped body immersed in a fluid stream that is flowing slowly parallel to the surface. Also, determine the local velocity distribution in the fluid as a function of position in the fluid.

NOTATION*

C_A	concentration of species A, $[M/L^3]$
c_v	specific heat at constant volume, [H/MT]
D_{AB}	diffusivity of species A in medium B, $[L/t^2]$
e	concentration of charge (electrical potential), $[C/L^2]$
i_y	current density, or flux of charge, in the y direction, $[C/L^2 t]$
k	thermal conductivity, [H/LTt]
L	dimension of length
n_{Ay}	flux of species A in the y direction, $[M/L^2 t]$
Q	"generic" notation for any conserved (transported) quantity
q_y	flux (i.e., rate of transport per unit area normal to direction of transport) of heat in the y direction, $[H/L^2 t]$
t	time, [t]
T	temperature, [K or °R]
v_x	local or point velocity in the x direction, [L/t]
V	spatial average velocity or velocity at boundary
y	coordinate direction, [L]
α_T	thermal diffusivity ($= \rho c_v T$), $[L^2/t]$
ρ	density, $[M/L^3]$

*Dimensions given in brackets: L = length, M = mass, t = time, T = temperature, C = charge, H = "heat" = thermal energy = ML^2/t^2. (See Chapter 2 for discussion of units and dimensions.)

μ	viscosity, [M/Lt]
$(\tau_{yx})_{mf}$	flux of x-momentum in the y direction ($= -\tau_{yx}$), [M/Lt^2]
τ_{yx}	shear stress [= (force in x direction)/(area of y surface) $= -(\tau_{yx})_{mf}$], [M/Lt^2]
ν	kinematic viscosity ($= \mu/\rho$), [L^2/t]

2

Dimensional Analysis and Scale-up

I. INTRODUCTION

In this chapter we consider the concepts of dimensions and units and the various systems in use for describing these quantities. In particular, the distinction between scientific and engineering systems of dimensions is explained, and the various metric and English units used in each system are discussed. It is important that the engineer be familiar with these systems, as they are all in common use in various fields of engineering and will continue to be for the indefinite future. It is common to encounter a variety of units in different systems during the analysis of a given problem, and the engineer must be adept at dealing with all of them.

The concept of "conservation of dimensions" will then be applied to the dimensional analysis and scale-up of engineering systems. It will be shown how these principles are used in the design and interpretation of laboratory experiments on "model" systems to predict the behavior of large-scale ("field") systems (this is also known as *similitude*). These concepts are presented early on, because we shall make frequent use of them in describing the results of both theoretical and experimental analyses of engineering systems in a form that is the most concise, general, and useful. These methods can also provide guidance in choosing the best approach to take in the solution of many complex problems.

II. UNITS AND DIMENSIONS

A. Dimensions

The dimensions of a quantity identify the physical charcter of that quantity, e.g., force (F), mass (M), length (L), time (t), temperature (T), electric charge (e), etc. On the other hand, "units" identify the reference scale by which the magnitude of the respective physical quantity is measured. Many different reference scales (units) can be defined for a given dimension; for example, the dimension of length can be measured in units of miles, centimeters, inches, meters, yards, angstroms, furlongs, light years, kilometers, etc.

Dimensions can be classified as either *fundamental* or *derived. Fundamental* dimensions cannot be expressed in terms of other dimensions and include length (L), time (t), temperature (T), mass (M), and/or force (F) (depending upon the *system* of dimensions used). *Derived* dimensions can be expressed in terms of fundamental dimensions, for example, area ($[A] = \mathrm{L}^2$), volume ($[V] = \mathrm{L}^3$), energy ($[E] = FL = \mathrm{ML}^2/\mathrm{t}^2$), power ($[\mathrm{HP}] = FL/\mathrm{t} = \mathrm{ML}^2/\mathrm{t}^3$), viscosity ($[\mu] = F\mathrm{t}/\mathrm{L}^2 = \mathrm{M}/\mathrm{Lt}$), etc.*

There are two systems of fundamental dimensions in use (with their associated units), which are referred to as *scientific* and *engineering* systems. These systems differ basically in the manner in which the dimensions of force is defined. In both systems, mass, length, and time are fundamental dimensions. Furthermore, Newton's second law provides a relation between the dimensions of force, mass, length, and time:

$$\text{Force} = \text{Mass} \times \text{Acceleration}$$

i.e.,

$$F = ma \tag{2-1}$$

or

$$[F] = [ma] = \mathrm{ML}/\mathrm{t}^2$$

In scientific systems, this is accepted as the definition of force; that is, force is a derived dimension, being identical to ML/t^2.

In engineering systems, however, force is considered in a more practical or "pragmatic" context as well. This is because the mass of a body is not usually measured directly but is instead determined by its "weight" (W), i.e., the gravitational force resulting from the mutual attraction between two bodies of mass m_1 and m_2:

$$W = G(m_1 m_2/r^2) \tag{2-2}$$

* The notation [ ] means "the dimensions of" whatever is in the brackets.

G is a constant having a value of $6.67 \times 10^{-11}\,\mathrm{N\,m^2/kg^2}$, and r is the distance between the centers of m_1 and m_2. If m_2 is the mass of the earth and r is its radius at a certain location on earth, then W is the "weight" of mass m_1 at that location:

$$W = m_1 g \tag{2-3}$$

The quantity g is called the *acceleration due to gravity* and is equal to m_2G/r^2. At sea level and 45° latitude on the Earth (i.e., the condition for "standard gravity," g_{std}) the value of g is $32.174\,\mathrm{ft/s^2}$ or $9.806\,\mathrm{m/s^2}$. The value of g is obviously different on the moon (different r and m_2) and varies slightly over the surface of the earth as well (since the radius of the earth varies with both elevation and latitude).

Since the mass of a body is determined indirectly by its weight (i.e., the gravitational force acting on the mass) under specified gravitational conditions, engineers decided that it would be more practical and convenient if a system of dimensions were defined in which "what you see is what you get"; that is, the numerical magnitudes of mass and weight are equal under standard conditions. This must not violate Newton's laws, however, so both Eqs. (2-1) and (2-3) are valid. Since the value of g is not unity when expressed in common units of length and time, the only way to have the numerical values of weight and mass be the same under any conditions is to introduce a "conversion factor" that forces this equivalence. This factor is designated g_c and is incorporated into Newton's second law for engineering systems (sometimes referred to as "gravitational systems") as follows:

$$F = \frac{ma}{g_c}, \qquad W = \frac{mg}{g_c} \tag{2-4}$$

This additional definition of force is equvialent to treating F as a fundamental dimension, the redundancy being accounted for by the conversion factor g_c. Thus, if a unit for the weight of mass m is defined so that the numerical values of F and m are identical under standard gravity conditions (i.e., $a = g_{std}$), it follows that the numerical magnitude of g_c must be identical to that of g_{std}. However, it is important to distinguish between g and g_c, because they are fundamentally different quantities. As explained above, g is not a constant; it is a *variable* that depends on both m_2 and r [Eq. (2-2)]. However, g_c is a *constant* because it is merely a conversion factor that is defined by the value of standard gravity. Note that these two quantities are also physically different, because they have different dimensions:

$$[g] = \frac{\mathrm{L}}{\mathrm{t}^2}, \qquad [g_c] = \frac{\mathrm{ML}}{\mathrm{Ft}^2} \tag{2-5}$$

The factor g_c is the conversion factor that relates equivalent force and mass ($\mathrm{ML/t^2}$) units in engineering systems. In these systems both force and mass can be considered fundamental dimensions, because they are related by two separate (but compatible) definitions: Newton's second law and the engineering definition of weight. The conversion factor g_c thus accounts for the redundancy in these two definitions.

B. Units

Several different sets of units are used in both scientific and engineering systems of dimensions. These can be classified as either metric (SI and cgs) or English (fps). Although the internationally accepted standard is the SI scientific system, English engineering units are still very common and will probably remain so for the foreseeable future. Therefore, the reader should at least master these two systems and become adept at converting between them. These systems are illustrated in Table 2-1. Note that there are two different English scientific systems, one in which M, L, and t are fundamental and F is derived, and another in which F, L, and t are fundamental and M is derived. In one system, mass (with the unit "slug") is fundamental; in the other, force (with the unit "poundal") is fundamental. However, these systems are archaic and rarely used in practice. Also, the metric engineering systems with units of kg_f and g_f have generally been replaced by the SI system, although they are still in use in some places. The most common systems in general use are the scientific metric (e.g., SI) and English engineering systems.

Since Newton's second law is satisfied identically in scientific units with no conversion factor (i.e., $g_c = 1$), the following identities hold:

$$g_c = 1\frac{\text{kg m}}{\text{N s}^2} = 1\frac{\text{g cm}}{\text{dyn s}^2} = 1\frac{\text{slug ft}}{\text{lb}_f\text{ s}^2} = 1\frac{\text{lb}_m\text{ ft}}{\text{poundal s}^2}$$

TABLE 2-1 Systems of Dimensions/Units

	Scientific				Engineering			
	L	M	F	g_c	L	M	F	g_c
English	ft	lb_m	poundal	1	ft	lb_m	lb_f	32.2
	ft	slug	lb_f	1				
Metric (SI)	m	kg	N	1	m	kg_m	kg_f	9.8
(cgs)	cm	g	dyn	1	cm	g_m	g_f	980

Conversion factors: $g_c[\text{ML/Ft}^2]$, $F = ma/g_c$

$g_c = 32.174\ \text{lb}_m\,\text{ft}/(\text{lb}_f\text{s}^2) = 9.806\ (\text{kg}_m\,\text{m}/(\text{kg}_f\text{s}^2) = 980.6\ \text{g}_m\,\text{cm}/(\text{g}_f\text{s}^2)$

$= 1\ \text{kg s}^2\ \text{m}/(\text{N s}^2) = 1\ \text{g cm}/(\text{dyn s}^2) = 1\ \text{slug ft}/(\text{lb}_f\,\text{s}^2) = 1\ \text{lb}_m\,\text{ft}/(\text{poundal s}^2)$

$= 12\ \text{in./ft} = 60\ \text{s/min} = 30.48\ \text{cm/ft} = 778\ \text{ft lb}_f/\text{Btu}) = \cdots = 1[0]$

In summary, for engineering systems both F and M can be considered fundamental because of the engineering definition of weight in addition to Newton's second law. However, this results in a redundancy, which needs to be recified by the conversion factor g_c. The value of this conversion factor in the various engineering units provides the following identities:

$$g_c = 9.806 \frac{\text{kg}_\text{m}\,\text{m}}{\text{kg}_\text{f}\,\text{s}^2} = 980.6 \frac{\text{g}_\text{m}\,\text{cm}}{\text{g}_\text{f}\,\text{s}^2} = 32.174 \frac{\text{lb}_\text{m}\,\text{ft}}{\text{lb}_\text{f}\,\text{s}^2}$$

C. Conversion Factors

Conversion factors relate the magnitudes of different units with common dimensions and are actually identities; that is, 1 ft is identical to 12 in., 1 Btu is identical to 778 ft lb_f, etc. Because any identity can be expressed as a ratio with a magnitude but no dimensions, the same holds for any conversion factor, i.e.,

$$12 \frac{\text{in.}}{\text{ft}} = 778 \frac{\text{ft, lb}_\text{f}}{\text{Btu}} = 30.48 \frac{\text{cm}}{\text{ft}} = 14.7 \frac{\text{psi}}{\text{atm}} = 10^5 \frac{\text{dyn}}{\text{N}} = 1(0), \qquad \text{etc.}$$

A table of commonly encountered conversion factors is included at the front of the book. The value of any quantity expressed in a given set of units can be converted to any other equivalent set of units by multiplying or dividing by the appropriate conversion factor to cancel the unwanted units.

Example 2-1: To convert a quantity X measured in feet to the equivalent in miles:

$$\frac{X\,\text{ft}}{5280\,\text{ft/mi}} = \frac{X}{5280}\,\text{mi}$$

Note that the conversion factor relating mass units in scientific systems to those in engineering systems can be obtained by equating the appropriate g_c values from the two systems, e.g.,

$$g_c = 1 \frac{\text{slug ft}}{\text{lb}_\text{f}\,\text{s}^2} = 32.174 \frac{\text{lb}_\text{m}\,\text{ft}}{\text{lb}_\text{f}\,\text{s}^2}$$

Thus, after canceling common units, the conversion factor relating slugs to lb_m is 32.174 lb_m/slug.

III. CONSERVATION OF DIMENSIONS

Physical laws, theories, empirical relations, etc., are normally expressed by equations relating the significant variables and parameters. These equations usually contain a number of terms. For example, the relation between the vertical elevation (z) and the horizontal distance (x) at any time for a projectile fired from a gun can be expressed in the form

$$z = ax + bx^2 \tag{2-6}$$

This equation can be derived from the laws of physics, in which case the parameteers a and b can be related to such factors as the muzzle velocity, projectile mass, angle of inclination of the gun, and wind resistance. The equation may also be empirical if measured values of z versus x are related by an equation of this form, with no reference to the laws of physics.

For any equation to be valid, every term in the equation must have the same physical character, i.e., the same net dimensifons (and consequently the same units in any consistent system of units). This is known as the *law of conservation of dimensions* (otherwise known as the "fruit salad law"—"you can't add apples and oranges, unless you are making fruit salad"). Let us look further at Eq. (2-6). Since both z and x have dimensifons of length, e.g., $[x] = \text{L}$, $[z] = \text{L}$, it follows from the fruit salad law that the dimensions of a and b must be

$$[a] = 0, \qquad [b] = 1/\text{L}$$

(i.e., a has no dimensions—it is dimensionless, and the dimensions of b are 1/length, or length^{-1}). For the sake of argument, let us assume that x and z are measured in feet and that the values of a and b in the equation are 5 and $10\,\text{ft}^{-1}$, respectively. Thus if $x = 1$ ft,

$$z = (5)(1\,\text{ft}) + (10\,\text{ft}^{-1})(1\,\text{ft})^2 = 15\,\text{ft}$$

On the other hand, if we choose to measure x and z in inches, the value of z for $x = 1$ in. is

$$z = (5)(1\,\text{in.}) + (10\,\text{ft}^{-1})\left(\frac{1}{12\,\text{in./ft}}\right)(1\,\text{in.})^2 = 5.83\,\text{in.}$$

This is still in the form of Eq. (2-6), i.e.,

$$z = ax + bx^2$$

but now $a = 5$ and $b = 10/12 = 0.833\,\text{in.}^{-1}$. Thus the magnitude of a has not changed, but the magnitude of b *has* changed. This simple example illustrates two important principles:

1. *Conservation of dimensions* ("fruit salad" law). All terms in a given equation must have the same net dimensions (and units) for the equation to be valid.
2. *Scaling.* The fact that the value of the dimensionless parameter a is the same regardless of the units (e.g., scale) used in the problem illustrates the universal nature of dimensionless quantities. That is, the magnitude of any dimensionless quantity will always be independent of the scale of the problem or the system of (consistent) units used. This is the basis for the application of dimensional analysis, which permits information and relationships determined in a small-scale system (e.g., a "model") to be applied directly to a similar system of a different size if the system variables are expressed in dimensionless form. This process is known as *scale-up*.

The universality of certain dimensionless quantities is often taken for granted. For example, the exponent 2 in the last term of Eq. (2-6) has no dimensions and hence has the same magnitude regardless of the scale or units used for measurement. Likewise, the kinetic energy per unit mass of a body moving with a velocity v is given by

$$\text{ke} = \tfrac{1}{2}v^2$$

Both of the numerical quantities in this equation, 1/2 and 2, are dimensionless, so they always have the same magnitude regardless of the units used to measure v.

A. Numerical Values

Ordinarily, any numerical quantities that appear in equations that have a theoretical basis (such as that for ke above) are dimensionless and hence "universal." However, many valuable engineering relations have an empirical rather than a theoretical basis, in which case this conclusion does not always hold. For example, a very useful expression for the (dimensionless) friction loss coefficient (K_f) for valves and fittings is

$$K_\text{f} = \frac{K_1}{N_\text{Re}} + K_\text{i}\left(1 + \frac{K_\text{d}}{\text{ID}}\right)$$

Here, N_Re is the Reynolds number,* which is dimensionless, as are K_f and the constants K_1 and K_i. However, the term ID is the internal diameter of

---

*We use the recommended notation of the AIChE for dimensionless groups that are named after their originator, i.e., a capital N with a subscript identifying the person the group is named for. However, a number of dimensionless quantities that are identified by other symbols; see, for example, Section IV.

the fitting, with dimensions of length. By the "fruit salad" law, the constant K_d in the term K_d/ID must also have dimensions of length and so is not independent of scale, i.e., its magnitude is defined only in specific units. In fact, its value is normally given in units of in., so ID must also be measured in inches for this value to be valid. If ID were to be measured in centimeters, for example, the value of K_d would be 2.54 times as large, because (1 in.)(2.54 cm/in.) = 2.54 cm.

B. Consistent Units

The conclusion that dimensionless numerical values are universal is valid only if a consistent system of units is used for all quantities in a given equation. If such is not the case, then the numerical quantities may include conversion factors relating the different units. For example, the velocity (V) of a fluid flowing in a pipe can be related to the volumetric flow rate (Q) and the internal pipe diameter (D) by any of the following equations:

$$V = 183.3Q/D^2 \tag{2-7}$$

$$V = 0.408Q/D^2 \tag{2-8}$$

$$V = 0.286Q/D^2 \tag{2-9}$$

$$V = 4Q/\pi D^2 \tag{2-10}$$

although the dimensions of V (i.e., L/t) are the same as those for Q/D^2 (i.e., $\text{L}^3/\text{tL}^2 = \text{L/t}$), it is evident that the numerical coefficient is not universal despite the fact that it must be dimensionless. This is because a consistent system of units is not used except in Eq. (2–10). In each equation, the units of V are ft/s. However, in Eq. (2-7), Q is in ft^3/s, whereas in Eq. (2-8), Q is in gallons per minute (gpm), and in Eq. (2-9) it is in barrels per hour (bbl/hr), with D in inches in each case. Thus, although the dimensions are consistent, the units are not, and thus the numerical coefficients include unit conversion factors. Only in Eq. (2-10) are all the units assumed to be from the same consistent system (i.e., Q in ft^3/s and D in ft) so that the factor $4/\pi$ is both dimensionless and unitless and is thus universal. It is always advisable to write equations in a universally valid from to avoid confusion; i.e., all quantities should be expressed in consistent units.

IV. DIMENSIONAL ANALYSIS

The law of conservation of dimensions can be applied to arrange the variables or parameters that are important in a given problem into a set of *dimensionless groups*. The original set of (dimensional) variables can then be replaced by the resulting set of dimensionless groups, and these can be used to completely define the system behavior. That is, any valid relationship (theoretical or empirical) between the original variables can be expressed in terms of these dimensionless groups. This has two important advantages:

1. Dimensionless quantities are universal (as we have seen), so any relationship involving dimensionless variables is independent of the size or scale of the system. Consequently, information obtained from a model (small-scale) system that is represented in dimensionless form can be applied directly to geometrically and dynamically similar systems of any size or scale. This allows us to translate information directly from laboratory models to large-scale equipment or plant operations (scale-up). Geometrical similarity requires that the two systems have the same shape (geometry), and dynamical similarity requries them to be operating in the same dynamic regime (i.e., both must be either laminar or turbulent). This will be expanded upon later.
2. The number of dimensionless groups is invariably less than the number of original variables involved in the problem. Thus the relations that define the behavior of a given system are much simpler when expressed in terms of the dimensionless variables, because fewer variables are required. In other words, the amount of effort required to represent a relationship between the dimensionless groups is much less than that required to relate each of the variables independently, and the resulting relation will thus be simpler in form. For example, a relation between two variables (x vs. y) requires two dimensions, whereas a relation between three variables (x vs. y vs z) requires three dimensions, or a family of two-dimensional "curves" (e.g., a set of x vs. y curves, each curve for a different z). This is equivalent to the difference between one page and a book of many pages. Relating four variables would obviously require many books or volumes. Thus, reducing the number of variables from, say, four to two would dramatically simplify any problem involving these variables.

It is important to realize that the process of dimensional analysis only replaces the set of original (dimensional) variables with an equivalent (smaller) set of dimensionless variables (i.e., the dimensionless groups). It does not tell how these variables are related—the relationship must be determined either theoretically by application of basic scientific principles or empirically by measurements and data analysis. However, dimensional analysis is a very powerful tool in that it can rovide a direct guide for experimental design and scale-up and for expressing operating relationships in the most general and useful form.

There are a number of different approaches to dimensional analysis. The classical method is the "Buckingham Π Theorem", so-called because Buckingham used the symbol Π to represent the dimensionless groups. Another classic approach, which involves a more direct application of the law of conservation of dimensions, is attributed to Lord Rayleigh. Numerous variations on these methods have also been presented in the literature. The one thing all of these methods have in common is that they require a knowledge of the variables and parameters that are important in the problem as a starting point. This can be determined through common sense, logic, intuition, experience, or physical reasoning or by asking someone who is more experienced or knowledgeable. They can also be determined from a knowledge of the physical principles that govern the system (e.g., the conservation of mass, energy, momentum, etc., as written for the specific system to be analyzed) along with the fundamental equations that describe these principles. These equations may be macroscopic or microscopic (e.g., coupled sets of partial differential equations, along with their boundary conditions). However, this knowledge often requires as much (or more) insight, intuition, and/or experience as is required to compose the list of variables from logical deduction or intuition. The analysis of any engineering problem requires key assumptions to distinguish those factors that are important in the problem from those that are insignificant. [This can be referred to as the "bathwater" rule—it is necessary to separate the "baby" from the "bathwater" in any problem, i.e., to retain the significant elements (the "baby") and discard the insignificant ones (the "bathwater"), and not vice versa!] The talent required to do this depends much more upon sound understanding of fundamentals and the exercise of good judgment than upon mathematical facility, and the best engineer is often the one who is able to make the most appropriate assumptions to simplify a problem (i.e., to discard the "bathwater" and retain the "baby"). Many problem statements, as well as solutions, involve assumptions that are implied but not stated. One should always be on the lookout for such implicit assumptions and try to identify them wherever possible, since they set corresponding limits on the applicability of the results.

The method we will use to illustrate the dimensional analysis process is one that involves a minimum of manipulations. It does require an initial knowledge of the variables (and parameters) that are important in the system and the dimensions of these variables. The objective of the process is to determine an appropriate set of dimensionless groups of these variables that can then be used in place of the original individual variables for the purpose of describing the behavior of the system. The process will be explained by means of an example, and the results will be used to illustrate the application of dimensional analysis to experimental design and scale-up.

A. PIPELINE ANALYSIS

The procedure for performing a dimensional analysis will be illustrated by means of an example concerning the flow of a liquid through a circular pipe. In this example we will determine an appropriate set of dimensionless groups that can be used to represent the relationship between the flow rate of an incompressible fluid in a pipeline, the properties of the fluid, the dimensions of the pipeline, and the driving force for moving the fluid, as illustrated in Fig. 2-1. The procedure is as follows.

Step 1: Identify the important variables in the system. Most of the important fundamental variables in this system should be obvious. The flow rate can be represented by either the total volumetric flow rate (Q) or the average velocity in the pipeline (V). However, these are related by the definition $Q = \pi D^2 V/4$, so if D is chosen as an important variable, then either V or Q can be chosen to represent the flow rate, but not both. We shall choose V. The driving force can be represented by ΔP, the difference between the pressure at the upstream end of the pipe (P_1) and that at the downstream end (P_2) ($\Delta P = P_1 - P_2$). The pipe dimensions are the diameter (D) and length (L), and the fluid properties are the density (ρ) and viscosity (μ). It is also possible that the "texture" of the pipe wall (i.e., the surface roughness ε) is important. This identification of the pertinent variables is the most important step in the process and can be done by using experience, judgment, brainstorming, and intuition or by examining the basic equations that describe the fundamental physical principles governing the system along with appropriate boundary conditions. It is also important to include only those "fundamental" variables, i.e., those that are not derivable from others through basic definitions. For example, as pointed out above, the fluid velocity (V), the pipe diameter (D),

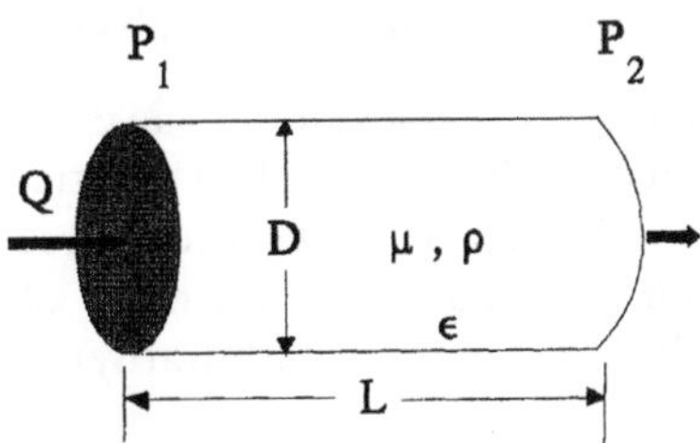

FIGURE 2-1 Flow in a pipeline.

and the volumetric flow rate (Q) are related by the definition $Q = \pi D^2 V/4$. Thus these three variables are not independent, since any one of them can be derived from the other two by this definition, it would therefore be necessary to include only two of the three.

Step 2: List all the problem variables and parameters, along with their dimensions. The procedure is simplest if the most fundamental dimensions in a scientific system (i.e., M, L, t) are used (e.g., energy should be converted to $FL = ML^2/t^2$, etc.):

Variable	Dimensions
V	L/t
ΔP	$F/L^2 = M/Lt^2$
D	L
L	L
ε	L
ρ	M/L^3
μ	M/Lt
7	3

The number of dimensionless groups that will be obtained is equal to the number of variables less the minimum number of fundamental dimensions involved in these variables (7-3 = 4 groups in this problem).

Step 3: Choose a set of reference variables. The choice of variables is arbitrary, except that the following criteria must be satisfied:

1. The number of reference variables must be equal to the minimum number of fundamental dimensions in the problem (in this case, three).
2. No two reference variables should have exactly the same dimensions.
3. All the dimensions that appear in the problem variables must also appear somewhere in the dimensions of the reference variables.

In general, the procedure is easiest if the reference variables chosen have the simplest combination of dimensions, consistent with the preceding criteria. In this problem we have three dimensions (M, L, t), so we need three reference variables. The variables D, ε, and L all have the dimension of length, so we can choose only one of these. We will choose D (arbitrarily) as one reference variable:

$$[D] = L$$

The dimension t appears in V, ΔP, and μ, but V has the simplest combination of dimensions, so we choose it as our second reference variable:

$$[V] = \mathrm{L/t}$$

We also need a reference variable containing the dimension M, which could be either ρ or μ. Since ρ has the simplest dimensions, we choose it for the third reference variable:

$$[\rho] = \mathrm{M/L^3}$$

Our three reference variables are therefore D, V, and ρ.

***Step 4: Solve the foregoing "dimensional equations" for the dimensions* (L, t, M) *in terms of the reference variables* (D, V, ρ), i.e.,**

$$\mathrm{L} = [D], \qquad \mathrm{t} = [D/V], \qquad \mathrm{M} = [\rho D^3]$$

Step 5: Write the dimensional equations for each of the remaining variables. Then substitute the results of step 4 for the dimensions in terms of the reference variables:

$$[\varepsilon] = \mathrm{L} = [D]$$

$$[L] = \mathrm{L} = [D]$$

$$[\mu] = \frac{\mathrm{M}}{\mathrm{Lt}} = \left[\frac{\rho D^3}{D(D/V)}\right] = [\rho V D]$$

$$[\Delta P] = \frac{\mathrm{M}}{\mathrm{Lt^2}} = \left[\frac{\rho D^3}{D(D/V)^2}\right] = [\rho V^2]$$

Step 6: These equations are each a dimensional identity, so dividing one side by the other results in one dimensionless group from each equation:

$$N_1 = \frac{\varepsilon}{D} \quad \text{or} \quad \frac{D}{\varepsilon}$$

$$N_2 = \frac{L}{D} \quad \text{or} \quad \frac{D}{L}$$

$$N_3 = \frac{\mu}{DV\rho} \quad \text{or} \quad \frac{DV\rho}{\mu}$$

$$N_4 = \frac{\Delta P}{\rho V^2} \quad \text{or} \quad \frac{\rho V^2}{\Delta P}$$

These four dimensionless groups can now be used as the primary variables to define the system behavior in place of the original seven variables.

B. Uniqueness

The results of the foregoing procedure are not unique, because the reciprocal of each group is just as valid as the initial group. In fact, any combination of these groups will be dimensionless and will be just as valid as any other combination as long as all of the original variables are represented among the groups. Thus these four groups can be replaced by any other four groups formed by a combination of these groups, and, indeed, a different set of groups would have resulted if we had used a different set of reference variables. However, any set of groups derived by forming a suitable combination of any other set would be just as valid. As we shall see, which set of groups is the most appropriate will depend on the particular problem to be solved, i.e., which of the variables are known (independent) and which are unknown (dependent). Specifically, it is most appropriate to arrange the groups so that the unknown variables each appear in only one group, if possible. It should be noted that the variables that were *not* chosen as the reference variables will each appear in only one group.

C. Dimensionless Variables

The original seven variables in this problem can now be replaced by an equivalent set of four dimensionless groups of variables. For example, if it is desired to determine the driving force required to transport a given fluid at a given rate through a given pipe, the relation could be represented as

$$\Delta P = \mathrm{fn}(V, D, L, \varepsilon, \rho, \mu)$$

or, in terms of the equivalent dimensionless variables (groups),

$$N_4 = \mathrm{fn}(N_1, N_2, N_3)$$

Note that the number of variables has been reduced from the original seven to four (groups). Furthermore, the relationship between these dimensionless variables or groups is independent of scale. That is, any two *similar* systems will be exactly equivalent, regardless of size or scale, if the values of all dimensionless variables or groups are the same in each. By "similar" we mean that both systems must have the same geometry or shape (which is actually another dimensionless variable), and both must be operating under comparable dynamic conditions (e.g., either laminar or turbulent—this will be expanded on later). Also, the fluids must be rheologically similar (e.g., Newtonian). The difference between Newtonian and non-Newtonian fluids will be discussed in Chapter 3. For the present, a Newtonian fluid is one that requires only one rheological property, the viscosity (μ), to determine flow behavior, whereas a non-Newtonian fluid requires a rheological "function" that contains two or more parameters. Each of these parameters is a rheological property, so in place of the viscosity for a Newtonian fluid, the non-Newtonian fluid would require two or more "rheological properties," depending upon the specific model that describes the fluid, with a corresponding increase in the number of dimensionless groups.

D. Problem Solution

It should be emphasized that the specific relationship between the variables or groups that is implied in the foregoing discussion is not determined by dimensional analysis. It must be determined from theoretical or experimental analysis. Dimensional analysis gives only an appropriate set of dimensionless groups that can be used as generalized variables in these relationships. However, because of the universal generality of the dimensionless groups, any functional relationship between them that is valid in any system must also be valid in any other similar system.

E. Alternative Groups

The preceding set of dimensionless groups is convenient for representing the behavior of a pipeline if it is desired to determine the driving force (ΔP) required to move a given fluid at a given rate through a given pipeline, because the unknown quantity (ΔP) appears in only one group (N_4), which can be considered the "dependent" group. However, the same variables apply to the case where the driving force is known and it is desired to determine the flow rate (Q or V) that would result for a given fluid through a given pipe. In this case, V is the dependent (unknown) variable, but it appears in more than one group (N_3 and N_4). Therefore, there is no single dependent group. However, this set of groups is not unique, so we can rearrange the groups into another equivalent set in which the unknown velocity appears in only one group. This can easily be done, for example, by combining groups N_3 and N_4 to form a group that does not contain V:

$$N_5 = (N_3)^2 N_4 = \left(\frac{DV\rho}{\mu}\right)^2 \left(\frac{\Delta P}{\rho V^2}\right) = \frac{\Delta P D^2 \rho}{\mu^2}$$

This new group can then be used in place of either N_3 or N_4, along with N_1 and N_2, to complete the required set of four groups in which the unknown V appears in only one group. If we replace N_4 by N_5, the implied relation can be expressed as

$$N_3 = \text{fn}(N_1, N_2, N_5) \qquad \text{or} \qquad \frac{DV\rho}{\mu} = \text{fn}\left(\frac{\varepsilon}{D}, \frac{L}{D}, \frac{\Delta P D^2 \rho}{\mu^2}\right)$$

in which the unknown (V) appears only in the group on the left.

Let us reexamine our original problem for a moment. If the pipeline is relatively long and is operating at steady state and the fluid is incompressible, then the conditions over any given length of the pipe will be the same as along any other segment of the same length, except for the regions very near the entrance and exit. If these regions are small relative to the rest of the pipe (e.g., $L \gg D$), their effect is negligible and the pressure drop per unit length of pipe should be the same over any given segment of the pipe. Thus the only significance of the pipe length is to spread the total pressure drop over the entire length, so that the two variables ΔP and L are not independent and can therefore be combined into one: the *pressure gradient*, $\Delta P/L$. This reduces the total number of variables from seven to six and the number of groups from four to three. These three groups can be derived by following the original procedure. However, because ΔP and L each appear in only one of the original groups (N_2 and N_4, respectively), dividing one of these by the other will automatically produce a group with the desired variable in the resulting group, which will then replace N_2 and N_4:

$$N_6 = \frac{N_4}{N_2} = \frac{D\,\Delta P/L}{\rho V^2}$$

The three groups are now N_1, N_3, and N_6:

$$N_1 = \frac{\varepsilon}{D}, \qquad N_3 = \frac{DV\rho}{\mu}, \qquad N_6 = \frac{D\,\Delta P/L}{\rho V^2}$$

Group N_6 (or some multiple thereof) is also known as a *friction factor* (f), because the driving force (ΔP) is required to overcome "friction" (i.e., the energy dissipated) in the pipeline (assuming it to be horizontal), and N_3 is known as the *Reynolds number* (N_{Re}). There are various definitions of the pipe friction factor, each of which is some multiple of N_6; e.g., the *Fanning friction factor* is $N_6/2$, and the *Darcy friction factor* is $2N_6$. The group N_4 is also known as the *Euler number*.

V. SCALE-UP

We have stated that dimensional analysis results in an appropriate set of groups that can be used to describe the behavior of a system, but it does not tell how these groups are related. In fact, dimensional analysis does not result in any *numbers* related to the groups (except for exponents on the variables). The relationship between the groups that represents the system behavior must be determined by either theoretical analysis or experimentation. Even when theoretical results are possible, however, it is often necessary to obtain data to evaluate or confirm the adequacy of the theory. Because relationships between dimensionless variables are independent of scale, the groups also provide a guide for the proper design of an experiment that is intended to simulate another (larger scale) similar system and for scaling up the results of model measurements to the full-scale system. For example, the operation of our pipeline can be described by a functional relationship of the form

$$N_6 = \text{fn}(N_1, N_3)$$

or

$$\frac{D\,\Delta P/L}{\rho V^2} = \text{fn}\left(\frac{\varepsilon}{D}, \frac{DV\rho}{\mu}\right)$$

This is valid for any Newtonian fluid in any (circular) pipe of any size (scale) under given dynamic conditions (e.g., laminar or turbulent). Thus, if the values of N_3 (i.e., the Reynolds number N_{Re}) and N_1 (ε/D) for an experimental model are identical to the values for a full-scale system, it follows that the value of N_6 (the friction factor) must also be the same in the two systems. In such a case the model is said to be dynamically similar to the full-scale (field) system, and measurements of the variables in N_6 can be translated (scaled) directly from the model to the field system. In other words, the equality between the groups N_3 (N_{Re}) and N_1 (ε/D) in the model and in the field is a necessary condition for the dynamic similarity of the two systems.

Example 2-2: Laminar Flow of a Newtonian Fluid in a Pipe. It turns out (for reasons that will be explained later) that if the Reynolds number in pipe flow has a value less than about 2000, the fluid elements follow a smooth, straight pattern called *laminar flow*. In this case, the "friction loss" (i.e., the pressure drop) does not depend upon the pipe wall roughness (ε) or the density (ρ) (the reason for this will become clear when we examine the mechanism of pipe flow in Chapter 6). With two fewer variables we would have two fewer groups, so that for a "long" pipe ($L \gg D$) the system can be described completely by only one group (that does not contain either ε or ρ). The form of this group can be determined by repeating the dimensional analysis procedure or simply by eliminating these two variables from the original three groups. This is easily done by multiplying the friction factor (f) by the Reynolds number (N_{Re}) to get the required group, i.e.,

$$N_7 = fN_{\text{Re}} = \frac{\Delta P\, D^2}{L\mu V} = \text{const.}$$

Because this is the only "variable" that is needed to describe this system, it follows that the value of this group must be the same, i.e., a constant, for the laminar flow of *any* Newtonian fluid at *any* flow rate in *any* pipe. This is in contrast to *turbulent pipe flow* (which occurs for $N_{Re} > 4000$) in long pipes, which can be described completely only by three groups (e.g., f, N_{Re}, and ε/D). That is, turbulent flow in two different pipes must satisfy the same functional relationship between these three groups even though the actual values of the individual groups may be quite different. However, for laminar pipe flow, since only one group (fN_{Re}) is required, the value of that group must be the same in *all* laminar pipe flows of Newtonian fluids, regardless of the values of the individual variables. The numerical value of this group will be derived theoretically in Chapter 6.

As an example of the application of dimensional analysis to experimental design and scale-up, consider the following example.

Example 2-3: Scale-Up of Pipe Flow. We would like to know the total pressure driving force (ΔP) required to pump oil ($\mu = 30$ cP, $\rho = 0.85$ g/cm^3) through a horizontal pipeline with a diameter (D) of 48 in. and a length (L) of 700 mi, at a flow rate (Q) of 1 million barrels per day. The pipe is to be of commercial steel, which has an equivalent roughness (ε) of 0.0018 in. To get this information, we want to design a laboratory experiment in which the laboratory model (m) and the full-scale field pipeline (f) are operating under dynamically similar conditions so that measurements of ΔP in the model can be scaled up directly to find ΔP in the field. The necessary conditions for dynamic similarity for this system are

$$(N_3)_m = (N_3)_f \quad \text{or} \quad \left(\frac{DV\rho}{\mu}\right)_m = \left(\frac{DV\rho}{\mu}\right)_f$$

and

$$(N_1)_m = (N_1)_f \quad \text{or} \quad \left(\frac{\varepsilon}{D}\right)_m = \left(\frac{\varepsilon}{D}\right)_f$$

from which it follows that

$$(N_6)_m = (N_6)_f \quad \text{or} \quad \left(\frac{\Delta PD}{L\rho V^2}\right)_m = \left(\frac{\Delta PD}{L\rho V^2}\right)_f$$

where the subscript m represents the experimental model and f represents the full-scale field system. Since the volumetric flow rate (Q) is specified instead of the velocity (V), we can make the substitution $V = 4Q/\pi D^2$ to get the following equivalent groups:

$$\left(\frac{\varepsilon}{D}\right)_{\mathrm{m}} = \left(\frac{\varepsilon}{D}\right)_{\mathrm{f}} \tag{2-11}$$

$$\left(\frac{4Q\rho}{\pi D\mu}\right)_{\mathrm{m}} = \left(\frac{4Q\rho}{\pi D\mu}\right)_{\mathrm{f}} \tag{2-12}$$

$$\left(\frac{\pi^2 \,\Delta P\, D^5}{16L\rho Q^2}\right)_{\mathrm{m}} = \left(\frac{\pi^2 \,\Delta P\, D^5}{16L\rho Q^2}\right)_{\mathrm{f}} \tag{2-13}$$

Note that all the numerical coefficients cancel out. By substituting the known values for the pipeline variables into Eq. (2-12), we find that the value of the Reynolds number for this flow is 5.4×10^4, which is turbulent. Thus all three of these groups are important.

We now identify the knowns and unknowns in the problem. The knowns obviously include all of the field variables except $(\Delta P)_{\mathrm{f}}$. Because we will measure the pressure drop in the lab model $(\Delta P)_{\mathrm{m}}$ after specifying the lab test conditions that simulate the field conditions, this will also be known. This value of $(\Delta P)_{\mathrm{m}}$ will then be scaled up to find the unknown pressure drop in the field, $(\Delta P)_{\mathrm{f}}$. Thus,

Knowns (7): $(D, L, \varepsilon, Q, \mu, \rho)_{\mathrm{f}}, (\Delta P)_{\mathrm{m}}$

Unknowns (7): $(D, L, \varepsilon, Q, \mu, \rho)_{\mathrm{m}}, (\Delta P)_{\mathrm{f}}$

There are seven unknowns but only three equations that relate these quantities. Therefore, four of the unknowns can be chosen "arbitrarily." This process is not really arbitrary, however, because we are constrained by certain practical considerations such as a lab model that must be smaller than the field pipeline, and test materials that are convenient, inexpensive, and readily available. For example, the diameter of the pipe to be used in the model could, in principle, be chosen arbitrarily. However, it is related to the field pipe diameter by Eq. (2-11):

$$D_{\mathrm{m}} = D_{\mathrm{f}}\left(\frac{\varepsilon_{\mathrm{m}}}{\varepsilon_{\mathrm{f}}}\right)$$

Thus, if we were to use the same pipe material (commercial steel) for the model as in the field, we would also have to use the same diameter (48 in.). This is obviously not practical, but a smaller diameter for the model would obviously require a much smoother material in the lab (because $D_{\mathrm{m}} \ll D_{\mathrm{f}}$ requires $\varepsilon_{\mathrm{m}} \ll \varepsilon_{\mathrm{f}}$). The smoothest material we can find would be glass or plastic or smooth drawn tubing such as copper or stainless steel, all of which have equivalent roughness values of the order of 0.00006 in. (see Table 6-1).

If we choose one of these (e.g., plastic), then the required lab diameter is set by Eq. (2-11):

$$D_m = D_f\left(\frac{\varepsilon_m}{\varepsilon_f}\right) = (48\text{ in.})\left(\frac{0.00006}{0.0018}\right) = 1.6\text{ in.}$$

Since the roughness values are only approximate, so is this value of D_m. Thus we could choose a convenient size pipe for the model with a diameter of the order of 1.6 in. (for example, from Appendix F, we see that a Schedule 40, $1\frac{1}{2}$ in. pipe has a diameter of 1.61 in., which is fortuitous).

We now have five remaining unknowns—Q_m, ρ_m, μ_m, L_m and $(\Delta P)_f$—and only two remaining equations, so we still have three "arbitrary" choices. Of course, we will choose a pipe length for the model that is much less than the 700 miles in the field, but it only has to be much longer than its diameter to avoid end effects. Thus we can choose any convenient length that will fit into the lab (say 50 ft), which still leaves two "arbitrary" unknowns to specify. Since there are two fluid properties to specify (μ and ρ), this means that we can choose (arbitrarily) any (Newtonian) fluid for the lab test. Water is the most convenient, available, and inexpensive fluid, and if we use it ($\mu = 1$ cP, $\rho = 1\text{ g/cm}^3$) we will have used up all our "arbitrary" choices. The remaining two unknowns, Q_m and $(\Delta P)_f$, are determined by the two remaining equations. From Eq. (2-12),

$$Q_m = Q_f\left(\frac{\rho_f}{\rho_m}\right)\left(\frac{D_m}{D_f}\right)\left(\frac{\mu_m}{\mu_f}\right) = \left(10^6\,\frac{\text{bbl}}{\text{day}}\right)\left(\frac{0.85}{1.0}\right)\left(\frac{1.6}{48}\right)\left(\frac{1.0}{30}\right)$$

$$= 944\text{ bbl/day}$$

or

$$Q_m = \left(\frac{994\text{ bbl}}{\text{day}}\right)\left(\frac{42\text{ gal}}{\text{bbl}}\right)\left(\frac{1}{1440\text{ min/day}}\right) = 27.5\text{ gal/min (gpm)}$$

Note that if the same units are used for the variables in both the model and the field, no conversion factors are needed, because only ratios are involved.

Now our experiment has been designed: We will use plastic pipe with an inside diameter of 1.6 in. and length of 50 ft and pump water through it at a rate of 27.5 gpm. Then we measure the pressure drop through this pipe and use our final equation to scaleup this value to find the field pressure drop. If the measured pressure drop with this system in the lab is, say, 1.2 psi, then the pressure drop in the field pipeline, from Eq. (2-13), would be

$$(\Delta P)_f = (\Delta P)_m \left(\frac{D_m}{D_f}\right)^5 \left(\frac{L_f}{L_m}\right)\left(\frac{\rho_r}{\rho_m}\right)\left(\frac{Q_f}{Q_m}\right)^2$$
$$= (1.2\,\text{psi})\left(\frac{1.6}{48}\right)^5 \left(\frac{700\,\text{mi} \times 5280\,\text{ft/mi}}{50\,\text{ft}}\right)\left(\frac{0.85}{1.0}\right)\left(\frac{10^6}{944}\right)^2$$
$$= 3480\,\text{psi}$$

This total pressure driving force would probably not be produced by a single pump but would be apportioned among several pumps spaced along the pipeline.

This example illustrates the power of dimensional analysis as an aid in experimental design and the scale-up of lab measurements to field conditions. We have actually determined the pumping requirements for a large pipeline by applying the results of dimensional analysis to select laboratory conditions and design a laboratory test model that simulates the field pipeline, making one measurement in the lab and scaling up this value to determine the field performance. We have not used any scientific principles or engineering correlations other than the principle of conservation of dimensions and the exercise of logic and judgment. However, we shall see later that information is available to us, based upon similar experiments that have been conducted by others (and presented in dimensionless form), that we can use to solve this and similar problems without conducting any additional experiments.

VI. DIMENSIONLESS GROUPS IN FLUID MECHANICS

Table 2-2 lists some dimensionless groups that are commonly encountered in fluid mechanics problems. The name of the group, and its symbol, definition, significance, and most common area of application are given in the table. Wherever feasible, it is desirable to express basic relations (either theoretical or empirical) in dimensionless form, with the variables being dimensionless groups, because this represents the most general way of presenting results and is independent of scale or specific system properties. We shall follow this guideline as far as is practical in this book.

VII. ACCURACY AND PRECISION

At this point, we digress slightly to make some observations about the accuracy and precision of experimental data. Since we, as engineers, continuously make use of data that represent measurements of various

Table 2-2 Dimensionless Groups in Fluid Mechanics

Name	Symbol	Formula	Notation	Significance	Application
Archimedes number	N_{Ar}	$N_{Ar} = \frac{\rho_f g \Delta\rho d^3}{\mu^2}$	$\rho_f =$ fluid density $\Delta\rho =$ solid density − fluid density	(Buoyant × inertial)/(viscous) forces	Settling particles, fluidization
Bingham number	N_{Bi}	$N_{Bi} = \frac{\tau_o D}{\mu_\infty V}$	$\tau_o =$ yield stress $\mu_\infty =$ limiting viscosity	(Yield/viscous) stresses	Flow of Bingham plastics
Bond number number	N_{Bo}	$N_{Bo} = \frac{\Delta\rho d^2 g}{\sigma}$	$\sigma =$ surface tension	(Gravity/surface tension) forces	Rise or fall of drops or bubbles
Cauchy number number	N_C	$N_C = \frac{\rho V^2}{K}$	$K =$ bulk modulus	(Inertial/compressible) forces	Compressible flow
Euler number	N_{Eu}	$N_{Eu} = \frac{\Delta P}{\rho V^2}$	$\Delta P =$ pressure drop in pipe	(Pressure energy)/(kinetic energy)	Flow in closed conduits
Drag coefficient	C_D	$C_D = \frac{F_D}{\frac{1}{2}\rho V^2 A}$	$F_D =$ drag force $A =$ area normal to flow	(Drag stress)/($\frac{1}{2}$ momentum flux)	External flows
Fanning (Darcy) friction factor	f (f or f_D)	$f = \frac{e_f D}{2V^2 L}$ $f_D = 4f$ $f = \frac{\tau_w}{\frac{1}{2}\rho V^2}$	$e_f =$ friction loss (energy/mass) $\tau_w =$ wall stress	(Energy dissipated)/(KE of flow × $4L/D$) or (Wall stress)/(momentum flux)	Flow in pipes, channels, fittings, etc.

Froude number	N_{Fr}	$N_{Fr} = V^2/gL$	L = characteristic length	(Inertial/gravity) forces	Free surface flows
Hedstrom number	N_{He}	$N_{He} = \frac{\tau_o D^2 \rho}{\mu_\infty^2}$	τ_o = yield stress μ_∞ = limiting viscosity	(Yield × inertia)/ viscous stresses	Flow of Bingham plastics
Reynolds number flows	N_{Re}	$N_{Re} = \frac{DV\rho}{\mu}$ $= \frac{\rho V^2}{\mu V/D}$ $= \frac{4Q\rho}{\pi D\mu}$ $= \frac{\rho V^2}{\tau_w/8}$	Pipe flow: τ_w = wall stress	(inertial momentum flux)/ (viscous momentum flux)	Pipe/internal flows (Equivalent forms for external flows)
Mach number	N_{Ma}	$N_{Ma} = \frac{V}{c}$	c = speed of sound	(Gas velocity)/(speed of sound)	High speed compressible flow

quantities, it is important that we understand and appreciate which of the numbers we deal with are useful and which are not.

First of all, we should make a clear distinction between accuracy and precision. *Accuracy* is a measure of how close a given value is to the "true" value, whereas *precision* is a measure of the uncertainty in the value or how "reproducible" the value is. For example, if we were to measure the width of a standard piece of paper using a ruler, we might find that it is 21.5 cm, give or take 0.1 cm. The "give or take" (i.e., the uncertainty) value of 0.1 cm is the precision of the measurement, which is determined by how close we are able to reproduce the measurement with the ruler. However, it is possible that when the ruler is compared with a "standard" unit of measure it is found to be in error by, say, 0.2 cm. Thus the "accuracy" of the ruler is limited, which contributes to the uncertainty of the measurement, although we may not know what this limitation is unless we can compare our "instrument" to one we know to be true.

Thus, the accuracy of a given value may be difficult to determine, but the precision of a measurement can be determined by the evaluation of reproducibility if multiple repetitions of the measurement are made. Unfortunately, when using values or data provided by others from handbooks, textbooks, journals, and so on, we do not usually have access to either the "true" value or information on the reproducibility of the measured values. However, we can make use of both common sense (i.e., reasonable judgment) and convention to estimate the implied precision of a given value. The number of decimal places when the value is represented in scientific notation, or the number of digits, should be indicative of its precision. For example, if the distance from Dallas to Houston is stated as being 250 miles and we drive at 60 miles/hr, should we say that it would take us 4.166667 (=250/60) hours for the trip? This number implies that we can determine the answer to a precision of 0.0000005 hr, which is one part in 10^7, or less than 2 milliseconds! This is obviously ludicrous, because the mileage value is nowhere near that precise (is it ± 1 mile, ± 5 miles?—*exactly* where did we start and end?), nor can we expect to drive at a speed having this degree of precision (e.g., 60 ± 0.000005 mph, or about $\pm 20\ \mu$m/s!). It is conventional to assume that the precision of a given number is comparable to the magnitude of the last digit to the right in that number. That is, we assume that the value of 250 miles implies 250 ± 1 mile (or perhaps $\pm$ 0.5 mile). However, unless the numbers are always given in scientific notation, so that the least significant digit can be associated with a specific decimal place, there will be some uncertainty, in which case common sense (judgment) should prevail.

For example, if the diameter of a tank is specified to be 10.32 ft, we could assume that this value has a precision (or uncertainty) of about

0.005 ft (or 0.06 in., or 1.5 mm). However, if the diameter is said to be 10 ft, the number of digits cannot provide an accurate guide to the precision of the number. It is unlikely that a tank of that size would be constructed to the precision of 1.5 mm, so we would probably assume (optimistically!) that the uncertainty is about 0.5 in or that the measurement is "roughly 10.0 ft." However, if I say that I have five fingers on my hand, this means *exactly* five, no more, no less (i.e., an "infinite" number of "significant digits").

In general, the number of decimal digits that are included in reported data, or the precision to which values can be read from graphs or plots, should be consistent with the precision of the data. Therefore, answers calculated from data with limited precision will likewise be limited in precision (computer people have an acronym for this—"GIGO," which stands for "garbage in, garbage out"). When the actual precision of data or other information is uncertain, a general rule of thumb is to report numbers to no more than three "significant digits," this corresponds to an uncertainty of somewhere between 0.05% and 0.5 % (which is actually much greater precision than can be justified by most engineering data). Inclusion of more that three digits in your answer implies a greater precision than this and should be justified. Those who report values with a large number of digits that cannot be justified are usually making the implied statement "I just wrote down the numbers—I really didn't think about it." This is most unfortunate, because if these people don't think about the numbers they write down, how can we be sure that they are thinking about other critical aspects of the problem?

Example 2-4: Our vacation time accrues by the hour, a certain number of hours of vacation time being credited per month worked. When we request leave or vacation, we are likewise expected to report it in increments of 1 hr. We received a statement from the accountants that we have accrued "128.00 hours of vacation time." What is the precision of this number?

The precision is implied by half of the digit furtherest to the right of the decimal point, i.e., 0.005 hr, or 18 s. Does this imply that we must report leave taken to the closest 18 s? (We think not. It takes at least a minute to fill out the leave request form—would this time be charged against our accrued leave? The accountant just "wasn't thinking" when the numbers were reported.)

When combining values, each of which has a finite precision or uncertainty, it is important to be able to estimate the corresponding uncertainty of the result. Although there are various "rigorous" ways of doing this, a very simple method that gives good results as long as the relative uncertainty is a small fraction of the value is to use the approximation (which is really just the first term of a Taylor series expansion)

$$A(1 \pm a)^x \cong A(1 \pm xa + \cdots)$$

which is valid for any value of x if $a < 0.1$ (about). This assumes that the relative uncertainty of each quantity is expressed as a fraction of the given value, e.g., the fractional uncertainty in the value A is a or, equivalently, the percentage error in A is $100a$.

Example 2-5: Suppose we wish to calculate the shear stress on the bob surface in a cup-and-bob viscometer from a measured value of the torque or moment on the bob. The equation for this is

$$\tau_{r\theta} = \frac{T}{2\pi R_i^2 L}$$

If the torque (T) can be measured to ±5%, the bob radius (R_i) is known to ±1%, and the length (L) is known to ±3%, the corresponding uncertainty in the shear stress can be determined as follows:

$$\begin{aligned}\tau_{r\theta} &= \frac{T(1 \pm 0.05)}{2\pi R_i^2 (1 \pm 0.01)^2 L(1 \pm 0.03)} \\ &= \frac{T}{2\pi R_i^2 L}[1 \pm (0.05) \pm (2)(0.01) \pm (0.03)] \\ &= \frac{T}{2\pi R_i^2 L}(1 \pm 0.1)\end{aligned}$$

That is, there would be a 10% error, or uncertainty, in the answer. Note that even though terms in the denominator have a negative exponent, the maximum error due to these terms is still cumulative, because a given error may be either positive or negative; i.e., errors may either accumulate (giving rise to the maximum possible error) or cancel out (we should be so lucky!).

PROBLEMS

Units and Dimensions

1. Determine the weight of 1 g mass at sea level in units of (a) dynes; (b) lb_f; (c) g_f; (d) poundals.

2. One cubic foot of water weighs 62.4 lb_f under conditions of standard gravity.
   (a) What is its weight in dynes, poundals, and g_f?
   (b) What is its density in lb_m/ft^3 and $slugs/ft^3$?
   (c) What is its weight on the moon ($g = 5.4\,ft^2$) in lb_f?
   (d) What is its density on the moon?
3. The acceleration due to gravity on the moon is about $5.4\,ft/s^2$. If your weight is 150 lb_f on earth:
   (a) What is your mass on the moon, in slugs?
   (b) What is your weight on the moon, in SI units?
   (c) What is your weight on earth, in poundals?
4. You weigh a body with a mass m on an electronic scale, which is calibrated with a known mass.
   (a) What does the scale actually measure, and what are its dimensions?
   (b) If the scale is calibrated in the appropriate system of units, what would the scale reading be if the mass of m is (1) 1 slug; (2) 1 lb_m (in scientific units); (3) 1 lb_m (in engineering units); (3) 1 g_m (in scientific units); (4) 1 g_m (in engineering units).
5. Explain why the gravitational "constant" (g) is different at Reykjavik, Iceland, than it is at La Paz, Bolivia. At which location is it greater, and why? If you could measure the value of g at these two locations, what would this tell you about the earth?
6. You have purchased a 5 oz. bar of gold (100% pure), at a cost of \$400/oz. Because the bar was weighed in air, you conclude that you got a bargain, because its true mass is greater than 5 oz due to the buoyancy of air. If the true density of the gold is $1.9000\ g/cm^3$, what is the actual value of the bar based upon its true mass?
7. You purchased 5 oz of gold in Quito, Ecuador ($g = 977.110\,cm/s^2$), for \$400/oz. You then took the gold and the same spring scale on which you weighed it in Quito to Reykjavik Iceland ($G = 983.06\,cm/s^2$), where you weighed it again and sold it for \$400/oz. How much money did you make or lose, or did you break even?
8. Calculate the pressure at a depth of 2 miles below the surface of the ocean. Explain and justify any assumptions you make. The physical principle that applies to this problem can be described by the equation

$$\Phi = \text{constant}$$

   where $\Phi = P + \rho g z$ and z is the vertical distance measured upward from any horizontal reference plane. Express your answer in units of (a) atm, (b) psi, (c) Pa, (d) $poundal/ft^2$, (e) dyn/cm^2.
9. (a) Use the principle in Problem 8 to calculate the pressure at a depth of 1000 ft below the surface of the ocean (in psi, Pa, and atm). Assume that the ocean water density is 64 lb_m/ft^3.
   (b) If this ocean were on the moon, what would be the answer to (a)?. Use the following information to solve this problem: The diameter of the moon is 2160 mi, the diameter of the earth is 8000 mi, and the density of the earth is 1.6 times that of the moon.

10. The following formula for the pressure drop through a valve was found in a design manual:

$$h_L = \frac{522Kq^2}{d^4}$$

where h_l = the "head loss" in feet of fluid flowing through the valve, K = dimensionless resistance coefficient for the valve, q = flow rate through the valve, in ft^3/s, and d = diameter of the valve, in inches.
   (a) Can this equation be used without changing anything if SI units are used for the variables? Explain.
   (b) What are the dimensions of "522" in this equation? What are its units?
   (c) Determine the pressure drop through a 2-in. valve with a K of 4 for water at 20°C flowing at a rate of 50 gpm (gal/min), in units of (1) feet of water, (2) psi, (3) atm, (4) Pa, (5) dyn/cm^2, and (6) inches of mercury.
11. When the energy balance on the fluid in a stream tube is written in the following form, it is known as Bernoulli's equation:

$$\frac{P_2 - P_1}{\rho} + g(z_2 - z_1) + \frac{\alpha}{2}(V_2^2 - V_1^2) + e_f + w = 0$$

where $-w$ is the work done on a unit mass of fluid, e_f is the energy dissipated by friction in the fluid per unit mass, including all thermal energy effects due to heat transfer or internal generation, and α is equal to either 1 or 2 for turbulent or laminar flow, respectively. If $P_1 = 25$ psig, $P_2 = 10$ psig, $z_1 = 5$ m, $z_2 = 8$ m, $V_1 = 20$ ft/s, $V_2 = 5$ ft/s, $\rho = 62.4\ lb_m/ft^3$, $\alpha = 1$, and $w = 0$, calculate the value of e_f in each of the following systems of units:
   (a) SI
   (b) mks engineering (e.g., metric engineering)
   (c) English engineering
   (d) English scientific (with M as a fundamental dimension)
   (e) English thermal units (e.g., Btu)
   (f) Metric thermal units (e.g., calories)

Conversion Factors, Precision

12. Determine the value of the gas constant R in units of ft^3 atm/lb mol °R), starting with the value of the standard molar volume of a perfect gas.
13. Calculate the value of the Reynolds number for sodium flowing at a rate of 50 gpm through a 1/2 in. ID tube at 400°F.
14. The conditions at two different positions along a pipeline (at points 1 and 2) are related by the Bernoulli equation (see Problem 11). For flow in a pipe,

$$e_f = \left(\frac{4fL}{D}\right)\left(\frac{V^2}{2}\right)$$

where D is the pipe diameter and L is the pipe length between points 1 and 2. If the flow is laminar ($N_{Re} < 2000$), the value of α is 2 and $f = 16/N_{Re}$, but for

turbulent flow in a smooth pipe $\alpha = 1$ and $f = 0.0791/N_{\text{Re}}^{1/4}$. The work done by a pump on the fluid $(-w)$ is related to the power delivered to the fluid (HP) and the mass flow rate of the fluid $(\dot{m})$ by HP $= -w\dot{m}$. Consider water ($\rho = 1\ \text{g/cm}^3$, $\mu = 1$ cP) being pumped at a rate of 150 gpm (gal/min) through a 2000 ft long, 3 in. diameter pipe. The water is transported from a reservoir ($z = 0$) at atmospheric pressure to a condenser at the top of a column that is at an elevation of 30 ft and a pressure of 5 psig. Determine:
(a) The value of the Reynolds number in the pipe
(b) The value of the friction factor in the pipe (assuming that it is smooth)
(c) The power that the pump must deliver to the water, in horsepower (hp)

15. The Peclet number (P_{Pe}) is defined as

$$N_{\text{Pe}} = N_{\text{Re}}N_{\text{Pr}} = \left(\frac{DV\rho}{\mu}\right)\left(\frac{c_p\mu}{k}\right) = \frac{DGc_p}{k}.$$

where D = pipe diameter, G = mass flux $= \rho V$, c_p = specific heat, k = thermal conductivity, μ = viscosity. Calculate the value of N_{Pe} for water at 60°F flowing through a 1 cm diameter tube at a rate of $100\ \text{lb}_m/\text{hr}$. (Use the most accurate data you can find, and state your answer in the appropriate number of digits consistent with the data you use.)

16. The heat transfer coefficient (h) for a vapor bubble rising through a boiling liquid is given by

$$h = A\left[\frac{kV\rho c_p}{d}\right]^{1/2} \qquad \text{where } V = \left[\frac{\Delta\rho g\sigma}{\rho_v^2}\right]^{1/4}$$

where h = heat transfer coefficient [e.g., Btu/(hr °F ft^2)], c_p = liquid heat capacity [e.g., cal/(g °C)], k = liquid thermal conductivity [e.g., J/(s K m)], σ = liquid/vapor surface tension [e.g., dyn/cm], $\Delta\rho = \rho_{\text{liquid}} - \rho_{\text{vapor}} = \rho_l - \rho_v$, d = bubble diameter, and g = acceleration due to gravity.
(a) What are the fundamental dimensions of V and A?
(b) If the value of h is 1000 Btu/(hr ft^2 °F) for a 5 mm diameter steam bubble rising in boiling water at atmospheric pressure, determine the corresponding values of V and A in SI units. You must look up values for the other quantities you need; be sure to cite the sources you use for these data.

17. Determine the value of the Reynolds number for SAE 10 lube oil at 100°F flowing at a rate of 2000 gpm through a 10 in. Schedule 40 pipe. The oil SG is 0.92, and its viscosity can be found in Appendix A. If the pipe is made of commercial steel ($\varepsilon = 0.0018$ in.), use the Moody diagram (see Fig. 6-4) to determine the friction factor f for this system. Estimate the precision of your answer, based upon the information and procedure you used to determine it (i.e., tell what the reasonable upper and lower bounds, or the corresponding percentage variation, should be for the value of f based on the information you used).

18. Determine the value of the Reynolds number for water flowing at a rate of 0.5 gpm through a 1 in. ID pipe. If the diameter of the pipe is doubled at the same flow rate, how much will each of the following change:

(a) The Reynolds number
(b) The pressure drop
(c) The friction factor

19. The pressure drop for a fluid with a viscosity of 5 cP and a density of 0.8 g/cm^3, flowing at a rate of 30 g/s in a 50 ft long 1/4 in. diameter pipe is 10 psi. Use this information to determine the pressure drop for water at 60°F flowing at 0.5 gpm in a 2 in. diameter pipe. What is the value of the Reynolds number for each of these cases?

Dimensional Analysis and Scale-Up

20. In the steady flow of a Newtonian fluid through a long uniform circular tube, if $N_{Re} < 2000$ the flow is laminar and the fluid elements move in smooth straight parallel lines. Under these conditions, it is known that the relationship between the flow rate and the pressure drop in the pipe does not depend upon the fluid density or the pipe wall material.
    (a) Perform a dimensional analysis of this system to determine the dimensionless groups that apply. Express your result in a form in which the Reynolds number can be identified.
    (b) If water is flowing at a rate of 0.5 gpm through a pipe with an ID of 1 in., what is the value of the Reynolds number? If the diameter is doubled at the same flow rate, what will be the effect on the Reynolds number and on the pressure drop?
21. Perform a dimensional analysis to determine the groups relating the variables that are important in determining the settling rate of very small solid particles falling in a liquid. Note that the driving force for moving the particles is gravity and the corresponding *net* weight of the particle. At very slow settling velocities, it is known that the velocity is independent of the fluid density. Show that this also requires that the velocity be inversely proportional to the fluid viscosity.
22. A simple pendulum consists of a small, heavy ball of mass m on the end of a long string of length L. The period of the pendulum should depend on these factors, as well as on gravity, which is the driving force for making it move. What information can you get about the relationship between these variables from a consideration of their dimensions? Suppose you measured the period, T_1, of a pendulum with mass m_1 and length L_1. How could you use this to determine the period of a different pendulum with a different mass and length? What would be the ratio of the pendulum period on the moon to that on the earth? How could you use the pendulum to determine the variation of g on the earth's surface?
23. An ethylene storage tank in your plant explodes. The distance that the blast wave travels from the blast site (R) depends upon the energy released in the blast (E), the density of the air (ρ), and time (t). Use dimensional analysis to determine:
    (a) The dimensionless group(s) that can be used to describe the relationship between the variables in the problem

(b) The ratio of the velocity of the blast wave at a distance of 2000 ft from the blast site to the velocity at a distance of 500 ft from the site
The pressure difference across the blast wave (ΔP) also depends upon the blast energy (E), the air density (ρ), and time (t). Use this information to determine:
(c) The ratio of the blast pressure at a distance of 500 ft from the blast site to that at a distance of 2000 ft from the site

24. It is known that the power required to drive a fan depends upon the impeller diameter (D), the impeller rotational speed (ω), the fluid density (ρ), and the volume flow rate (Q). (Note that the fluid viscosity is not important for gases under normal conditions.)
    (a) What is the minimum number of fundamental dimensions required to define all of these variables?
    (b) How many dimensionless groups are required to determine the relationship between the power and all the other variables? Find these groups by dimensional analysis, and arrange the results so that the power and the flow rate each appear in only one group.
25. A centrifugal pump with an 8 in. diameter impeller operating at a rotational speed of 1150 rpm requires 1.5 hp to deliver water at a rate of 100 gpm and a pressure of 15 psi. Another pump for water, which is geometrically similar but has an impeller diameter of 13 in., operates at a speed of 1750 rpm. Estimate the pump pressure, flow capacity, and power requirements of this second pump. (Under these conditions, the performance of both pumps is independent of the fluid viscosity.)
26. A gas bubble of diameter d rises with velocity V in a liquid of density ρ and viscosity μ.
    (a) Determine the dimensionless groups that include the effects of all the significant variables, in such a form that the liquid viscosity appears in only one group. Note that the driving force for the bubble motion is buoyancy, which is equal to the weight of the displaced fluid.
    (b) You want to know how fast a 5 mm diameter air bubble will rise in a liquid with a viscosity of 20 cP and a density of 0.85 g/cm^3. You want to simulate this system in the laboratory using water ($\mu = 1$ cP, $\rho = 1$ g/cm^3) and air bubbles. What size air bubble should you use?
    (c) You perform the experiment, and measure the velocity of the air bubble in water (V_m). What is the ratio of the velocity of the 5 mm bubble in the field liquid (V_f) to that in the lab (V_m)?
27. You must predict the performance of a large industrial mixer under various operating conditions. To obtain the necessary data, you decide to run a laboratory test on a small-scale model of the unit. You have deduced that the power (P) required to operate the mixer depends upon the following variables:

Tank diameter D	Impeller diameter d
Impeller rotational speed N	Fluid density ρ
Fluid viscosity μ	

(a) Determine the minimum number of fundamental dimensions involved in these variables and the number of dimensionless groups that can be defined by them.
(b) Find an appropriate set of dimensionless groups such that D and N each appear in only one group. If possible, identify one or more of the groups with groups commonly encountered in other systems.
(c) You want to know how much power would be required to run a mixer in a large tank 6 ft in diameter, using an impeller with a diameter of 3 ft operating at a speed of 10 rpm, when the tank contains a fluid with a viscosity of 25 cP and a specific gravity of 0.85. To do this, you run a lab test on a model of the system, using a scale model of the impeller that is 10 in. in diameter. The only appropriate fluid you have in the lab has a viscosity of 15 cP and a specific gravity of 0.75. Can this fluid be used for the test? Explain.
(d) If the preceding lab fluid is used, what size tank should be used in the lab, and how fast should the lab impeller be rotated?
(e) With the lab test properly designed and the proper operating conditions chosen, you run the test and find that it takes 150 W to operate the lab test model. How much power would be required to operate the larger field mixer under the plant operating conditions?

28. When an open tank with a free surface is stirred with an impeller, a vortex will form around the shaft. It is important to prevent this vortex from reaching the impeller, because entrainment of air in the liquid tends to cause foaming. The shape of the free surface depends upon (among other things) the fluid properties, the speed and size of the impeller, the size of the tank, and the depth of the impeller below the free surface.
(a) Perform a dimensional analysis of this system to determine an appropriate set of dimensionless groups that can be used to describe the system performance. Arrange the groups so that the impeller speed appears in only one group.
(b) In your plant you have a 10 ft diameter tank containing a liquid that is 8 ft deep. The tank is stirred by an impeller that is 6 ft in diameter and is located 1 ft from the tank bottom. The liquid has a viscosity of 100 cP and a specific gravity of 1.5. You need to know the maximum speed at which the impeller can be rotated without entraining the vortex. To find this out, you design a laboratory test using a scale model of the impeller that is 8 in. in diameter. What, if any, limitations are there on your freedom to select a fluid for use in the lab test?
(c) Select an appropriate fluid for the lab test and determine how large the tank used in the lab should be and where in the tank the impeller should be located.
(d) The lab impeller is run at such a speed that the vortex just reaches the impeller. What is the relation between this speed and that at which entrainment would occur in the tank in the plant?

29. The variables involved in the performance of a centrifugal pump include the fluid properties (μ and ρ), the impeller diameter (d), the casing diameter (D), the impeller rotational speed (N), the volumetric flow rate of the fluid (Q), the head

(H) developed by the pump ($\Delta P = \rho g H$), and the power required to drive the pump (HP).

(a) Perform a dimensional analysis of this system to determine an appropriate set of dimensionless groups that would be appropriate to characterize the pump. Arrange the groups so that the fluid viscosity and the pump power each appear in only one group.

(b) You want to know what pressure a pump will develop with a liquid that has a specific gravity of 1.4 and a viscosity of 10 cP, at a flow rate of 300 gpm. The pump has an impeller with a diameter of 12 in., which is driven by a motor running at 1100 rpm. (It is known that the pump performance is independent of fluid viscosity unless the viscosity is greater than about 50 cP.) You want to run a lab test that simulates the operation of the larger field pump using a similar (scaled) pump with an impeller that has a diameter of 6 in. and a 3600 rpm motor,. Should you use the same liquid in the lab as in the field, or can you use a different liquid? Why?

(c) If you use the same liquid, what flow rate should be used in the lab to simulate the operating conditions of the field pump?

(d) If the lab pump develops a pressure of 150 psi at the proper flow rate, what pressure will the field pump develop with the field fluid?

(e) What pressure would the field pump develop with water at a flow rate of 300 gpm?

30. The purpose of a centrifugal pump is to increase the pressure of a liquid in order to move it through a piping system. The pump is driven by a motor, which must provide sufficient power to operate the pump at the desired conditions. You wish to find the pressure developed by a pump operating at a flow rate of 300 gpm with an oil having a specific gravity (SG) of 0.8 and a viscosity of 20 cP, and the required horsepower for the motor to drive the pump. The pump has an impeller diameter of 10 in., and the motor runs at 1200 rpm.

(a) Determine the dimensionless groups that would be needed to completely describe the performance of the pump.

(b) You want to determine the pump pressure and motor horsepower by measuring these quantities in the lab on a smaller scale model of the pump that has a 3 in. diameter impeller and a 1800 rpm motor, using water as the test fluid. Under the operating conditions for both the lab model and the field pump, the value of the Reynolds number is very high, and it is known that the pump performance is independent of the fluid viscosity under these conditions. Determine the proper flow rate at which the lab pump should be tested and the ratio of the pressure developed by the field pump to that of the lab pump operating at this flow rate as well as the ratio of the required motor power in the field to that in the lab.

(c) The pump efficiency (η_e) is the ratio of the power delivered by the pump to the fluid (as determined by the pump pressure and flow rate) to the power delivered to the pump by the motor. Because this is a dimensionless number, it should also have the same value for both the lab and field pumps when they are operating under equivalent conditions. Is this condition satisfied?

31. When a ship moves through the water, it causes waves. The energy and momentum in these waves must come from the ship, which is manifested as a "wave drag" force on the ship. It is known that this drag force (F) depends upon the ship speed (V), the fluid properties (ρ, μ), the length of the waterline (L), and the beam width (W) as well as the shape of the hull, among other things. (There is at least one important "other thing" that relates to the "wave drag," i.e., the energy required to create and sustain the waves from the bow and the wake. What is this additional variable?) Note that "shape" is a dimensionless parameter, which is implied by the requirement of geometrical similarity. If two geometries have the same shape, the ratio of each corresponding dimension of the two will also be the same.
   (a) Perform a dimensional analysis of this system to determine a suitable set of dimensionless groups that could be used to describe the relationship between all the variables. Arrange the groups such that viscous and gravitational parameters each appear in separate groups.
   (b) It is assumed that "wave drag" is independent of viscosity and that "hull drag" is independent of gravity. You wish to determine the drag on a ship having a 500 ft long waterline moving at 30 mph through seawater (SG = 1.1). You can make measurements on a scale model of the ship, 3 ft long; in a towing tank containing fresh water. What speed should be used for the model to simulate the wave drag and the hull drag?
32. You want to find the force exerted on an undersea pipeline by a 10 mph current flowing normal to the axis of the pipe. The pipe is 30 in. in diameter; the density of seawater is 64 $\mathrm{lb_m/ft^3}$ and its viscosity is 1.5 cP. To determine this, you test a $1\frac{1}{2}$ in. diameter model of the pipe in a wind tunnel at 60°F. What velocity should you use in the wind tunnel to scale the measured force to the conditions in the sea? What is the ratio of the force on the pipeline in the sea to that on the model measured in the wind tunnel?
33. You want to determine the thickness of the film when a Newtonian fluid flows uniformly down an inclined plane at an angle θ with the horizontal at a specified flow rate. To do this, you design a laboratory experiment from which you can scale up measured values to any other Newtonian fluid under corresponding conditions.
   (a) List all the independent variables that are important in this problem, with their dimensions. If there are any variables that are not independent but act only in conjunction with one another, list only the net combination that is important.
   (b) Determine an appropriate set of dimensionless groups for this system, in such a way that the fluid viscosity and the plate inclination each appear in only one group.
   (c) Decide what variables you would choose for convenience, what variables would be specified by the analysis, and what you would measure in the lab.
34. You would like to know the thickness of a syrup film as it drains at a rate of 1 gpm down a flat surface that is 6 in. wide and is inclined at an angle of 30° from the vertical. The syrup has a viscosity of 100 cP and an SG of 0.9. In the

laboratory, you have a fluid with a viscosity of 70 cP and an SG of 1.0 and a 1 ft wide plane inclined at an angle of 45° from the vertical.

(a) At what flow rate, in gpm, would the laboratory conditions simulate the specified conditions?

(b) If the thickness of the film in the laboratory is 3 mm at the proper flow rate, what would the thickness of the film be for the 100 cP fluid at the specified conditions?

35. The size of liquid droplets produced by a spray nozzle depends upon the nozzle diameter, the fluid velocity, and the fluid properties (which may, under some circumstances, include surface tension).
    (a) Determine an appropriate set of dimensionless groups for this system.
    (b) You want to know what size droplets will be generated by a fuel oil nozzle with a diameter of 0.5 mm at an oil velocity of 10 m/s. The oil has a viscosity of 10 cP, an SG of 0.82, and a surface tension of 35 dyn/cm. You have a nozzle in the lab with a nozzle diameter of 0.2 mm that you want to use in a lab experiment to find the answer. Can you use the same fuel oil in the lab test as in the field? If not, why not?
    (c) If the only fluid you have is water, tell how you would design the lab experiment. *Note*: Water has a viscosity of 1 cP and an SG of 1, but its surface tension can be varied by adding small amounts of surfactant, which does not affect the viscosity or density.
    (d) Determine what conditions you would use in the lab, what you would measure, and the relationship between the measured and the unknown droplet diameters.
36. Small solid particles of diameter d and density ρ_s are carried horizontally by an air stream moving at velocity V. The particles are initially at a distance h above the ground, and you want to know how far they will be carried horizontally before they settle to the ground. To find this out, you decide to conduct a lab experiment using water as the test fluid.
    (a) Determine what variables you must set in the lab and how the value of each of these variables is related to the corresponding variable in the air system. You should note that two forces act on the particle: the drag force due to the moving fluid, which depends on the fluid and solid properties, the size of the particle, and the relative velocity; and the gravitational force, which is directly related to the densities of both the solid and the fluid in a known manner.
    (b) Is there any reason why this experiment might not be feasible in practice?
37. You want to find the wind drag on a new automobile design at various speeds. To do this, you test a 1/30 scale model of the car in the lab. You must design an experiment whereby the drag force measured in the lab can be scaled up directly to find the force on the full-scale car at a given speed.
    (a) What is the minimum number of (dimensionless) variables required to completely define the relationship between all the important variables in the problem? Determine the appropriate variables (e.g., the dimensionless groups).

(b) The only fluids you have available in the lab are air and water. Could you use either one of these, if you wanted to? Why (or why not)?
(c) Tell which of these fluids you would use in the lab, and then determine what the velocity of this fluid past the model car would have to be so that the experiment would simulate the drag on the full-scale car at 40 mph. If you decide that it is possible to use either fluid, determine the answer for each of them.
(d) What is the relationship between the measured drag force on the model and the drag force on the full scale car? If possible, determine this relationship for the other fluid as well. Repeat this for a speed of 70 mph.
(e) It turns out that for very high values of the Reynolds number, the drag force is independent of the fluid viscosity. Under these conditions, if the speed of the car doubles, by what factor does the power required to overcome wind drag change?

38. The power required to drive a centrifugal pump and the pressure that the pump will develop depend upon the size (diameter) and speed (angular velocity) of the impeller, the volumetric flow rate through the pump, and the fluid properties. However, if the fluid is not too viscous (e.g. less than about 100 cP), the pump performance is essentially independent of the fluid viscosity. Under these conditions:
(a) Perform a dimensional analysis to determine the dimensionless groups that would be required to define the pump performance. Arrange the groups so that the power and pump pressure each appear in only one group.

You have a pump with an 8 in. diameter impeller that develops a pressure of 15 psi and requires 1.5 hp to operate when running at 1150 rpm with water at a flow rate of 100 gpm. You also have a similar pump with a 13 in. diameter impeller, driven by a 1750 rpm motor, and you would like to know what pressure this pump would develop with water and what power would be required to drive it.
(b) If the second pump is to be operated under equivalent (similar) conditions to the first one, what should the flow rate be?
(c) If this pump is operated at the proper flow rate, what pressure will it develop, and what power will be required to drive it when pumping water?

39. In a distillation column, vapor is bubbled through the liquid to provide good contact between the two phases. The bubbles are formed when the vapor passes upward through a hole (orifice) in a plate (tray) that is in contact with the liquid. The size of the bubbles depends upon the diameter of the orifice, the velocity of the vapor through the orifice, the viscosity and density of the liquid, and the surface tension between the vapor and the liquid.
(a) Determine the dimensionless groups required to completely describe this system, in such a manner that the bubble diameter and the surface tension do not appear in the same group.
(b) You want to find out what size bubbles would be formed by a hydrocarbon vapor passing through a 1/4 in. orifice at a velocity of 2 ft/s, in contact with a liquid having a viscosity of 4 cP and a density of 0.95 g/cm^3 (the surface tension is 30 dyn/cm). To do this, you run a lab experiment using air and

water (surface tension 60 dyn/cm). (1) What size orifice should you use, and what should the air velocity through the orifice be? (2) You design and run this experiment and find that the air bubbles are 0.1 in. in diameter. What size would the vapor bubbles be in the organic fluid above the 1/4 in. orifice?

40. A flag will flutter in the wind at a frequency that depends upon the wind speed, the air density, the size of the flag (length and width), gravity, and the "area density" of the cloth (i.e. the mass per unit area). You have a very large flag (40 ft long and 30 ft wide) which weighs 240 lb, and you want to find the frequency at which it will flutter in a wind of 20 mph.
    (a) Perform a dimensional analysis to determine an appropriate set of dimensionless groups that could be used to describe this problem.
    (b) To find the flutter frequency you run a test in a wind tunnel (at normal atmospheric temperature and pressure) using a flag made from a cloth that weighs 0.05 lb/ft^2. Determine (1) the size of the flag and the wind speed that you should use in the wind tunnel and (2) the ratio of the flutter frequency of the big flag to that which you observe for the model flag in the wind tunnel.
41. If the viscosity of the liquid is not too high (e.g., less than about 100 cP), the performance of many centrifugal pumps is not very sensitive to the fluid viscosity. You have a pump with an 8 in. diameter impeller that develops a pressure of 15 psi and consumes 1.5 hp when running at 1150 rpm pumping water at a rate of 100 gpm. You also have a similar pump with a 13 in. diameter impeller driven by a 1750 rpm motor, and you would like to know what pressure that pump would develop with water and how much power it would take to drive it.
    (a) If the second pump is to be operated under conditions similar to that of the first, what should the flow rate be?
    (b) When operated at this flow rate with water, (1) what pressure should it develop and (2) what power would be required to drive it?
42. The pressure developed by a centrifugal pump depends on the fluid density, the diameter of the pump impeller, the rotational speed of the impeller, and the volumetric flow rate through the pump (centrifugal pumps are not recommended for highly viscous fluids, so viscosity is not commonly an important variable). Furthermore, the pressure developed by the pump is commonly expressed as the "pump head," which is the height of a column of the fluid in the pump that exerts the same pressure as the pump pressure.
    (a) Perform a dimensional analysis to determine the minimum number of variables required to represent the pump performance characteristic in the most general (dimensionless) form.
    (b) The power delivered to the fluid by the pump is also important. Should this be included in the list of important variables, or can it be determined from the original set of variables? Explain.

    You have a pump in the field that has a 1.5 ft diameter impeller that is driven by a motor operating at 750 rpm. You want to determine what head the pump will develop when pumping a liquid with a density of 50 lb$_m$/ft^3 at a rate of 1000 gpm. You do this by running a test in the lab on a scale model of the pump that has a 0.5 ft diameter impeller using water (at 70°F) and a motor that runs at 1200 rpm.

(c) At what flow rate of water (in gpm) should the lab pump be operated?

(d) If the lab pump develops a head of 85 ft at this flow rate, what head would the pump in the field develop with the operating fluid at the specified flow rate?

(e) How much power (in horsepower) is transferred to the fluid in both the lab and the field cases?

(f) The pump efficiency is defined as the ratio of the power delivered to the fluid to the power of the motor that drives the pump. If the lab pump is driven by a 2 hp motor, what is the efficiency of the lab pump? If the efficiency of the field pump is the same as that of the lab pump, what power motor (horsepower) would be required to drive it?

NOTATION

D	diameter, [L]
f	friction factor, [—]
F	dimension of force
G	gravitational constant 6.67×10^{-11} N m^2/kg^2, Eq. (2-2), [FL^2/M^2]=[L^3/Mt^2]
g	acceleration due to gravity, [L/t^2]
g_c	conversion factor, [ML/(Ft^2]
ID	inside diameter of pipe, [L]
K_1	loss parameter (see Chapter 7), [—]
K_d	loss parameter (see Chapter 7), [—]
K_f	loss coefficient (see Chapter 7), [—]
K_i	loss parameter (see Chapter 7), [—]
ke	kinetic energy/mass, [FL/M = L^2/t^2]
L	dimension of length
L	length, [L]
M	dimension of mass
m	mass, [M]
N_{Re}	Reynolds number, [—]
N_x	dimensionless group x [—]
P	pressure, [F/L^2 = M/Lt^2]
Q	volume flow rate, [L^3/t]
R	radius, [L]
T	torque, [FL = ML^2/t^2]
t	dimension of time
V	spatial average velocity, [L/t]
v	local velocity, [L/t]
W	weight, [F = ML/t2]
x	coordinate direction, [L]
z	coordinate direction (measured upward), [L]

ε	roughness, [L]
μ	viscosity, [M/Lt]
ρ	density, $[M/L^3]$
τ_{yx}	shear stress, force in x direction on y surface, $[F/L^2 = ML/t^2]$

Subscripts

1	reference point 1
2	reference point 2
m	model
f	field
x, y, r, θ	coordinate directions

3

Fluid Properties in Perspective

I. CLASSIFICATION OF MATERIALS AND FLUID PROPERTIES

What is a fluid? It isn't a solid, but what is a solid? Perhaps it is easier to define these materials in terms of how they respond (i.e., deform or flow) when subjected to an applied force in a specific situation such as the *simple shear* situation illustrated in Fig. 3-1 (which is virtually identical to Fig. 1-1). We envision the material contained between two infinite parallel plates, the bottom one being fixed and the top one subject to an applied force parallel to the plate, which is free to move in its plane. The material is assumed to adhere to the plates, and its properties can be classified by the way the top plate responds when the force is applied.

The mechanical behavior of a material, and its corresponding mechanical or rheological* properties, can be defined in terms of how the *shear stress* (τ_{yx}) (force per unit area) and the *shear strain* (γ_{yx}) (which is a relative displacement) are related. These are defined, respectively, in terms of the total force (F_x) acting on area A_y of the plate and the displacement (U_x) of the plate:

$$\tau_{yx} = F_x / A_y \tag{3-1}$$

*Rheology is the study of the deformation and flow behavior of materials, both fluids and solids. See, e.g., Barnes et al. (1989).

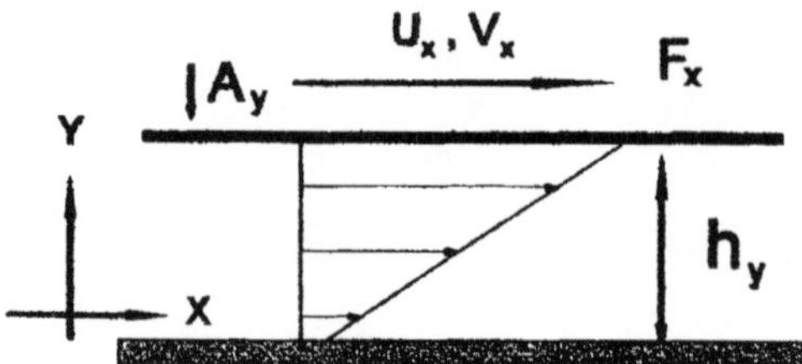

FIGURE 3-1 Simple shear.

and

$$\gamma_{yx} = \frac{U_x}{h_y} = \frac{du_x}{dy} \tag{3-2}$$

The manner in which the shear strain responds to the shear stress (or vice versa) in this situation defines the mechanical or rheological classification of the material. The parameters in any quantitative functional relation between the stress and strain are the *rheological properties* of the material. It is noted that the shear stress has dimensions of force per unit area (with units of, e.g., Pa, dyn/cm^2, lb$_f$/ft^2) and that shear strain is dimensionless (it has no units).

For example, if the material between the plates is a perfectly *rigid solid* (e.g., a brick), it will not move at all no matter how much force is applied (unless it breaks). Thus, the quantitative relation that defines the behavior of this material is

$$\gamma_{yx} = 0 \tag{3-3}$$

However, if the top plate moves a distance which is in proportion to the applied force and then stops, the material is called a *linear elastic (Hookean) solid* (e.g., rubber). The quantitative relation that defines such a material is

$$\tau_{yx} = G\gamma_{yx} \tag{3-4}$$

where G is a constant called the *shear modulus*. Note also that if the force (stress) is removed, the strain (displacement) also goes to zero, i.e., the material reverts to its original undeformed state. Such an ideal elastic material is thus said to have a "perfect memory."

On the other hand, if the top plate moves but its displacement is not directly proportional to the applied force (it may be either more or less than proportional to the force), the material is said to be a *nonlinear* (i.e., *non-Hookean*) *elastic solid*. It can be represented by an equation of the form

$$G = \tau_{yx}/\gamma_{yx} = \text{fn}(\tau \text{ or } \gamma) \tag{3-5}$$

Here G is still the shear modulus, but it is no longer a constant. It is, instead, a *function* of either how far the plate moves (γ_{yx}) or the magnitude of the applied force (τ_{yx}), i.e., $G(\gamma)$ or $G(\tau)$. The particular form of the function will depend upon the specific nature of the material. Note, however, that such a material still exhibits a "perfect memory," because it returns to its undeformed state when the force (stress) is removed.

At the other extreme, if the molecules of the material are so far apart that they exhibit negligible attraction for each other (e.g., a gas under very low pressure), the plate can be moved by the application of a negligible force. The equation that describes this material is

$$\tau_{yx} = 0 \tag{3-6}$$

Such an ideal material is called an *inviscid (Pascalian) fluid.* However, if the molecules do exhibit a significant mutual attraction such that the force (e.g., the shear stress) is proportional to the *relative rate* of movement (i.e., the velocity gradient), the material is known as a *Newtonian fluid.* The equation that describes this behavior is

$$\tau_{yx} = \mu \dot{\gamma}_{yx} \tag{3-7}$$

where $\dot{\gamma}_{yx}$ is the *rate of shear strain* or *shear rate*:

$$\dot{\gamma}_{yx} = \frac{d\gamma_{yx}}{dt} = \frac{dv_x}{dy} = \frac{V_x}{h_y} \tag{3-8}$$

and μ is the fluid viscosity. Note that Eq. (3-7) defines the viscosity, i.e., $\mu \equiv \tau_{yx}/\dot{\gamma}_{yx}$, which has dimensions of $\mathrm{Ft/L^2}$ [with units of Pa s, dyn s/cm^2 = g/(cm s) = poise, $\mathrm{lb_f\ s/ft^2}$, etc.]. Note that when the stress is removed from this fluid, the shear rate goes to zero, i.e., the motion stops, but there is no "memory" or tendency to return to any past state.

If the properties of the fluid are such that the shear stress and shear rate are not directly proportional but are instead related by some more complex function, the fluid is said to be *non-Newtonian.* For such fluids the viscosity may still be defined as $\tau_{yx}/\dot{\gamma}_{yx}$, but it is no longer a constant. It is, instead, a function of either the shear rate or shear stress. This is called the *apparent viscosity (function)* and is designated by η:

$$\eta = \frac{\tau_{yx}}{\dot{\gamma}_{yx}} = \text{fn}(\tau \text{ or } \dot{\gamma}) \tag{3-9}$$

The actual mathematical form of this function will depend upon the nature (i.e., the "constitution") of the particular material. Most common fluids of simple structure water, air, glycerine, oils, etc.) are Newtonian. However, fluids with complex structure (polymer melts or solutions, suspensions, emulsions, foams, etc.) are generally non-Newtonian. Some very common

TABLE 3-1 Classification of Materials

Rigid solid (Euclidian)	Linear elastic solid (Hookean)	Nonlinear elastic solid (non-Hookean)	Viscoelastic	Nonlinear viscous fluid (non-Newtonian)	Linear viscous fluid (Newtonian)	Inviscid fluid (Pascalian)
$\gamma = 0$	$\tau = G\gamma$ or $G = \tau/\gamma$	$\tau = \text{fn}(\gamma)$ or $G = \tau/\gamma$	fluids and solids	$\tau = \text{fn}(\dot{\gamma})$ or $\eta = \tau/\dot{\gamma}$	$\tau = \mu\dot{\gamma}$ or $\mu = \tau/\dot{\gamma}$	$\tau = 0$
	Shear modulus—constant	Modulus—function of γ or τ	(Nonlinear)	Viscosity—function of $\dot{\gamma}$ or τ	Viscosity—constant	
← Purely elastic solids →			$\tau = \text{fn}(\gamma, \dot{\gamma}, \textit{dots})$	← Purely viscous fluids →		
Elastic deformations store energy.				Viscous deformations dissipate energy		

examples of non-Newtonian fluids are mud, paint, ink, mayonnaise, shaving cream, dough, mustard, toothpaste, and sludge.

Actually, some fluids and solids have both elastic (solid) properties and viscous (fluid) properties. These are said to be *viscoelastic* and are most notably materials composed of high polymers. The complete description of the rheological properties of these materials may involve a function relating the stress and strain as well as derivatives or integrals of these with respect to time. Because the elastic properties of these materials (both fluids and solids) impart "memory" to the material (as described previously), which results in a tendency to recover to a preferred state upon the removal of the force (stress), they are often termed "memory materials" and exhibit time-dependent properties.

This classification of material behavior is summarized in Table 3-1 (in which the subscripts have been omitted for simplicity). Since we are concerned with fluids, we will concentrate primarily on the flow behavior of Newtonian and non-Newtonian fluids. However, we will also illustrate some of the unique characteristics of viscoelastic fluids, such as the ability of solutions of certain high polymers to flow through pipes in turbulent flow with much less energy expenditure than the solvent alone.

II. DETERMINATION OF FLUID VISCOUS (RHEOLOGICAL) PROPERTIES

As previously discussed, the flow behavior of fluids is determined by their rheological properties, which govern the relationship between shear stress and shear rate. In principle these properties could be determined by measurements in a "simple shear" test as illustrated in Fig. 3-1. One would put the "unknown" fluid in the gap between the plates, subject the upper plate to a specified velocity (V), and measure the required force (F) (or vice versa). The shear stress (τ) would be determined by F/A, the shear rate ($\dot{\gamma}$) is given by V/h, and the viscosity (η) by $\tau/\dot{\gamma}$. The experiment is repeated for different combinations of V and F to determine the viscosity at various shear rates (or shear stresses). However, this geometry is not convenient to work with, because it is hard to keep the fluid in the gap with no confining walls, and correction for the effect of the walls is not simple. However, there are more convenient geometries for measuring viscous properties. The working equations used to obtain viscosity from measured quantities will be given here, although the development of these equations will be delayed until after the appropriate fundamental principles have been discussed.

A. Cup-and-Bob (Couette) Viscometer

As the name implies, the cup-and-bob viscometer consists of two concentric cylinders, the outer "cup" and the inner "bob," with the test fluid in the annular gap (see Fig. 3-2). One cylinder (preferably the cup) is rotated at a fixed angular velocity (Ω). The force is transmitted to the sample, causing it to deform, and is then transferred by the fluid to the other cylinder (i.e., the bob). This force results in a torque (T) that can be measured by a torsion spring, for example. Thus, the known quantities are the radii of the inner bob (R_i) and the outer cup (R_o), the length of surface in contact with the sample (L), and the measured angular velocity (Ω) and torque (T). From these quantities, we must determine the corresponding shear stress and shear rate to find the fluid viscosity. The shear stress is determined by a balance of moments on a cylindrical surface within the sample (at a distance r from the center), and the torsion spring:

$$T = \text{Force} \times \text{Lever arm} = \text{Shear stress} \times \text{Surface area} \times \text{Radius}$$

or

$$T = \tau_{r\theta}(2\pi r L)(r)$$

where the subscripts on the shear stress (r, θ) represent the force in the θ direction acting on the r surface (the cylindrical surface perpendicular to r). Solving for the shear stress, we have

$$\tau_{r\theta} = \frac{T}{2\pi r^2 L} = \tau \tag{3-10}$$

Setting $r = R_i$ gives the stress on the bob surface (τ_i), and $r = R_o$ gives the stress on the cup (τ_o). If the gap is small [i.e., $(R_o - R_i)/R_i \leq 0.02$], the

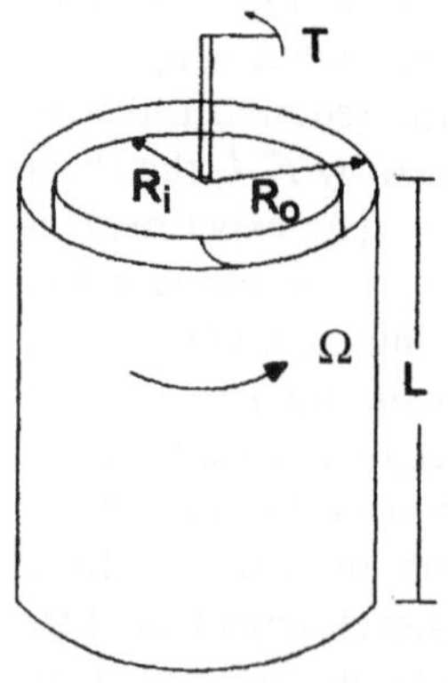

FIGURE 3-2 Cup-and-bob (Couette) viscometer.

curvature can be neglected and the flow in the gap is equivalent to flow between parallel plates. In this case, an average shear stress should be used [i.e., $(\tau_i + \tau_o)/2$], and the average shear rate is given by

$$\dot{\gamma} = \frac{dv_\theta}{dr} \cong \frac{\Delta V}{\Delta r} = \frac{V_o - V_i}{R_o - R_i} = \frac{R_o \Omega}{R_o - R_i} = \frac{\Omega}{1 - R_i/R_o}$$

or

$$\dot{\gamma}_{r\theta} = \frac{dv_\theta}{dr} = \frac{\Omega}{1 - \beta} = \dot{\gamma} \tag{3-11}$$

where $\beta = R_i/R_o$. However, if the gap is not small, the shear rate must be corrected for the curvature in the velocity profile. This can be done by applying the following approximate expression for the shear rate at the bob [which is accurate to ±1% for most conditions and is better than ±5% for the worst case; see, e.g., Darby (1985)]:

$$\dot{\gamma}_i = \frac{2\Omega}{n'(1 - \beta^{2/n'})} \tag{3-12}$$

where

$$n' = \frac{d(\log T)}{d(\log \Omega)} \tag{3-13}$$

is the point slope of the plot of log T versus log Ω, at the value of Ω (or T) in Eq. (3-12). Thus a series of data points of T versus Ω must be obtained in order to determine the value of the slope (n') at each point, which is needed to determine the corresponding values of the shear rate. If $n' = 1$ (i.e, $T \propto \Omega$), the fluid is Newtonian. The viscosity at each shear rate (or shear stress) is then determined by dividing the shear stress at the bob [Eq. (3-10) with $r = R_i$] by the shear rate at the bob [Eq. (3-12)], for each data point.

Example 3-1: The following data were taken in a cup-and-bob viscometer with a bob radius of 2 cm, a cup radius of 2.05 cm, and a bob length of 15 cm. Determine the viscosity of the sample and the equation for the model that best represents this viscosity.

Torque (dyn cm)	Speed (rpm)
2,000	2
3,500	4
7,200	10
12,500	20
20,000	40

The viscosity is the shear stress at the bob, as given by Eq. (3-10), divided by the shear rate at the bob, as given by Eq. (3-12). The value of n' in Eq. (3-12) is determined from the *point slope* of the log T versus log *rpm* plot at each data point. Such a plot is shown Fig. 3-3a. The line through the data is the best fit of all data points by linear least squares (this is easily found by using a spreadsheet) and has a slope of 0.77 (with $r^2 = 0.999$). In general, if the

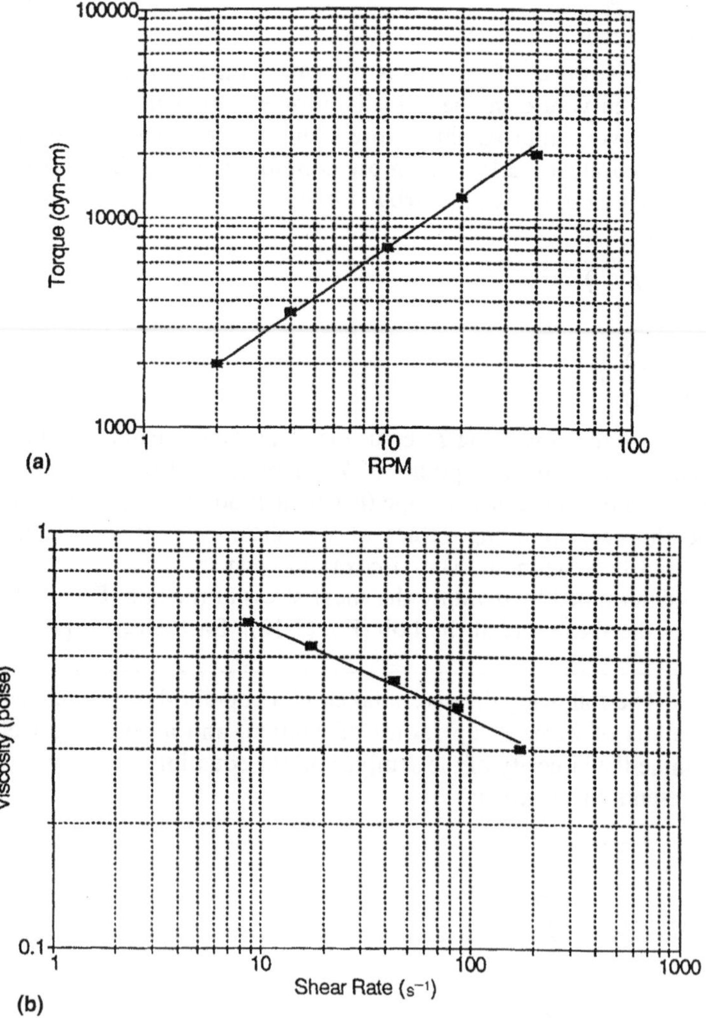

FIGURE 3-3 Examples of (a) cup-and-bob viscosity (torque vs. speed) and (b) viscosity versus shear rate. (Lines are least-mean-square fit.)

data do not fall on a straight line on this plot, the point slope (tangent) must be determined at each data point, giving a different value of n' for each data point. Using 0.77 for n' in Eq. (3-12) for the shear rate and Eq. (3-10) for the shear stress, gives the following values:

Shear stress at bob (dyn/cm^2)	Shear rate at bob (1/s)	Viscosity (poise)
5.31	8.74	0.61
9.28	17.5	0.53
19.1	43.7	0.43
33.2	87.4	0.37
53.1	175	0.31

The plot of viscosity versus shear rate is shown in Fig. 3-3b, in which the line represents Eq. (3-24), with $n = 0.77$ and $m = 1.01$ dyn s^n/cm^2 (or "poise"). In this case the power law model represents the data quite well over the entire range of shear rate, so that $n = n'$ is the same for each data point. If this were not the case, the local slope of log T versus log rpm would determine a different value of n' for each data point, and the power law model would not give the best fit over the entire range of shear rate. The shear rate and viscosity would still be determined as above (using the local value of n' for each data point), but the viscosity curve could probably be best fit by some other model, depending upon the trend of the data (see Section III).

B. Tube Flow (Poiseuille) Viscometer

Another common method of determining viscosity is by measuring the total pressure drop ($\Delta\Phi = \Delta P + \rho g \, \Delta z$) and flow rate ($Q$) in steady laminar flow through a uniform circular tube of length L and diameter D (this is called *Poiseuille flow*). This can be done by using pressure taps through the tube wall to measure the pressure difference directly or by measuring the total pressure drop from a reservoir to the end of the tube, as illustrated in Fig. 3-4. The latter is more common, because tubes of very small diameter are usually used, but this arrangement requires that correction factors be applied for the static head of the fluid in the reservoir and the pressure loss from the reservoir to the tube.

As will be shown later, a momentum (force) balance on the fluid in the tube provides a relationships between the shear stress at the tube wall (τ_w) and the measured pressure drop:

$$\tau_w = \frac{-\Delta\Phi}{4L/D} \tag{3-14}$$

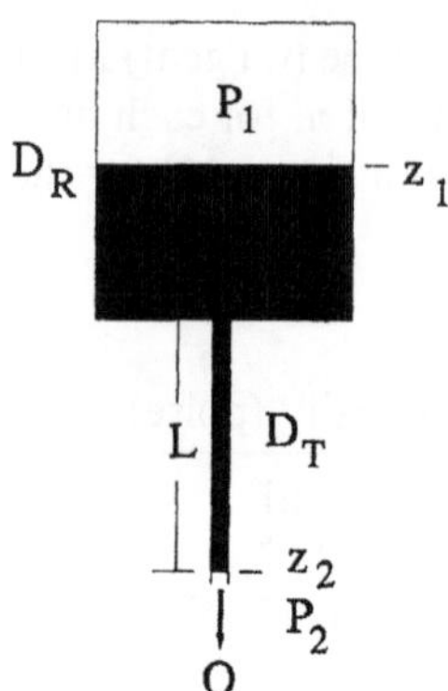

FIGURE 3-4 Tube flow (Poiseuille) viscometer.

The corresponding shear rate at the tube wall ($\dot{\gamma}_w$) is given by

$$\dot{\gamma}_w = \Gamma\left(\frac{3n' + 1}{4n'}\right) \tag{3-15}$$

where

$$\Gamma = \frac{32Q}{\pi D^3} = \frac{8V}{D} \tag{3-16}$$

is the wall shear rate for a Newtonian ($n' = 1$) fluid, and

$$n' = \frac{d \log \tau_w}{d \log \Gamma} = \frac{d \log(-\Delta\Phi)}{d \log Q} \tag{3-17}$$

is the point slope of the log-log plot of $\Delta\Phi$ versus Q, evaluated at each data point. This n' is the same as that determined from the cup-and-bob viscometer for a given fluid. As before, if $n' = 1$ (i.e., $\Delta\Phi \propto Q$), the fluid is Newtonian. The viscosity is given by $\eta = \tau_w/\dot{\gamma}_w$.

III. TYPES OF OBSERVED FLUID BEHAVIOR

When the measured values of shear stress or viscosity are plotted versus shear rate, various types of behavior may be observed depending upon the fluid properties, as shown in Figs. 3-5 and 3-6. It should be noted that the shear stress and shear rate can both be either positive or negative, depending upon the direction of motion or the applied force, the reference frame, etc. (however, by our convention they are always the same sign). Because the viscosity must always be positive, the shear rate (or shear stress) argument in

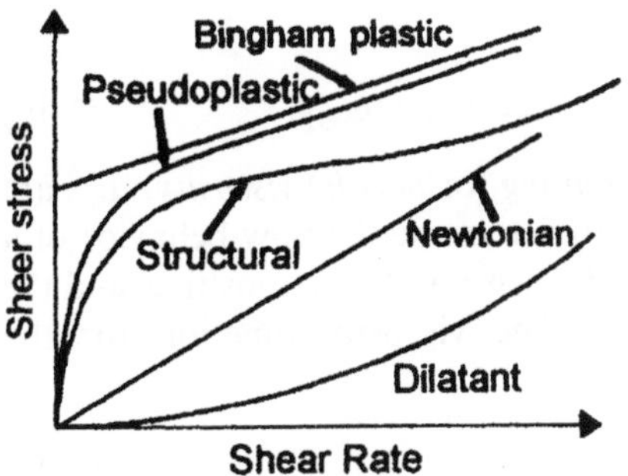

FIGURE 3-5 Shear stress versus shear rate for various fluids.

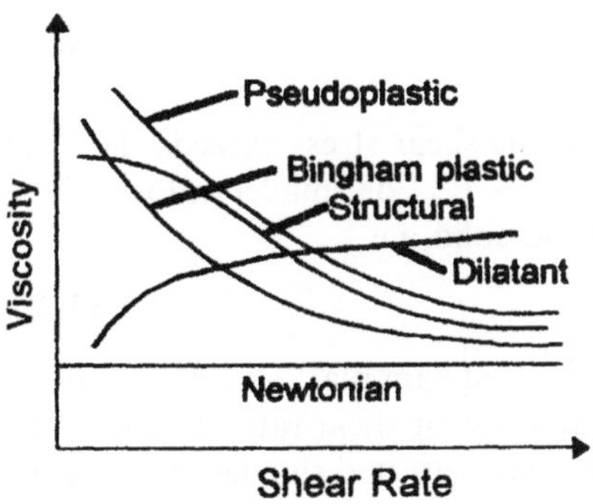

FIGURE 3-6 Viscosity versus shear rate for fluids in Fig. 3-5.

the viscosity function for a non-Newtonian fluid should be the absolute magnitude regardless of the actual sign of the shear rate and shear stress.

A. Newtonian Fluid

If the shear stress versus shear rate plot is a straight line through the origin (or a straight line with a slope of unity on a log-log plot), the fluid is Newtonian:

$$\text{Newtonian:} \qquad \tau = \mu \dot{\gamma} \tag{3-18}$$

where μ is the viscosity.

B. Bingham Plastic Model

If the data appear to be linear but do not extrapolate through the origin, intersecting the τ axis instead at a shear stress value of τ_o, the material is called a Bingham plastic:

$$\text{Bingham plastic:} \qquad \text{For } |\tau| > \tau_o, \quad \tau = \pm\tau_o + \mu_\infty\dot{\gamma} \tag{3-19}$$

The *yield stress*, τ_o, and the *high shear limiting* (or *plastic*) *viscosity*, μ_∞, are the two rheological properties required to determine the flow behavior of a Bingham plastic. The positive sign is used when τ and $\dot{\gamma}$ are positive, and the negative sign when they are negative. The viscosity function for the Bingham plastic is

$$\eta(\dot{\gamma}) = \frac{\tau_o}{|\dot{\gamma}|} + \mu_\infty \tag{3-20}$$

or

$$\eta(\tau) = \frac{\mu_\infty}{1 - \tau_o/|\tau|} \tag{3-21}$$

Because this material will not flow unless the shear stress exceeds the yield stress, these equations apply only when $|\tau| > \tau_o$. For smaller values of the shear stress, the material behaves as a rigid solid, i.e.,

$$\text{For } |\tau| < \tau_o: \quad \dot{\gamma} = 0 \tag{3-22}$$

As is evident from Eq. (3-20) or (3-21), the Bingham plastic exhibits a *shear thinning* viscosity; i.e., the larger the shear stress or shear rate, the lower the viscosity. This behavior is typical of many concentrated slurries and suspensions such as muds, paints, foams, emulsions (e.g., mayonnaise), ketchup, or blood.

C. Power Law Model

If the data (either shear stress or viscosity) exhibit a straight line on a log-log plot, the fluid is said to follow the *power law* model, which can be represented as

$$\text{Power law:} \qquad \tau = m|\dot{\gamma}|^{n-1}\dot{\gamma} \tag{3-23}$$

or

$$\tau = m\dot{\gamma}^n \qquad \text{if } \tau \text{ and } \dot{\gamma} \text{ are } (+)$$

$$\tau = -m(-\dot{\gamma})^n \qquad \text{if } \tau \text{ and } \dot{\gamma} \text{ are } (-)$$

The two viscous rheological properties are m, the *consistency coefficient*, and n, the *flow index*. The apparent viscosity function for the power law model in terms of shear rate is

$$\eta(\dot{\gamma}) = m|\dot{\gamma}|^{n-1} \tag{3-24}$$

or, in terms of shear stress,

$$\eta(\tau) = m^{1/n}|\tau|^{(n-1)/n} \tag{3-25}$$

Note that n is dimensionless but m has dimensions of Ft^n/L^2. However, m is also equal to the viscosity of the fluid at a shear rate of $1\,s^{-1}$, so it is a "viscosity" parameter with equivalent units. It is evident that if $n = 1$ the power law model reduces to a Newtonian fluid with $m = \mu$. If $n < 1$, the fluid is shear thinning (or *pseudoplastic*); and if $n > 1$, the model represents shear thickening (or *dilatant*) behavior, as illustrated in Figs. 3-5 and 3-6. Most non-Newtonian fluids are shear thinning, whereas shear thickening behavior is relatively rare, being observed primarily for some concentrated suspensions of very small particles (e.g., starch suspensions) and some unusual polymeric fluids. The power law model is very popular for curve fitting viscosity data for many fluids over one to three decades of shear rate. However, it is dangerous to extrapolate beyond the range of measurements using this model, because (for $n < 1$) it predicts a viscosity that increases without bound as the shear rate decreases and a viscosity that decreases without bound as the shear rate increases, both of which are physically unrealistic.

D. Structural Viscosity Models

The typical viscous behavior for many non-Newtonian fluids (e.g., polymeric fluids, flocculated suspensions, colloids, foams, gels) is illustrated by the curves labeled "structural" in Figs. 3-5 and 3-6. These fluids exhibit Newtonian behavior at very low and very high shear rates, with shear thinning or pseudoplastic behavior at intermediate shear rates. In some materials this can be attributed to a reversible "structure" or network that forms in the "rest" or equilibrium state. When the material is sheared, the structure breaks down, resulting in a shear-dependent (shear thinning) behavior. Some real examples of this type of behavior are shown in Fig. 3-7. These show that structural viscosity behavior is exhibited by fluids as diverse as polymer solutions, blood, latex emulsions, and mud (sediment). Equations (i.e., models) that represent this type of behavior are described below.

1. Carreau Model

The *Carreau model* (Carreau, 1972) is very useful for describing the viscosity of structural fluids:

$$\text{Carreau:} \qquad \eta(\dot{\gamma}) = \eta_\infty + \frac{\eta_0 - \eta_\infty}{[1 + (\lambda^2\dot{\gamma}^2)]^p} \tag{3-26}$$

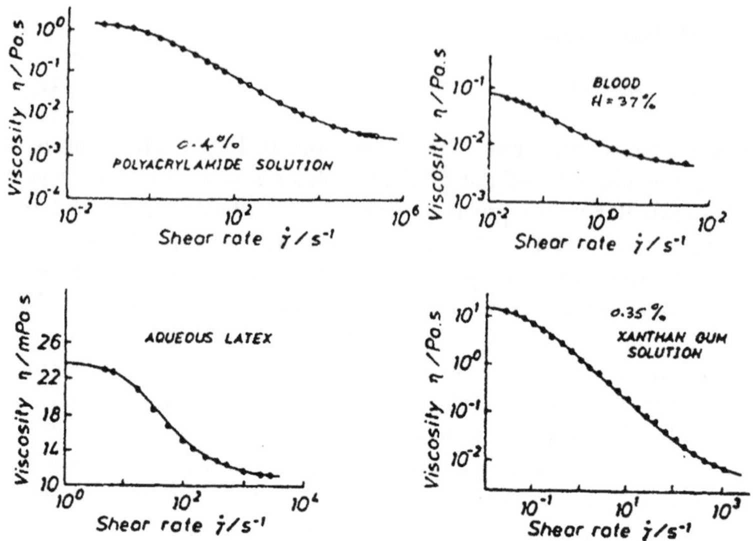

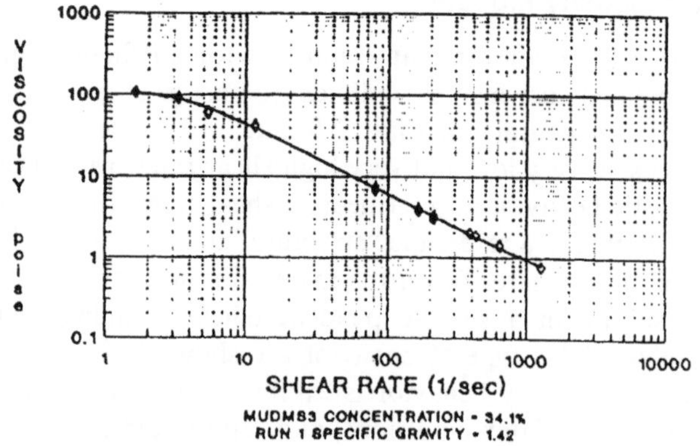

Figure 3-7 Some examples of structural viscosity behavior.

This model contains four rheological parameters: the *low shear limiting viscosity* (η_0), the *high shear limiting viscosity* (η_∞), a *time constant* (λ), and the *shear thinning index* (p). This is a very general viscosity model, and it can represent the viscosity function for a wide variety of materials. However, it may require data over a range of six to eight *decades* of shear

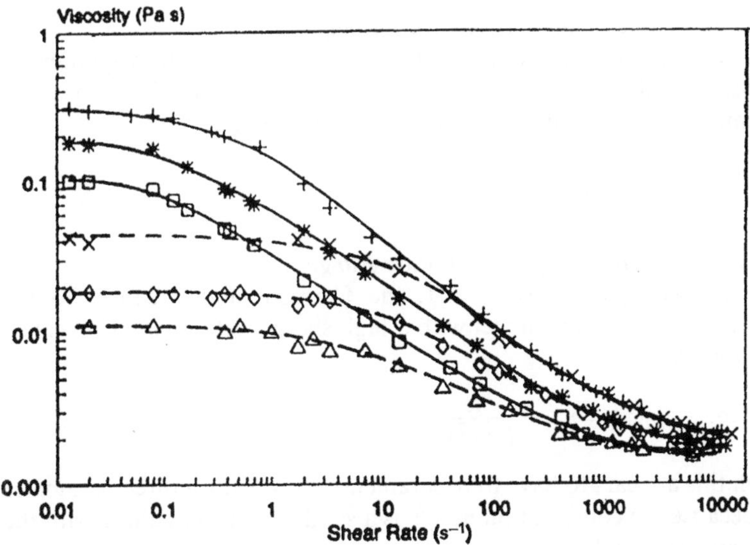

FIGURE 3-8 Viscosity data and Carreau model fit of polyacrylamide solutions. (From Darby and Pivsa-Art, 1991.)

rate to completely define the shape of the curve (and hence to determine all four parameters). As an example, Fig. 3-8 shows viscosity data for several polyacrylamide solutions over a range of about 10^6 in shear rate, with the curves through the data showing the Carreau model fit of the data. The corresponding values of the Carreau parameters for each of the curves are given in Table 3-2. In fact, over certain ranges of shear rate, the Carreau model reduces to various other popular models as special cases (including the Bingham plastic and power law models), as shown below.

TABLE 3-2 Values of Carreau Parameters for Model Fit in Fig. 3-8

Solution	η_0[(Pa s) × 10]	λ(s)	p	η_∞[(Pa s) × 1000]
100 mg/kg (fresh)	1.113	11.89	0.266	1.30
250 mg/kg (fresh)	1.714	6.67	0.270	1.40
500 mg/kg (fresh)	3.017	3.53	0.300	1.70
100 mg/kg (sheared)	0.098	0.258	0.251	1.30
250 mg/kg (sheared)	0.169	0.167	0.270	1.40
500 mg/kg (sheared)	0.397	0.125	0.295	1.70

Source: Darby and Pivsa-Art (1991).

a. Low to Intermediate Shear Rate Range

If $\eta_\infty \ll (\eta, \eta_0)$, the Carreau model reduces to a three-parameter model (η_0, λ, and p) that is equivalent to a power law model with a low shear limiting viscosity, also known as the *Ellis model*:

$$\text{Ellis:} \qquad \eta(\dot{\gamma}) = \frac{\eta_0}{[1 + \lambda^2\dot{\gamma}^2]^p} \tag{3-27}$$

b. Intermediate to High Shear Rate Range

If $\eta_0 \gg (\eta, \eta_\infty)$ and $(\lambda\dot{\gamma})^2 \gg 1$, the Carreau model reduces to the equivalent of a power law model with a high shear limiting viscosity, called the *Sisko model*:

$$\text{Sisko:} \qquad \eta(\dot{\gamma}) = \eta_\infty + \frac{\eta_0}{(\lambda^2\dot{\gamma}^2)^p} \tag{3-28}$$

Although this appears to have four parameters, it is really a three-parameter model, because the combination η_0/λ^{2p} is a single parameter, along with the two parameters p and η_∞.

c. Intermediate Shear Rate Behavior

For $\eta_\infty \ll \eta \ll \eta_0$ and $(\lambda\dot{\gamma})^2 \gg 1$, the Carreau model reduces to the *power law model*:

$$\text{Power law:} \qquad \eta(\dot{\gamma}) = \frac{\eta_0}{(\lambda^2\dot{\gamma}^2)^p} \tag{3-29}$$

where the power law parameters m and n are equivalent to the following combinations of Carreau parameters:

$$m = \frac{\eta_0}{\lambda^{2p}}, \qquad n = 1 - 2p \tag{3-30}$$

d. Bingham Plastic Behavior

If the value of p is set equal to 1/2 in the Sisko model, the result is equivalent to the *Bingham plastic* model:

$$\text{Bingham:} \quad \eta(\dot{\gamma}) = \eta_\infty + \frac{\eta_0}{\lambda|\dot{\gamma}|} \tag{3-31}$$

where the yield stress τ_o is equivalent to η_0/λ, and η_∞ is the limiting (high shear) viscosity.

2. Other Models

A variety of more complex models have been proposed to fit a wider range and variety of viscosity data. Three of these are presented here.

a. Meter Model

A stress-dependent viscosity model, which has the same general characteristics as the Carreau model, is the *Meter model* (Meter, 1964):

$$\text{Meter:} \qquad \eta(\tau) + \eta_\infty + \frac{\eta_0 - \eta_\infty}{1 + \tau^2/\sigma^2)^a} \tag{3-32}$$

where σ is a characteristic stress parameter and a is the shear thinning index.

b. Yasuda Model

The *Yasuda model* (Yasuda et al., 1981) is a modification of the Carreau model with one additional parameter a (a total of five parameters):

$$\text{Yasuda:} \qquad \eta = \eta_\infty + \frac{\eta_0 - \eta_\infty}{[1 + (\lambda^2\dot{\gamma}^2)^a]^{p/a}} \tag{3-33}$$

which reduces to the Carreau model for $a = 1$. (This is also sometimes called the *Carreau–Yasuda* model). This model is particularly useful for representing melt viscosity data for polymers with a broad molecular weight distribution, for which the zero-shear viscosity is approached very gradually. Both of these models reduce to Newtonian behavior at very low and very high shear rates and to power law behavior at intermediate shear rates.

IV. TEMPERATURE DEPENDENCE OF VISCOSITY

All fluid properties are dependent upon temperature. For most fluids the viscosity is the property that is most sensitive to temperature changes.

A. Liquids

For liquids, as the temperature increases, the degree of molecular motion increases, reducing the short-range attractive forces between molecules and lowering the viscosity. The viscosity of various liquids is shown as a function of temperature in Appendix A. For many liquids, this temperature dependence can be represented reasonably well by the Arrhenius equation:

$$\mu = A \exp(B/T) \tag{3-34}$$

where T is the absolute temperature. If the viscosity of a liquid is known at two different temperatures, this information can be used to evaluate the parameters A and B, which then permits the calculation of the viscosity at any other temperature. If the viscosity is known at only one temperature, this value can be used as a reference point to establish the temperature scale for Fig. A-2 of Appendix A, which can then be used to estimate the viscosity at any other temperature. Viscosity data for 355 liquids have been fit by

Yaws et al. (1994) by the equation

$$\log_{10} \mu = A + B/T + CT + DT^2 \tag{3-35}$$

where T is in kelvin and the viscosity μ is in centipoise. The values of the correlation parameters A, B, C, and D are tabulated by Yaws et al. (1994).

For non-Newtonian fluids, any model parameter with the dimensions or physical significance of viscosity (e.g., the power law consistency, m, or the Carreau parameters η_∞ and η_0) will depend on temperature in a manner similar to the viscosity of a Newtonian fluid [e.g., Eq. (3-34)].

B. Gases

In contrast to the behavior of liquids, the viscosity of a gas increases with increasing temperature. This is because gas molecules are much farther apart, so the short-range attractive forces are very small. However, as the temperature is increased, the molecular kinetic energy increases, resulting in a greater exchange of momentum between the molecules and consequently a higher viscosity. The viscosity of gases is not as sensitive to temperature as that of liquids, however, and can often be represented by the equation

$$\mu = aT^b \tag{3-36}$$

The parameters a and b can be evaluated from a knowledge of the viscosity at two different temperatures, and the equation can then be used to calculate the viscosity at any other temperature. The value of the parameter b is often close to 1.5. In fact, if the viscosity (μ_1) of a gas is known at only one temperature (T_1), the following equation can be used to estimate the viscosity at any other temperature:

$$\mu = \mu_1 \left(\frac{T}{T_1}\right)^{3/2} \left(\frac{T_1 + 1.47T_B}{T + 1.47T_B}\right) \tag{3-37}$$

where the temperatures are in degrees Rankine and T_B is the boiling point of the gas.

V. DENSITY

In contrast to viscosity, the density of both liquids and gases decreases with increasing temperature, and the density of gases is much more sensitive to temperature than that of liquids. If the density of a liquid and its vapor are

TABLE 3-3 Parameter N in Eq. (3-38)

Liquid	N
Water and alcohols	4
Hydrocarbons and ethers	3.45
Organics	3.23
Inorganics	3.03

known at 60°F, the density at any other temperature can be estimated with the equation

$$\frac{(\rho - \rho_v)_T}{(\rho - \rho_v)_{60°F}} = \left(\frac{T_c - T}{T_0 - 519.67}\right)^{1/N} \tag{3-38}$$

where the temperatures are in degrees Rankine and T_c is the critical temperature. The value of N is given in Table 3-3 for various liquids.

The specific gravity of hydrocarbon liquids at 60°F is also often represented by the API gravity:

$$SG_{60°F} = \frac{141.5}{131.5 + °API} \tag{3-39}$$

For gases, if the temperature is well above the critical temperature and the pressure is below the critical pressure, the ideal gas law usually applies:

$$\rho = \frac{PM}{RT} = \frac{M}{\tilde{V}_m}\left(\frac{T_{ref}}{T}\right)\left(\frac{P}{P_{ref}}\right) \tag{3-40}$$

where M is the gas molecular weight, temperatures and pressures are absolute, and $\tilde{V}_m$ is the "standard molar volume" [22.4 m^3/(kg mol) at 273 K and 1 atm, 359 ft^3/(lb mol) at 492°R and 1 atm, or 379.4 ft^3/(lb mol) at 520°R (60°F) and 1 atm]. The notation SCF (which stands for "standard cubic feet") is often used for hydrocarbon gases to represent the volume in ft^3 that would be occupied by the gas at 60°F and 1 atm pressure which is really a measure of the mass of the gas.

For other methods of predicting fluid properties and their temperature dependence, the reader is referred to the book by Reid et al. (1977).

PROBLEMS

Rheological Properties

1. (a) Using tabulated data for the viscosity of water and SAE 10 lube oil as a function of temperature, plot the data in a form that is consistent with each of the following equations: (1) $\lambda = A \exp(B/T)$; (2) $\mu = aT^b$,

(b) Arrange the equations in (a) in such a form that you can use linear regression analysis to determine the values of A, B and a, b that give the best fit to the data for each fluid (a spreadsheet is useful for this). (Note that T is absolute temperature.)

2. The viscosity of a fluid sample is measured in a cup-and-bob viscometer. The bob is 15 cm long with a diameter of 9.8 cm, and the cup has a diameter of 10 cm. The cup rotates, and the torque is measured on the bob. The following data were obtained:

Ω (rpm)	T (dyn cm)
2	3.6×10^5
4	3.8×10^5
10	4.4×10^5
20	5.4×10^5
40	7.4×10^5

(a) Determine the viscosity of the sample.
(b) What viscosity model equation would be the most appropriate for describing the viscosity of this sample? Convert the data to corresponding values of viscosity versus shear rate, and plot them on appropriate axes consistent with the data and your equation. Use linear regression analysis in a form that is consistent with the plot to determine the values of each of the parameters in your equation.
(c) What is the viscosity of this sample at a cup speed of 100 rpm in the viscometer?

3. A fluid sample is contained between two parallel plates separated by a distance of 2 ± 0.1 mm. The area of the plates is $100 \pm 0.01\ \text{cm}^2$. The bottom plate is stationary, and the top plate moves with a velocity of 1 cm/s when a force of 315 ± 25 dyn is applied to it, and at 5 cm/s when the force is 1650 ± 25 dyn.
(a) Is the fluid Newtonian?
(b) What is its viscosity?
(c) What is the range of uncertainty to your answer to (b)?

4. The following materials exhibit flow properties that can be described by models that include a yield stress (e.g., Bingham plastic): (a) catsup; (b) toothpaste; (c) paint; (d) coal slurries; (e) printing ink. In terms of typical applications of these materials, describe how the yield stress is beneficial to their behavior, in contrast to how they would behave if they were Newtonian.

5. Consider each of the fluids for which the viscosity is shown in Fig. 3-7, all of which exhibit a "structural viscosity" characteristic. Explain how the "structure" of each of these fluids influences the nature of the viscosity curve for that fluid.

6. Starting with the equations for $\tau = \text{fn}(\dot{\gamma})$ that define the power law and Bingham plastic fluids, derive the equations for the viscosity functions for these models as a function of shear stress, i.e., $\eta = \text{fn}(\tau)$.

7. A paint sample is tested in a Couette (cup-and-bob) viscometer that has an outer radius of 5 cm, an inner radius of 4.9 cm, and a bob length of 10 cm. When the outer cylinder is rotated at a speed of 4 rpm, the torque on the bob is 0.0151 N m, and at a speed of 20 rpm the torque is 0.0226 N m.
   (a) What are the corresponding values of shear stress and shear rate for these two data points (in cgs units)?
   (b) What can you conclude about the viscous properties of the paint sample?
   (c) Which of the following models could be used to describe the paint: (1) Newtonian; (2) Bingham plastic; (3) power law? Explain why.
   (d) Determine the values of the fluid properties (i.e., parameters) of the models in (c) that could be used.
   (e) What would the viscosity of the paint be at a shear rate of 500 s^{-1} (in poise)?
8. The quantities that are measured in a cup-and-bob viscometer are the rotation rate of the cup (rpm) and the corresponding torque transmitted to the bob. These quantities are then converted to corresponding values of shear rate ($\dot{\gamma}$) and shear stress (τ), which in turn can be converted to corresponding values of viscosity (η).
   (a) Show what a log-log plot of τ vs. $\dot{\gamma}$ and η vs. $\dot{\gamma}$ would look like for materials that follow the following models: (1) Newtonian; (2) power law (shear thinning); (3) power law (shear thickening); (4) Bingham plastic; (5) structural.
   (b) Show how the values of the parameters for each of the models listed in (a) can be evaluated from the respective plot of η vs. $\dot{\gamma}$. That is, relate each model parameter to some characteristic or combination of characteristics of the plot such as the slope, specific values read from the plot, or intersection of tangent lines, etc.
9. What is the difference between shear stress and momentum flux? How are they related? Illustrate each one in terms of the angular flow in the gap in a cup-and-bob viscometer, in which the outer cylinder (cup) is rotated and the torque is measured at the stationary inner cylinder (bob).
10. A fluid is contained in the annulus in a cup-and-bob viscometer. The bob has a radius of 50 mm and a length of 10 cm and is made to rotate inside the cup by application of a torque on a shaft attached to the bob. If the cup's inside radius is 52 mm and the applied torque is 0.03 ft lb_f, what is the shear stress in the fluid at the bob surface and at the cup surface? If the fluid is Newtonian with a viscosity of 50 cP, how fast will the bob rotate (in rpm) with this applied torque?
11. You measure the viscosity of a fluid sample in a cup-and-bob viscometer. The radius of the cup is 2 in. and that of the bob is 1.75 in., and the length of the bob is 3 in. At a speed of 10 rpm, the measured torque is 500 dyn cm, and at 50 rpm it is 1200 dyn cm. What is the viscosity of the fluid? What can you deduce about the properties of the fluid?
12. A sample of a coal slurry is tested in a Couette (cup-and-bob) viscometer. The bob has a diameter of 10.0 cm and a length of 8.0 cm, and the cup has a diameter of 10.2 cm. When the cup is rotated at a rate of 2 rpm, the torque on the bob is 6.75×10^4 dyn cm, and at a rate of 50 rpm it is 2.44×10^6 dyn cm. If the slurry follows the power law model, what are the values of the flow index

and consistency coefficient? If the slurry follows the Bingham plastic model, what are the values of the yield stress and the limiting viscosity? What would the viscosity of this slurry be at a shear rate of 500 s^{-1} as predicted by each of these models? Which number would you be more likely to believe, and why?

13. You must analyze the viscous properties of blood. Its measured viscosity is 6.49 cP at a shear rate of 10 s^{-1} and 4.66 cP at a shear rate of 80 s^{-1}.
    (a) How would you describe these viscous properties?
    (b) If the blood is subjected to a shear stress of 50 dyn/cm^2, what would its viscosity be if it is described by: (1) the power law model? (2) the Bingham plastic model? Which answer do you think would be better, and why?
14. The following data were measured for the viscosity of a 500 ppm polyacrylamide solution in distilled water:

Shear rate (s^{-1})	Viscosity (cP)	Shear rate (s^{-1})	Viscosity (cP)
0.015	300	15	30
0.02	290	40	22
0.05	270	80	15
0.08	270	120	11
0.12	260	200	8
0.3	200	350	6
0.4	190	700	5
0.8	180	2,000	3.3
2	100	4,500	2.2
3.5	80	7,000	2.1
8	50	20,000	2

Find the model that best represents these data, and determine the values of the model parameters by fitting the model to the data. (This can be done most easily by trial and error, using a spreadsheet.)

15. What viscosity model best represents the following data? Determine the values of the parameters in the model. Show a plot of the data together with the line that represents the model, to show how well the model works. (*Hint:* The easiest way to do this is by trial and error, fitting the model equation to the data using a spreadsheet.)

Shear rate (s^{-1})	Viscosity (poise)	Shear rate (s^{-1})	Viscosity (poise)
0.007	7,745	20	270
0.01	7,690	50	164
0.02	7,399	100	113
0.05	6,187	200	77.9
0.07	5,488	500	48.1
0.1	4,705	700	40.4
0.2	3,329	1,000	33.6

Shear rate (s^{-1})	Viscosity (poise)	Shear rate (s^{-1})	Viscosity (poise)
0.5	2,033	2,000	23.8
0.7	1,692	5,000	15.3
1	1,392	7,000	13.2
2	952	10,000	11.3
5	576	20,000	8.5
7	479	50,000	6.1
10	394	70,000	5.5

16. You would like to determine the pressure drop in a slurry pipeline. To do this, you need to know the rheological properties of the slurry. To evaluate these properties, you test the slurry by pumping it through a $\frac{1}{8}$ in. ID tube that is 10 ft long. You find that it takes a 5 psi pressure drop to produce a flow rate of 100 cm^3/s in the tube and that a pressure drop of 10 psi results in a flow rate of 300 cm^3/s. What can you deduce about the rheological characteristics of the slurry from these data? If it is assumed that the slurry can be adequately described by the power law model, what would be the values of the appropriate fluid properties (i.e., the flow index and consistency parameter) for the slurry?
17. A film of paint, 3 mm thick, is applied to a flat surface that is inclined to the horizontal by an angle θ. If the paint is a Bingham plastic, with a yield stress of 150 dyn/cm^2, a limiting viscosity of 65 cP, and an SG of 1.3, how large would the angle θ have to be before the paint would start to run? At this angle, what would the shear rate be if the paint follows the power law model instead, with a flow index of 0.6 and a consistency coefficient of 215 (in cgs units)?
18. A thick suspension is tested in a Couette (cup-and-bob) viscometer that has having a cup radius of 2.05 cm, a bob radius of 2.00 cm, and a bob length of 15 cm. The following data are obtained:

Cup speed (rpm)	Torque on bob (dyn cm)
2	2,000
4	6,000
10	19,000
20	50,000
50	150,000

What can you deduce about (a) the viscous properties of this material and (b) the best model to use to represent these data?
19. You have obtained data for a viscous fluid in a cup-and-bob viscometer that has the following dimensions: cup radius = 2 cm, bob radius = 1.5 cm, bob length = 3 cm. The data are tabulated below, where n is the point slope of the log T versus log N curve.

N (rpm)	T (dyn cm)	n	N (rpm)	T (dyn cm)	n
1	1.13×10^4	0.01	100	1.25×10^4	0.6
2	1.13×10^4	0.02	200	1.42×10^4	0.7
5	1.13×10^4	0.05	500	1.93×10^4	0.8
10	1.13×10^4	0.1	1000	2.73×10^4	0.9
20	1.14×10^4	0.2	2000	4.31×10^4	1.0
50	1.16×10^4	0.5			

(a) Determine the viscosity of the fluid. How would you describe its viscosity?
(b) What kind of viscous model (equation) would be appropriate to describe this fluid?
(c) Use the data to determine the values of the fluid properties that are defined by the model.

20. A sample of a viscous fluid is tested in a cup-and-bob viscometer that has a cup radius of 2.1 cm, a bob radius of 2.0 cm, and a bob length of 5 cm. When the cup is rotated at 10 rpm, the torque measured at the bob is 6000 dyn cm, and at 100 rpm the torque is 15,000 dyn cm.
    (a) What is the viscosity of this sample?
    (b) What can you conclude about the viscous properties of the sample?
    (c) If the cup is rotated at 500 rpm, what will be the torque on the bob and the fluid viscosity? Clearly explain any assumptions you make to answer this question, and tell how you might check the validity of these assumptions.
21. You have a sample of a sediment that is non-Newtonian. You measure its viscosity in a cup-and-bob viscometer that has a cup radius of 3.0 cm, a bob radius of 2.5 cm, and a length of 5 cm. At a rotational speed of 10 rpm the torque transmitted to the bob is 700 dyn cm, and at 100 rpm the torque is 2500 dyn cm.
    (a) What is the viscosity of the sample?
    (b) Determine the values of the model parameters that represent the sediment viscous properties if it is represented by (1) the power law model or (2) the Bingham plastic model.
    (c) What would the flow rate of the sediment be (in cm^3/s) in a 2 cm diameter tube, 50 m long, that is subjected to a differential pressure driving force of 25 psi (1) assuming that the power law model applies? (2) assuming that the Bingham plastic model applies? Which of these two answers do you think is best, and why?
22. The Bingham plastic model can describe acrylic latex paint, with a yield stress of 112 dyn/cm^2, a limiting viscosity of 80 cP, and a density of 0.95 g/cm^3. What is the maximum thickness of this paint that can be applied to a vertical wall without running?
23. Santa Claus and his loaded sleigh are sitting on your roof, which is covered with snow. The sled's two runners each have a length L and width W, and the roof is inclined at an angle θ to the horizontal. The thickness of the snow between the runners and the roof is H. If the snow has properties of a Bingham plastic, derive an expression for the total mass (m) of the loaded sleigh at which it will

just start to slide on the roof if it is pointed straight downhill. If the actual mass is twice this minimum mass, determine an expression for the speed at which the sled will slide. (*Note*: Snow does *not* actually behave as a Bingham plastic!)

24. You must design a piping system to handle a sludge waste product. However, you don't know the properties of the sludge, so you test it in a cup-and-bob viscometer with a cup diameter of 10 cm, a bob diameter of 9.8 cm, and a bob length of 8 cm. When the cup is rotated at 2 rpm, the torque on the bob is 2.4×10^4 dyn cm, and at 20 rpm, it is 6.5×10^4 dyn cm.
    (a) If you use the power law model to describe the sludge, what are the values of the flow index and consistency?
    (b) If you use the Bingham plastic model instead, what are the values of the yield stress and limiting viscosity?

25. A fluid sample is tested in a cup-and-bob viscometer that has a cup diameter of 2.25 in., a bob diameter of 2 in., and length of 3 in. The following data are obtained:

Rotation rate (rpm)	Torque (dyn cm)
20	2,500
50	5,000
100	8,000
200	10,000

    (a) Determine the viscosity of this sample.
    (b) What model would provide the best representation of this viscosity function, and why?

26. You test a sample in a cup-and-bob viscometer to determine the viscosity. The diameter of the cup is 55 mm, that of the bob is 50 mm, and the length is 65 mm. The cup is rotated, and the torque on the bob is measured, giving the following data:

Cup speed (rpm)	Torque on bob (dyn cm)
2	3,000
4	6,000
10	11,800
20	14,500
40	17,800

    (a) Determine the viscosity of this sample.
    (b) How would you describe the viscosity of this material?
    (c) What model would be the most appropriate to represent this viscosity?
    (d) Determine the values of the parameters in the model that fit the model to the data.

27. Consider each of the fluids for which the viscosity is shown in Fig. 3-7, all of which exhibit a typical "structural viscosity" characteristic. Explain why this is

a logical consequence of the composition or "structural makeup" for each of these fluids.

28. You are asked to measure the viscosity of an emulsion, so you use a tube flow viscometer similar to that illustrated in Fig. 3-4, with the container open to the atmosphere. The length of the tube is 10 cm, its diameter is 2 mm, and the diameter of the container is 3 in. When the level of the sample is 10 cm above the bottom of the container the emulsion drains through the tube at a rate of 12 cm^3/min, and when the level is 20 cm the flow rate is 30 cm^3/min. The emulsion density is 1.3 g/cm^3.
    (a) What can you tell from the data about the viscous properties of the emulsion?
    (b) Determine the viscosity of the emulsion.
    (c) What would the sample viscosity be at a shear rate of 500 s^{-1}?
29. You must determine the horsepower required to pump a coal slurry through an 18 in. diameter pipeline, 300 mi long, at a rate of 5 million tons/yr. The slurry can be described by the Bingham plastic model, with a yield stress of 75 dyn/cm^2, a limiting viscosity of 40 cP, and a density of 1.4 g/cm^3. For non-Newtonian fluids, the flow is not sensitive to the wall roughness.
    (a) Determine the dimensionless groups that characterize this system. You want to use these to design a lab experiment from which you can scale up measurements to find the desired horsepower.
    (b) Can you use the same slurry in the lab as in the pipeline?
    (c) If you use a slurry in the lab that has a yield stress of 150 dyn/cm^2, a limiting viscosity of 20 cP, and a density of 1.5 g/cm^3, what size pipe and what flow rate (in gpm) should you use in the lab?
    (d) If you run the lab system as designed and measure a pressure drop ΔP (psi) over a 100 ft length of pipe, show how you would use this information to determine the required horsepower for the pipeline.
30. You want to determine how fast a rock will settle in mud, which behaves like a Bingham plastic. The first step is to perform a dimensional analysis of the system.
    (a) List the important variables that have an influence on this problem, with their dimensions (give careful attention to the factors that cause the rock to fall when listing these variables), and determine the appropriate dimensionless groups.
    (b) Design an experiment in which you measure the velocity of a solid sphere falling in a Bingham plastic in the lab, and use the dimensionless variables to scale the answer to find the velocity of a 2 in. diameter rock, with a density of 3.5 g/cm^3, falling in a mud with a yield stress of 300 dyn/cm^2, a limiting viscosity of 80 cP, and a density of 1.6 g/cm^3. Should you use this same mud in the lab, or can you use a different material that is also a Bingham plastic but with a different yield stress and limiting viscosity?
    (c) If you use a suspension in the lab with a yield stress of 150 dyn/cm^2, a limiting viscosity of 30 cP, and a density of 1.3 g/cm^3 and a solid sphere, how big should the sphere be, and how much should it weigh?

(d) If the sphere in the lab falls at a rate of 4 cm/s, how fast will the 2 in. diameter rock fall in the other mud?

31. A pipeline has been proposed to transport a coal slurry 1200 mi from Wyoming to Texas, at a rate of 50 million tons/yr, through a 36 in. diameter pipeline. The coal slurry has the properties of a Bingham plastic, with a yield stress of 150 dyn/cm^2, a limiting viscosity of 40 cP, and an SG of 1.5. You must conduct a lab experiment in which the measured pressure gradient can be used to determine the total pressure drop in the pipeline.
    (a) Perform a dimensional analysis of the system to determine an appropriate set of dimensionless groups to use (you may neglect the effect of wall roughness for this fluid).
    (b) For the lab test fluid, you have available a sample of the above coal slurry and three different muds with the following properties:

	Yield stress (dyn/cm^2)	Limiting viscosity (cP)	Density (g/cm^3)
Mud 1	50	80	1.8
Mud 2	100	20	1.2
Mud 3	250	10	1.4

    Which of these would be the best to use in the lab, and why?
    (c) What size pipe and what flow rate (in lb_m/min) should you use in the lab?
    (d) If the measured pressure gradient in the lab is 0.016 psi/ft, what is the total pressure drop in the pipeline?

32. A fluid sample is subjected to a "sliding plate" (simple shear) test. The area of the plates is 100 ± 0.01 cm^2, and the spacing between them is 2 ± 0.1 mm. When the moving plate travels at a speed of 0.5 cm/s, the force required to move it is measured to be 150 dyn, and at a speed of 3 cm/s the force is 1100 dyn. The force transducer has a sensitivity of 50 dyn. What can you deduce about the viscous properties of the sample?

33. You want to predict how fast a glacier that is 200 ft thick will flow down a slope inclined 25° to the horizontal. Assume that the glacier ice can be described by the Bingham plastic model with a yield stress of 50 psi, a limiting viscosity of 840 poise, and an SG of 0.98. The following materials are available to you in the lab, which also may be described by the Bingham plastic model:

	Yield stress (dyn/cm^2)	Limiting viscosity (cP)	SG
Mayonnaise	300	130	0.91
Shaving cream	175	15	0.32
Catsup	130	150	1.2
Paint	87	95	1.35

You want to set up a lab experiment to measure the velocity at which the model fluid flows down an inclined plane and scale this value to find the velocity of the glacier.

(a) Determine the appropriate set of dimensionless groups.
(b) Which of the above materials would be the best to use in the lab? Why?
(c) What is the film thickness that you should use in the lab, and at what angle should the plane be inclined?
(d) What would be the scale factor between the measured velocity in the lab and the glacier velocity?
(e) What problems might you encounter when conducting this experiment?

34. Your boss gives you a sample of "gunk" and asks you to measure its viscosity. You do this in a cup-and-bob viscometer that has an outer (cup) diameter of 2 in., an inner (bob) diameter of 1.75 in., and a bob length of 4 in. You run the viscometer at three speeds, and record the following data:

Rotational velocity Ω (rpm)	Torque on bob T (dyn cm)
1	10,500
10	50,000
100	240,000

(a) How would you classify the viscous properties of this material?
(b) Calculate the viscosity of the sample in cP.
(c) What viscosity model best represents these data, and what are the values of the viscous properties (i.e., the model parameters) for the model?

35. The dimensions and measured quantities in the viscometer in Problem 34 are known to the following precision:

T: $\pm 1\%$ of full scale (full scale $= 500,000$ dyn cm)

Ω: $\pm 1\%$ of reading

D_o, D_i, and L: ± 0.002 in.

Estimate the maximum percentage uncertainty in the measured viscosity of the sample for each of the three data points.

36. A concentrated slurry is prepared in an open 8 ft diameter mixing tank, using an impeller with a diameter of 6 ft located 3 ft below the surface. The slurry is non-Newtonian and can be described as a Bingham plastic with a yield stress of 50 dyn/cm^2, a limiting viscosity of 20 cP, and a density of 1.5 g/cm^3. A vortex is formed above the impeller, and if the speed is too high the vortex can reach the blades of the impeller, entraining air and causing problems. Since this condition is to be avoided, you need to know how fast the impeller can be rotated without entraining the vortex. To do this, you conduct a lab experiment using a scale model of the impeller that is 1 ft in diameter. You must design the experiment so that the critical impeller speed can be measured in the lab and scaled up to determine the critical speed in the larger mixer.

(a) List all the variables that are important in this system, and determine an appropriate set of dimensionless groups.
(b) Determine the diameter of the tank that should be used in the lab and the depth below the surface at which the impeller should be located.
(c) Should you use the same slurry in the lab model as in the field? If not, what properties should the lab slurry have?
(d) If the critical speed of the impeller in the lab system is ω (rpm), what is the critical speed of the impeller in the large tank?

37. You would like to know the thickness of a paint film as it drains at a rate of 1 gpm down a flat surface that is 6 in. wide and is inclined at an angle of 30° to the vertical. The paint is non-Newtonian and can be described as a Bingham plastic with a limiting viscosity of 100 cP, a yield stress of 60 dyn/cm^2, and a density of 0.9 g/cm^3. You have data from the laboratory for the film thickness of a Bingham plastic that has a limiting viscosity of 70 cP, a yield stress of 40 dyn/cm^2, and a density of 1 g/cm^3 flowing down a plane 1 ft wide inclined at an angle of 45° to the vertical, at various flow rates.
(a) At what flow rate (in gpm) will the laboratory system correspond to the conditions of the other system?
(b) If the film thickness of the laboratory fluid is 3 mm at these conditions, what would the film thickness be for the other system?

NOTATION

A_y	area whose outward normal vector is in the y direction, [L^2]
F_x	force component in the x direction, [F = ML/t^2]
fn( )	a function of whatever is in the ( )
G	shear modulus, [F/L^2 = M/Lt^2]
g	acceleration due to gravity, [L/t^2]
h_y	distance between plates in the y direction, [L]
L	length, [L]
m	power law consistency parameter, [M/Lt^{2-a}]
n	power law flow index, [—]
n'	variable defined by Eq. (3-13) or (3-17), [—]
P	pressure, [F/L^2 = M/Lt^2]
p	parameter in Carreau model, [—]
Q	volumetric flow rate, [L^3/t]
R	radius, [L]
r	radial coordinate, [L]
SG	specific gravity, [—]
T	temperature, [T]
T	torque or moment, [FL = ML^2/t^2]
U_x	displacement of boundary in the x direction, [L]
u_x	local displacement in the x direction, [L]
V	bulk or average velocity, [L/t]
z	vertical direction measured upward, [L]

β	$(= R_i/R_o)$ ratio of inner to outer radius [—]
Γ	shear rate at tube wall for Newtonian fluid, Eq. (3-16), [1/t]
γ_{yx}	gradient of x displacement in the y direction (shear strain, or γ), [—]
$\dot{\gamma}_{yx}$	gradient of x velocity in the y direction (shear rate, or γ), [1/t]
$\Delta(\,)$	value of $(\,)_2 - (\,)_1$
λ	fluid time constant parameter, [t]
μ	viscosity (constant), [M/Lt]
μ_∞	Bingham limiting viscosity [M/Lt]
η	viscosity (function), [M/Lt]
ρ	density, $[M/L^3]$
τ_o	yield stress, $[F/L^2 = M/Lt^2]$
τ_{yx}	force in the x direction on y surface (shear stress, or τ), $[F/L^2 = M/Lt^2]$
Φ	potential $(= P + \rho gz)$, $[F/L^2 = M/Lt^2]$
Ω	angular velocity of cylinder, [1/t]

Subscripts

1	reference point 1
2	reference point 2
0	zero shear rate parameter
i	inner
o	outer
w	value at wall
x, y, r, θ	coordinate directions
∞	high shear limiting parameter

REFERENCES

Barnes HA, JF Hutton, K Walters An Introduction to Rheology. New York: Elsevier, 1989.

Carreau PJ. Trans Soc Rheol. 16:99, 1972.

Darby R. J Rheol., 29:359, 1985.

Darby R, and S Pivsa-Art. Can J Chem Eng. 69:1395, 1991.

Meter DM. AIChE J 10:881, 1964.

Reid RC, JM Prausnitz, TK Sherwood. Properties of Gases and Liquids. 3rd ed. New York: McGraw-Hill, 1977.

Yasuda K, RC Armstrong, and RE Cohen. Rheol Acta 20:163, 1981.

Yaws CL, X Lin, and L Bu, Chem Eng April: 119, 1994.

4

Fluid Statics

I. STRESS AND PRESSURE

The forces that exist within a fluid at any point may arise from various sources. These include gravity, or the "weight" of the fluid, an external driving force such as a pump or compressor, and the internal resistance to relative motion between fluid elements or inertial effects resulting from variation in local velocity and the mass of the fluid (e.g., the transport or rate of change of momentum).

Any or all of these forces may result in local stresses within the fluid. "Stress" can be thought of as a (local) "concentration of force," or the force per unit area that bounds an infinitesimal volume of the fluid. Now both force and area are vectors, the direction of the area being defined by the normal vector that points outward relative to the volume bounded by the surface. Thus, each stress component has a magnitude and *two* directions associated with it, which are the characteristics of a "second-order tensor" or "dyad." If the direction in which the local force acts is designated by subscript j (e.g., $j = x$, y, or z in Cartesian coordinates) and the orientation (normal) of the local area element upon which it acts is designated by subscript i, then the corresponding stress component (σ_{ij}) is given by

$$\sigma_{ij} = \frac{F_j}{A_i}, \qquad i, j = 1, 2, \text{ or } 3 \text{ (e.g., } x, y, \text{ or } z) \tag{4-1}$$

Note that since i and j each represent any of three possible directions, there are a total of nine possible components of the stress tensor (at any given point in a fluid). However, it can readily be shown that the stress tensor is symmetrical (i.e., the ij components are the same as the ji components), so there are at most six independent stress components.

Because of the various origins of these forces, as mentioned above, there are different "types" of stresses. For example, the only stress that can exist in a fluid at rest is pressure, which can result from gravity (e.g., hydrostatic head) or various other forces acting on the fluid. Although pressure is a stress (e.g., a force per unit area), it is *isotropic*, that is, the force acts uniformly in all directions normal to any local surface at a given point in the fluid. Such a stress has no directional character and is thus a scalar. (Any isotropic tensor is, by definition, a scalar, because it has magnitude only and no direction.) However, the stress components that arise from the fluid motion do have directional characteristics, which are determined by the relative motion in the fluid. These stresses are associated with the local resistance to motion due to viscous or inertial properties and are *anisotropic* because of their directional character. We shall designate them by τ_{ij}, where the i and j have the same significance as in Eq. (4-1).

Thus the total stress, σ_{ij}, at any point within a fluid is composed of both the isotropic pressure and anisotropic stress components, as follows:

$$\sigma_{ij} = -P\delta_{ij} + \tau_{ij} \tag{4-2}$$

where P is the (isotropic) pressure. By convention, pressure is considered a "negative" stress because it is compressive, whereas tensile stresses are positive (i.e., a positive F_j acting on a positive A_i or a negative F_j on a negative A_i). The term δ_{ij} in Eq. (4-2) is a "unit tensor" (or Kronecker delta), which has a value of zero if $i \neq j$ and a value of unity if $i = j$. This is required, because the isotropic pressure acts only in the normal direction (e.g., $i = j$) and has only one component. As mentioned above, the anisotropic shear stress components τ_{ij} in a fluid are associated with relative motion within the fluid and are therefore zero in any fluid at rest. It follows that the only stress that can exist in a fluid at rest or in a state of uniform motion in which there is no relative motion between fluid elements is pressure. (This is a major distinction between a fluid and a solid, as solids can support a shear stress in a state of rest.) It is this situation with which we will be concerned in this chapter.

II. THE BASIC EQUATION OF FLUID STATICS

Consider a cylindrical region of arbitrary size and shape within a fluid, as shown in Fig. 4-1. We will apply a momentum balance to a "slice" of the

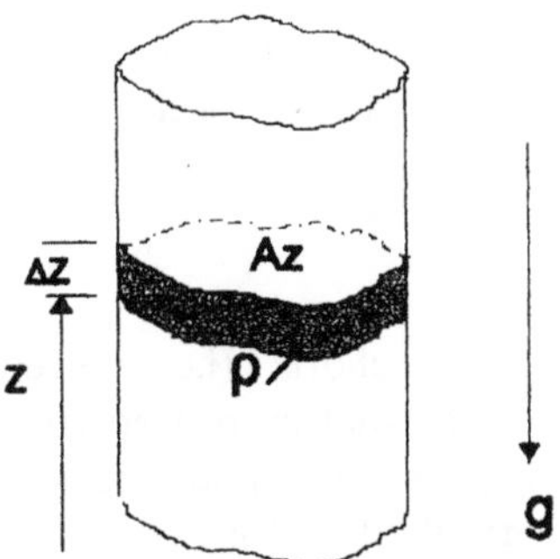

FIGURE 4-1 Arbitrary region within a fluid.

fluid that has a "z area" A_z and a thickness Δz and is located a vertical distance z above some horizontal reference plane. The density of the fluid in the slice is ρ, and the force of gravity (g) acts in the $-z$ direction. A momentum balance on a "closed" system (e.g., the slice) is equivalent to Newton's second law of motion, i.e.,

$$\sum F_z = ma_z \tag{4-3}$$

Because this is a vector equation, we apply it to the z vector components. $\sum F_z$ is the sum of all of the forces acting *on the system* (the "slice") in the z direction, m is the mass of the system, and a_z is the acceleration in the z direction. Because the fluid is not moving, $a_z = 0$, and the momentum balance reduces to a force balance. The z forces acting on the system include the (−) pressure on the bottom (at z) times the (−) z area, the (−) pressure on the top (at $z + \Delta z$) times the (+) z area, and the z component of gravity, i.e., the "weight" of the fluid ($-\rho g A\,\Delta z$). The first force is positive, and the latter two are negative because they act in the $-z$ direction. The momentum (force) balance thus becomes

$$(A_z P)_z - (A_z P)_{z+\Delta z} - \rho g A_z \Delta z = 0 \tag{4-4}$$

If we divide through by $A_z \Delta z$, then take the limit as the slice shrinks to zero ($\Delta z \to 0$), the result is

$$\frac{dP}{dz} = -\rho g \tag{4-5}$$

which is the *basic equation of fluid statics*. This equation states that at any point within a given fluid the pressure decreases as the elevation (z) increases, at a local rate that is equal to the product of the fluid density and the gravitational acceleration at that point. This equation is valid at all points within any given static fluid regardless of the nature of the fluid. We shall now show how the equation can be applied to various special situations.

A. Constant Density Fluids

If density (ρ) is constant, the fluid is referred to as "isochoric" (i.e., a given mass occupies a *constant volume*), although the somewhat more restrictive term "incompressible" is commonly used for this property (liquids are normally considered to be incompressible or isochoric fluids). If gravity (g) is also constant, the only variables in Eq. (4-5) are pressure and elevation, which can then be integrated between any two points (1 and 2) in a given fluid to give

$$P_1 - P_2 = \rho g(z_2 - z_1) \tag{4-6}$$

This can also be written

$$\Phi_1 = \Phi_2 = \text{constant} \tag{4-7}$$

where

$$\Phi = P + \rho g z$$

This says that the sum of the local pressure (P) and static head ($\rho g z$), which we call the *potential* (Φ), is constant at all points within a given isochoric (incompressible) fluid. This is an important result for such fluids, and it can be applied directly to determine how the pressure varies with elevation in a static liquid, as illustrated by the following example.

Example 4-1: Manometer. The pressure difference between two points in a fluid (flowing or static) can be measured by using a manometer. The manometer contains an incompressible liquid (density ρ_m) that is immiscible with the fluid flowing in the pipe (density ρ_f). The legs of the manometer are connected to taps on the pipe where the pressure difference is desired (see Fig. 4-2). By applying Eq. (4-7) to any two points within either one of the fluids within the manometer, we see that

$$(\Phi_1 = \Phi_3, \Phi_2 = \Phi_4)_f, \qquad (\Phi_3 = \Phi_4)_m \tag{4-8}$$

or

$$\begin{aligned} P_1 + \rho_f g z_1 &= P_3 + \rho_f g z_3 \\ P_3 + \rho_m g z_3 &= P_4 + \rho_m g z_4 \\ P_4 + \rho_f g z_4 &= P_2 + \rho_f g z_2 \end{aligned} \tag{4-9}$$

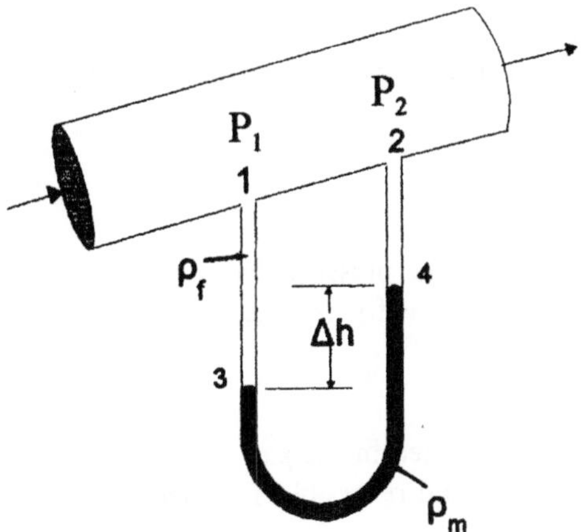

FIGURE 4-2 Manometer attached to pressure taps on a pipe carrying a flowing fluid.

When these three equations are added, P_3 and P_4 cancel out. The remaining terms can be collected to give

$$\Delta\Phi = -\Delta\rho\, g\, \Delta h \tag{4-10}$$

where $\Phi = P + \rho g z$ and $\Delta\Phi = \Phi_2 - \Phi_1$, $\Delta\rho = \rho_m - \rho_f$, $\Delta h = z_4 - z_3$. Equation (4-10) is the basic *manometer equation* and can be applied to a manometer in any orientation. Note that the manometer reading (Δh) is a direct measure of the *potential difference* ($\Phi_2 - \Phi_1$), which is identical to the pressure difference ($P_2 - P_1$) only if the pipe is horizontal (i.e., $z_2 = z_1$). It should be noted that these static fluid equations cannot be applied within the pipe, since the fluid in the pipe is not static.

B. Ideal Gas—Isothermal

If the fluid can be described by the ideal gas law (e.g., air, under normal atmospheric conditions), then

$$\rho = \frac{PM}{RT} \tag{4-11}$$

and Eq. (4-5) becomes

$$\frac{dP}{dz} = -\frac{PMg}{RT} \tag{4-12}$$

Now if the temperature is constant for all z (i.e., isothermal conditions), Eq. (4-12) can be integrated from (P_1, z_1) to (P_2, z_2) to give the pressure as a function of elevation:

$$P_2 = P_1 \exp\left(-\frac{Mg\Delta z}{RT}\right) \tag{4-13}$$

where $\Delta z = z_2 - z_1$. Note that in this case the pressure drops exponentially as the elevation increases instead of linearly as for the incompressible fluid.

C. Ideal Gas—Isentropic

If there is no heat transfer or energy dissipated in the gas when going from state 1 to state 2, the process is adiabatic and reversible, i.e., *isentropic*. For an ideal gas under these conditions,

$$\frac{P}{\rho^k} = \text{constant} = \frac{P_1}{\rho_1^k} \tag{4-14}$$

where $k = c_p/c_v$ is the specific heat ratio for the gas (for an ideal gas, $c_p = c_v + R/M$). If the density is eliminated from Eqs. (4-14) and (4-11), the result is

$$\frac{T}{T_1} = \left(\frac{P}{P_1}\right)^{(k-1)/k} \tag{4-15}$$

which relates the temperature and pressure at any two points in an isentropic ideal gas. If Eq. (4-15) is used to eliminate T from Eq. (4-12), the latter can be integrated to give the pressure as a function of elevation:

$$P_2 = P_1\left[1 - \left(\frac{k-1}{k}\right)\left(\frac{gM\,\Delta z}{RT_1}\right)\right]^{k/(k-1)} \tag{4-16}$$

which is a nonlinear relationship between pressure and elevation. Equation (4-15) can be used to eliminate P_2/P_1 from this equation to give an expression for the temperature as a function of elevation under isentropic conditions:

$$T_2 = T_1\left[1 - \left(\frac{k-1}{k}\right)\left(\frac{gM\,\Delta z}{RT_1}\right)\right] \tag{4-17}$$

That is, the temperature drops linearly as the elevation increases.

D. The Standard Atmosphere

Neither Eq. (4-13) nor Eq. (4-16) would be expected to provide a very good representation of the pressure and temperature in the real atmosphere, which is neither isothermal nor isentropic. Thus, we must resort to the use of observations (i.e., empiricism) to describe the real atmosphere. In fact, atmospheric conditions vary considerably from time to time and from place to place over the earth. However, a reasonable representation of atmospheric conditions "averaged" over the year and over the earth based on observations results in the following:

$$\begin{aligned} &\text{For } 0 < z < 11\text{ km:} \quad \frac{dT}{dz} = -6.5^{\circ}\text{C/km} = -G \\ &\text{For } z > 11\text{ km:} \quad T = -56.5^{\circ}\text{C} \end{aligned} \tag{4-18}$$

where the average temperature at ground level ($z = 0$) is assumed to be 15°C (288 K). These equations describe what is known as the "standard atmosphere," which represents an average state. Using Eq. (4-18) for the temperature as a function of elevation and incorporating this into Eq. (4-12) gives

$$\frac{dP}{dz} = \frac{PMg}{R(T_0 - Gz)} \tag{4-19}$$

where $T_0 = 288$ K and $G = 6.5^{\circ}$C/km. Integrating Eq. (4-19) assuming that g is constant gives the pressure as a function of elevation:

$$P_2 = P_1\left[1 - \frac{G\,\Delta z}{T_0 - Gz_1}\right]^{Mg/RG} \tag{4-20}$$

which applies for $0 < z < 11$ km.

III. MOVING SYSTEMS

We have stated that the only stress that can exist in a fluid at rest is pressure, because the shear stresses (which resist motion) are zero when the fluid is at rest. This also applies to fluids in motion provided there is no *relative* motion within the fluid (because the shear stresses are determined by the velocity *gradients*, e.g., the shear rate). However, if the motion involves an acceleration, this can contribute an additional component to the pressure, as illustrated by the examples in this section.

A. Vertical Acceleration

Consider the vertical column of fluid illustrated in Fig. 4-1, but now imagine it to be on an elevator that is accelerating upward with an acceleration of a_z, as illustrated in Fig. 4-3. Application of the momentum balance to the "slice" of fluid, as before, gives

$$\sum F_z = ma_z \tag{4-21}$$

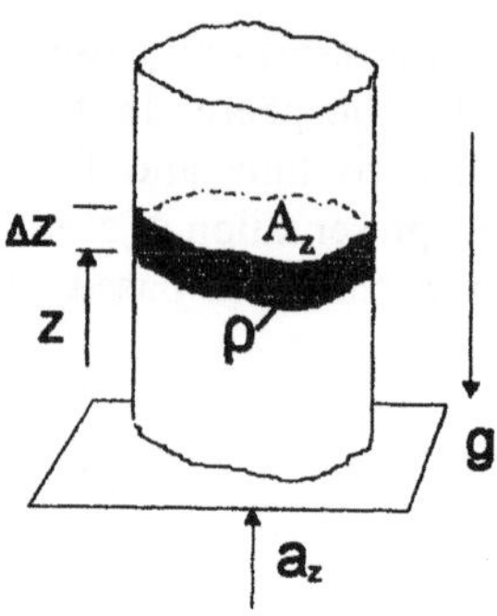

FIGURE 4-3 Vertically accelerating column of fluid.

which is the same as Eq. (4-3), except that now $a_z \neq 0$. The same procedure that led to Eq. (4-5) now gives

$$\frac{dP}{dz} = -\rho(g + a_z) \tag{4-22}$$

which shows that the effect of a superimposed vertical acceleration is equivalent to increasing the gravitational acceleration by an amount a_z (which is why you feel "heavier" on a rapidly accelerating elevator). In fact, this result may be generalized to any direction; an acceleration in the i direction will result in a pressure gradient within the fluid in the $-i$ direction, of magnitude ρa_a:

$$\frac{\partial P}{\partial x_i} = -\rho a_i \tag{4-23}$$

Two applications of this are illustrated next.

B. Horizontally Accelerating Free Surface

Consider a pool of water in the bed of your pickup truck. If you accelerate from rest, the water will slosh toward the rear, and you want to know how fast you can accelerate (a_x) without spilling the water out of the back of the truck (see Fig. 4-4). That is, you must determine the slope ($\tan\theta$) of the water surface as a function of the rate of acceleration (a_x). Now at any point within the liquid there is a vertical pressure gradient due to gravity [Eq. (4-5)] and a horizontal pressure gradient due to the acceleration a_x [Eq. (4-23)]. Thus at any location within the liquid the total differential pressure

FIGURE 4-4 Horizontally accelerating tank.

dP between two points separated by dx in the horizontal direction and dz in the vertical direction is given by

$$\begin{aligned} dP &= \frac{\partial P}{\partial x}\,dx + \frac{\partial P}{\partial z}\,dz \\ &= -\rho a_x\,dx - \rho g\,dz \end{aligned} \tag{4-24}$$

Since the surface of the water is open to the atmosphere, where $P =$ constant (1 atm),

$$(dP)_s = 0 = -\rho g(dz)_s - \rho a_x(dx)_s \tag{4-25}$$

or

$$\left(\frac{dz}{dx}\right)_s = -\frac{a_x}{g} = \tan\theta \tag{4-26}$$

which is the slope of the surface and is seen to be independent of fluid properties. A knowledge of the initial position of the surface plus the surface slope determines the elevation at the rear of the truck bed and hence whether or not the water will spill out.

C. Rotating Fluid

Consider an open bucket of water resting on a turntable that is rotating at an angular velocity ω (see Fig. 4-5). The (inward) radial acceleration due to the rotation is $\omega^2 r$, which results in a corresponding radial pressure gradient at all points in the fluid, in addition to the vertical pressure gradient due to gravity. Thus the pressure differential between any two points within the fluid separated by dr and dz is

$$dP = \left(\frac{\partial P}{\partial z}\right)dz + \left(\frac{\partial P}{\partial r}\right)dr = \rho(-g\,dz + \omega^2 r\,dr) \tag{4-27}$$

Just like the accelerating tank, the shape of the free surface can be determined from the fact that the pressure is constant at the surface, i.e.,

$$(dP)_s = 0 = -g(dz)_s + \omega^2 r(dr)_s \tag{4-28}$$

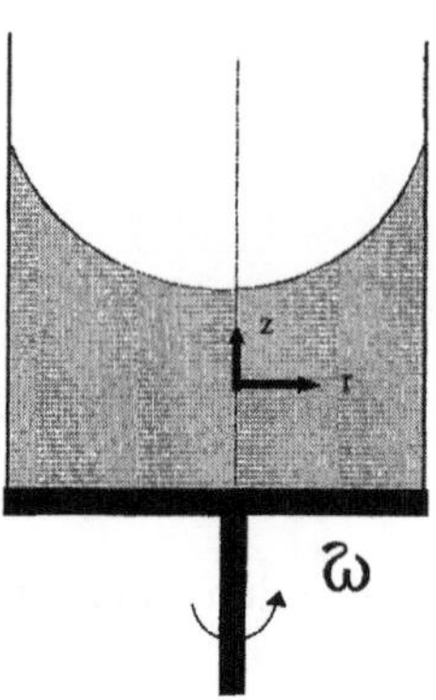

FIGURE 4-5 Rotating fluid.

This can be integrated to give an equation for the shape of the surface:

$$z = z_0 + \frac{\omega^2 r^2}{2g} \tag{4-29}$$

which shows that the shape of the rotating surface is parabolic.

IV. BUOYANCY

As a consequence of Archimedes' principle, the buoyant force exerted on a submerged body is equal to the weight of the displaced fluid and acts in a direction opposite to the acceleration vector. Thus the "effective net weight" of a submerged body is its actual weight less the weight of an equal volume of the fluid. The result is equivalent to replacing the density of the body (ρ_s) in the expression for the weight ($\rho_s g \tilde{V}_s$, where $\tilde{V}_s$ is the volume of the body) by the difference between the density of the body and that of the fluid (i.e., $\Delta\rho g \tilde{V}_s$, where $\Delta\rho = \rho_s - \rho_f$).

This also applies to a body submerged in a fluid that is subject to any acceleration. For example, a solid particle of volume $\tilde{V}_s$ submerged in a fluid within a centrifuge at a point r where the angular velocity is ω is subjected to a net radial force equal to $\Delta\rho\, \omega^2 r \tilde{V}_s$. Thus, the effect of buoyancy is to effectively reduce the density of the body by an amount equal to the density of the surrounding fluid.

V. STATIC FORCES ON SOLID BOUNDARIES

The force exerted on a solid boundary by a static pressure is given by

$$\vec{F} = \int_A P\, d\vec{A} \tag{4-30}$$

Note that both force and area are vectors, whereas pressure is a scalar. Hence the directional character of the force is determined by the orientation of the surface on which the pressure acts. That is, the component of force acting in a given direction on a surface is the integral of the pressure over the *projected component* area of the surface, where the surface vector (normal to the surface component) is parallel to the direction of the force [recall that pressure is a *negative* isotropic stress and the *outward* normal to the (fluid) system boundary represents a positive area]. Also, from Newton's third law ("action equals reaction"), the force exerted *on* the fluid system boundary is of opposite sign to the force exerted *by* the system on the solid boundary.

Example 4-2: Consider the force within the wall of a pipe resulting from the pressure of the fluid inside the pipe, as illustrated in Fig. 4-6. The pressure P acts equally in all directions on the inside wall of the pipe. The resulting force exerted within the pipe wall normal to a plane through the pipe axis is simply the product of the pressure and the projected area of the wall on this plane, e.g., $F_x = PA_x = 2PRL$. This force acts to pull the metal in the wall apart and is resisted by the internal stress within the metal holding it together. This is the effective *working stress*, σ, of the particular material of which the pipe is made. If we assume a thin-walled pipe (i.e., we neglect the radial variation of the stress from point to point within the wall), a force balance between the "disruptive" pressure force and the "restorative" force due to the internal stress in the metal gives

$$2PRL = 2\sigma tL \tag{4-31}$$

or

$$\frac{t}{R} \cong \frac{P}{\sigma} \tag{4-32}$$

This relation determines the pipe wall thickness required to withstand a fluid pressure P in a pipe of radius R made of a material with a working stress σ.

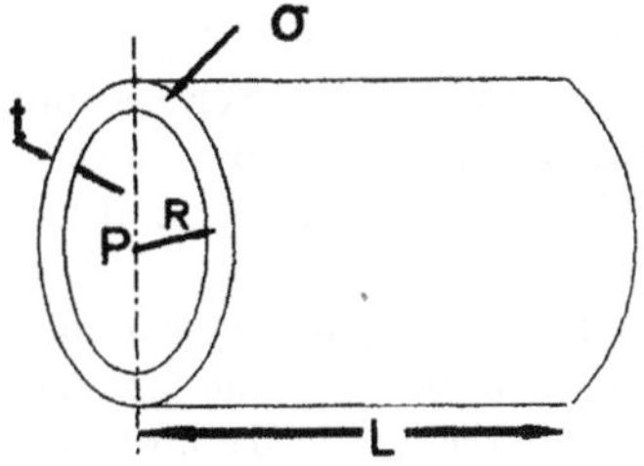

FIGURE 4-6 Fluid pressure inside a pipe.

The dimensionless pipe wall thickness (times 1000) is known as the *Schedule number* of the pipe:

$$\text{Schedule No.} \cong \frac{1000t}{R} = \frac{1000P}{\sigma} \tag{4-33}$$

This expression is only approximate, as it does not make any allowance for the effects of such things as pipe threads, corrosion, or wall damage. To compensate for these factors, an additional allowance is made for the wall thickness in the working definition of the "schedule thickness," t_s:

$$\text{Schedule No.} = \frac{1000P}{\sigma} = \frac{1750t_s - 200}{D_o} \tag{4-34}$$

where both t_s and D_o (the pipe outside diameter) are measured in inches. This relation between schedule number and pipe dimensions can be compared with the actual dimensions of commercial pipe for various schedule pipe sizes, as tabulated in Appendix F.

PROBLEMS

Statics

1. The manometer equation is $\Delta\Phi = -\Delta\rho g\,\Delta h$, where $\Delta\Phi$ is the difference in the total pressure plus static head $(P + \rho gz)$ between the two points to which the manometer is connected, $\Delta\rho$ is the difference in the densities of the two fluids in the manometer, Δh is the manometer reading, and g is the acceleration due to gravity. If $\Delta\rho$ is 12.6 g/cm^3 and Δh is 6 in. for a manometer connected to two points on a horizontal pipe, calculate the value of ΔP in the following units: (a) dyn/cm^2; (b) psi; (c) pascals; (d) atmospheres.
2. A manometer containing an oil with a specific gravity (SG) of 0.92 is connected across an orifice plate in a horizontal pipeline carrying seawater (SG = 1.1). If the manometer reading is 16.8 cm, what is the pressure drop across the orifice in psi? What is it in inches of water?
3. A mercury manometer is used to measure the pressure drop across an orifice that is mounted in a vertical pipe. A liquid with a density of 0.87 g/cm^3 is flowing upward through the pipe and the orifice. The distance between the manometer taps is 1 ft. If the pressure in the pipe at the upper tap is 30 psig, and the manometer reading is 15 cm, what is the pressure in the pipe at the lower manometer tap in psig?
4. A mercury manometer is connected between two points in a piping system that contains water. The downstream tap is 6 ft higher than the upstream tap, and the manometer reading is 16 in. If a pressure gage in the pipe at the upstream tap reads 40 psia, what would a pressure gage at the downstream tap read in (a) psia, (b) dyn/cm^2; (c) Pa; (d) kg$_f$/m^2?

5. An inclined tube manometer with a reservoir is used to measure the pressure gradient in a large pipe carrying oil (SG = 0.91) (see Fig. 4-P5). The pipe is inclined at an angle of 60° to the horizontal, and flow is uphill. The manometer tube is inclined at an angle of 20° to the horizontal, and the pressure taps on the pipe are 5 in. apart. The manometer reservoir diameter is eight times as large as the manometer tube diameter, and the manometer fluid is water. If the manometer reading (L) is 3 in. and the displacement of the interface in the reservoir is neglected, what is the pressure drop in the pipe in (a) psi; (b) Pa; (c) in. H_2O? What is the percentage error introduced by neglecting the change in elevation of the interface in the reservoir?

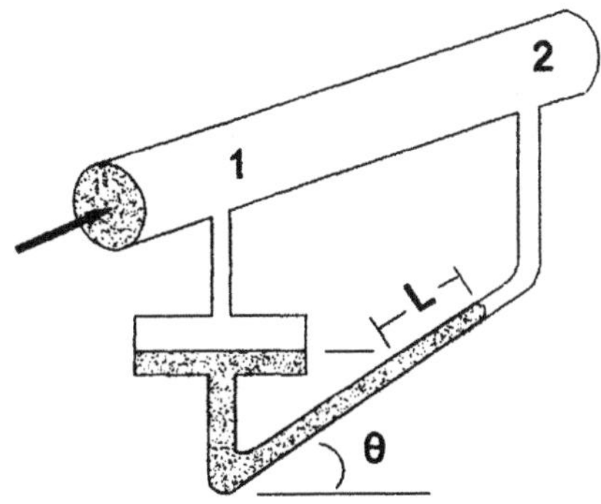

FIGURE 4-P5

6. Water is flowing downhill in a pipe that is inclined 30° to the horizontal. A mercury manometer is attached to pressure taps 5 cm apart on the pipe. The interface in the downstream manometer leg is 2 cm higher than the interface in the upstream leg. What is the pressure gradient ($\Delta P/L$) in the pipe in (a) Pa/m, (b) dyn/cm^3, in. (c) H_2O/ft, (d) psi/mi?
7. Repeat Problem 6 for the case in which the water in the pipe is flowing uphill instead of downhill, all other conditions remaining the same.
8. Two horizontal pipelines are parallel, with one carrying salt water ($\rho = 1.988$ slugs/ft^3) and the other carrying fresh water ($\rho = 1.937$ slugs/ft^3). An inverted manometer using linseed oil ($\rho = 1.828$ slugs/ft^3) as the manometer fluid is connected between the two pipelines. The interface between the oil and the fresh water in the manometer is 38 in. above the centerline of the freshwater pipeline, and the oil/salt water interface in the manometer is 20 in. above the centerline of the salt water pipeline. If the manometer reading is 8 in., determine the difference in the pressures between the pipelines (a) in Pa and (b) in psi.
9. Two identical tanks are 3 ft in diameter and 3 ft high, and they are both vented to the atmosphere. The top of tank B is level with the bottom of tank A, and they are connected by a line from the bottom of A to the top of B with a valve in it. Initially A is full of water and B is empty. The valve is opened for a short

time, letting some of the water drain into B. An inverted manometer having an oil with SG = 0.7 is connected between taps on the bottom of each tank. The manometer reading is 6 in., and the oil/water interface in the leg connected to tank A is higher. What is the water level in each of the tanks?

10. An inclined tube manometer is used to measure the pressure drop in an elbow through which water is flowing (see Fig. 4-P10). The manometer fluid is an oil with SG = 1.15. The distance L is the distance along the inclined tube that the interface has moved from its equilibrium (no pressure differential) position. If $h = 6\,\text{in.}$, $L = 3\,\text{in.}$, $\theta = 30°$, the reservoir diameter is 2 in., and the tubing diameter is 0.25 in., calculate the pressure drop $(P_1 - P_2)$ in (a) atm; (b) Pa; (c) cm H_2O; (d) dyn/cm^2. What would be the percentage error in pressure difference as read by the manometer if the change in level in the reservoir were neglected?

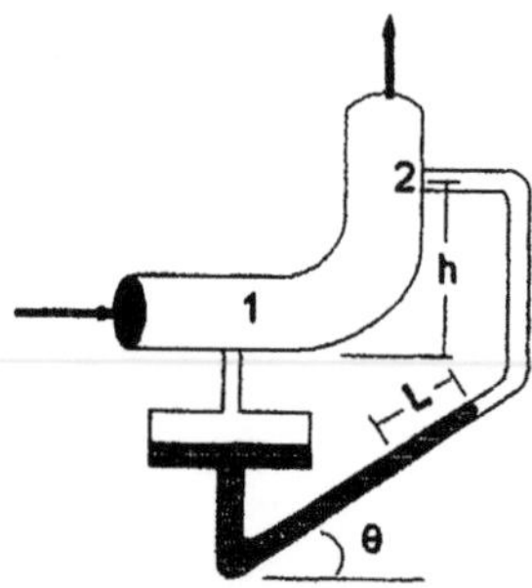

FIGURE 4-P10

11. The three-fluid manometer illustrated in Fig. 4-P11 is used to measure a very small pressure difference $(P_1 - P_2)$. The cross-sectional area of each of the reservoirs is A, and that of the manometer legs is a. The three fluids have densities ρ_a, ρ_b, and ρ_c, and the difference in elevation of the interfaces in the reservoir is x. Derive the equation that relates the manometer reading h to the pressure difference $P_1 - P_2$. How would the relation be simplified if $A \gg a$?

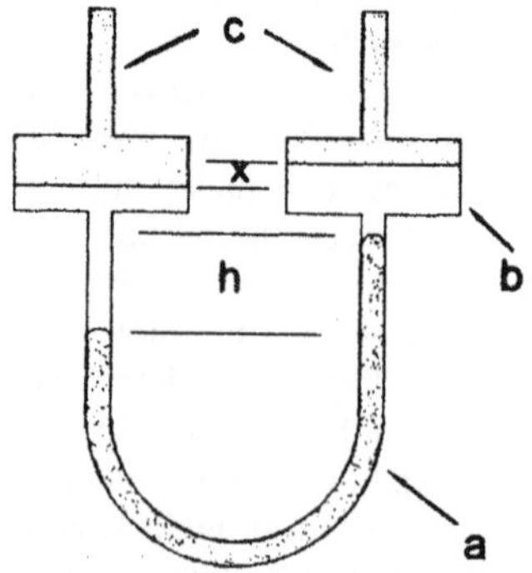

FIGURE 4-P11

12. A tank that is vented to the atmosphere contains a liquid with a density of $0.9\ g/cm^3$. A dip tube inserted into the top of the tank extends to a point 1 ft from the bottom of the tank. Air is bubbled slowly through the dip tube, and the air pressure in the tube is measured with a mercury (SG = 13.6) manometer. One leg of the manometer is connected to the air line feeding the dip tube, and the other leg is open to the atmosphere. If the manometer reading is 5 in., what is the depth of the liquid in the tank?
13. An inclined manometer is used to measure the pressure drop between two taps on a pipe carrying water, as shown in Fig. 4-P13. The manometer fluid is an oil with SG = 0.92, and the manometer reading (L) is 8 in. The manometer reservoir is 4 in. in diameter, the tubing is $\frac{1}{4}$ in. in diameter, and the manometer tube is inclined at an angle of 30° to the horizontal. The pipe is inclined at 20° to the horizontal, and the pressure taps are 40 in. apart.
    (a) What is the pressure difference between the two pipe taps that would be indicated by the difference in readings of two pressure gages attached to the taps, in (1) psi, (2) Pa, and (3) in. H_2O?
    (b) Which way is the water flowing?
    (c) What would the manometer reading be if the valve were closed?

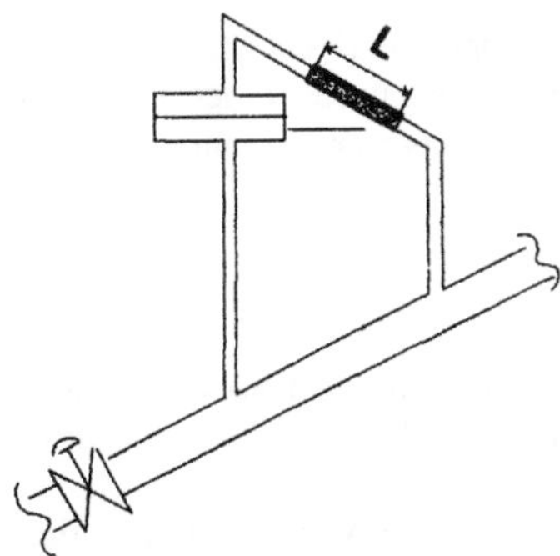

FIGURE 4-P13

14. The pressure gradient required to force water through a straight horizontal $\frac{1}{4}$ in. ID tube at a rate of 2 gpm is 1.2 psi/ft. Consider this same tubing coiled in an expanding helix with a vertical axis. Water enters the bottom of the coil and flows upward at a rate of 2 gpm. A mercury manometer is connected between two pressure taps on the coil, one near the bottom where the coil radius is 6 in., and the other near the top where the coil radius is 12 in. The taps are 2 ft apart in the vertical direction, and there is a total of 5 ft of tubing between the two taps. Determine the manometer reading, in cm.
15. It is possible to achieve a weightless condition for a limited time in an airplane by flying in a circular arc above the earth (like a rainbow). If the plane flies at 650 mph, what should the radius of the flight path be (in miles) to achieve weightlessness?

16. Water is flowing in a horizontal pipe bend at a velocity of 10 ft/s. The radius of curvature of the inside of the bend is 4 in., and the pipe ID is 2 in. A mercury manometer is connected to taps located radically opposite each other on the inside and outside of the bend. Assuming that the water velocity is uniform over the pipe cross section, what would be the manometer reading in centimeters? What would it be if the water velocity were 5 ft/s? Convert the manometer reading to equivalent pressure difference in psi and Pa.
17. Calculate the atmospheric pressure at an elevation of 3000 m, assuming (a) air is incompressible, at a temperature of 59°F; (b) air is isothermal at 59°F and an ideal gas; (c) the pressure distribution follows the model of the standard atmosphere, with a temperature of 59°F at the surface of the earth.
18. One pound mass of air (MW = 29) at sea level and 70°F is contained in a balloon, which is then carried to an elevation of 10,000 ft in the atmosphere. If the balloon offers no resistance to expansion of the gas, what is its volume at this elevation?
19. A gas well contains hydrocarbon gases with an average molecular weight of 24, which can be assumed to be an ideal gas with a specific heat ratio of 1.3. The pressure and temperature at the top of the well are 250 psig and 70°F, respectively. The gas is being produced at a slow rate, so conditions in the well can be considered to be isentropic.
    (a) What are the pressure and temperature at a depth of 10,000 ft?
    (b) What would the pressure be at this depth if the gas were isothermal?
    (c) What would the pressure be at this depth if the gas were incompressible?
20. The adiabatic atmosphere obeys the equation

$$P/\rho^k = \text{constant}$$

    where k is a constant and ρ is density. If the temperature decreases 0.3°C for every 100 ft increase in altitude, what is the value of k? [*Note*: Air is an ideal gas; $g = 32.2\ \text{ft/s}^2$; $R = 1544\ \text{ft lb}_f/(^\circ\text{R lbmol})$].
21. Using the actual dimensions of commercial steel pipe from Appendix F, plot the pipe wall thickness versus the pipe diameter for both Schedule 40 and Schedule 80 pipe, and fit the plot with a straight line by linear regression analysis. Rearrange your equation for the line in a form consistent with the given equation for the schedule number as a function of wall thickness and diameter:

$$\text{Schedule No} = (1750t_s - 200)/D$$

    and use the results of the regression to calculate values corresponding to the 1750 and 200 in this equation. Do this using (for D) (a) the nominal pipe diameter and (b) the outside pipe diameter. Explain any discrepancies or differences in the numerical values determined from the data fit compared to those in the equation.
22. The "yield stress" for carbon steel is 35,000 psi, and the "working stress" is one-half of this value. What schedule number would you recommend for a pipe carrying ethylene at a pressure of 2500 psi if the pipeline design calls for a pipe

of 2 in. ID? Give the dimensions of the pipe that you would recommend. What would be a safe maximum pressure to recommend for this pipe?

23. Consider a 90° elbow in a 2 in. pipe (all of which is in the horizontal plane). A pipe tap is drilled through the wall of the elbow on the inside curve of the elbow, and another through the outer wall of the elbow directly across from the inside tap. The radius of curvature of the inside of the bend is 2 in., and that of the outside of the bend is 4 in. The pipe is carrying water, and a manometer containing an immiscible oil with SG of 0.90 is connected across the two taps on the elbow. If the reading of the manometer is 7 in., what is the average velocity of the water in the pipe, assuming that the flow is uniform across the pipe inside the elbow?
24. A pipe carrying water is inclined at an angle of 45° to the horizontal. A manometer containing a fluid with an SG of 1.2 is attached to taps on the pipe, which are 1 ft apart. If the liquid interface in the manometer leg that is attached to the lower tap is 3 in. below the interface in the other leg, what is the pressure gradient in the pipe ($\Delta P/L$), in units of (a) psi/ft and (b) Pa/m? Which direction is the water flowing?
25. A tank contains a liquid of unknown density (see the Fig. 4-P25). Two dip tubes are inserted into the tank, each to a different level in the tank, through which air is bubbled very slowly through the liquid. A manometer is used to measure the difference in pressure between the two dip tubes. If the difference in level of the ends of the dip tubes (H) is 1 ft, and the manometer reads 1.5 ft (h) with water as the manometer fluid, what is the density of the liquid in the tank?

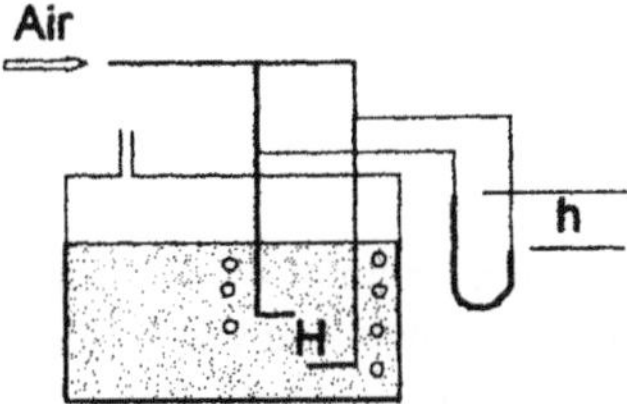

FIGURE 4-P25

26. The tank shown in the Fig. 4-P26 has a partition that separates two immiscible liquids. Most of the tank contains water, and oil is floating above the water on the right of the partition. The height of the water in the standpipe (h) is 10 cm, and the interface between the oil and water is 20 cm below the top of the tank and 25 cm above the bottom of the tank. If the specific gravity of the oil is 0.82, what is the height of the oil in the standpipe (H)?
27. A manometer that is open to the atmosphere contains water, with a layer of oil floating on the water in one leg (see Fig. 4-P27). If the level of the water in the left leg is 1 cm above the center of the leg, the interface between the water and oil is 1 cm below the center in the right leg, and the oil layer on the right extends 2 cm above the center, what is the density of the oil?

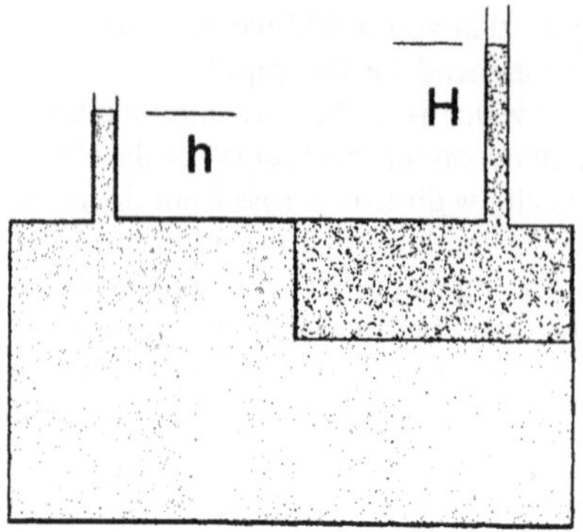

FIGURE 4-P26

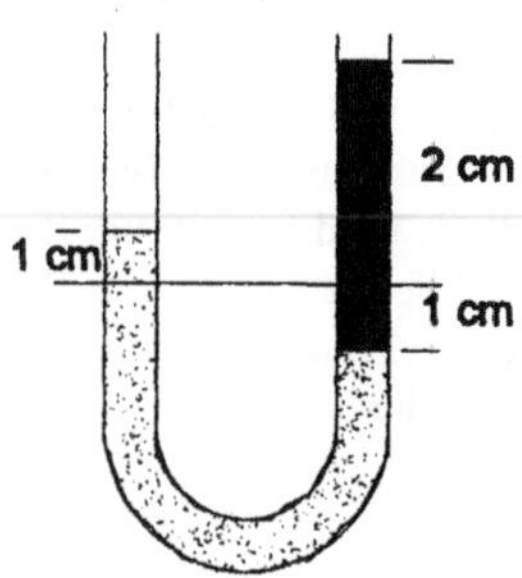

FIGURE 4-P27

28. An open cylindrical drum, with a diameter of 2 ft and a length of 4 ft, is turned upside down in the atmosphere and then submerged in a liquid so that it floats partially submerged upside down, with air trapped inside. If the drum weighs 150 lb_f, and it floats with 1 ft extending above the surface of the liquid, what is the density of the liquid? How much additional weight must be added to the drum to make it sink to the point where it floats just level with the liquid?
29. A solid spherical particle with a radius of 1 mm and a density of 1.3 g/cm^3 is immersed in water in a centrifuge. If the particle is 10 cm from the axis of the centrifuge, which is rotating at a rate of 100 rpm, what direction will the particle be traveling relative to a horizontal plane?
30. A manometer with mercury as the manometer fluid is attached to the wall of a closed tank containing water (see Fig. 4-P30). The entire system is rotating about the axis of the tank at N rpm. The radius of the tank is r_1, the distances from the tank centerline to the manometer legs are r_2 and r_3 (as shown), and the manometer reading is h. If $N = 30$ rpm, $r_1 = 12$ cm, $r_2 = 15$ cm, $r_3 = 18$ cm, and $h = 2$ cm, determine the gauge pressure at the wall of the tank and also at the centerline at the elevation of the pressure tap on the tank.

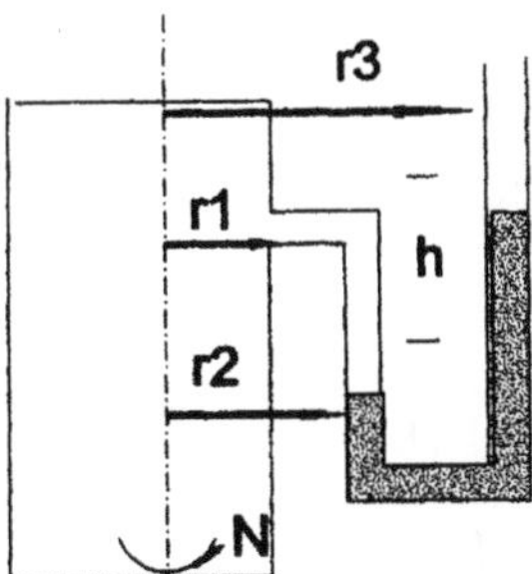

FIGURE 4-P30

31. With reference to the figure for Prob. 30, the manometer contains water as the manometer fluid and is attached to a tank that is empty and open to the atmosphere. When the tank is stationary, the water level is the same in both legs of the manometer. If the entire system is rotated about the centerline of the tank at a rate of N (rpm):
    (a) What happens to the water levels in the legs of the manometer?
    (b) Derive an equation for the difference in elevation of the levels (h) in the legs of the manometer as a function of known quantities.
32. You want to measure the specific gravity of a liquid. To do this, you first weigh a beaker of the liquid on a scale (W_{Lo}). You then attach a string to a solid body that is heavier than the liquid, and while holding the string you immerse the solid body in the liquid and measure the weight of the beaker containing the liquid with the solid submerged (W_{Ls}). You then repeat the procedure using the same weight but with water instead of the "unknown" liquid. The corresponding weight of the water without the weight submerged is W_{wo}, and with the solid submerged it is W_{ws}. Show how the specific gravity of the "unknown" liquid can be determined from these four weights, and show that the result is independent of the size, shape, or weight of the solid body used (provided, of course, that it is heavier than the liquids and is large enough that the difference in the weights can be measured precisely).
33. A vertical U-tube manometer is open to the atmosphere and contains a liquid that has an SG of 0.87 and a vapor pressure of 450 mmHg at the operating temperature. The vertical tubes are 4 in. apart, and the level of the liquid in the tubes is 6 in. above the bottom of the manometer. The manometer is then rotated about a vertical axis through its centerline. Determine what the rotation rate would have to be (in rpm) for the liquid to start to boil.
34. A spherical particle with SG = 2.5 and a diameter of 2 mm is immersed in water in a cylindrical centrifuge with has a diameter of 20 cm. If the particle is initially 8 cm above the bottom of the centrifuge and 1 cm from the centerline, what is the speed of the centrifuge (in rpm) if this particle strikes the wall of the centrifuge just before it hits the bottom?

NOTATION

A	area, $[L^2]$
A_i	area with outward normal in the i direction, $[L^2]$
a_z	acceleration in the z direction, $[L/t^2]$
D_o	pipe diameter (outer), [L]
F_j	force in the j direction, $[F = ML/t^2]$
G	atmospheric temperature gradient (= 6.5°C/km), [T/L]
g	acceleration due to gravity, $[L/t^2]$
h	vertical displacement of manometer interface, [L]
k	isentropic exponent $(= c_p/c_v)$ for ideal gas, [—]
M	molecular weight [M/mol]
P	pressure, $[F/L^2 = M/Lt^2]$
R	gas constant, $[FL/(\text{mol T}) = ML^2 (\text{mol } t^2T)]$
r	radial direction, [L]
T	temperature [T]
t	pipe thickness, [L]
$\tilde{V}$	volume $[L^3]$
z	vertical direction, measured upward, [L]
$\Delta(\,)$	difference of two values $[=(\,)_2 - (\,)_1]$
ρ	density, $[M/L^3]$
Φ	potential $(= P + \rho gz)$, $[F/L^2 = M/Lt^2]$
σ	working stress of metal, $[F/L^2 = M/Lt^2]$
σ_{ij}	ij total stress component, force in j direction on i surface, $[F/L^2 = M/Lt^2]$
ω	angular velocity, [1/t]

Subscripts

1	reference point 1
2	reference point 2
i, j, x, y, z	coordinate directions
o	outer

5

Conservation Principles

I. THE SYSTEM

As discussed in Chapter 1, the basic principles that apply to the analysis and solution of flow problems include the conservation of mass, energy, and momentum in addition to appropriate transport relations for these conserved quantities. For flow problems, these conservation laws are applied to a *system*, which is defined as any clearly specified region or volume of fluid with either macroscopic or microscopic dimensions (this is also sometimes referred to as a "control volume"), as illustrated in Fig. 5-1. The general conservation law is

$$\begin{matrix}\text{Rate of } X \\ \text{into the system}\end{matrix} - \begin{matrix}\text{Rate of } X \\ \text{out of the system}\end{matrix} = \begin{matrix}\text{Rate of accumulation} \\ \text{of } X \text{ in the system}\end{matrix}$$

where X is the conserved quantity, i.e., mass, energy, or momentum. In the case of momentum, because a "rate of momentum" is equivalent to a force (by Newton's second law), the "rate in" term must also include any (net) forces acting on the system. It is emphasized that the *system* is *not* the "containing vessel" (e.g., a pipe, tank, or pump) but is the *fluid* contained within the designated boundary. We will show how this generic expression is applied for each of the these conserved quantities.

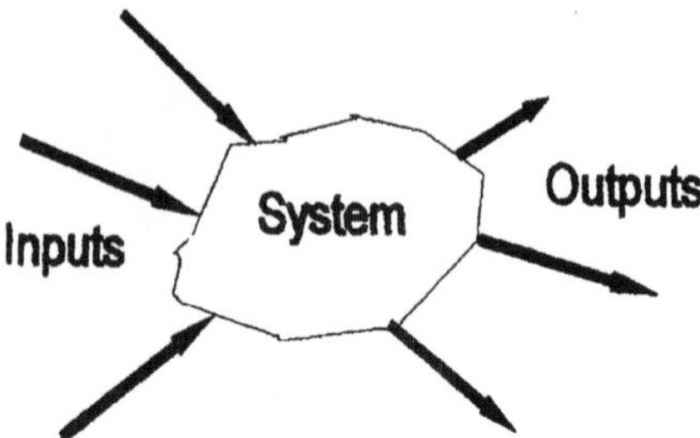

FIGURE 5-1 A system with inputs and outputs.

II. CONSERVATION OF MASS

A. Macroscopic Balance

For a given system (e.g., Fig. 5-1), each entering stream (subscript i) will carry mass into the system (at rate $\dot{m}_i$), and each exiting stream (subscript o) carries mass out of the system (at rate $\dot{m}_o$). Hence, the conservation of mass, or "continuity," equation for the system is

$$\sum_{\text{in}} \dot{m}_i - \sum_{\text{out}} \dot{m}_o = \frac{dm_s}{dt} \tag{5-1}$$

where m_s is the mass of the system. For each stream,

$$\dot{m} = \int_A d\dot{m} = \int_A \rho\vec{v} \cdot d\vec{A} = \rho\vec{V} \cdot \vec{A} \tag{5-2}$$

that is, the total mass flow rate through a given area for any stream is the integrated value of the local mass flow rate over that area. Note that mass flow rate is a scalar, whereas velocity and area are vectors. Thus it is the scalar (or dot) product of the velocity and area vectors that is required. (The "direction" or orientation of the area is that of the unit vector that is normal to the area.) The corresponding definition of the average velocity through the conduit is

$$V = \frac{1}{A}\int \vec{v} \cdot d\vec{A} = \frac{Q}{A} \tag{5-3}$$

where $Q = \dot{m}/\rho$ is the volumetric flow rate and the area A is the projected component of $\vec{A}$ that is normal to $\vec{V}$ (i.e., the component of $\vec{A}$ whose normal is in the same direction as $\vec{V}$). For a system at steady state, Eq. (5-1) reduces to

$$\sum_{\text{in}} \dot{m}_i = \sum_{\text{out}} \dot{m}_o \tag{5-4}$$

or

$$\sum_{\text{in}} (\rho VA)_i = \sum_{\text{out}} (\rho VA)_o \tag{5-5}$$

B. Microscopic Balance

The conservation of mass can be applied to an arbitrarily small fluid element to derive the "microscopic continuity" equation, which must be satisfied at all points within any continuous fluid. This can be done by considering an arbitrary (cubical) differential element of dimensions dx, dy, dz, with mass

---

Example 5-1: Water is flowing at a velocity of 7 ft/s in both 1 in. and 2 in. ID pipes, which are joined together and feed into a 3 in. ID pipe, as shown in Fig. 5-2. Determine the water velocity in the 3 in. pipe.

Solution. Because the system is at steady state, Eq. (5-5) applies:

$$(\rho VA)_1 + (\rho VA)_2 = (\rho VA)_3$$

For constant density, this can be solved for V_3:

$$V_3 = V_1 \frac{A_1}{A_3} + V_2 \frac{A_2}{A_3}$$

Since $A = \pi D^2/4$, this gives

$$V_3 = \left(7 \frac{\text{ft}}{\text{s}}\right)\left(\frac{1}{9} + \frac{4}{9}\right) = 3.89 \text{ ft/s}$$

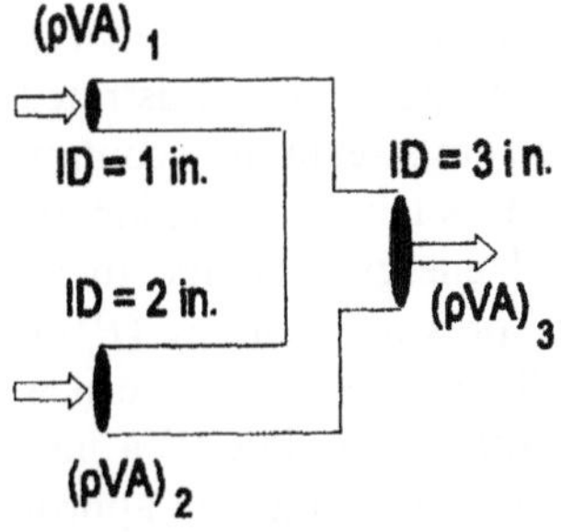

FIGURE 5-2 Continuity.

---

flow components into or out of each surface, e.g.,

$$\begin{aligned}\dot{m}_{\text{out}} - \dot{m}_{\text{in}} &= dy\,dz[(\rho v_x)_{x+dx} - (\rho v_x)_x] \\ &\quad + dx\,dz[(\rho v_y)_{y+dy} - (\rho v_y)_y] + dx\,dy[(\rho v_z)_{z+dz} - (\rho v_z)_z] \\ &= -\frac{\partial \rho}{\partial t}\,dx\,dy\,dz \end{aligned} \tag{5-6}$$

Dividing by the volume of the element $(dx \;\; dy \;\; dz)$ and taking the limit as the size of the element shrinks to zero gives

$$\frac{\partial(\rho v_x)}{\partial x} + \frac{\partial(\rho v_y)}{\partial y} + \frac{\partial(\rho v_z)}{\partial z} = -\frac{\partial \rho}{\partial t} \tag{5-7}$$

This is the microscopic (local) continuity equation and must be satisfied at all points within any flowing fluid. If the fluid is incompressible (i.e., constant ρ), Eq. (5-7) reduces to

$$\frac{\partial v_x}{\partial x} + \frac{\partial v_y}{\partial y} + \frac{\partial v_z}{\partial z} = 0 \tag{5-8}$$

We will make use of this equation in subsequent chapters.

III. CONSERVATION OF ENERGY

Energy can take a wide variety of forms, such as internal (thermal), mechanical, work, kinetic, potential, surface, electrostatic, electromagnetic, and nuclear energy. Also, for nuclear reactions or velocities approaching the speed of light, the interconversion of mass and energy can be significant. However, we will not be concerned with situations involving nuclear reactions or velocities near that of light, and some other possible forms of energy will usually be negligible as well. Our purposes will be adequately served if we consider only internal (thermal), kinetic, potential (due to gravity), mechanical (work), and heat forms of energy. For the system illustrated in Fig. 5-1, a unit mass of fluid in each inlet and outlet stream may contain a certain amount of internal energy (u) by virtue of its temperature, kinetic energy ($V^2/2$) by virtue of its velocity, potential energy (gz) due to its position in a (gravitational) potential field, and "pressure" energy (P/ρ). The "pressure" energy is sometimes called the "flow work," because it is associated with the amount of work or energy required to "inject" a unit mass of fluid into the system or "eject" it out of the system at the appropriate pressure. In addition, energy can cross the boundaries of the system other than with the flow streams, in the form of heat (Q) resulting from a temperature difference and "shaft work" (W). Shaft work is so named because it is normally associated with work transmitted to or from the system by a shaft, such as that of a pump, compressor, mixer, or turbine.

The sign conventions for heat (Q) and work (W) are arbitrary and consequently vary from one authority to another. Heat is usually taken to be positive when it is added to the system, so it would seem to be consistent to use this same convention for work (which is the convention in most "scientific" references). However, engineers, being pragmatic, use a sign convention that is directly associated with "value." That is, if work can be *extracted* from the system (e.g., to drive a turbine) then it is *positive*, because a positive asset can be sold to produce revenue. However, if work must be *put into* the system (such as from a pump), then it is *negative*, because it must be purchased (a negative asset). This convention is also more consistent with the "driving force" interpretation of the terms in the energy balance, as will be shown later.

With this introduction, we can write the rate form of the conservation of energy equation for any system as follows:

$$\sum_{\text{in}} \left(h + gz + \frac{V^2}{2} \right)_{\text{i}} \dot{m}_{\text{i}} - \sum_{\text{out}} \left(h + gz + \frac{V^2}{2} \right)_{\text{o}} \dot{m}_{\text{o}} + \dot{Q} - \dot{W} = \frac{d}{dt} \left[\left(u + gz + \frac{V^2}{2} \right) m \right]_{\text{sys}} \tag{5-9}$$

Here, $h = u + P/\rho$ is the *enthalpy* per unit mass of fluid. Note that the inlet and exit streams include enthalpy (i.e., both internal energy, u, and flow work, P/ρ), whereas the "system energy" includes only the internal energy but no P/ρ flow work (for obvious reasons). If there are only one inlet stream and one exit stream ($\dot{m}_{\text{i}} = \dot{m}_{\text{o}} = \dot{m}$) and the system is at steady state, the energy balance becomes

$$\Delta h + g\,\Delta z + \tfrac{1}{2}\,\Delta V^2 = q - w \tag{5-10}$$

where $\Delta =$ ("out") — ("in"), and $q = \dot{Q}/\dot{m}$, $w = \dot{W}/\dot{m}$ are the heat added to the system and work done by the system, respectively, per unit mass of fluid. This expression also applies to a system comprising the fluid between any two points along a streamline (a "stream tube") within a flow field. Specifically, if these two points are only an infinitesimal distance apart, the result is the differential form of the energy balance:

$$dh + g\,dz + v\,dv = \delta q - \delta w \tag{5-11}$$

where $dh = du + d(P/\rho)$. The $d(\ )$ notation represents a total or "exact" differential and applies to those quantities that are determined only by the state (T, P) of the system and are thus "point" properties. The $\delta(\ )$ notation represents quantities that are inexact differentials and depend upon the path taken from one point to another.

Note that the energy balance contains several different forms of energy, which may be generally classified as either *mechanical* energy, associated with motion or position, or *thermal* energy, associated with temperature. Mechanical energy is "useful," in that it can be converted directly into useful work, and includes potential energy, kinetic energy, "flow work," and shaft work. The thermal energy terms, i.e., internal energy and heat, are not directly available to do useful work unless they are transformed into mechanical energy, in which case it is the mechanical energy that does the work.

In fact, the total amount of energy represented by a relatively small temperature change is equivalent to a relatively large amount of "mechanical energy." For example, 1 Btu of thermal energy is equivalent to 778 ft lb_f of mechanical energy. This means that the amount of energy required to raise the temperature of 1 lb of water by 1°F (the definition of the Btu) is equivalent to the amount of energy required to raise the elevation of that same pound of water by 778 ft (e.g., an 80 story building!). Thus, for systems that involve significant temperature changes, the mechanical energy terms (e.g., pressure, potential and kinetic energy, and work) may be negligible compared with the thermal energy terms (e.g., heat transfer, internal energy). In such cases the energy balance equation reduces to a "heat balance," i.e., $\Delta h = q$. However, the reader should be warned that "heat" is not a conserved quantity and that the inherent assumption that other forms of energy are negligible when a "heat balance" is being written should always be confirmed.

Before proceeding further, we will take a closer look at the significance of enthalpy and internal energy, because these cannot be measured directly but are determined indirectly by measuring other properties such as temperature and pressure.

A. Internal Energy

An infinitesimal change in internal energy is an exact differential and is a unique function of temperature and pressure (for a given composition). Since the density of a given material is also uniquely determined by temperature and pressure (e.g., by an equation of state for the material), the internal energy may be expressed as a function of any two of the three terms T, P, or ρ (or $\nu = 1/\rho$). Hence, we may write:

$$du = \left(\frac{\partial u}{\partial T}\right)_\nu dT + \left(\frac{\partial u}{\partial \nu}\right)_T d\nu \tag{5-12}$$

By making use of classical thermodynamic identities, this is found to be equivalent to

$$du = c_v dT + \left[T\left(\frac{\partial P}{\partial T}\right)_v - P \right] dv \tag{5-13}$$

where

$$c_v = \left(\frac{\partial u}{\partial T}\right)_v \tag{5-14}$$

is the specific heat at constant volume (e.g., constant density). We will now consider several special cases for various materials.

1. Ideal Gas

For an ideal gas,

$$\rho = \frac{PM}{RT} \qquad \text{so that} \qquad T\left(\frac{\partial P}{\partial T}\right)_v = P \tag{5-15}$$

Thus Eq. (5-13) reduces to

$$du = c_v\, dT \qquad \text{or} \qquad \Delta u = \int_{T_1}^{T_2} c_v\, dT = \bar{c}_v(T_2 - T_1) \tag{5-16}$$

which shows that the internal energy for an ideal gas is a function of temperature only.

2. Non-Ideal Gas

For a non-ideal gas, Eq. (5-15) is not valid, so

$$T\left(\frac{\partial P}{\partial T}\right)_v \neq P \tag{5-17}$$

Consequently, the last term in Eq. (5-13) does not cancel as it did for the ideal gas, which means that

$$\Delta u = \text{fn}(T, P) \tag{5-18}$$

The form of the implied function, fn(T, P), may be analytical if the material is described by a non-ideal equation of state, or it could be empirical such as for steam, for which the properties are expressed as data tabulated in steam tables.

3. Solids and Liquids

For solids and liquids, $\rho \approx$ constant (or $dv = 0$), so

$$du = c_v\,dT \qquad \text{or} \qquad \Delta u = \int_{T_1}^{T_2} c_v\,dT = \bar{c}_v(T_2 - T_1) \tag{5-19}$$

This shows that the internal energy depends upon temperature only (just as for the ideal gas, but for an entirely different reason).

B. Enthalpy

The enthalpy can be expressed as a function of temperature and pressure:

$$dh = \left(\frac{\partial h}{\partial T}\right)_P dT + \left(\frac{\partial h}{\partial P}\right)_T dP \tag{5-20}$$

which, from thermodynamic identities, is equivalent to

$$dh = c_p\,dT + \left[v - T\left(\frac{\partial v}{\partial T}\right)_P\right] dP \tag{5-21}$$

Here

$$c_p = \left(\frac{\partial h}{\partial T}\right)_P \tag{5-22}$$

is the specific heat of the material at constant pressure. We again consider some special cases.

1. Ideal Gas

For an ideal gas,

$$T\left(\frac{\partial v}{\partial T}\right)_P = v \qquad \text{and} \qquad c_p = c_v + \frac{R}{M} \tag{5-23}$$

Thus Eq. (5-21) for the enthalpy becomes

$$dh = c_p\,dT \qquad \text{or} \qquad \Delta h = \int_{T_1}^{T_2} c_p\,dT = \bar{c}_p(T_2 - T_1) \tag{5-24}$$

which shows that the enthalpy for an ideal gas is a function of temperature only (as is the internal energy).

2. Non-Ideal Gas

For a non-ideal gas,

$$T\left(\frac{\partial v}{\partial T}\right)_P \neq v \qquad \text{so that} \qquad \Delta h = \text{fn}(T, P) \tag{5-25}$$

which, like Δu, may be either an analytical or an empirical function. All gases can be described as ideal gases under appropriate conditions (i.e., far enough from the critical point) and become more nonideal as the critical point is approached. That is, under conditions that are sufficiently far from the critical point that the enthalpy at constant temperature is essentially independent of pressure, the gas should be adequately described by the ideal gas law.

3. Solids and Liquids

For solids and liquids, $v = 1/\rho \approx \text{constant}$, so that $(\partial v/\partial T)_p = 0$ and $c_p \approx c_v$. Therefore,

$$dh = c_p\, dT + v\, dP \tag{5-26}$$

or

$$\Delta h = \int_{T_1}^{T_2} c_p\, dT + \int_{P_1}^{P_2} \frac{dP}{\rho} = \bar{c}_p(T_2 - T_1) + \frac{P_2 - P_1}{\rho} \tag{5-27}$$

This shows that for solids and liquids the enthalpy depends upon both temperature and pressure. This is in contrast to the internal energy, which depends upon temperature only. Note that for solids and liquids $c_p = c_v$.

The thermodynamic properties of a number of compounds are shown in Appendix D as pressure–enthalpy diagrams with lines of constant temperature, entropy, and specific volume. The vapor, liquid, and two-phase regions are clearly evident on these plots. The conditions under which each compound may exhibit ideal gas properties are identified by the region on the plot where the enthalpy is independent of pressure at a given temperature (i.e., the lower the pressure and the higher the temperature relative to the critical conditions, the more nearly the properties can be described by the ideal gas law).

IV. IRREVERSIBLE EFFECTS

We have noted that if there is a significant change in temperature, the thermal energy terms (i.e., q and u) may represent much more energy than the mechanical terms (i.e., pressure, potential and kinetic energy, and work). On the other hand, if the temperature difference between the system and its surroundings is very small, the only source of "heat" (thermal energy) is the internal (irreversible) dissipation of mechanical energy into thermal energy, or "friction." The origin of this "friction loss" is the irreversible work required to overcome intermolecular forces, i.e., the attractive forces between the "fluid elements," under dynamic (nonequilibrium) conditions. This can be quantified as follows.

For a system at equilibrium (i.e., in a reversible or "static" state), thermodynamics tells us that

$$du = T\,ds - P\,d(1/\rho) \qquad \text{and} \qquad T\,ds = \delta q \tag{5-28}$$

That is, the total increase in entropy (which is a measure of "disorder") comes from heat transferred across the system boundary (δq). However, a flowing fluid is in a "dynamic," or irreversible, state. Because entropy is proportional to the degree of departure from the most stable (equilibrium) conditions, this means that the further the system is from equilibrium, the greater the entropy, so for a dynamic (flow) system

$$T\,ds > \delta q \qquad \text{or} \qquad T\,ds = \delta q + \delta e_f \tag{5-29}$$

i.e.,

$$du = \delta q + \delta e_f - P\,d(1/\rho) \tag{5-30}$$

where δe_f represents the "irreversible energy" associated with the departure of the system from equilibrium, which is extracted from mechanical energy and transformed (or "dissipated") into thermal energy. The farther from equilibrium (e.g., the faster the motion), the greater this irreversible energy. The origin of this energy (or "extra entropy") is the mechanical energy that drives the system and is thus reduced by e_f. This energy ultimately appears as an increase in the temperature of the system (δu), heat transferred from the system (δq), and/or expansion energy [$P\,d(1/\rho)$] (if the fluid is compressible). This mechanism of transfer of useful mechanical energy to low grade (non-useful) thermal energy is referred to as "energy dissipation." Although e_f is often referred to as the friction loss, it is evident that this energy is not really lost, but is transformed (dissipated) from useful high level mechanical energy to non-useful low grade thermal energy. It should be clear that e_f must always be positive, because energy can be transformed spontaneously only from a higher state (mechanical) to a lower state (thermal) and not in the reverse direction, as a consequence of the second law of thermodynamics.

When Eq. (5-30) is introduced into the definition of enthalpy, we get

$$dh = du + d\left(\frac{P}{\rho}\right) = \delta q + \delta e_f + \frac{dP}{\rho} \tag{5-31}$$

Substituting this for the enthalpy in the differential energy balance, Eq. (5-11), gives

$$\frac{dP}{\rho} + g\,dz + V\,dV + \delta w + \delta e_f = 0 \tag{5-32}$$

This can be integrated along a streamline from the inlet to the outlet of the system to give

$$\int_{P_i}^{P_o} \frac{dP}{\rho} + g(z_o - z_i) + \frac{1}{2}(V_o^2 - V_i^2) + e_f + w = 0 \tag{5-33}$$

where, from Eq. (5-30),

$$e_f = (u_o - u_i) - q + \int_{\rho_i}^{\rho_o} P\, d\left(\frac{1}{\rho}\right) \tag{5-34}$$

Equations (5-33) and (5-34) are simply rearrangements of the steady-state energy balance equation [Eq. (5-10)], but are in much more useful forms. Without the friction loss (e_f) term (which includes all of the thermal energy effects), Eq. (5-33) represents a *mechanical energy balance*, although mechanical energy is not a conserved quantity. Equation (5-33) is referred to as the *engineering Bernoulli equation* or simply the *Bernoulli equation.* Along with Eq. (5-34), it accounts for all of the possible thermal and mechanical energy effects and is the form of the energy balance that is most convenient when mechanical energy dominates and thermal effects are minor. It should be stressed that the first three terms in Eq. (5-33) are *point* functions—they depend only on conditions at the inlet and outlet of the system—whereas the w and e_f terms are *path* functions, which depend on what is happening to the system *between* the inlet and outlet points (i.e., these are rate-dependent and can be determined from an appropriate rate or transport model).

If the fluid is incompressible (constant density), Eq. (5-33) can be written

$$\frac{\Delta\Phi}{\rho} + \frac{1}{2}\Delta(V^2) + e_f + w = 0 \tag{5-35}$$

where $\Phi = P + \rho g z$. For a fluid at rest, $e_f = V = w = 0$, and Eq. (5-35) reduces to the basic equation of fluid statics for an incompressible fluid (i.e., $\Phi = \text{const.}$), Eq. (4-7). For any static fluid, Eq. (5-32) reduces to the more general basic equation of fluid statics, Eq. (4-5). For gases, if the pressure change is such that the density does not change more than about 30%, the incompressible equation can be applied with reasonable accuracy by assuming the density to be constant at a value equal to the average density in the system (a general consideration of compressible fluids is given in Chapter 9).

Note that if each term of Eq. (5-35) is divided by g, then all terms will have the dimension of length. The result is called the "head" form of the Bernoulli equation, and each term then represents the equivalent amount of potential energy in a static column of the system fluid of the specified height. For example, the pressure term becomes the "pressure head ($-\Delta P/\rho g = H_p$)," the potential energy term becomes the "static head ($-\Delta z = H_z$)," the kinetic energy term becomes the "velocity head ($\Delta V^2/2g = H_v$)," the friction loss becomes the "head loss ($e_f/g = H_f$)," and the work term is, typically, the "work (or pump) head ($-w/g = H_w$)."

A. Kinetic Energy Correction

In the foregoing equations, we assumed that the fluid velocity (V) at a given point in the system (e.g., in a tube) is the same for all fluid elements at a given cross section of the flow stream. However, this is not true in conduits, because the fluid velocity is zero at a stationary boundary or wall and thus increases with the distance from the wall. The total rate at which kinetic energy is transported by a fluid element moving with local velocity $\vec{v}$ at a mass flow rate $d\dot{m}$ through a differential area $d\vec{A}$ is ($v^2\, d\dot{m}/2$), where $d\dot{m} = \rho\vec{v}\cdot d\vec{A}$. Thus, the total rate of transport of kinetic energy through the cross section A is

$$\int \frac{1}{2} v^2\, d\dot{m} = \frac{\rho}{2}\int v^3\, dA \tag{5-36}$$

If the fluid velocity is uniform over the cross section at a value equal to the average velocity V (i.e., "plug flow"), then the rate at which kinetic energy is transported would be

$$\tfrac{1}{2}\rho V^3 A \tag{5-37}$$

Therefore, a *kinetic energy correction factor*, α, can be defined as the ratio of the true rate of kinetic energy transport relative to that which would occur if the fluid velocity is everywhere equal to the average (plug flow) velocity, e.g.,

$$\alpha = \frac{\text{True KE transport rate}}{\text{Plug flow KE transport rate}} = \frac{1}{A}\int_A \left(\frac{v}{V}\right)^3 dA \tag{5-38}$$

The Bernoulli equation should therefore include this kinetic energy correction factor, i.e.,

$$\frac{\Delta\Phi}{\rho} + \frac{1}{2}\Delta(\alpha V^2) + e_f + w = 0 \tag{5-39}$$

As will be shown later, the velocity profile for a Newtonian fluid in laminar flow in a circular tube is parabolic. When this is introduced into Eq. (5-38), the result is $\alpha = 2$. For highly turbulent flow, the profile is much flatter and $\alpha \approx 1.06$, although for practical applications it is usually assumed that $\alpha = 1$ for turbulent flow.

Example 5-2: Kinetic Energy Correction Factor for Laminar Flow of a Newtonian Fluid. We will show later that the velocity profile for the laminar flow of a Newtonian fluid in fully developed flow in a circular tube is parabolic. Because the velocity is zero at the wall of the tube and maximum in the center, the equation for the profile is

$$v(r) = V_{\max}\left(1 - \frac{r^2}{R^2}\right)$$

This can be used to calculate the kinetic energy correction factor from Eq. (5-38) as follows. First we must calculate the average velocity, V, using Eq. (5-3):

$$V = \frac{1}{\pi R^2}\int_0^R v(r) 2\pi r \, dr$$

$$= 2V_{\max}\int_0^1 (1 - x^2)x \, dx = \frac{V_{\max}}{2}$$

which shows that the average velocity is simply one-half of the maximum (centerline) velocity. Thus, replacing V in Eq. (5-38) by $V_{\max}/2$ and then integrating the cube of the parabolic velocity profile over the tube cross section gives $\alpha = 2$. (The details of the manipulation are left as an exercise for the reader.)

Example 5-3: Diffuser. A diffuser is a section in a conduit over which the flow area increases gradually from upstream to downstream, as illustrated in Fig. 5-3. If the inlet and outlet areas (A_1 and A_2) are known, and the upstream pressure and velocity (P_1 and V_1) are given, we would like to find the downstream pressure and velocity (P_2 and V_2). If the fluid is incompressible, the continuity equation gives V_2:

$$(\rho VA)_1 = (\rho VA)_2 \qquad \text{or} \qquad V_2 = V_1 \frac{A_1}{A_2}$$

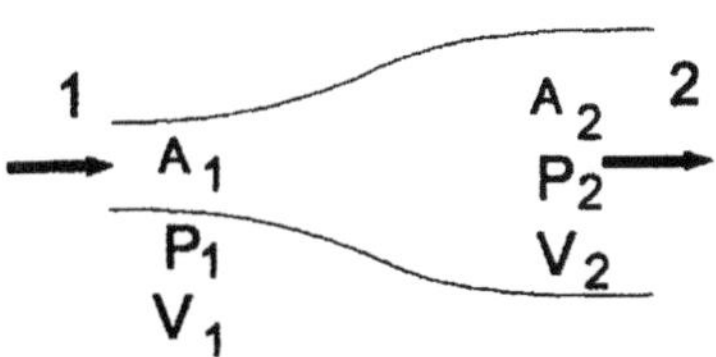

FIGURE 5-3 Diffuser.

The pressure P_2 is determined by Bernoulli's equation. If the diffuser is horizontal, there is no work done between the inlet and outlet, and the friction loss is small (which is a good assumption for a well designed diffuser), the Bernoulli equation gives

$$P_2 = P_1 + \frac{\rho}{2}(V_1^2 - V_2^2) - \rho e_f \cong P_1 + \frac{\rho V_1^2}{2}\left(1 - \frac{A_1^2}{A_2^2}\right)$$

Because $A_1 < A_2$ and the losses are small, this shows that $P_2 > P_1$, i.e., the pressure increases downstream. This occurs because the decrease in kinetic energy is transformed into an increase in "pressure energy." The diffuser is said to have a "high pressure recovery."

Example 5-4: Sudden Expansion. We now consider an incompressible fluid flowing from a small conduit through a sudden expansion into a larger conduit, as illustrated in Fig. 5-4. The objective, as in the previous example, is to determine the exit pressure and velocity (P_2 and V_2), given the upstream conditions and the dimensions of the ducts. The conditions are all identical to those of the above diffuser example, so the continuity and Bernoulli equations are also identical. The major difference is that the friction loss is not as small as for the diffuser. Because of inertia, the fluid cannot follow the sudden 90° change in direction of the boundary, so considerable turbulence is generated after the fluid leaves the small duct and before it can expand to fill the large duct, resulting in much greater friction loss. The equation for P_2 is the same as before:

$$P_2 = P_1 + \frac{\rho V_1^2}{2}\left(1 - \frac{A_1^2}{A_2^2}\right) - \rho e_f$$

The "pressure recovery" is reduced by the friction loss, which is relatively high for the sudden expansion. The pressure recovery is therefore relatively low.

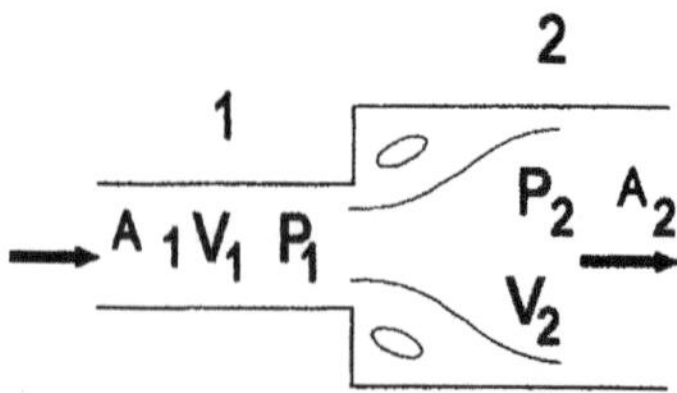

FIGURE 5-4 Sudden expansion.

Example 5-5: The Toricelli Problem. Consider an open vessel with diameter D_1 containing a fluid at a depth h that is draining out of a hole of diameter D_2 in the bottom of the tank. We would like to determine the velocity of the fluid flowing out of the hole in the bottom. As a first approximation, we neglect the friction loss in the tank and through the hole. Point 1 is taken at the surface of the fluid in the tank, and point 2 is taken at the exit from the hole, because the pressure is known to be atmospheric at both points. The velocity in the tank is related to that through the hole by the continuity equation

$$(\rho VA)_1 = (\rho VA)_2 \qquad \text{or} \qquad V_1 = V_2 \frac{A_2}{A_1} = V_2 \beta^2$$

where $\beta = D_2/D_1$. The Bernoulli equation for an incompressible fluid between points 1 and 2 is

$$\frac{P_2 - P_1}{\rho} + g(z_2 - z_1) + \frac{1}{2}(\alpha_2 V_2^2 - \alpha_1 V_1^2) + w + e_f = 0$$

Because points 1 and 2 are both at atmospheric pressure, $P_2 = P_1$. We assume that $w = 0$, $\alpha = 1$, and we neglect friction, so $e_f = 0$ (actually a very poor assumption in many cases). Setting $z_2 - z_1 = -h$, eliminating V_1 from these two equations, and solving for V_2 gives

$$V_2 = \left(\frac{2gh}{1 - \beta^4}\right)^{1/2}$$

This is known as the *Toricelli equation.* We now consider what happens as the hole gets larger. Specifically, as $D_2 \to D_1$ (i.e., as $\beta \to 1$), the equation says that $V_2 \to \infty$! This is obviously an unrealistic limit, so there must be

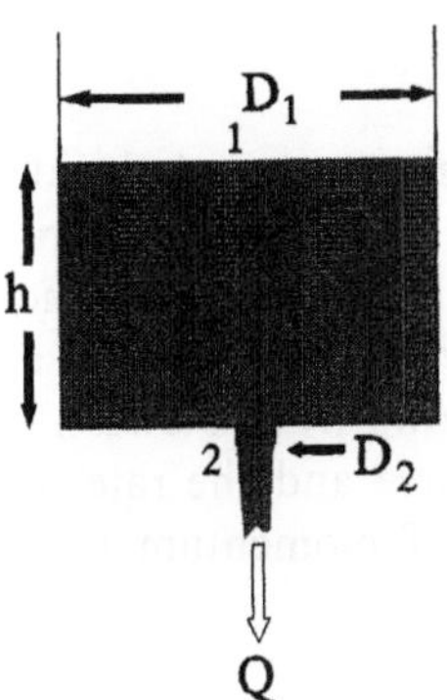

FIGURE 5-5 Draining tank. The Toricelli problem.

something wrong. Of course, our assumption that friction is negligible may be valid at low velocities, but as the velocity increases it becomes less valid and is obviously invalid long before this condition is reached.

Upon examining the equation for V_2, we see that it is independent of the properties of the fluid in the tank. We might suspect that this is not accurate, because if the tank were to be filled with CO_2 we intuitively expect that it would drain more slowly than if it were filled with water. So, what is wrong? In this case, it is our assumption that $P_2 = P_1$. Of course, the pressure is atmospheric at both points 1 and 2, but we have neglected the static head of air between these points, which is the actual difference in the pressure. This results in a buoyant force due to the air and can have a significant effect on the drainage of CO_2 although it will be negligible for water. Thus, if we account for the static head of air, i.e., $P_2 - P_1 = \rho_a gh$, in the Bernoulli equation and then solve for V_2, we get

$$V_2 = \left(\frac{2gh(1 - \rho_a/\rho)}{1 - \beta^4} \right)^{1/2}$$

where ρ is the density of the fluid in the tank. This also shows that as $\rho \to \rho_a$, the velocity goes to zero, as we would expect.

These examples illustrate the importance of knowing what can and cannot be neglected in a given problem and the necessity for matching the appropriate assumptions to the specific problem conditions in order to arrive at a valid solution. They also illustrate the importance of understanding what is happening within the system as well as knowing the inlet and outlet conditions.

V. CONSERVATION OF MOMENTUM

A macroscopic momentum balance for a flow system must include all equivalent forms of momentum. In addition to the rate of momentum convected into and out of the system by the entering and leaving streams, the sum of all the forces that act *on* the system (the system being defined as a specified volume of *fluid*) must be included. This follows from Newton's second law, which provides an equivalence between force and the rate of momentum. The resulting macroscopic conservation of momentum thus becomes

$$\sum_{\text{on system}} \vec{F} + \sum_{\text{in}} (\dot{m}\vec{V})_i - \sum_{\text{out}} (\dot{m}\vec{V})_o = \frac{d}{dt}(m\vec{V})_{sys} \tag{5-40}$$

Note that because momentum is a vector, this equation represents three component equations, one for each direction in three-dimensional space. If there is only one entering and one leaving stream, then $\dot{m}_i = \dot{m}_o = \dot{m}$. If the system is also at steady state, the momentum balance becomes

$$\sum_{\text{on fluid}} \vec{F} = \dot{m}\left(\sum_{\text{out}} \vec{V}_o - \sum_{\text{in}} \vec{V}_i\right) \tag{5-41}$$

Note that the vector (directional) character of the "convected" momentum terms (i.e., $\dot{m}\vec{V}$) is that of the velocity, because $\dot{m}$ is a scalar (i.e., $\dot{m} = \rho\vec{V} \cdot \vec{A}$ is a scalar product).

A. One-Dimensional Flow in a Tube

We will apply the steady state momentum balance to a fluid in plug flow in a tube, as illustrated in Fig. 5-6. (The "stream tube" may be bounded by either solid or imaginary boundaries; the only condition is that no fluid crosses the boundaries other that through the "inlet" and "outlet" planes.) The shape of the cross section does not have to be circular; it can be any shape. The fluid element in the "slice" of thickness dx is our system, and the momentum balance equation on this system is

$$\sum_{\text{on fluid}} F_x + \dot{m}V_x - \dot{m}(V_x + dV_x) = \sum_{\text{on fluid}} F_x - \dot{m}\, dV_x$$

$$= \frac{d}{dt}(\rho V_x A\, dx) = 0 \tag{5-42}$$

The forces acting on the fluid result from pressure (dF_P), gravity (dF_g), wall drag (dF_w), and external "shaft" work ($\delta W = -F_{ext}\, dx$, not shown in Fig. 5-6):

$$\sum_{\text{on fluid}} F_x = dF_p + dF_g + F_{ext} + dF_w \tag{5-43}$$

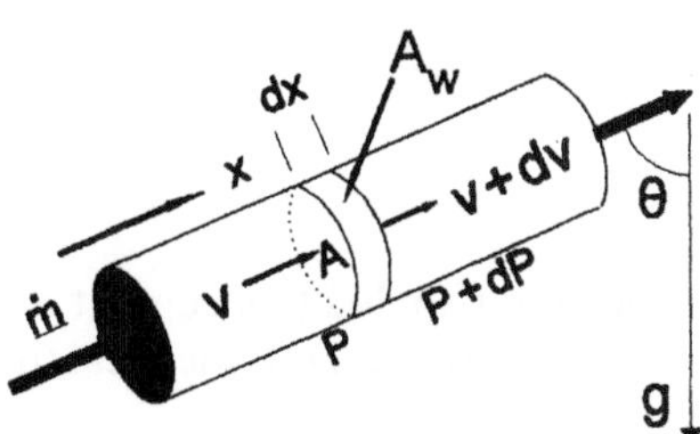

FIGURE 5-6 Momentum balance on a "slice" in a stream tube.

where

$$dF_p = A_x[P - (P + dP)] = -A_x \, dP$$

$$dF_g = \rho g_x A_x \, dx = -\rho A_x \, dx \, g \cos\theta = -\rho g A_x \, dx$$

$$dF_w = -\tau_w \, dA_w = -\tau_w W_p \, dx$$

$$-\delta W = F_{ext} \, dx \qquad \text{or} \qquad F_{ext} = -\frac{\delta W}{dx}$$

Here, τ_w is the stress exerted *by* the fluid *on* the wall (the reaction to the stress exerted *on* the fluid *by* the wall), and W_p is the perimeter of the wall in the cross section that is wetted by the fluid (the "wetted perimeter"). After substituting the expressions for the forces from Eq. (5-43) into the momentum balance equation, Eq. (5-42), and dividing the result by $-\rho A$, where $A = A_x$, the result is

$$\frac{dP}{\rho} + g \, dz + \frac{\tau_w W_p}{\rho A} \, dx + \delta w + V \, dV = -\frac{dV}{dt} \, dx \tag{5-44}$$

where $\delta w = \delta W/(\rho A \, dx)$ is the work done per unit mass of fluid. Integrating this expression from the inlet (i) to the outlet (o) and assuming steady state gives

$$\int_{P_i}^{P_o} \frac{dP}{\rho} + g(z_o - z_i) + \frac{1}{2}(V_o^2 - V_i^2) + \int_L \frac{\tau_w W_p}{\rho A} \, dx + w = 0 \tag{5-45}$$

Comparing this with the Bernoulli equation [Eq. (5-33)] shows that they are identical, provided

$$e_f = \int_L \left(\frac{\tau_w W_p}{\rho A} \right) dx \tag{5-46}$$

or, for steady flow in a uniform conduit

$$e_f \cong \frac{\tau_w W_p L}{\rho A} = \frac{\tau_w}{\rho} \left(\frac{4L}{D_h} \right) \tag{5-47}$$

where

$$D_h = 4 \frac{A}{W_p} \tag{5-48}$$

is called the *hydraulic diameter*. Note that this result applies to a conduit of any cross-sectional shape. For a circular tube, for example, D_h is identical to the tube diameter D.

We see that there are several ways of interpreting the term e_f. From the Bernoulli equation, it represents the "lost" (i.e., dissipated) energy associated with irreversible effects. From the momentum balance, e_f is also seen to be directly related to the stress between the fluid and the tube wall (τ_w), i.e., it can be interpreted as the work required to overcome the resistance to flow in the conduit. These interpretations are both correct and are equivalent.

Although the energy and momentum balances lead to equivalent results for this special case of one-dimensional fully developed flow in a straight uniform tube, this is an exception and not the rule. In general, the momentum balance gives additional information relative to the forces exerted on and/or by the fluid in the system through the boundaries, which is not given by the energy balance or Bernoulli equation. This will be illustrated shortly.

B. The Loss Coefficient

Looking at the Bernoulli equation, we see that the friction loss (e_f) can be made dimensionless by dividing it by the kinetic energy per unit mass of fluid. The result is the dimensionless *loss coefficient*, K_f:

$$K_f \equiv \frac{e_f}{V^2/2} \tag{5-49}$$

A loss coefficient can be defined for any element that offers resistance to flow (i.e., in which energy is dissipated), such as a length of conduit, a valve, a pipe fitting, a contraction, or an expansion. The total friction loss can thus be expressed in terms of the sum of the losses in each element, i.e., $e_f = \sum_i (K_{fi} V_i^2/2)$. This will be discussed further in Chapter 6.

As can be determined from Eqs. (5-47) and (5-49), the pipe wall stress can also be made dimensionless by dividing by the kinetic energy per unit volume of fluid. The result is known as the pipe *Fanning friction factor*, *f*:

$$f = \frac{\tau_w}{\rho V^2/2} \tag{5-50}$$

Although $\rho V^2/2$ represents kinetic energy per unit volume, ρV^2 is also the *flux of momentum* carried by the fluid along the conduit. The latter interpretation is more logical in Eq. (5-50), because τ_w is also a flux of momentum from the fluid to the tube wall. However, the conventional definition includes the (arbitrary) factor $\frac{1}{2}$. Other definitions of the pipe friction factor are in use that are some multiple of the Fanning friction factor. For example, the *Darcy friction factor*, which is equal to $4f$, is used frequently by mechanical and civil engineers. Thus, it is important to know which definition is implied when data for friction factors are used.

Because the friction loss and wall stress are related by Eq. (5-47), the loss coefficient for pipe flow is related to the pipe Fanning friction factor as follows:

$$K_f = \frac{4fL}{D_h} \qquad \text{(pipe)} \tag{5-51}$$

Example 5-6: Friction Loss in a Sudden Expansion. Figure 5-7 shows the flow in a sudden expansion from a small conduit to a larger one. We assume that the conditions upstream of the expansion (point 1) are known, as well as the areas A_1 and A_2. We desire to find the velocity and pressure downstream of the expansion (V_2 and P_2) and the loss coefficient, K_f. As before, V_2 is determined from the mass balance (continuity equation) applied to the system (the fluid in the shaded area). Assuming constant density,

$$V_2 = V_1 \frac{A_1}{A_2}$$

For plug flow, the Bernoulli equation for this system is

$$\frac{P_2 - P_1}{\rho} + \frac{1}{2}(V_2^2 - V_1^2) + e_f = 0$$

which contains two unknowns, P_2 and e_f. Thus, we need another equation, the steady-state momentum balance:

$$\sum F_x = \dot{m}(V_{2x} - V_{1x})$$

where $V_{1x} = V_1$ and $V_{2x} = V_2$, because all velocities are in the x direction. Accounting for all the forces that can act on the system through each section of the boundary, this becomes

$$P_1 A_1 + P_{1a}(A_2 - A_1) - P_2 A_2 + F_{wall} = \rho V_1 A_1 (V_2 - V_1)$$

where P_{1a} is the pressure on the left-hand boundary of the system (i.e., the "washer-shaped" surface), and F_{wall} is the force due to the drag of the wall on the fluid at the horizontal boundary of the system. The fluid pressure cannot change discontinuously, so $P_{1a} \simeq P_1$. Also, because the contact area

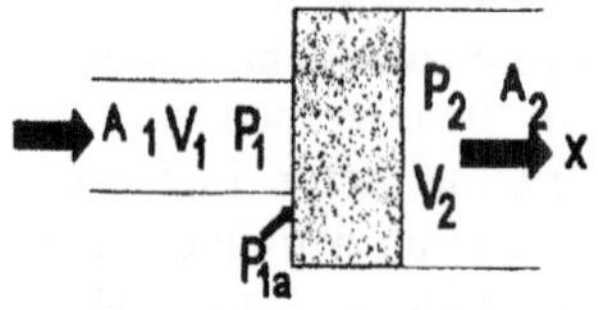

FIGURE 5-7 Sudden expansion.

with the wall bounding the system is relatively small, we can neglect F_{wall} with no serious consequences. The result is

$$(P_1 - P_2)A_2 = \rho V_1^2 A_1\left(\frac{A_1}{A_2} - 1\right)$$

This can be solved for $(P_2 - P_1)$, which, when inserted into the Bernoulli equation, allows us to solve for e_f:

$$e_f = \frac{V_1^2}{2}\left(1 - \frac{A_1}{A_2}\right)^2 = \frac{K_f V_1^2}{2}$$

Thus,

$$K_f = \left(1 - \frac{A_1}{A_2}\right)^2 = (1 - \beta^2)^2$$

where $\beta = D_1/D_2$.

The loss coefficient is seen to be a function only of the geometry of the system (note that the assumption of plug flow implies that the flow is highly turbulent). For most systems (i.e., flow in valves, fittings, etc.), the loss coefficient cannot be determined accurately from simple theoretical concepts (as in this case) but must be determined empirically. For example, the friction loss in a sudden contraction cannot be calculated by this simple method due to the occurrence of the vena contracta just downstream of the contraction (see Table 7-5 in Chapter 7 and the discussion in Section IV of Chapter 10). For a sharp 90° contraction, the contraction loss coefficient is given by

$$K_f = 0.5(1 - \beta^2)$$

where β is the ratio of the small to the large tube diameter.

Example 5-7: Flange Forces on a Pipe Bend. Consider an incompressible fluid flowing through a pipe bend, as illustrated in Fig. 5-8. We would like to determine the forces in the bolts in the flanges that hold the bend in the pipe, knowing the geometry of the bend, the flow rate through the bend, and the exit pressure (P_2) from the bend. Taking the system to be the fluid within the pipe bend, a steady-state "x-momentum" balance is

$$\sum (F_x)_{\text{on sys}} = \dot{m}(V_{2x} - V_{1x})$$

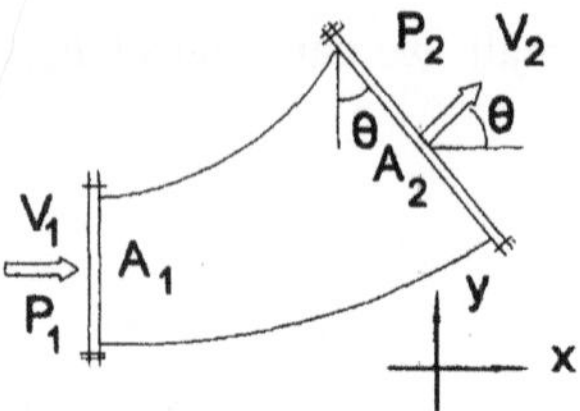

FIGURE 5-8 Flange forces in a pipe bend.

Various factors contribute to the forces on the left-hand side of this equation:

$$\sum (F_x)_{\text{on sys}} = P_1 A_{1x} + P_2 A_{2x} - (F_x)_{\text{on wall by fluid}}$$
$$= P_1 A_1 - P_2 A_2 \cos\theta - (F_x)_{\text{on bolts}}$$

[The sign of the force resulting from the pressure acting on any area element is intuitive, because pressure acts on any system boundary from the outside, i.e., pressure on the left-hand boundary acts to the right on the system, and vice versa. This is also consistent with previous definitions, because the sign of a surface element corresponds to the direction of the normal vector that points *outward* from the bounded volume, and pressure is a compressive (negative) stress. Thus $P_1 A_{x1}$ is (+) because it is a negative stress acting on a negative area, and $P_2 A_{x2}$ is (−) because it is a negative stress acting on a positive area. These signs have been accounted for intuitively in the equation.]

The right-side of the momentum balance reduces to

$$\dot{m}(V_{2x} - V_{1x}) = \dot{m}(V_2 \cos\theta - V_1)$$

Equating these two expressions and solving for $(F_x)_{\text{on wall}}$ gives

$$(F_x)_{\text{on wall}} = (F_x)_{\text{on bolts}} = P_1 A_1 - P_2 A_2 \cos\theta - \dot{m}(V_2 \cos\theta - V_1)$$

Similarly, the "y-momentum" balance is

$$\sum (F_y)_{\text{on sys}} = \dot{m}(V_{2y} - V_{1y})$$

which becomes

$$(F_y)_{\text{on wall}} = (F_y)_{\text{on bolts}} = -P_2 A_2 \sin\theta - \dot{m} V_2 \sin\theta$$

This assumes that the xy plane is horizontal. If the y direction is vertical, the total weight of the bend, including the fluid inside, could be included as an additional (negative) force component due to gravity. The magnitude and direction of the net force are

$$\bar{F} = \sqrt{F_x^2 + F_y^2}, \qquad \varphi = \tan^{-1}\left(\frac{F_y}{F_x}\right)$$

where φ is the direction of the net force vector measured counterclockwise from the $+x$ direction. Note that either P_1 or P_2 must be known, but the other is determined by the Bernoulli equation if the loss coefficient is known:

$$\frac{P_2 - P_1}{\rho} + g(z_2 - z_1) + \frac{1}{2}(V_2^2 - V_1^2) + e_f = 0$$

where

$$e_f = \tfrac{1}{2} K_f V_1^2$$

Methods for evaluating the loss coefficient K_f will be discussed in Chapter 6.

It should be noted that in evaluating the forces acting on the system, the effect of the external pressure transmitted through the boundaries to the system from the surrounding atmosphere was not included. Although this pressure does result in forces that act on the system, these forces all cancel out, so the pressure that appears in the momentum balance equation is the *net* pressure in excess of atmospheric, e.g., *gage pressure.*

C. Conservation of Angular Momentum

In addition to linear momentum, angular momentum (or the moment of momentum) may be conserved. For a fixed mass (m) moving in the x direction with a velocity V_x, the linear x-momentum (M_x) is mV_x. Likewise, a mass m rotating counterclockwise about a center of rotation at an angular velocity $\omega = d\theta/dt$ has an angular momentum (L_θ) equal to $mV_\theta R = m\omega R^2$, where R is the distance from the center of rotation to m. Note that the angular momentum has dimensions of "length times momentum," and is thus also referred to as the "moment of momentum." If the mass is not a point but a rigid distributed mass (M) rotating at a uniform angular velocity, the total angular momentum is given by

$$L_\theta = \int_M \omega r^2 \, dm = \omega \int_M r^2 \, dm = \omega I \tag{5-52}$$

where I is the moment of inertia of the body with respect to the center of rotation.

For a fixed mass, the conservation of linear momentum is equivalent to Newton's second law:

$$\sum \vec{F} = m\vec{a} = \frac{d(m\vec{V})}{dt} = m\,\frac{d\vec{V}}{dt} \tag{5-53}$$

The corresponding expression for the conservation of angular momentum is

$$\sum \Gamma_\theta = \sum F_\theta R = \frac{d(I\omega)}{dt} = I\,\frac{d\omega}{dt} = I\alpha \tag{5-54}$$

where Γ_θ is the moment (torque) acting on the system and $d\omega/dt = \alpha$ is the angular acceleration.

For a flow system, streams with curved streamlines may carry angular momentum into and/or out of the system by convection. To account for this, the general macroscopic angular momentum balance applies:

$$\sum_{\text{in}} (\dot{m}RV_\theta)_\text{i} - \sum_{\text{out}} (\dot{m}RV_\theta)_\text{o} + \sum \Gamma_\theta = \frac{d(I\omega)}{dt} = I\alpha \tag{5-55}$$

For a steady-state system with only one inlet and one outlet stream, this becomes

$$\sum \Gamma_\theta = \dot{m}[(RV_\theta)_\text{o} - (RV_\theta)_\text{i}] = \dot{m}[(R^2\omega)_\text{o} - (R^2\omega)_\text{i}] \tag{5-56}$$

This is known as the *Euler turbine equation*, because it applies directly to turbines and all rotating fluid machinery. We will find it useful later in the analysis of the performance of centrifugal pumps.

D. Moving Boundary Systems and Relative Motion

We sometimes encounter a system that is in contact with a moving boundary, such that the fluid that composes the system is carried along with the boundary while streams carrying momentum and/or energy may flow into and/or out of the system. Examples of this include the flow impinging on a turbine blade (with the system being the fluid in contact with the moving blade) and the flow of exhaust gases from a moving rocket motor. In such cases, we often have direct information concerning the velocity of the fluid relative to the moving boundary (i.e., relative to the system), V_r, and so we must also consider the velocity of the system, V_s, to determine the absolute velocity of the fluid that is required for the conservation equations.

For example, consider a system that is moving in the x direction with a velocity V_s a fluid stream entering the system with a velocity in the x direction relative to the system of V_ri, and a stream leaving the system with a velocity V_ro relative to the system. The absolute stream velocity in the x

direction V_x is related to the relative velocity V_{rx} and the system velocity V_{sx} by

$$V_x = V_{sx} + V_{rx} \tag{5-57}$$

The linear momentum balance equation becomes

$$\begin{aligned}\sum \vec{F} &= \dot{m}_o \vec{V}_o - \dot{m}_i \vec{V}_i + \frac{d(m\vec{V}_s)}{dt} \\ &= \dot{m}_o(\vec{V}_{ro} + \vec{V}_s) - \dot{m}_i(\vec{V}_{ri} + \vec{V}_s) + \frac{d(m\vec{V}_s)}{dt}\end{aligned} \tag{5-58}$$

Example 5-8: Turbine Blade. Consider a fluid stream impinging on a turbine blade that is moving with a velocity V_s. We would like to know what the velocity of the impinging stream should be in order to transfer the maximum amount of energy to the blade. The system is the fluid in contact with the blade, which is moving at velocity V_s. The impinging stream velocity is V_i, and the stream leaves the blade at velocity V_o. Since $V_o = V_{ro} + V_s$ and $V_i = V_{ri} + V_s$, the system velocity cancels out of the momentum equation:

$$F_x = \dot{m}(V_o - V_i) = \dot{m}(V_{ro} - V_{ri})$$

If the friction loss is negligible, the energy balance (Bernoulli's equation) becomes

$$w = \tfrac{1}{2}(V_i^2 - V_o^2)$$

which shows that the maximum energy or work transferred from the fluid to the blade occurs when $V_o = 0$. Now from continuity at steady state, recognizing that V_i and V_s are of opposite sign

$$|V_i| = |V_o| \qquad \text{or} \qquad V_{ri} = -V_{ro}$$

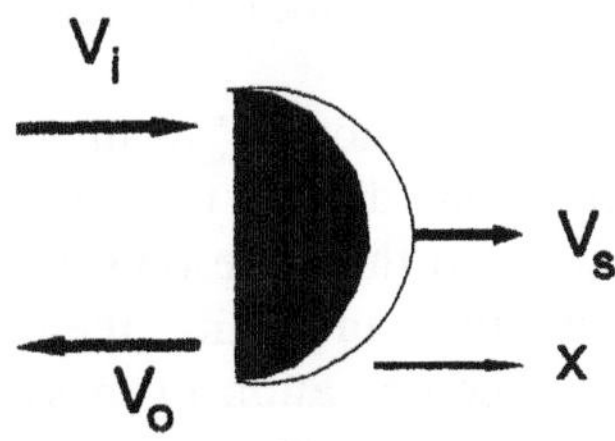

Figure 5-9 Turbine blade.

That is,

$$V_i - V_s = -(V_o - V_s)$$

Rearranging this for V_s gives

$$V_s = \tfrac{1}{2}(V_i + V_o)$$

Since the maximum energy is transferred when $V_o = 0$, this reduces to

$$V_s = \tfrac{1}{2} V_i$$

That is, the maximum efficiency for energy transfer from the fluid to the blade occurs when the velocity of the impinging fluid is twice that of the moving blade.

E. Microscopic Momentum Balance

The conservation of momentum principle can be applied to a system comprising the fluid within an arbitrary (differential) cubical volume within any flow field. This is done by accounting for convection of momentum through all six surfaces of the cube, all possible stress components acting on each of the six surfaces, and any body forces (e.g., gravity) acting on the mass as a whole. Dividing the result by the volume of the cube and taking the limit as the volume shrinks to zero results in a general microscopic form of the momentum equation that is valid at all points within any fluid. This is done in a manner similar to the earlier derivation of the microscopic mass balance (continuity) equation, Eq. (5-7), for each of the three vector components of momentum. The result can be expressed in general vector notation as

$$\rho\left(\frac{\partial \vec{v}}{\partial t} + \vec{v} \cdot \vec{\nabla} \vec{v}\right) = -\vec{\nabla} P + \vec{\nabla} \cdot \vec{\tau} + \rho \vec{g} \tag{5-59}$$

The three components of this momentum equation, expressed in Cartesian, cylindrical, and spherical coordinates, are given in detail in Appendix E. Note that Eq. (5-59) is simply a microscopic ("local") expression of the conservation of momentum, e.g., Eq. (5-40), and it applies locally at any and all points in any flowing stream.

Note that there are 11 dependent variables, or "unknowns" in these equations (three v_i's, six τ_{ij}'s, P, and ρ), all of which may depend on space and time. (For an incompressible fluid, ρ is constant so there are only 10 "unknowns.") There are four conservation equations involving these unknowns (the three momentum equations plus the conservation of mass or continuity equation), which means that we still need six more equations (seven, if the fluid is compressible). These additional equations are the "constitutive" equations that relate the local stress components to the flow or deformation of the particular fluid in laminar flow (i.e., these are determined by the constitution or structure of the material) or equations for the local turbulent stress components (the "Reynolds stresses" see Chapter 6). These equations describe the deformation or flow properties of the specific fluid of interest and relate the six shear stress components (τ_{ij}) to the deformation rate (i.e., the velocity gradient components). [Note there are only six independent components of the shear stress tensor (τ_{ij}) because it is symmetrical, i.e., $\tau_{ij} = \tau_{ji}$, which is a result of the conservation of angular momentum.] For a compressible fluid, the density is related to the pressure through an appropriate equation of state. When the equations for the six τ_{ij} components are coupled with the four conservation equations, the result is a set of differential equations for the 10 (or 11) unknowns that can be solved (in principle) with appropriate boundary conditions for the velocity components as a function of time and space. In laminar flows, the constitutive equation gives the shear stress components as a unique function of the velocity gradient components. For example, the constitutive equation for a Newtonian fluid, generalized from the one dimensional form (i.e., $\tau = \mu\dot{\gamma}$), is

$$\tau = \mu[(\vec{\nabla}\vec{v} + (\vec{\nabla}\vec{v})^t] \tag{5-60}$$

where $(\vec{\nabla}\vec{v})^t$ represents the transpose of the matrix of the $\vec{\nabla}\vec{v}$ components. The component forms of this equation are also given in Appendix E for Cartesian, cylindrical, and spherical coordinate systems. If these equations are used to eliminate the stress components from the momentum equations, the result is called the *Navier–Stokes equations*, which apply to the laminar flow of any Newtonian fluid in any system and are the starting point for the detailed solution of many fluid flow problems. Similar equations can be developed for non-Newtonian fluids, based upon the appropriate rheological (constitutive) model for the fluid. For turbulent flows, additional equations are required to describe the momentum transported by the fluctuating ("eddy") components of the flow (see Chapter 6). However, the number of flow problems for which closed analytical solutions are possible is rather limited, so numerical computer techniques are required for many problems of practical interest. These procedures are beyond the scope of this book, but we will illustrate the application of the momentum equations to the solution of an example problem.

Example 5-9: Flow Down an Inclined Plane. Consider the steady laminar flow of a thin layer or film of liquid down a flat plate that is inclined at an angle θ to the vertical, as illustrated in Fig. 5-10. The width of the plate is W (normal to the plane of the figure). Flow is only in the x direction (parallel to

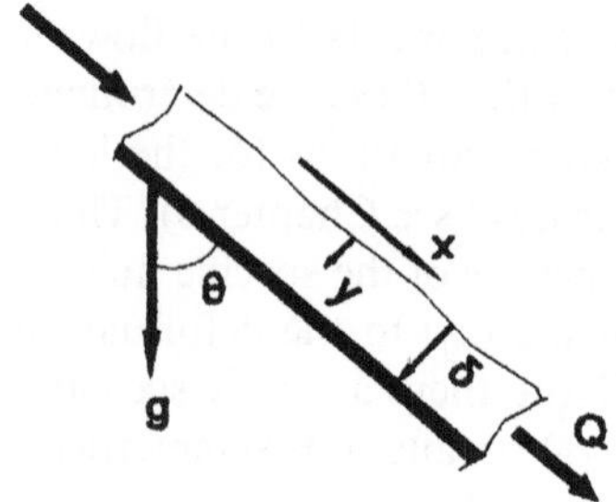

FIGURE 5-10 Flow down an inclined plane.

the surface), and the velocity varies only in the y direction (normal to the surface). These prescribed conditions constitute the definition of the problem to be solved. The objective is to determine the film thickness, δ, as a function of the flow rate per unit width of plate (Q/W), the fluid properties (μ, ρ), and other parameters in the problem. Since $v_y = v_z = 0$, the microscopic mass balance (continuity equation) reduces to

$$\frac{\partial v_x}{\partial x} = 0$$

which tells us that the velocity v_x must be independent of x. Hence, the only independent variable is y. Considering the x component of the momentum equation (see Appendix E), and discarding all y and z velocity and stress components and all derivatives except those with respect to the y direction, the result is

$$0 = \frac{\partial \tau_{yx}}{\partial y} + \rho g \cos\theta$$

The pressure gradient term has been discarded, because the system is open to the atmosphere and thus the pressure is constant (or, at most, hydrostatic) everywhere. The above equation can be integrated to give the shear stress distribution in the film:

$$\tau_{yx} = -\rho g y \cos\theta$$

where the constant of integration is zero, because there is zero (negligible) stress at the free surface of the film ($y = 0$). Note that this result is valid for any fluid (Newtonian or non-Newtonian) under any flow conditions (laminar or turbulent), because it is simply a statement of the conservation of momentum. If the fluid is Newtonian fluid and the flow is laminar, the shear stress is

$$\tau_{yx} = \mu \frac{\partial v_x}{\partial y}$$

Eliminating the stress between the last two equations gives a differential equation for $v_x(y)$ that can be integrated to give the velocity distribution:

$$v_x = \frac{\rho g \delta^2 \cos\theta}{2\mu}\left(1 - \frac{y^2}{\delta^2}\right)$$

where the boundary condition that $v_x = 0$ at $y = \delta$ (the wall) has been used to evaluate the constant of integration.

The volumetric flow rate can now be determined from

$$Q = W \int_0^\delta v_x \, dy = \frac{W \rho g \delta^3 \cos\theta}{3\mu}$$

The film thickness is seen to be proportional to the cube root of the flow rate and the fluid viscosity. The shear stress exerted on the plate is

$$\tau_w = (-\tau_{yx})_{y=\delta} = \rho g \delta \cos\theta$$

which is just the component of the weight of the fluid on the plate acting parallel to the plate.

It is also informative to express these results in dimensionless form, i.e., in terms of appropriate dimensionless groups. Because this is a noncircular conduit, the appropriate flow "length" parameter is the hydraulic diameter defined by Eq. (5.48):

$$D_h = 4\frac{xA}{W_p} = \frac{4W\delta}{W} = 4\delta$$

The appropriate form for the Reynolds number is thus

$$N_{Re} = \frac{D_h V \rho}{\mu} = \frac{4\delta V \rho}{\mu} = \frac{4\rho Q/W}{\mu}$$

because $V = Q/A = Q/W\delta$. The wall stress can also be expressed in terms of the Fanning friction factor [Eq. (5-50)]:

$$\tau_w = f\frac{\rho V^2}{2} = \rho g \delta \cos\theta$$

Substituting $V = Q/W\delta$ and eliminating $\rho g \cos\theta$ from the solution for Q gives

$$f\frac{\rho}{2}\left(\frac{Q}{W\delta}\right)^2 = \left(\frac{Q}{W}\right)\left(\frac{3\mu}{\delta^2}\right)$$

or

$$f = \frac{24}{N_{Re}}$$

i.e., $fN_{Re} = 24 = \text{constant}$.

This can be compared with the results of the dimensional analysis for the laminar flow of a Newtonian fluid in a pipe (Chapter 2, Section V), for which we deduced that $fN_{Re} = \text{constant}$. In this case, we have determined the value of the constant analytically, using first principles rather than by experiment.

The foregoing procedure can be used to solve a variety of steady, fully developed laminar flow problems, such as flow in a tube or in a slit between parallel walls, for Newtonian or non-Newtonian fluids. However, if the flow is turbulent, the turbulent eddies transport momentum in three dimensions within the flow field, which contributes additional momentum flux components to the shear stress terms in the momentum equation. The resulting equations cannot be solved exactly for such flows, and methods for treating turbulent flows will be considered in Chapter 6.

PROBLEMS

Conservation of Mass and Energy

1. Water is flowing into the top of a tank at a rate of 200 gpm. The tank is 18 in. in diameter and has a 3 in. diameter hole in the bottom, through which the water flows out. If the inflow rate is adjusted to match the outflow rate, what will the height of the water be in the tank if friction is negligible?
2. A vacuum pump operates at a constant volumetric flow rate of 10 liters/min (l/min) based upon pump inlet conditions. How long will it take to pump down a 100 L tank containing air from 1 atm to 0.01 atm, assuming that the temperature is constant?
3. Air is flowing at a constant mass flow rate into a tank that has a volume of 3 ft^3. The temperature of both the tank and the air is constant at 70°F. If the pressure in the tank is observed to increase at a rate of 5 psi/min, what is the flow rate of air into the tank?
4. A tank contains water initially at a depth of 3 ft. The water flows out of a hole in the bottom of the tank, and air at a constant pressure of 10 psig is admitted to the top of the tank. If the water flow rate is directly proportional to the square root of the gage pressure inside the bottom of the tank, derive expressions for the water flow rate and air flow rate as a function of time. Be sure to define all symbols you use in your equations.

5. The flow rate of a hot coal/oil slurry in a pipeline is measured by injecting a small side stream of cool oil and measuring the resulting temperature change downstream in the pipeline. The slurry is initially at 300°F and has a density of 1.2 g/cm^3 and a specific heat of 0.7 Btu/(lb$_m$ °F). With no side stream injected, the temperature downstream of the mixing point is 298°F. With a side stream at 60°F and a flow rate of 1 lb$_m$/s, the temperature at this point is 295°F. The side stream has a density of 0.8 g/cm^3 and a c_p of 0.6 Btu/(lb$_m$ °F). What is the mass flow rate of the slurry?
6. A gas enters a horizontal 3 in. Schedule 40 pipe at a constant rate of 0.5 lb$_m$/s, with temperature of 70°F, and pressure of 1.15 atm. The pipe is wrapped with a 20 kW heating coil covered with a thick layer of insulation. At the point where the gas is discharged, the pressure is 1.05 atm. What is the gas temperature at the discharge point, assuming it to be ideal with a MW of 29 and a c_p of 0.24 Btu/(lb$_m$ °F)?
7. Water is flowing into the top of an open cylindrical tank (diameter D) at a volume flow rate of Q_i and out of a hole in the bottom at a rate of Q_o. The tank is made of wood that is very porous, and the water is leaking out through the wall uniformly at a rate of q per unit of wetted surface area. The initial depth of water in the tank is Z_1. Derive an equation for the depth of water in the tank at any time. If $Q_i = 10$ gpm, $Q_o = 5$ gpm, $D = 5$ ft, $q = 0.1$ gpm/ft^2, and $Z_1 = 3$ ft, is the level in the tank rising or falling?
8. Air is flowing steadily through a horizontal tube at a constant temperature of 32°C and a mass flow rate of 1 kg/s. At one point upstream where the tube diameter is 50 mm, the pressure is 345 kPa. At another point downstream the diameter is 75 mm and the pressure is 359 kPa. What is the value of the friction loss (e_f) between these two points? [$c_p = 1005$ J/(kg K).]
9. Steam is flowing through a horizontal nozzle. At the inlet the velocity is 1000 ft/s and the enthalpy is 1320 Btu/lb$_m$. At the outlet the enthalpy is 1200 Btu/lb$_m$. If heat is lost through the nozzle at a rate of 5 Btu/lb$_m$ of steam, what is the outlet velocity?
10. Oil is being pumped from a large storage tank, where the temperature is 70°F, through a 6 in. ID pipeline. The oil level in the tank is 30 ft above the pipe exit. If a 25 hp pump is required to pump the oil at a rate of 600 gpm through the pipeline, what would the temperature of the oil at the exit be if no heat is transferred across the pipe wall? State any assumptions that you make. Oil properties: SG = 0.92, $\mu = 35$ cP, $c_p = 0.5$ Btu/(lb$_m$ °F).
11. Ethylene 12 enters a 1 in. Schedule (sch) 80 pipe at 170°F and 1000 psia and a velocity of 10 ft/s. At a point somewhere downstream, the temperature has dropped to 140°F and the pressure to 15 psia. Calculate the velocity at the downstream conditions and the Reynolds number at both the upstream and downstream conditions.
12. Number 3 fuel oil (30° API) is transferred from a storage tank at 60°F to a feed tank in a power plant at a rate of 2000 bbl/day. Both tanks are open to the atmosphere, and they are connected by a pipeline containing 1200 ft equivalent length of $1\frac{1}{2}$ in. sch 40 steel pipe and fittings. The level in the feed tank is 20 ft higher than that in the storage tank, and the transfer pump is 60% efficient.

The Fanning friction factor is given by $f = 0.0791/N_{Re}^{1/4}$.

(a) What horsepower motor is required to drive the pump?
(b) If the specific heat of the oil is 0.5 Btu/(lb_m °F) and the pump and transfer line are perfectly insulated, what is the temperature of the oil entering the feed tank?

13. Oil with a viscosity of 35 cP, SG of 0.9, and a specific heat of 0.5 Btu/(lb_m °F) is flowing through a straight pipe at a rate of 100 gpm. The pipe is 1 in. sch 40, 100 ft long, and the Fanning friction factor is given by $f = 0.0791/N_{Re}^{1/4}$. If the temperature of the oil entering the pipe is 150°F, determine:
    (a) The Reynolds number.
    (b) The pressure drop in the pipe, assuming that it is horizontal.
    (c) The temperature of the oil at the end of the pipe, assuming the pipe to be perfectly insulated.
    (d) The rate at which heat must be removed from the oil (in Btu/hr) to maintain it at a constant temperature if there is no insulation on the pipe.
14. Water is pumped at a rate of 90 gpm by a centrifugal pump driven by a 10 hp motor. The water enters the pump through a 3 in. sch 40 pipe at 60°F and 10 psig and leaves through a 2 in. sch 40 pipe at 100 psig. If the water loses 0.1 Btu/lb_m while passing through the pump, what is the water temperature leaving the pump?
15. A pump driven by a 7.5 hp motor, takes water in at 75°F and 5 psig and discharges it at 60 psig, at a flow rate of 600 lb_m/min. If no heat is transferred to or from the water while it is in the pump, what will the temperature of the water be leaving the pump?
16. A high pressure pump takes water in at 70°F, 1 atm, through a 1 in. ID suction line and discharges it at 1000 psig through a 1/8 in. ID line. The pump is driven by a 20 hp motor and is 65% efficient. If the flow rate is 500 g/s and the temperature of the discharge is 73°F, how much heat is transferred between the pump casing and the water, per pound of water? Does the heat go into or out of the water?

Bernoulli's Equation

17. Water is flowing from one large tank to another through a 1 in. diameter pipe. The level in tank A is 40 ft above the level in tank B. The pressure above the water in tank A is 5 psig, and in tank B it is 20 psig. Which direction is the water flowing?
18. A pump that is driven by a 7.5 hp motor takes water in at 75°F and 5 psig and discharges it at 60 psig at a flow rate of 600 lb_m/min. If no heat is transferred between the water in the pump and the surroundings, what will be the temperature of the water leaving the pump?
19. A 90% efficient pump driven by a 50 hp motor is used to transfer water at 70°F from a cooling pond to a heat exchanger through a 6 in. sch 40 pipeline. The heat exchanger is located 25 ft above the level of the cooling pond, and the water pressure at the discharge end of the pipeline is 40 psig. With all valves in the line wide open, the water flow rate is 650 gpm. What is the rate of energy dissipation (friction loss) in the pipeline, in kilowatts (kW)?

20. A pump takes water from the bottom of a large tank where the pressure is 50 psig and delivers it through a hose to a nozzle that is 50 ft above the bottom of the tank at a rate of 100 lb_m/s. The water exits the nozzle into the atmosphere at a velocity of 70 ft/s. If a 10 hp motor is required to drive the pump, which is 75% efficient, find:
    (a) The friction loss in the pump
    (b) The friction loss in the rest of the system
    Express your answer in units of ft lb_f/lb_m.
21. You have purchased a centrifugal pump to transport water at a maximum rate of 1000 gpm from one reservoir to another through an 8 in. sch 40 pipeline. The total pressure drop through the pipeline is 50 psi. If the pump has an efficiency of 65% at maximum flow conditions and there is no heat transferred across the pipe wall or the pump casing, calculate:
    (a) The temperature change of the water through the pump
    (b) The horsepower of the motor that would be required to drive the pump
22. The hydraulic turbines at Boulder Dam power plant are rated at 86,000 kW when water is supplied at a rate of 66.3 m^3/s. The water enters at a head of 145 m at 20°C and leaves through a 6 m diameter duct.
    (a) Determine the efficiency of the turbines.
    (b) What would be the rating of these turbines if the dam power plant was on Jupiter ($g = 26\, m/s^2$)?
23. Water is draining from an open conical funnel at the same rate at which it is entering at the top. The diameter of the funnel is 1 cm at the top and 0.5 cm at the bottom, and it is 5 cm high. The friction loss in the funnel per unit mass of fluid is given by $0.4V^2$, where V is the velocity leaving the funnel. What is (a) the volumetric flow rate of the water and (b) the value of the Reynolds number entering and leaving the funnel?
24. Water is being transferred by pump between two open tanks (from A to B) at a rate of 100 gpm. The pump receives the water from the bottom of tank A through a 3 in. sch 40 pipe and discharges it into the top of tank B through a 2 in. sch 40 pipe. The point of discharge into B is 75 ft higher than the surface of the water in A. The friction loss in the piping system is 8 psi, and both tanks are 50 ft in diameter. What is the head (in feet) which must be delivered by the pump to move the water at the desired rate? If the pump is 70% efficient, what horsepower motor is required to drive the pump?
25. A 4 in. diameter open can has a 1/4 in. diameter hole in the bottom. The can is immersed bottom down in a pool of water, to a point where the bottom is 6 in. below the water surface and is held there while the water flows through the hole into the can. How long will it take for the water in the can to rise to the same level as that outside the can? Neglect friction, and assume a "pseudo steady state," i.e., time changes are so slow that at any instant the steady state Bernoulli equation applies.
26. Carbon tetrachloride (SG = 1.6) is pumped at a rate of 2 gpm through a pipe that is inclined upward at an angle of 30°. An inclined tube manometer (with a 10° angle of inclination) using mercury as the manometer fluid (SG = 13.6) is connected between two taps on the pipe that are 2 ft apart. The manometer

reading is 6 in. If no heat is lost through the tube wall, what is the temperature rise of the CCl_4 over a 100 ft length of the tube?

27. A pump that is taking water at 50°F from an open tank at a rate of 500 gpm is located directly over the tank. The suction line entering the pump is a nominal 6 in. sch 40 straight pipe 10 ft long and extends 6 ft below the surface of the water in the tank. If friction in the suction line is neglected, what is the pressure at the pump inlet (in psi)?
28. A pump is transferring water from tank A to tank B, both of which are open to the atmosphere, at a rate of 200 gpm. The surface of the water in tank A is 10 ft above ground level, and that in tank B is 45 ft above ground level. The pump is located at ground level, and the discharge line that enters tank B is 50 ft above ground level at its highest point. All piping is 2 in. ID, and the tanks are 20 ft in diameter. If friction is neglected, what would be the required pump head rating for this application (in ft), and what size motor (horsepower) would be needed to drive the pump if it is 60% efficient? (Assume the temperature is constant at 77°F.)
29. A surface effect (air cushion) vehicle measures 10 ft by 20 ft and weighs 6000 lb_f. The air is supplied by a blower mounted on top of the vehicle, which must supply sufficient power to lift the vehicle 1 in. off the ground. Calculate the required blower capacity in scfm (standard cubic feet per minute), and the horsepower of the motor required to drive the blower if it is 80% efficient. Neglect friction, and assume that the air is an ideal gas at 80°F with properties evaluated at an average pressure.
30. The air cushion car in Problem 29 is equipped with a 2 hp blower that is 70% efficient.
    (a) What is the clearance between the skirt of the car and the ground?
    (b) What is the air flow rate, in scfm?
31. An ejector pump operates by injecting a high speed fluid stream into a slower stream to increase its pressure. Consider water flowing at a rate of 50 gpm through a 90° elbow in a 2 in. ID pipe. A stream of water is injected at a rate of 10 gpm through a 1/2 in. ID pipe through the center of the elbow in a direction parallel to the downstream flow in the larger pipe. If both streams are at 70°F, determine the increase in pressure in the larger pipe at the point where the two streams mix.
32. A large tank containing water has a 51 mm diameter hole in the bottom. When the depth of the water is 15 m above the hole, the flow rate through the hole is found to be 0.0324 m^3/s. What is the head loss due to friction in the hole?
33. Water at 68°F is pumped through a 1000 ft length of 6 in. sch 40 pipe. The discharge end of the pipe is 100 ft above the suction end. The pump is 90% efficient, and it is driven by a 25 hp motor. If the friction loss in the pipe is 70 ft lb_f/lb_m, what is the flow rate through the pipe in gpm? ($P_{in} = P_{out} =$ 1 atm.)
34. You want to siphon water out of a large tank using a 5/8 in. ID hose. The highest point of the hose is 10 ft above the water surface in the tank, and the hose exit outside the tank is 5 ft below the inside surface level. If friction

is neglected, (a) what would be the flow rate through the hose (in gpm), and (b) what is the minimum pressure in the hose (in psi)?

35. It is desired to siphon a volatile liquid out of a deep open tank. If the liquid has a vapor pressure of 200 mmHg and a density of 45 lb_m/ft^3 and the surface of the liquid is 30 ft below the top of the tank, is it possible to siphon the liquid? If so, what would the velocity be through a frictionless siphon, 1/2 in. in diameter, if the exit of the siphon tube is 3 ft below the level in the tank?
36. The propeller of a speedboat is 1 ft in diameter and 1 ft below the surface of the water. At what speed (rpm) will cavitation occur? The vapor pressure of the water is 18.65 mmHg at 70°F.
37. A conical funnel is full of liquid. The diameter of the top (mouth) is D_1, that of the bottom (spout) is D_2 (where $D_2 \ll D_1$), and the depth of the fluid above the bottom is H_0. Derive an expression for the time required for the fluid to drain by gravity to a level of $H_0/2$, assuming frictionless flow.
38. An open cylindrical tank of diameter D contains a liquid of density ρ at a depth H. The liquid drains through a hole of diameter d in the bottom of the tank. The velocity of the liquid through the hole is $C\sqrt{h}$, where h is the depth of the liquid at any time t. Derive an equation for the time required for 90% of the liquid to drain out of the tank.
39. An open cylindrical tank that is 2 ft in diameter and 4 ft high is full of water. If the tank has a 2 in. diameter hole in the bottom, how long will it take for half of the water to drain out, if friction is neglected?
40. A large tank has a 5.1 mm diameter hole in the bottom. When the depth of liquid in the tank is 1.5 m above the hole, the flow rate through the hole is found to be 324 cm^3/s. What is the head loss due to friction in the hole (in ft)?
41. A window is left slightly open while the air conditioning system is running. The air conditioning blower develops a pressure of 2 in.H_2O (gage) inside the house, and the window opening measures 1/8 in. × 20 in. Neglecting friction, what is the flow rate of air through the opening, in scfm (ft^3/min at 60°F, 1 atm)? How much horsepower is required to move this air?
42. Water at 68°F is pumped through a 1000 ft length of 6 in. sch 40 pipe. The discharge end of the pipe is 100 ft above the suction end. The pump is 90% efficient and is driven by a 25 hp motor. If the friction loss in the pipe is 70 ft lb_f/lb_m, what is the flow rate through the pipe (in gpm)?
43. The plumbing in your house is 3/4 in. sch 40 galvanized pipe, and it is connected to an 8 in. sch 80 water main in which the pressure is 15 psig. When you turn on a faucet in your bathroom (which is 12 ft higher than the water main), the water flows out at a rate of 20 gpm.
    (a) How much energy is lost due to friction in the plumbing?
    (b) If the water temperature in the water main is 60°F, and the pipes are well insulated, what would the temperature of the water be leaving the faucet?
    (c) If there were no friction loss in the plumbing, what would the flow rate be (in gpm)?
44. A 60% efficient pump driven by a 10 hp motor is used to transfer bunker C fuel oil from a storage tank to a boiler through a well-insulated line. The pressure in the tank is 1 atm, and the temperature is 100°F. The pressure at the burner in

the boiler is 100 psig, and it is 100 ft above the level in the tank. If the temperature of the oil entering the burner is 102°F, what is the oil flow rate, in gpm? [Oil properties: SG = 0.8, $c_p = 0.5$ Btu/(lb$_m$ °F).]

Fluid Forces, Momentum Transfer

45. You have probably noticed that when you turn on the garden hose it will whip about uncontrollably if it is not restrained. This is because of the unbalanced forces developed by the change of momentum in the tube. If a 1/2 in. ID hose carries water at a rate of 50 gpm, and the open end of the hose is bent at an angle of 30° to the rest of the hose, calculate the components of the force (magnitude and direction) exerted by the water on the bend in the hose. Assume that the loss coefficient in the hose is 0.25.
46. Repeat Problem 46, for the case in which a nozzle is attached to the end of the hose and the water exits the nozzle through a 1/4 in. opening. The loss coefficient for the nozzle is 0.3 based on the velocity through the nozzle.
47. You are watering your garden with a hose that has a 3/4 in. ID, and the water is flowing at a rate of 10 gpm. A nozzle attached to the end of the hose has an ID of 1/4 in. The loss coefficient for the nozzle is 20 based on the velocity in the hose. Determine the force (magnitude and direction) that you must apply to the nozzle in order to deflect the free end of the hose (nozzle) by an angle of 30° relative to the straight hose.
48. A 4 in. ID fire hose discharges water at a rate of 1500 gpm through a nozzle that has a 2 in. ID exit. The nozzle is conical and converges through a total included angle of 30°. What is the total force transmitted to the bolts in the flange where the nozzle is attached to the hose? Assume the loss coefficient in the nozzle is 3.0 based on the velocity in the hose.
49. A 90° horizontal reducing bend has an inlet diameter of 4 in. and an outlet diameter of 2 in. If water enters the bend at a pressure of 40 psig and a flow rate of 500 gpm, calculate the force (net magnitude and direction) exerted on the supports that hold the bend in place. The loss coefficient for the bend may be assumed to be 0.75 based on the highest velocity in the bend.
50. A fireman is holding the nozzle of a fire hose that he is using to put out a fire. The hose is 3 in. in diameter, and the nozzle is 1 in. in diameter. The water flow rate is 200 gpm, and the loss coefficient for the nozzle is 0.25 (based on the exit velocity). How much force must the fireman use to restrain the nozzle? Must he push or pull on the nozzle to apply the force? What is the pressure at the end of the hose where the water enters the nozzle?
51. Water flows through a 30° pipe bend at a rate of 200 gpm. The diameter of the entrance to the bend is 2.5 in., and that of the exit is 3 in. The pressure in the pipe is 30 psig, and the pressure drop in the bend is negligible. What is the total force (magnitude and direction) exerted by the fluid on the pipe bend?
52. A nozzle with a 1 in. ID outlet is attached to a 3 in. ID fire hose. Water pressure inside the hose is 100 psig, and the flow rate is 100 gpm. Calculate the force (magnitude and direction) required to hold the nozzle at an angle of 45° relative to the axis of the hose. (Neglect friction in the nozzle.)

53. Water flows through a 45° expansion pipe bend at a rate of 200 gpm, exiting into the atmosphere. The inlet to the bend is 2 in. ID, the exit is 3 in. ID, and the loss coefficient for the bend is 0.3 based on the inlet velocity. Calculate the force (magnitude and direction) exerted by the fluid on the bend relative to the direction of the entering stream.
54. A patrol boat is powered by a water jet engine, which takes water in at the bow through a 1 ft diameter duct and pumps it out the stern through a 3 in. diameter exhaust jet. If the water is pumped at a rate of 5000 gpm, determine:
    (a) The thrust rating of the engine
    (b) The maximum speed of the boat, if the drag coefficient is 0.5 based on an underwater area of 600 ft^2
    (c) the horsepower required to operate the motor (neglecting friction in the motor, pump, and ducts)
55. A patrol boat is powered by a water jet pump engine. The engine takes water in through a 3 ft. diameter duct in the bow and discharges it through a 1 ft. diameter duct in the stern. The drag coefficient of the boat has a value of 0.1 based on a total underwater area of 1500 ft^2. Calculate the pump capacity in gpm and the engine horsepower required to achieve a speed of 35 mph, neglecting friction in the pump and ducts.
56. Water is flowing through a 45° pipe bend at a rate of 200 gpm and exits into the atmosphere. The inlet to the bend is $1\frac{1}{2}$ in. inside diameter, and the exit is 1 in. in diameter. The friction loss in the bend can be characterized by a loss coefficient of 0.3 (based on the inlet velocity). Calculate the net force (magnitude and direction) transmitted to the flange holding the pipe section in place.
57. The arms of a lawn sprinkler are 8 in. long and 3/8 in. ID. Nozzles at the end of each arm direct the water in a direction that is 45° from the arms. If the total flow rate is 10 gpm, determine:
    (a) The moment developed by the sprinkler if it is held stationary and not allowed to rotate.
    (b) The angular velocity (in rpm) of the sprinkler if there is no friction in the bearings.
    (c) The trajectory of the water from the end of the rotating sprinkler (i.e., the radial and angular velocity components)
58. A water sprinkler contains two 1/4 in. ID jets at the ends of a rotating hollow (3/8 in. ID) tube, which direct the water 90° to the axis of the tube. If the water leaves at 20 ft/s, what torque would be necessary to hold the sprinkler in place?
59. An open container 8 in. high with an inside diameter of 4 in. weighs 5 lb_f when empty. The container is placed on a scale, and water flows into the top of the container through a 1 in. diameter tube at a rate of 40 gpm. The water flows horizontally out into the atmosphere through two 1/2 in. holes on opposite sides of the container. Under steady conditions, the height of the water in the tank is 7 in.
    (a) Determine the reading on the scale.
    (b) Determine how far the holes in the sides of the container should be from the bottom so that the level in the container will be constant.

60. A boat is tied to a dock by a line from the stern of the boat to the dock. A pump inside the boat takes water in through the bow and discharges it out the stern at the rate of 3 ft^3/s through a pipe running through the hull. The pipe inside area is 0.25 ft^2 at the bow and 0.15 ft^2 at the stern. Calculate the tension on the line, assuming inlet and outlet pressures are equal.
61. A jet ejector pump is shown in Fig. 5-P61. A high speed stream (Q_A) is injected at a rate of 50 gpm through a small tube 1 in. in diameter into a stream (Q_B) in a larger, 3 in. diameter, tube. The energy and momentum are transferred from the small stream to the larger stream, which increases the pressure in the pump. The fluids come in contact at the end of the small tube and become perfectly mixed a short distance downstream (the flow is turbulent). The energy dissipated in the system is significant, but the wall force between the end of the small tube and the point where mixing is complete can be neglected. If both streams are water at 60°F, and $Q_B = 100$ gpm, calculate the pressure rise in the pump.

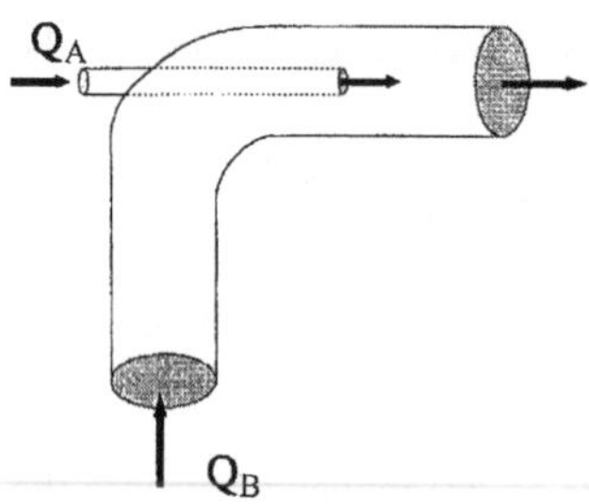

FIGURE 5-P61

62. Figure 5-P62 illustrates two relief valves. The valve disk is designed to lift when the upstream pressure in the vessel (P_1) reaches the valve set pressure. Valve A has a disk that diverts the fluid leaving the valve by 90° (i.e., to the horizontal direction), whereas the disk in valve B diverts the fluid to a direction that is 60° downward from the horizontal. The diameter of the valve nozzle is 3 in., and

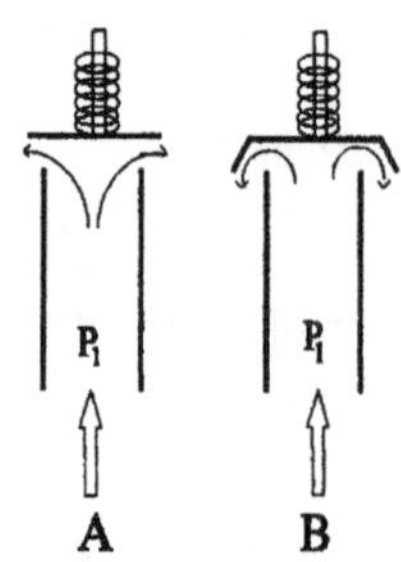

FIGURE 5-P62

the clearance between the end of the nozzle and the disk is 1 in., for both valves. If the fluid is water at 200°F, $P_1 = 100$ psig, and the discharge pressure is atmospheric, determine the force exerted on the disk for both cases A and B. The loss coefficient for the valve in both cases is 2.4 based on the velocity in the nozzle.

63. A relief valve is mounted on the top of a large vessel containing hot water. The inlet diameter to the valve is 4 in., and the outlet diameter is 6 in. The valve is set to open when the pressure in the vessel reaches 100 psig, which happens when the water is at 200°F. The liquid flows through the open valve and exits to the atmosphere on the side of the valve, 90° from the entering direction. The loss coefficient for the valve has a value of 5, based on the exit velocity from the valve.
    (a) Determine the net force (magnitude and direction) acting on the valve.
    (b) You want to attach a cable to the valve to brace it such that the tensile force in the cable balances the net force on the valve. Show exactly where you would attach the cable at both ends.
64. A relief valve is installed on the bottom of a pressure vessel. The entrance to the valve is 4.5 in. diameter, and the exit (which discharges in the horizontal direction, 90° from the entrance) is 5 in. in diameter. The loss coefficient for the valve is 4.5 based on the inlet velocity. The fluid in the tank is a liquid with a density of 0.8 g/cm^3. If the valve opens when the pressure at the valve reaches 150 psig, determine:
    (a) The mass flow rate through the valve, in lb_m/s
    (b) The net force (magnitude and direction) exerted on the valve
    (c) Determine the location (orientation) of a cable that is to be attached to the valve to balance the net force. (Note that a cable can support only a tensile force.)
65. An emergency relief valve is installed on a reactor to relieve excess pressure in case of a runaway reaction. The lines upstream and downstream of the valve are 6 in. sch 40 pipe. The valve is designed to open when the tank pressure reaches 100 psig, and the vent exhausts to the atmosphere at 90° to the direction entering the valve. The fluid can be assumed to be incompressible, with an SG of 0.95, a viscosity of 3.5 cP, and a specific heat of 0.5 Btu/(lb_m °F). If the sum of the loss coefficients for the valve and the vent line is 6.5, determine:
    (a) The mass flow rate of the fluid through the valve in lb_m/s and the value of the Reynolds number in the pipe when the valve opens.
    (b) The rise in temperature of the fluid from the tank to the vent exit, if the heat transferred through the walls of the system is negligible.
    (c) The force exerted on the valve supports by the fluid flowing through the system. If you could install only one support cable to balance this force, show where you would put it.
66. Consider the "tank on wheels" shown in Fig. 5-P66. Water is draining out of a hole in the side of the open tank, at a rate of 10 gpm. If the tank diameter is 2 ft and the diameter of the hole is 2 in., determine the magnitude and direction of the force transmitted from the water to the tank.

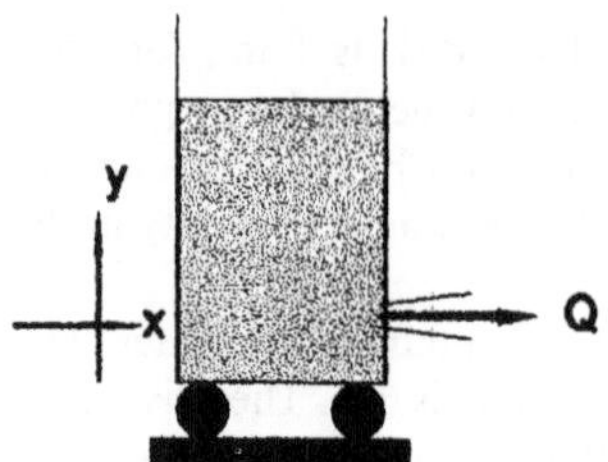

FIGURE 5-P66

67. The tank in Problem 66 is 6 in. in diameter and contains water at a depth of 3 ft. On the side of the tank near the bottom is a 1.5 in. ID outlet to which is attached a ball valve, which has a loss coefficient of 1.2. When the valve is opened, the water flows out in a horizontal stream. Calculate:
    (a) The flow rate of the water (in gpm)
    (b) The thrust exerted on the tank by the escaping water, and the direction that the tank will move. If the diameter of the outlet and valve are increased, will the thrust on the tank increase or decrease? Why?

Laminar Flow

68. Use the microscopic equations of motion in Appendix E as a starting point to derive a relationship between the volumetric flow rate and the pressure gradient for a Newtonian fluid in a pipe that is valid for any orientation of the pipe axis. (*Hint*: The critical starting point requires that you identify which velocity and velocity gradients are nonzero, and hence the corresponding nonzero stress components, for this problem. This allows you to tailor the differential equations to suit the problem, and the resulting equations can be integrated, with appropriate boundary conditions, to get the answer.)
69. A viscous molten polymer is pumped through a thin slit between two flat surfaces. The slit has a depth H, width W, and length L and is inclined upward at an angle θ to the horizontal ($H \ll W$). The flow is laminar, and the polymer is non-Newtonian, with properties that can be represented by the power law model.
    (a) Derive an equation relating the volume flow rate of the polymer (Q) to the applied pressure difference along the slit, the slit dimensions, and the fluid properties.
    (b) Using the definition of the Fanning friction factor (f), solve your equation for f in terms of the remaining quantities. The corresponding solution for a Newtonian fluid can be written $f = 24/N_{Re}$. Use your solution to obtain an equivalent expression for the power law Reynolds number (i.e., $N_{RePL} = 24/f$). Use the hydraulic diameter as the length scale in the Reynolds number. (*Note*: It is easiest to take the origin of your coordinates at the center of the slit, then calculate the flow rate for one-half the slit and double this to get the answer. Why is this the easiest way?)

70. Acrylic latex paint can be described as a Bingham plastic with a yield stress of 200 dyn/cm^2, a limiting viscosity of 50 cP, and a density of 0.95 g/cm^3.
    (a) What is the maximum thickness at which a film of this paint could be spread on a vertical wall without running?
    (b) If the power law model were used to describe this paint, such that the apparent viscosity predicted by both the power law and Bingham plastic models is the same at shear rates of 1 and 100 s^{-1}, what would the flow rate of the film be if it has the thickness predicted in (a)?
71. A vertical belt is moving upward continuously through a liquid bath, at a velocity V. A film of the liquid adheres to the belt, which tends to drain downward due to gravity. The equilibrium thickness of the film is determined by the steady-state condition at which the downward drainage velocity of the surface of the film is exactly equal to the upward velocity of the belt. Derive an equation for the film thickness if the fluid is (a) Newtonian; (b) a Bingham plastic.
72. Water at 70°F is draining by gravity down the outside of a 4 in. OD vertical tube at a rate of 1 gpm. Determine the thickness of the film. Is the flow laminar or turbulent?
73. For laminar flow of a Newtonian fluid in a tube:
    (a) Show that the average velocity over the cross section is half the maximum velocity in the tube.
    (b) Derive the kinetic energy correction factor for laminar flow of a Newtonian fluid in a tube (i.e., $\alpha = 2$).
74. A slider bearing can be described as one plate moving with a velocity V parallel to a stationary plate, with a viscous lubricant in between the plates. The force applied to the moving plate is F, and the distance between the plates is H. If the lubricant is a grease with properties that can be described by the power law model, derive an equation relating the velocity V to the applied force F and the gap clearance H, starting with the general microscopic continuity and momentum equations. If the area of the plate is doubled, with everything else staying the same, how will the velocity V change?
75. Consider a fluid flowing in a conical section, as illustrated in Fig. 5-P75. The mass flow rate is the same going in (through point 1) as it is coming out (point 2), but the velocity changes because the area changes. They are related by

    $$(\rho VA)_1 = (\rho VA)_2$$

    where ρ is the fluid density (assumed to be constant here). Because the velocity changes, the transport of momentum will be different going in than going out, which results in a net force in the fluid.

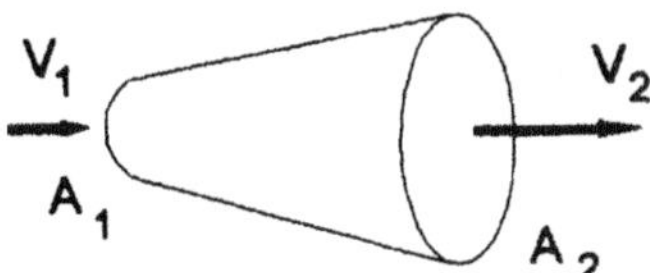

FIGURE 5-P75

(a) Derive an expression for the magnitude of this force associated with the change in momentum.
(b) Which direction will the force transmitted from the fluid to the cone act.

NOTATION

A	cross sectional area, [L^2]
A_w	area of wall, [L^2]
c_v	specific heat at constant volume, [$FL/MT = L^2/Mt^2$]
c_p	specific heat at constant pressure, [$FL/MT = L^2/Mt^2$]
d	diameter, [L]
D	diameter, [L]
D_h	hydraulic diameter, [L]
f	Fanning friction factor, [—]
g	acceleration due to gravity, [L/t^2]
e_f	energy dissipated per unit mass of fluid, [$FL/M = L^2/t^2$]
F_x	force component in the x direction, [$F = ML/t^2$]
h	enthalpy per unit mass, [$FL/M = L^2/t^2$]
H_f	friction loss head, [L]
H_p	pressure head, [L]
H_w	work (pump) head, [L]
H_v	velocity head, [L]
H_z	static head, [L]
I	moment of inertia, [$FLt^2 = ML^2$]
K_f	loss coefficient, [—]
L_θ	angular momentum in the θ direction, [ML^2/t]
M	molecular weight, [M/mol]
m	mass, [M]
$\dot{m}$	mass flow rate, [M/t]
P	pressure, [$F/L^2 = M/Lt^2$]
Q	volumetric flow rate, [L^3/t]
$\dot{Q}$	rate of heat transfer into the system, [$FL/t = ML^2/t^3$]
q	heat transferred into the system per unit mass of fluid, [$FL/M = L^2/t^2$]
R	gas constant, [$FL/(mol\ T) = ML^2\ (mol\ t^2T)$]
R	radius, [L]
s	entropy per unit mass, [$FL/M = L^2/t^2$]
T	temperature, [T]
t	time, [t]
u	internal energy per unit mass, [$FL/M = L^2/t^2$]
v	local velocity, [L/t]
V	spatial average velocity, [L/t]
W	width of plate, [L]
$\dot{W}$	rate of work done by fluid system, [$FL/t = ML^2/t^3$]
W_p	wetted perimeter, [L]

w	work done by fluid system per unit mass of fluid, $[FL/M = L^2/t^2]$
x, y, z	coordinate directions, [L]
α	kinetic energy correction factor, [—]
β	the ratio d/D, where $d < D$, [—]
Γ	moment or torque, $[FL = ML^2/t^2]$
$\Delta()$	$()_2 - ()_1$
$\vec{\nabla}$	gradient vector operator, [1/L]
δ	film thickness, [L]
μ	viscosity (constant), [M/Lt]
ν	specific volume $[L^3/M]$
Φ	potential $(= P + \rho g z)$, $[F/L^2 = M/Lt^2]$
ρ	density, $[M/L^3]$
τ_{yx}	shear stress component, force in x direction on y area component, $[F/L^2 = M/Lt^2]$
τ	shear stress tensor, $[F/L^2 = M/Lt^2]$
τ_w	stress exerted on wall by fluid, $[F/L^2 = M/Lt^2]$
ω	angular velocity, [1/t]

Subscripts

1	reference point 1
2	reference point 2
i	input
o	output
s	system
x, y, z	coordinate directions, [L]

6

Pipe Flow

I. FLOW REGIMES

In 1883, Osborn Reynolds conducted a classical experiment, illustrated in Fig. 6-1, in which he measured the pressure drop as a function of flow rate for water in a tube. He found that at low flow rates the pressure drop was directly proportional to the flow rate, but as the flow rate was increased a point was reached where the relation was no longer linear and the "noise" or scatter in the data increased considerably. At still higher flow rates the data became more reproducible, but the relationship between pressure drop and flow rate became almost quadratic instead of linear.

To investigate this phenomenon further, Reynolds introduced a trace of dye into the flow to observe what was happening. At the low flow rates where the linear relationship was observed, the dye was seen to remain a coherent, rather smooth thread throughout most of the tube. However, where the data scatter occurred, the dye trace was seen to be rather unstable, and it broke up after a short distance. At still higher flow rates, where the quadratic relationship was observed, the dye dispersed almost immediately into a uniform "cloud" throughout the tube. The stable flow observed initially was termed *laminar flow*, because it was observed that the fluid elements moved in smooth layers or "lamella" relative to each other with no mixing. The unstable flow pattern, characterized by a high degree of

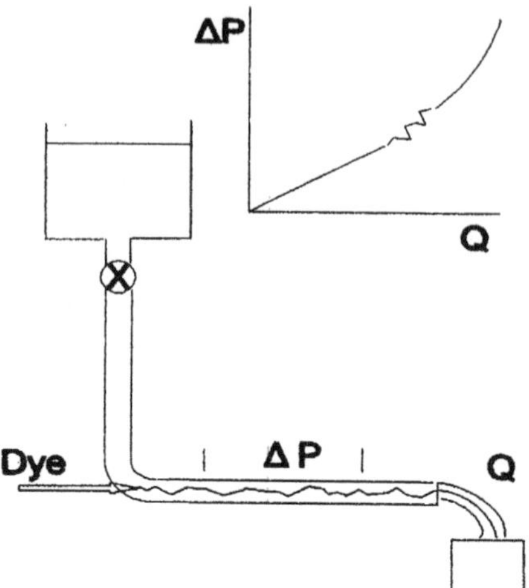

FIGURE 6-1 Reynolds' experiment.

mixing between the fluid elements, was termed *turbulent flow*. Although the transition from laminar to turbulent flow occurs rather abruptly, there is nevertheless a *transition region* where the flow is unstable but not thoroughly mixed.

Careful study of various fluids in tubes of different sizes has indicated that laminar flow in a tube persists up to a point where the value of the Reynolds number ($N_{Re} = DV\rho/\mu$) is about 2000, and turbulent flow occurs when N_{Re} is greater than about 4000, with a transition region in between. Actually, unstable flow (turbulence) occurs when disturbances to the flow are amplified, whereas laminar flow occurs when these disturbances are damped out. Because turbulent flow cannot occur unless there are disturbances, studies have been conducted on systems in which extreme care has been taken to eliminate any disturbances due to irregularities in the boundary surfaces, sudden changes in direction, vibrations, etc. Under these conditions, it has been possible to sustain laminar flow in a tube to a Reynolds number of the order of 100,000 or more. However, under all but the most unusual conditions there are sufficient natural disturbances in all practical systems that turbulence begins in a pipe at a Reynolds number of about 2000.

The physical significance of the Reynolds number can be appreciated better if it is rearranged as

$$N_{Re} = \frac{DV\rho}{\mu} = \frac{\rho V^2}{\mu V/D} \tag{6-1}$$

The numerator is the flux of "inertial" momentum carried by the fluid along the tube in the axial direction. The denominator is proportional to the viscous shear stress in the tube, which is equivalent to the flux of "viscous" momentum normal to the flow direction, i.e., in the radial direction. Thus, the Reynolds number is a ratio of the inertial momentum flux in the flow direction to the viscous momentum flux in the transverse direction. Because viscous forces are a manifestation of intermolecular attractive forces, they are stabilizing, whereas inertial forces tend to pull the fluid elements apart and are therefore destabilizing. It is thus quite logical that stable (laminar) flow should occur at low Reynolds numbers where viscous forces dominate, whereas unstable (turbulent) flow occurs at high Reynolds numbers where inertial forces dominate. Also, laminar flows are dominated by viscosity and are independent of the fluid density, whereas fully turbulent flows are dominated by the fluid density and are independent of the fluid viscosity at high turbulence levels. For fluids flowing near solid boundaries (e.g., inside conduits), viscous forces dominate in the immediate vicinity of the boundary, whereas for turbulent flows (high Reynolds numbers) inertial forces dominate in the region far from the boundary. We will consider both the laminar and turbulent flow of Newtonian and non-Newtonian fluids in pipes in this chapter.

II. GENERAL RELATIONS FOR PIPE FLOWS

For steady, uniform, fully developed flow in a pipe (or any conduit), the conservation of mass, energy, and momentum equations can be arranged in specific forms that are most useful for the analysis of such problems. These general expressions are valid for both Newtonian and non-Newtonian fluids in either laminar or turbulent flow.

A. Energy Balance

Consider a section of uniform cylindrical pipe of length L and radius R, inclined upward at an angle θ to the horizontal, as shown in Fig. 6-2. The steady-state energy balance (or Bernoulli equation) applied to an incompressible fluid flowing in a uniform pipe can be written

$$\frac{-\Delta\Phi}{\rho} = e_f = K_f \frac{V^2}{2} \tag{6-2}$$

where $\Phi = P + \rho g z$, $K_f = 4fL/D$, and f is the Fanning friction factor.

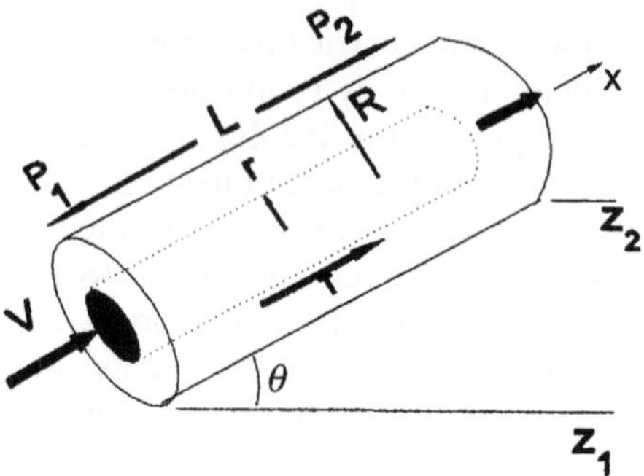

FIGURE 6-2 Pipe flow.

B. Momentum Balance

We can write a momentum balance on a cylindrical volume of fluid of radius r, length L, centered on the pipe centerline (see Fig. 6-2) as follows:

$$\sum F_x = (P_1 - P_2)\pi r^2 - \pi r^2 L\rho g \sin\theta + 2\pi r L \tau_{rx} = 0 \tag{6-3}$$

where τ_{rx} is the force in the x direction acting on the r surface of the fluid system. Solving Eq. (6-3) for τ_{rx} gives

$$\tau_{rx} = \frac{\Delta\Phi r}{2L} = -\tau_w \frac{r}{R} \tag{6-4}$$

where $\Delta\Phi = \Delta P + \rho g L \sin\theta = \Delta P + \rho g\, \Delta z$, and τ_w is the stress exerted *by* the fluid *on* the tube wall [i.e., $\tau_w = (-\tau_{rx})_{r=R}$]. Note that Eq. (6-4) also follows directly from integrating the axial component of the microscopic momentum equation of motion in cylindrical coordinates (i.e., the z-component equation in Appendix E).

Equation (6-4) is equivalent to Eq. (6-2), because

$$f = \frac{\tau_w}{\frac{1}{2}\rho V^2} = \frac{K_f}{4L/D} = \frac{e_f}{(4L/D)(V^2/2)} \tag{6-5}$$

Note that from Eq. (6-4) the shear stress is negative (i.e., the fluid outside the cylindrical system of radius r is moving more slowly than that inside the system and hence exerts a force in the $-x$ direction on the fluid in the system, which is bounded by the r surface). However, the stress at the wall (τ_w) is defined as the force exerted in the $+x$ direction *by* the fluid *on* the wall (which is positive).

C. Continuity

Continuity provides a relationship between the volumetric flow rate (Q) passing through a given cross section in the pipe and the local velocity (v_x), i.e.,

$$Q = \int_A v_x\, dA = \pi \int_0^R 2rv_x\, dr = \pi \int_A v_x\, dr^2 \tag{6-6}$$

This can be integrated by parts, as follows:

$$Q = \pi \int_A v_x\, d(r^2) = -\pi \int_A r^2\, dv_x = -\pi \int_0^R r^2 \frac{dv_x}{dr}\, dr \tag{6-7}$$

Thus, if the radial dependence of the shear rate (dv_x/dr) is known or can be found, the flow rate can be determined directly from Eq. (6-7). Application of this is shown below.

D. Energy Dissipation

A different, but related, approach to pipe flow that provides additional insight involves consideration of the rate at which energy is dissipated per unit volume of fluid. In general, the rate of energy (or power) expended in a system subjected to a force $\vec{F}$ and moving at a velocity $\vec{V}$ is simply $\vec{F} \cdot \vec{V}$. With reference to the "simple shear" deformation shown in Fig. 3-1, the corresponding rate of energy dissipation per unit volume of fluid is $\vec{F} \cdot \vec{V}/Ah = \tau\, dv_x/dy$. This can be generalized for any system as follows:

$$\dot{e}_f = e_f \dot{m} = e_f \rho Q = \int_{\text{vol}} \tau\colon \nabla\vec{v}\, d\tilde{V} \tag{6-8}$$

where τ was defined in Eq. (5-60) and $\tilde{V}$ is the volume of the fluid in the pipe. The ":" operator represents the scalar product of two dyads. Thus, integration of the local rate of energy dissipation throughout the flow volume, along with the Bernoulli equation, which relates the energy dissipated per unit mass (e_f) to the driving force ($\Delta\Phi$), can be used to determine the flow rate. All of the equations to this point are general, because they apply to any fluid (Newtonian or non-Newtonian) in any type of flow (laminar or turbulent) in steady, fully developed flow in a uniform cylindrical tube with any orientation. The following section will illustrate the application of these relations to laminar flow in a pipe.

III. NEWTONIAN FLUIDS

A. Laminar Flow

For a Newtonian fluid in laminar flow,

$$\tau_{rx} = \mu \frac{dv_x}{dr} \quad \text{or} \quad \frac{dv_x}{dr} = \frac{\tau_{rx}}{\mu} \tag{6-9}$$

When the velocity gradient from Eq. (6-9) is substituted into Eq. (6-7), and Eq. (6-4) is used to eliminate the shear stress, Eq. (6-7) becomes

$$Q = -\pi \int_0^R r^2 \frac{dv_x}{dr}\, dr = -\frac{\pi}{\mu}\int_0^R r^2 \tau_{rx}\, dr = \frac{\pi\tau_w}{\mu R}\int_0^R r^3\, dr \tag{6-10}$$

or

$$Q = \frac{\pi\tau_w R^3}{4\mu} = -\frac{\pi\,\Delta\Phi\, R^4}{8\mu L} = -\frac{\pi\,\Delta\Phi\, D^4}{128\mu L} \tag{6-11}$$

Equation (6-11) is known as the *Hagen–Poiseuille equation.*

This result can also be derived by equating the shear stress for a Newtonian fluid, Eq. (6-9), to the expression obtained from the momentum balance for tube flow, Eq. (6-4), and integrating to obtain the velocity profile:

$$v_x(r) = \frac{\tau_w R}{2\mu}\left(1 - \frac{r^2}{R^2}\right) \tag{6-12}$$

Inserting this into Eq. (6-6) and integrating over the tube cross section gives Eq. (6-11) for the volumetric flow rate.

Another approach is to use the Bernoulli equation [Eq. (6-2)] and Eq. (6-8) for the friction loss term e_f. The integral in the latter equation is evaluated in a manner similar to that leading to Eq. (6-10) as follows. Eliminating e_f between Eq. (6-8) and the Bernoulli equation [Eq. (6-2), i.e., $\rho e_f = -\Delta\Phi$] leads directly to

$$\dot{e}_f = \rho Q e_f = -\Delta\Phi\, Q = \int_{\text{vol}} \tau : \nabla\vec{v}\, d\tilde{V} = L\int_0^R \tau \frac{dv}{dr} 2\pi r\, dr$$

$$= \frac{2\pi L}{\mu}\int_0^R \tau^2 r\, dr = \frac{2\pi\tau_w^2}{\mu R^2}\int_0^R r^3\, dr = \frac{\pi L R^2 \tau_w^2}{2\mu} = \frac{\pi(-\Delta\Phi)^2 D^4}{128\mu L} \tag{6-13}$$

which is, again, the Hagen–Poiseuille equation [Eq. (6-11)].

If the wall stress (τ_w) in Eq. (6-11) is expressed in terms of the Fanning friction factor (i.e., $\tau_w = f\rho V^2/2$) and the result solved for f, the dimensionless form of the Hagen–Poiseuille equation results:

$$f = \frac{4\pi D\mu}{Q\rho} = \frac{16\mu}{DV\rho} = \frac{16}{N_{Re}} \tag{6-14}$$

It may be recalled that application of dimensional analysis (Chapter 2) showed that the steady fully developed laminar flow of a Newtonian fluid in a cylindrical tube can be characterized by a single dimensionless group that is equivalent to the product fN_{Re} (note that this group is independent of the fluid density, which cancels out). Since there is only one dimensionless variable, it follows that this group must be the same (i.e., constant) for all such flows, regardless of the fluid viscosity or density, the size of the tube, the flow rate, etc. Although the magnitude of this constant could not be obtained from dimensional analysis, we have shown from basic principles that this value is 16, which is also in agreement with experimental observations. Equation (6-14) is valid for $N_{Re} < 2000$, as previously discussed.

It should be emphasized that these results are applicable only to "fully developed" flow. However, if the fluid enters a pipe with a uniform ("plug") velocity distribution, a minimum hydrodynamic entry length (L_e) is required for the parabolic velocity flow profile to develop and the pressure gradient to become uniform. It can be shown that this (dimensionless) "hydrodynamic entry length" is approximately $L_e/D = N_{Re}/20$.

B. Turbulent Flow

As previously noted, if the Reynolds number in the tube is larger than about 2000, the flow will no longer be laminar. Because fluid elements in contact with a stationary solid boundary are also stationary (i.e., the fluid sticks to the wall), the velocity increases from zero at the boundary to a maximum value at some distance from the boundary. For uniform flow in a symmetrical duct, the maximum velocity occurs at the centerline of the duct. The region of flow over which the velocity varies with the distance from the boundary is called the *boundary layer* and is illustrated in Fig. 6-3.

1. The Boundary Layer

Because the fluid velocity at the boundary is zero, there will always be a region adjacent to the wall that is laminar. This is called the *laminar sub-*

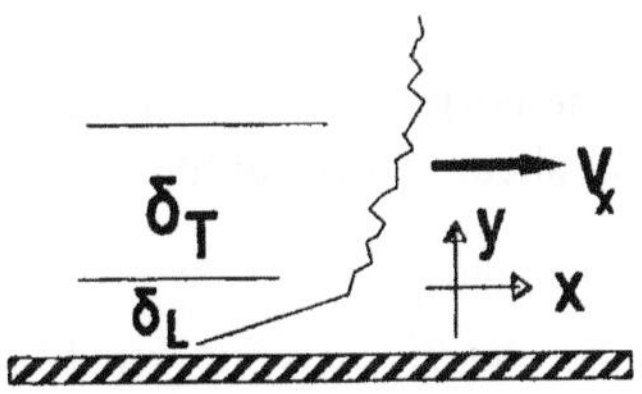

FIGURE 6-3 The boundary layer.

layer and is designated δ_L in Fig. 6-3. Note that for tube flow if $N_{Re} < 2000$ the entire flow is laminar and $\delta_L = R$. The turbulent boundary layer (δ_T) includes the region in the vicinity of the wall in which the flow is turbulent and in which the velocity varies with the distance from the wall (y). Beyond this region the fluid is almost completely mixed in what is often called the *turbulent core*, and the velocity is independent of y. The transition from the laminar sublayer to the turbulent boundary layer is gradual, not abrupt, and the transition region is called the *buffer zone*.

2. Turbulent Momentum Flux

The velocity field in turbulent flow can be described by a local "mean" (or time-average) velocity, upon which is superimposed a time-dependent fluctuating component or "eddy." Even in "one-dimensional" flow, in which the overall average velocity has only one directional component (as illustrated in Fig. 6-3), the turbulent eddies have a three-dimensional structure. Thus, for the flow illustrated in Fig. 6-3, the local velocity components are

$$\begin{aligned} v_x(y,t) &= \bar{v}_x(y) + v_x'(y,t) \\ v_y(y,t) &= 0 + v_y'(y,t) \\ v_z(y,t) &= 0 + v_z'(y,t) \end{aligned} \tag{6-15}$$

The time-average velocity ($\bar{v}$) obviously has zero components in the y and z directions, but the eddy velocity components are nonzero in all three directions. The time-average velocity is defined as

$$\bar{v}_x = \frac{1}{T}\int_0^T v_x\,dt \tag{6-16a}$$

so

$$\int_0^T v_v'\,dt = 0 \tag{6-16b}$$

The average in Eq. (6-16a) is taken over a time T that is long compared to the period of the eddy fluctuation.

Now the eddies transport momentum and the corresponding momentum flux components are equivalent to (negative) shear stress components:

$$\begin{aligned} &\tau_{xx}' = -\rho(v_x')^2, && \tau_{xy}' = -\rho v_x' v_y' && \tau_{xz}' = -\rho v_x' v_z' \\ &\tau_{yx}' = \tau_{xy}' && \tau_{yy}' = -\rho(v_y')^2 && \tau_{yz}' = -\rho v_y' v_z' \\ &\tau_{zx}' = \tau_{xz}' && \tau_{zy}' = \tau_{yz}' && \tau_{zz}' = -\rho(v_z')^2 \end{aligned} \tag{6-17}$$

These "turbulent momentum flux components" are also called *Reynolds stresses*. Thus, the total stress in a Newtonian fluid in turbulent flow is composed of both viscous and turbulent (Reynolds) stresses:

$$\tau_{ij} = \mu\left(\frac{\partial \bar{v}_i}{\partial x_j} + \frac{\partial \bar{v}_j}{\partial x_i}\right) - \rho v_i' v_j' \tag{6-18}$$

Although Eq. (6-18) can be used to eliminate the stress components from the general microscopic equations of motion, a solution for the turbulent flow field still cannot be obtained unless some information about the spatial dependence and structure of the eddy velocities or turbulent (Reynolds) stresses is known. A classical (simplified) model for the turbulent stresses, attributed to Prandtl, is outlined in the following subsection.

3. Mixing Length Theory

Turbulent eddies (with velocity components v_x', v_y', v_z') are continuously being generated, growing and dying out. During this process, there is an exchange of momentum between the eddies and the mean flow. Considering a two-dimensional turbulent field near a smooth wall, Prandtl assumed that $v_x' \approx v_y'$ (a gross approximation) so that

$$\tau_{yx}' = -\rho v_x' v_y' \cong -\rho (v_x')^2 \tag{6-19}$$

He also assumed that each eddy moves a distance l (the "mixing length") during the time it takes to exchange its momentum with the mean flow, i.e.,

$$\frac{v_x'}{l} \cong \frac{d\bar{v}_x}{dy} \tag{6-20}$$

Using Eq. (6-20) to eliminate the eddy velocity from Eq. (6-19) gives

$$\tau_{yx}' = \mu_e \frac{d\bar{v}_x}{dy} \tag{6-21}$$

where μ_e,

$$\mu_e = \rho l^2 \left|\frac{d\bar{v}_x}{dy}\right| \tag{6-22}$$

is called the *eddy viscosity*. Note that the eddy viscosity is *not* a fluid property; it is a function of the eddy characteristics (e.g., the mixing length or the degree of turbulence) and the mean velocity gradient. The only fluid property involved is the density, because turbulent momentum transport is an inertial (i.e., mass-dominated) effect. Since turbulence (and all motion) is zero at the wall, Prandtl further assumed that the mixing length should be proportional to the distance from the wall, i.e.,

$$l = \kappa y \tag{6-23}$$

Because these relations apply only in the vicinity of the wall, Prandtl also assumed that the eddy (Reynolds) stress must be of the same order as the wall stress, i.e.,

$$\tau'_{yx} \cong \tau_w = \rho \kappa^2 y^2 \left(\frac{d\bar{v}_x}{dy} \right)^2 \tag{6-24}$$

Integrating Eq. (6-24) over the turbulent boundary layer (from y_1, the edge of the buffer layer, to y) gives

$$\bar{v}_x = \frac{1}{\kappa} \left(\frac{\tau_w}{\rho} \right)^{1/2} \ln y + C_1 \tag{6-25}$$

This equation is called the *von Karman equation* (or, sometimes, the "law of the wall"), and can be written in the following dimensionless form

$$v^+ = \frac{1}{\kappa} \ln y^+ + A \tag{6-26}$$

where

$$v^+ = \frac{\bar{v}_x}{v_*} = \frac{\bar{v}_x}{V} \sqrt{\frac{2}{f}}, \qquad y^+ = \frac{y v_* \rho}{\mu} = \frac{y V \rho}{\mu} \sqrt{\frac{f}{2}} \tag{6-27}$$

The term

$$v_* = \sqrt{\frac{\tau_w}{\rho}} = V \sqrt{\frac{f}{2}} \tag{6-28}$$

is called the *friction velocity*, because it is a wall stress parameter with dimensions of velocity. The parameters κ and A in the von Karman equation have been determined from experimental data on Newtonian fluids in smooth pipes to be $\kappa = 0.4$ and $A = 5.5$. Equation (6-26) applies only within the turbulent boundary layer (outside the buffer region), which has been found empirically to correspond to $y^+ \geq 26$.

Within the laminar sublayer the turbulent eddies are negligible, so

$$\tau_{yx} \cong \tau_w = \mu \frac{d\bar{v}}{dy} \tag{6-29}$$

The corresponding dimensionless form of this equation is

$$\frac{dv^+}{dy^+} = 1 \tag{6-30}$$

or

$$v^+ = y^+ \tag{6-31}$$

Equation (6-31) applies to the laminar sublayer region in a Newtonian fluid, which has been found to correspond to $0 \leq y^+ \leq 5$. The intermediate region, or "buffer zone," between the laminar sublayer and the turbulent boundary layer can be represented by the empirical equation

$$v^+ = -3.05 + 5.0 \ln y^+ \tag{6-32}$$

which applies for $5 < y^+ < 26$.

4. Friction Loss in Smooth Pipe

For a Newtonian fluid in a smooth pipe, these equations can be integrated over the pipe cross section to give the average fluid velocity, e.g.,

$$V = \frac{2}{R^2} \int_0^R \bar{v}_x r \, dr = 2v_* \int_0^1 v^+(1 - x) \, dx \tag{6-33}$$

where $x = y/R = 1 - r/R$. If the von Karman equation [Eq. (6-26)] for v^+ is introduced into this equation and the laminar sublayer and buffer zones are neglected, the integral can be evaluated and the result solved for $1/\sqrt{f}$ to give

$$\frac{1}{\sqrt{f}} = 4.1 \log(N_{Re}\sqrt{f}) - 0.60 \tag{6-34}$$

The constants in this equation were modified by Nikuradse from observed data taken in smooth pipes as follows:

$$\frac{1}{\sqrt{f}} = 4.0 \log(N_{Re}\sqrt{f}) - 0.40 \tag{6-35}$$

Equation (6-35) is also known as the *von Karman–Nikuradse equation* and agrees well with observations for friction loss in smooth pipe over the range $5 \times 10^3 < N_{Re} < 5 \times 10^6$.

An alternative equation for smooth tubes was derived by Blasius based on observations that the mean velocity profile in the tube could be represented approximately by

$$\bar{v}_x = v_{max}\left(1 - \frac{r}{R}\right)^{1/7} \tag{6-36}$$

A corresponding expression for the friction factor can be obtained by writing this expression in dimensionless form and substituting the result into Eq. (6-33). Evaluating the integral and solving for f gives

$$f = \frac{0.0791}{N_{\text{Re}}^{1/4}} \tag{6-37}$$

Equation (6-37) represents the friction factor for Newtonian fluids in smooth tubes quite well over a range of Reynolds numbers from about 5000 to 10^5. The Prandtl mixing length theory and the von Karman and Blasius equations are referred to as "semiempirical" models. That is, even though these models result from a process of logical reasoning, the results cannot be deduced solely from first principles, because they require the introduction of certain parameters that can be evaluated only experimentally.

5. Friction Loss in Rough Tubes

All models for turbulent flows are semiempirical in nature, so it is necessary to rely upon empirical observations (e.g., data) for a quantitative description of friction loss in such flows. For Newtonian fluids in long tubes, we have shown from dimensional analysis that the friction factor should be a unique function of the Reynolds number and the relative roughness of the tube wall. This result has been used to correlate a wide range of measurements for a range of tube sizes, with a variety of fluids, and for a wide range of flow rates in terms of a generalized plot of f versus N_{Re}, with ε/D as a parameter. This correlation, shown in Fig. 6-4, is called a *Moody diagram.*

The laminar region (for $N_{\text{Re}} < 2000$) is described by the theoretical Hagen–Poiseuille equation [Eq. (6-14)], which is plotted in Fig. 6-4. In this region, the only fluid property that influences friction loss is the viscosity (because the density cancels out). Furthermore, the roughness has a negligible effect in laminar flow, as will be explained shortly. The "critical zone" is the range of transition from laminar to turbulent flow, which corresponds to values of N_{Re} from about 2000 to 4000. Data are not very reproducible in this range, and correlations are unreliable. The so-called transition zone in Fig. 6-4 is the region where the friction factor depends strongly on both the Reynolds number and relative roughness. The region in the upper right of the diagram where the lines of constant roughness are horizontal is called "complete turbulence, rough pipes" or "fully turbulent." In this region the friction factor is independent of Reynolds number (i.e., independent of viscosity) and is a function only of the relative roughness.

For turbulent flow in smooth tubes, the semiempirical Prandtl–von Karman/Nikuradse or Blasius models represent the friction factor quite well. Whether a tube is hydraulically "smooth" or "rough" depends upon

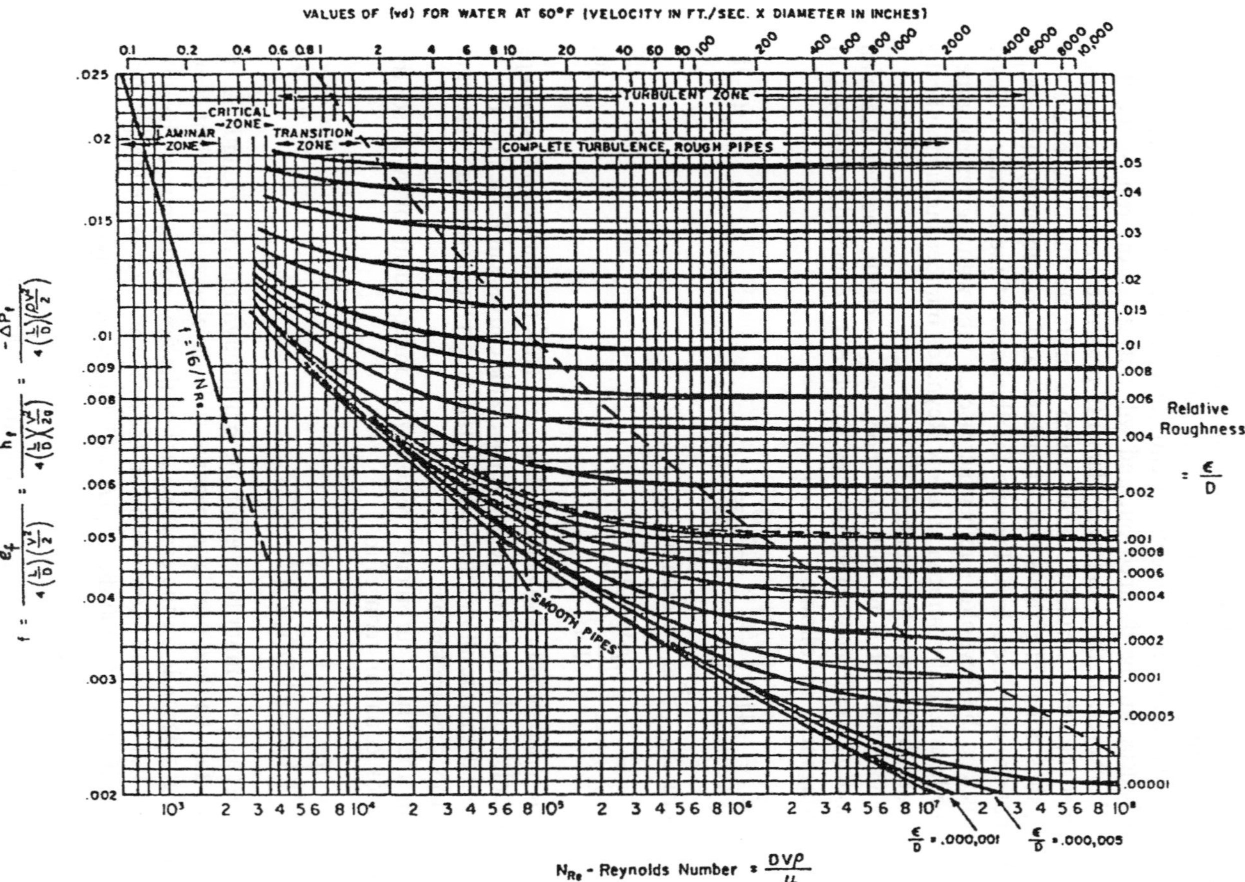

$$f = \frac{e_f}{4\left(\frac{L}{D}\right)\left(\frac{V^2}{2}\right)} = \frac{h_f}{4\left(\frac{L}{D}\right)\left(\frac{V^2}{2g}\right)} = \frac{-\Delta P_f}{4\left(\frac{L}{D}\right)\left(\frac{\rho V^2}{2}\right)}$$

FIGURE 6-4 Moody diagram.

the size of the wall roughness elements relative to the thickness of the laminar sublayer. Because laminar flow is stable, if the flow perturbations due to the roughness elements lie entirely within the laminar region the disturbances will be damped out and will not affect the rest of the flow field. However, if the roughness elements protrude through the laminar sublayer into the turbulent region, which is unstable, the disturbance will grow, thus enhancing the Reynolds stresses and consequently the energy dissipation or friction loss. Because the thickness of the laminar sublayer decreases as the Reynolds number increases, a tube with a given roughness may be hydraulically smooth at a low Reynolds number but hydraulically rough at a high Reynolds number.

For rough tubes in turbulent flow ($N_{Re} > 4000$), the von Karman equation was modified empirically by Colebrook to include the effect of wall roughness, as follows:

$$\frac{1}{\sqrt{f}} = -4 \log \left[\frac{\varepsilon/D}{3.7} + \frac{1.255}{N_{Re}\sqrt{f}} \right] \tag{6-38}$$

The term $N_{Re}\sqrt{f}$ is, by definition,

$$N_{Re}\sqrt{f} = \left(\frac{e_f D^3 \rho^2}{2L\mu^2} \right)^{1/2} \tag{6-39}$$

which is independent of velocity or flow rate. Thus the dimensionless groups in the Colebrook equation are in a form that is convenient if the flow rate is to be found and the allowable friction loss (e.g., driving force), tube size, and fluid properties are known.

In the "fully turbulent" region, f is independent of N_{Re}, so the Colebrook equation reduces to

$$f = \left(\frac{1}{4 \log[3.7/(\varepsilon/D)]} \right)^2 \tag{6-40}$$

Just as for laminar flow, a minimum hydrodynamic entry length (L_e) is required for the flow profile to become fully developed in turbulent flow. This length depends on the exact nature of the flow conditions at the tube entrance but has been shown to be on the order of $L_e/D = 0.623 N_{Re}^{0.25}$. For example, if $N_{Re} = 50{,}000$ then $L_e/D = 10$ (approximately).

6. Wall Roughness

The actual size of the roughness elements on the conduit wall obviously varies from one material to another, with age and usage, and with the amount of dirt, scale, etc. Characteristic values of wall roughness have been determined for various materials, as shown in Table 6-1. The most

TABLE 6–1 Equivalent Roughness of Various Surfaces

Material	Condition	Roughness range	Recommended
Drawn brass, copper, stainless	New	0.01–0.0015 mm (0.0004–0.00006 in.)	0.002 mm (0.00008 in.)
Commercial steel	New	0.1–0.02 mm (0.004–0.0008 in.)	0.045 mm (0.0018 in.)
	Light rust	1.0–0.15 mm (0.04–0.006 in.)	0.3 mm (0.015 in.)
	General rust	3.0–1.0 mm (0.1–0.04 in.)	2.0 mm (0.08 in.)
Iron	Wrought, new	0.045 mm (0.002 in.)	0.045 mm (0.002 in.)
	Cast, new	1.0–0.25 mm (0.04–0.01 in.)	0.30 mm (0.025 in.)
	Galvanized	0.15–0.025 mm (0.006–0.001 in.)	0.15 mm (0.006 in.)
	Asphalt-coated	1.0–0.1 mm (0.04–0.004 in.)	0.15 mm (0.006 in.)
Sheet metal	Ducts Smooth joints	0.1–0.02 mm (0.004–0.0008 in.)	0.03 mm (0.0012 in.)
Concrete	Very smooth	0.18–0.025 mm (0.007–0.001 in.)	0.04 mm (0.0016 in.)
	Wood floated, brushed	0.8–0.2 mm (0.03–0.007 in.)	0.3 mm (0.012 in.)
	Rough, visible form marks	2.5–0.8 mm (0.1–0.03 in.)	2.0 mm (0.08 in.)
Wood	Stave, used	1.0–0.25 mm (0.035–0.01 in.)	0.5 mm (0.02 in.)
Glass or plastic	Drawn tubing	0.01–0.0015 mm (0.0004–0.00006 in.)	0.002 mm (0.00008 in.)
Rubber	Smooth tubing	0.07–0.006 mm (0.003–0.00025 in.)	0.01 mm (0.0004 in.)
	Wire-reinforced	4.0–0.3 mm (0.15–0.01 in.)	1.0 mm (0.04 in.)

common pipe material—clean, new commercial steel or wrought iron—has been found to have an effective roughness of about 0.0018 in. (0.045 mm). Other surfaces, such as concrete, may vary by as much as several orders of magnitude, depending upon the nature of the surface finish. These roughness values are not measured directly but have been determined indirectly. Conduit surfaces artificially roughened by sand grains of various sizes were studied initially by Nikuradse, and measurements of f and N_{Re} were plotted to establish the reference curves for various known values of ε/D for these surfaces, as shown on the Moody diagram. The equivalent roughness factors for other materials are determined from similar measurements in conduits made of the material, by plotting the data on the Moody diagram and comparing the results with the reference curves (or by using the Colebrook equation). For this reason, such roughness values are sometimes termed the *equivalent sand grain roughness.*

C. All Flow Regimes

The expressions for the friction factor in both laminar and turbulent flow were combined into a single expression by Churchill (1977) as follows:

$$f = 2\left[\left(\frac{8}{N_{Re}}\right)^{12} + \frac{1}{(A+B)^{3/2}}\right]^{1/12} \tag{6-41}$$

where

$$A = \left[2.457 \ln\left(\frac{1}{(7/N_{Re})^{0.9} + 0.27\varepsilon/D}\right)\right]^{16}$$

and

$$B = \left(\frac{37{,}530}{N_{Re}}\right)^{16}$$

Equation (6-41) adequately represents the Fanning friction factor over the entire range of Reynolds numbers within the accuracy of the data used to construct the Moody diagram, including a reasonable estimate for the intermediate or transition region between laminar and turbulent flow. Note that it is explicit in f.

IV. POWER LAW FLUIDS

Corresponding expressions for the friction loss in laminar and turbulent flow for non-Newtonian fluids in pipes, for the two simplest (two-parameter) models—the power law and Bingham plastic—can be evaluated in a similar manner. The power law model is very popular for representing the viscosity of a wide variety of non-Newtonian fluids because of its simplicity and versatility. However, extreme care should be exercised in its application, because any application involving extrapolation beyond the range of shear stress (or shear rate) represented by the data used to determine the model parameters can lead to misleading or erroneous results.

Both laminar and turbulent pipe flow of highly loaded slurries of fine particles, for example, can often be adequately represented by either of these two models over an appreciable shear rate range, as shown by Darby et al. (1992).

A. Laminar Flow

Because the shear stress and shear rate are negative in pipe flow, the appropriate form of the power law model for laminar pipe flow is

$$\tau_{rx} = m\dot{\gamma}_{rx}^{n} = -m\left(-\frac{dv_x}{dr}\right)^{n} \tag{6-42}$$

By equating the shear stress from Eqs. (6-42) and (6-4), solving for the velocity gradient, and introducing the result into Eq. (6-7) (as was done for the Newtonian fluid), the flow rate is found to be

$$Q = \pi\left(\frac{\tau_w}{mR}\right)^{1/n}\int_0^R r^{2+1/n}\,dr = \pi\left(\frac{\tau_w}{mr}\right)^{1/n}\left(\frac{n}{3n+1}\right)R^{(3n+1)/n} \tag{6-43}$$

This is the power law equivalent of the Hagen–Poiseuille equation. It can be written in dimensionless form by expressing the wall stress in terms of the friction factor using Eq. (6-5), solving for f, and equating the result to $16/N_{Re}$ (i.e., the form of the Newtonian result). The result is an expression that is identical to the dimensionless Hagen–Poiseuille equation:

$$fN_{Re,pl} = 16 \tag{6-44}$$

if the Reynolds number for the power law fluid is defined as

$$N_{Re,pl} = \frac{8D^n V^{2-n}\rho}{m[2(3n+1)/n]^n} \tag{6-45}$$

It should be noted that a dimensional analysis of this problem results in one more dimensionless group than for the Newtonian fluid, because there is one more fluid rheological property (e.g., m and n for the power law fluid, versus μ for the Newtonian fluid). However, the parameter n is itself dimensionless and thus constitutes the additional "dimensionless group," even though it is integrated into the Reynolds number as it has been defined. Note also that because n is an empirical parameter and can take on any value, the units in expressions for power law fluids can be complex. Thus, the calculations are simplified if a scientific system of dimensional units is used (e.g., SI or cgs), which avoids the necessity of introducing the conversion factor g_c. In fact, the evaluation of most dimensionless groups is usually simplified by the use of such units.

B. Turbulent Flow

Dodge and Metzner (1959) modified the von Karman equation to apply to power law fluids, with the following result:

$$\frac{1}{\sqrt{f}} = \frac{4}{n^{0.75}} \log[N_{\text{Re,pl}} f^{1-n/2}] - \frac{0.4}{n^{1.2}} \tag{6-46}$$

Like the von Karman equation, this equation is implicit in f. Equation (6-46) can be applied to any non-Newtonian fluid if the parameter n is interpreted to be the point slope of the shear stress versus shear rate plot from (laminar) viscosity measurements, at the wall shear stress (or shear rate) corresponding to the conditions of interest in turbulent flow. However, it is not a simple matter to acquire the needed data over the appropriate range or to solve the equation for f for a given flow rate and pipe diameter, in turbulent flow.

Note that there is no effect of pipe wall roughness in Eq. (6-46), in contrast to the case for Newtonian fluids. There are insufficient data in the literature to provide a reliable estimate of the effect of roughness on friction loss for non-Newtonian fluids in turbulent flow. However, the evidence that does exist suggests that the roughness is not as significant for non-Newtonian fluids as for Newtonian fluids. This is partly due to the fact that the majority of non-Newtonian turbulent flows lie in the low Reynolds number range and partly due to the fact that the laminar boundary layer tends to be thicker for non-Newtonian fluids than for Newtonian fluids (i.e., the flows are generally in the "hydraulically smooth" range for common pipe materials).

C. All Flow Regimes

An expression that represents the friction factor for the power law fluid over the entire range of Reynolds numbers (laminar through turbulent) and encompasses Eqs. (6-44) and (6-46) has been given by Darby et al. (1992):

$$f = (1 - \alpha) f_{\text{L}} + \frac{\alpha}{[f_{\text{T}}^{-8} + f_{\text{Tr}}^{-8}]^{1/8}} \tag{6-47}$$

where

$$f_{\text{L}} = \frac{16}{N_{\text{Re,pl}}} \tag{6-48}$$

$$f_{\text{T}} = \frac{0.0682 n^{-1/2}}{N_{\text{Re,pl}}^{1/(1.87+2.39n)}} \tag{6-49}$$

$$f_{\text{Tr}} = 1.79 \times 10^{-4} \exp[-5.24n] N_{\text{Re,pl}}^{0.414+0.757n} \tag{6-50}$$

The parameter α is given by

$$\alpha = \frac{1}{1 + 4^{-\Delta}} \tag{6-51}$$

where

$$\Delta = N_{\text{Re,pl}} - N_{\text{Re,plc}} \tag{6-52}$$

and $N_{\text{Re,plc}}$ is the critical power law Reynolds number at which laminar flow ceases:

$$N_{\text{Re,plc}} = 2100 + 875(1 - n) \tag{6-53}$$

Equation (6-48) applies for $N_{\text{Re,pl}} < N_{\text{Re,plc}}$, Eq. (6-49) applies for $4000 < N_{\text{Re,pl}} < 10^5$, Eq. (6-50) applies for $N_{\text{Re,plc}} < N_{\text{Re,pl}} < 4000$, and all are encompassed by Eq. (6-47) for all $N_{\text{Re,pl}}$.

V. BINGHAM PLASTICS

The Bingham plastic model usually provides a good representation for the viscosity of concentrated slurries, suspensions, emulsions, foams, etc. Such materials often exhibit a yield stress that must be exceeded before the material will flow at a significant rate. Other examples include paint, shaving cream, and mayonnaise. There are also many fluids, such as blood, that may have a yield stress that is not as pronounced.

It is recalled that a "plastic" is really two materials. At low stresses below the critical or yield stress (τ_o) the material behaves as a solid, whereas for stresses above the yield stress the material behaves as a fluid. The Bingham model for this behavior is

$$\begin{aligned} &\text{For } |\tau| < \tau_o: \quad \dot{\gamma} = 0 \\ &\text{For } |\tau| > \tau_o: \quad \tau = \pm\tau_o + \mu_\infty \dot{\gamma} \end{aligned} \tag{6.54}$$

Because the shear stress and shear rate can be either positive or negative, the plus/minus sign in Eq. (6-54) is plus in the former case and minus in the latter. For tube flow, because the shear stress and shear rate are both negative, the appropriate form of the model is

$$\begin{aligned} &\text{For } |\tau_{rx}| < \tau_o: \quad \frac{dv_x}{dr} = 0 \\ &\text{For } |\tau_{rx}| > \tau_o: \quad \tau_{rx} = -\tau_o + \mu_\infty \frac{dv_x}{dr} \end{aligned} \tag{6-55}$$

A. Laminar Flow

Because the shear stress is always zero at the centerline in pipe flow and increases linearly with distance from the center toward the wall [Eq. (6-4)], there will be a finite distance from the center over which the stress is always less than the yield stress. In this region, the material has solid-like properties and does not yield but moves as a rigid plug. The radius of this plug (r_o) is, from Eq. (6-4),

$$r_o = R\frac{\tau_o}{\tau_w} \tag{6-56}$$

Because the stress outside of this plug region exceeds the yield stress, the material will deform or flow as a fluid between the plug and the wall. The flow rate must thus be determined by combining the flow rate of the "plug" with that of the "fluid" region:

$$Q = \int^A v_x\, dA = Q_{plug} + \pi\int_{r_o^2}^{R^2} v_x\, dr^2 \tag{6-57}$$

Evaluating the integral by parts and noting that the Q_{plug} term cancels with $\pi r_o^2 V_{plug}$ from the lower limit, the result is

$$Q = -\pi\int_{r_o}^{R} r^2\dot{\gamma}\, dr \tag{6-58}$$

When Eq. (6-55) is used for the shear rate in terms of the shear stress and Eq. (6-4) is used for the shear stress as a function of r, the integral can be evaluated to give

$$Q = \frac{\pi R^3\tau_w}{4\mu_\infty}\left[1 - \frac{4}{3}\left(\frac{\tau_o}{\tau_w}\right) + \frac{1}{3}\left(\frac{\tau_o}{\tau_w}\right)^4\right] \tag{6-59}$$

This equation is known as the *Buckingham–Reiner equation*. It can be cast in dimensionless form and rearranged as follows:

$$f_L = \frac{16}{N_{Re}}\left[1 + \frac{1}{6}\left(\frac{N_{He}}{N_{Re}}\right) - \frac{1}{3}\left(\frac{N_{He}^4}{f^3 N_{Re}^7}\right)\right] \tag{6-60}$$

where the Reynolds number is given by

$$N_{Re} = DV\rho/\mu_\infty \tag{6-61}$$

and

$$N_{He} = D^2\rho\tau_o/\mu_\infty^2 \tag{6-62}$$

is the *Hedstrom number*. Note that the Bingham plastic reduces to a Newtonian fluid if $\tau_o = 0 = N_{He}$. In this case Eq. (6-60) reduces to the Newtonian result, i.e., $f = 16/N_{Re}$ [see Eq. (6-14)]. Note that there are actually only two independent dimensionless groups in Eq. (6-60) (consistent with the results of dimensional analysis for a fluid with two rheological properties, τ_o and μ_∞), which are the combined groups fN_{Re} and N_{He}/N_{Re}. The ratio N_{He}/N_{Re} is also called the *Bingham number*, $N_{Bi} = D\tau_o/\mu_\infty V$. Equation (6-60) is implicit in f, so it must be solved by iteration for known values of N_{Re} and N_{He}. This is not difficult, however, because the last term in Eq. (6-60) is usually much smaller than the other terms, in which case neglecting this term provides a good first estimate for f. Inserting this first estimate into the neglected term to revise f and repeating the procedure usually results in rapid convergence.

B. Turbulent Flow

For the Bingham plastic, there is no abrupt transition from laminar to turbulent flow as is observed for Newtonian fluids. Instead, there is a gradual deviation from purely laminar flow to fully turbulent flow. For turbulent flow, the friction factor can be represented by the empirical expression of Darby and Melson (1981) [as modified by Darby et al. (1992)]:

$$f_T = 10^a / N_{Re}^{0.193} \tag{6-63}$$

where

$$a = -1.47[1 + 0.146\exp(-2.9 \times 10^{-5} N_{He})] \tag{6-64}$$

C. All Reynolds Numbers

The friction factor for a Bingham plastic can be calculated for any Reynolds number, from laminar through turbulent, from the equation

$$f = (f_L^m + f_T^m)^{1/m} \tag{6-65}$$

where

$$m = 1.7 + \frac{40{,}000}{N_{Re}} \tag{6-66}$$

In Eq. (6-65), f_T is given by Eq. (6-63) and f_L is given by Eq. (6-60).

VI. PIPE FLOW PROBLEMS

There are three typical problems encountered in pipe flows, depending upon what is known and what is to be found. These are the "unknown driving force," "unknown flow rate," and "unknown diameter" problems, and we will outline here the procedure for the solution of each of these for both Newtonian and non-Newtonian (power law and Bingham plastic) fluids. A fourth problem, perhaps of even more practical interest for piping system design, is the "most economical diameter" problem, which will be considered in Chapter 7.

We note first that the Bernoulli equation can be written

$$\mathrm{DF} = e_f + \tfrac{1}{2}(\alpha_2 V_2^2 - \alpha_1 V_1^2) \tag{6-67}$$

where

$$e_f = \left(\frac{4fL}{d}\right)\left(\frac{V^2}{2}\right) = \frac{32fLQ^2}{\pi^2 D^2} \tag{6-68}$$

and

$$\mathrm{DF} = -\left(\frac{\Delta\Phi}{\rho} + w\right) \tag{6-69}$$

DF represents the net energy input into the fluid per unit mass (or the net "driving force") and is the combination of static head, pressure difference, and pump work. When any of the terms in Eq. (6-69) are negative, they represent a positive "driving force" for moving the fluid through the pipe (positive terms represent forces resisting the flow, e.g., an increase in elevation, pressure, etc. and correspond to a negative driving force). In many applications the kinetic energy terms are negligible or cancel out, although this should be verified for each situation.

We will use the Bernoulli equation in the form of Eq. (6-67) for analyzing pipe flows, and we will use the total volumetric flow rate (Q) as the flow variable instead of the velocity, because this is the usual measure of capacity in a pipeline. For Newtonian fluids, the problem thus reduces to a relation between the three dimensionless variables:

$$N_{\mathrm{Re}} = \frac{4Q\rho}{\pi D\mu}, \qquad f = \frac{e_f \pi^2 D^5}{32LQ^2}, \qquad \frac{\varepsilon}{D} \tag{6-70}$$

A. Unknown Driving Force

For this problem, we want to know the net driving force (DF) that is required to move a given fluid (μ, ρ) at a specified rate (Q) through a specified pipe (D, L, ε). The Bernoulli equation in the form $\mathrm{DF} = e_f$ applies.

1. Newtonian Fluid

The "knowns" and "unknowns" in this case are

Given: $Q, \mu, \rho, D, L, \varepsilon$ *Find*: DF

All the relevant variables and parameters are uniquely related through the three dimensionless variables f, N_{Re}, and ε/D by the Moody diagram or the Churchill equation. Furthermore, the unknown ($DF = e_f$) appears in only one of these groups (f). The procedure is thus straightforward:

1. Calculate the Reynolds number from Eq. (6-70).
2. Calculate ε/D.
3. Determine f from the Moody diagram or Churchill equation [Eq. (6-41)]; (if $N_{Re} < 2000$, use $f = 16/N_{Re}$).
4. Calculate e_f (hence DF) from the Bernoulli Equation, Eq. (6-68).

From the resulting value of DF, the required pump head ($-w/g$) can be determined, for example, from a knowledge of the upstream and downstream pressures and elevations using Eq. (6-69).

2. Power Law Fluid

The equivalent problem statement is

Given: Q, m, n, ρ, D, L *Find*: DF

Note that we have an additional fluid property (m and n instead of μ), but we also assume that pipe roughness has a negligible effect, so the total number of variables is the same. The corresponding dimensionless variables are f, $N_{Re,pl}$, and n [which are related by Eq. (6-47)], and the unknown ($DF = e_f$) appears in only one group (f). The procedure just followed for a Newtonian fluid can thus also be applied to a power law fluid if the appropriate equations are used, as follows.

1. Calculate the Reynolds number ($N_{Re,pl}$), using Eq. (6-45) and the volumetric flow rate instead of the velocity, i.e.,

$$N_{Re,pl} = \frac{2^{7-3n} \rho Q^{2-n}}{m\pi^{2-n} D^{4-3n}} \left(\frac{n}{3n+1} \right)^n \tag{6-71}$$

2. Calculate f from Eq. (6-47).
3. Calculate e_f (hence DF) from Eq. (6-68).

3. Bingham Plastic

The problem statement is

Given: $Q, \mu_\infty, \tau_o, \rho, D, L$ *Find*: DF

The number of variables is the same as in the foregoing problems; hence the number of groups relating these variables is the same. For the Bingham plastic, these are f, N_{Re}, and N_{He}, which are related by Eq. (6-65) [along with Eqs. (6-60) and (6-63)]. The unknown ($DF = e_f$) appears only in f, as before. The solution procedure is similar to that followed for Newtonian and power law fluids.

1. Calculate the Reynolds number.

$$N_{Re} = \frac{4Q\rho}{\pi D \mu_\infty} \tag{6-72}$$

2. Calculate the Hedstrom number:

$$N_{He} = \frac{D^2 \rho \tau_o}{\mu_\infty^2} \tag{6-73}$$

3. Determine f from Eqs. (6-65), (6-63), and (6-60). [Note that an iteration is required to determine f_L from Eq. (6-60).]
4. Calculate e_f, hence DF, from Eq. (6-68).

B. Unknown Flow Rate

In this case, the flow rate is to be determined when a given fluid is transported in a given pipe with a known net driving force (e.g., pump head, pressure head, and/or hydrostatic head). The same total variables are involved, and hence the dimensionless variables are the same and are related in the same way as for the unknown driving force problems. The main difference is that now the unknown (Q) appears in two of the dimensionless variables (f and N_{Re}), which requires a different solution strategy.

1. Newtonian Fluid

The problem statement is

Given: DF, $D, L, \varepsilon, \mu, \rho$ *Find*: Q

The strategy is to redefine the relevant dimensionless variables by combining the original groups in such a way that the unknown variable appears in one group. For example, f and N_{Re} can be combined to cancel the unknown (Q) as follows:

$$fN_{Re}^2 = \left(\frac{\text{DF}\,\pi^2 D^5}{32LQ^2}\right)\left(\frac{4Q\rho}{\pi D \mu}\right)^2 = \frac{\text{DF}\,\rho^2 D^3}{2L\mu^2} \tag{6-74}$$

Thus, if we work with the three dimensionless variables fN_{Re}^2, N_{Re}, and ε/D, the unknown (Q) appears in only N_{Re}, which then becomes the unknown (dimensionless) variable.

There are various approaches that we can take to solve this problem. Since the Reynolds number is unknown, an explicit solution is not possible using the established relations between the friction factor and Reynolds number (e.g., the Moody diagram or Churchill equation). We can, however proceed by a trial-and-error method that requires an initial guess for an unknown variable, use the basic relations to solve for this variable, revise the guess accordingly, and repeat the process (iterating) until agreement between calculated and guessed values is achieved.

Note that in this context, either f or N_{Re} can be considered the unknown dimensionless variable, because they both involve the unknown Q. As an aid in making the choice between these, a glance at the Moody diagram shows that the practical range of possible values of f is approximately one order of magnitude, whereas the corresponding possible range of N_{Re} is over five orders of magnitude! Thus, the chances of our initial guess being close to the final answer are greatly enhanced if we choose to iterate on f instead of N_{Re}. Using this approach, the procedure is as follows.

1. A reasonable guess might be based on the assumption that the flow conditions are turbulent, for which the Colebrook equation, Eq. (6-38), applies.
2. Calculate the value of fN_{Re}^2 from given values.
3. Calculate f using the Colebrook equation, Eq. (6-38).
4. Calculate $N_{\text{Re}} = (fN_{\text{Re}}^2/f)^{1/2}$, using f from step 3.
5. Using the N_{Re} value from step 4 and the known value of ε/D, determine f from the Moody diagram or Churchill equation (if $N_{\text{Re}} < 2000$, use $f = 16/N_{\text{Re}}$).
6. If this value of f does not agree with that from step 3, insert the value of f from step 5 into step 4 to get a revised value of N_{Re}.
7. Repeat steps 5 and 6 until f no longer changes.
8. Calculate $Q = \pi D\mu N_{\text{Re}}/4\rho$.

2. Power Law Fluid

The problem statement is

Given: DF, D, L, m, n, ρ *Find*: Q

The simplest approach for this problem is also an iteration procedure based on an assumed value of f:

1. A reasonable starting value for f is 0.005, based on a "dart throw" at the (equivalent) Moody diagram.

2. Calculate Q from Eq. (6-68), i.e.,

$$Q = \pi \left(\frac{D^5 \, \mathrm{DF}}{32 f L} \right)^{1/2} \tag{6-75}$$

3. Calculate the Reynolds number from Eq. (6-71), i.e.,

$$N_{\mathrm{Re,pl}} = \frac{2^{7-3n} \rho Q^{2-n}}{m \pi^{2-n} D^{4-3n}} \left(\frac{n}{3n+1} \right)^n \tag{6-76}$$

4. Calculate f from Eq. (6-47).
5. Compare the values of f from step 4 and step 1. If they do not agree, use the result of step 4 in step 2 and repeat steps 2–5 until agreement is reached. Convergence usually requires only two or three trials at most, unless very unusual conditions are encountered.

3. Bingham Plastic

The procedure is very similar to the one above.

Given: DF, D, L, μ_∞, τ_o, ρ *Find*: Q

1. Assume $f = 0.005$.
2. Calculate Q from Eq. (6-75).
3. Calculate the Reynolds and Hedstrom numbers:

$$N_{\mathrm{Re}} = \frac{4 Q \rho}{\pi D \mu_\infty}, \qquad N_{\mathrm{He}} = \frac{D^2 \rho \tau_o}{\mu_\infty^2} \tag{6-77}$$

4. Calculate f from Eq. (6-65).
5. Compare the value of f from step 4 with the assumed value in step 1. If they do not agree, use the value of f from step 4 in step 2 and repeat steps 2–5 until they agree. Note that an iteration is required to determine f_L in Eq. (6-65), but this procedure normally converges rapidly unless unusual conditions are encountered.

C. Unknown Diameter

In this problem, it is desired to determine the size of the pipe (D) that will transport a given fluid (Newtonian or non-Newtonian) at a given flow rate (Q) over a given distance (L) with a given driving force (DF). Because the unknown (D) appears in each of the dimensionless variables, it is appropriate to regroup these variables in a more convenient form for this problem.

1. Newtonian Fluid

The problem statement is

Given: DF, $Q, L, \varepsilon, \rho, \mu$ *Find*: D

We can eliminate the unknown (D) from two of the three basic groups (N_{Re}, ε/D, and f) as follows:

$$fN_{Re}^5 = \left(\frac{\text{DF}\,\pi^2 D^5}{32LQ^2}\right)\left(\frac{4Q\rho}{\pi D\mu}\right)^5 = \frac{32\,\text{DF}\,\rho^5 Q^3}{\pi^3 L\mu^5} \tag{6-78}$$

$$N_R = \frac{N_{Re}}{\varepsilon/D} = \frac{4Q\rho}{\pi\mu\varepsilon} \tag{6-79}$$

Thus, the three basic groups for this problem are fN_{Re}^5, N_R, and N_{Re}, with N_{Re} being the dimensionless "unknown" (because it is now the only group containing the unknown D). D is unknown, so no initial estimate for f can be obtained from the equations, because ε/D is also unknown. Thus the following procedure is recommended for this problem:

1. Calculate fN_{Re}^5 from known quantities using Eq. (6-78).
2. Assume $f = 0.005$.
3. Calculate N_{Re}:

$$N_{Re} = \left(\frac{fN_{Re}^5}{0.005}\right)^{1/5} \tag{6-80}$$

4. Calculate D from N_{Re}:

$$D = \frac{4Q\rho}{\pi\mu N_{Re}} \tag{6-81}$$

5. Calculate ε/D.
6. Determine f from the Moody diagram or Churchill equation using the above values of N_{Re} and ε/D (if $N_{Re} < 2000$, use $f = 16/N_{Re}$).
7. Compare the value of f from step 6 with the assumed value in step 2. If they do not agree, use the result of step 6 for f in step 3 in place of 0.005 and repeat steps 3–7 until they agree.

2. Power Law Fluid

The problem statement is

Given: DF, Q, m, n, ρ, L *Find*: D

The procedure is analogous to that for the Newtonian fluid. In this case, the combined group $fN_{\text{Re,pl}}^{5/(4-3n)}$ (which we shall call K, for convenience) is independent of D:

$$fN_{\text{Re,pl}}^{5/(4-3n)} = \left(\frac{\pi^2 \text{DF}}{32LQ^2}\right)\left[\frac{2^{7-3n}Q^{2-n}}{m\pi^{2-n}}\left(\frac{n}{3n+1}\right)^n\right]^{5/(4-3n)} = K \tag{6-82}$$

The following procedure can be used to find D:

1. Calculate K from Eq. (6-82).
2. Assume $f = 0.005$.
3. Calculate $N_{\text{Re,pl}}$ from

$$N_{\text{Re,pl}} = \left(\frac{K}{f}\right)^{(4-3n)/5} \tag{6-83}$$

4. Calculate f from Eq. (6-47), using the value of $N_{\text{Re,pl}}$ from step 3.
5. Compare the result of step 4 with the assumed value in step 2. If they do not agree, use the value of f from step 4 in step 3, and repeat steps 3–5 until they agree.

The diameter D is obtained from the last value of $N_{\text{Re,pl}}$ from step 3:

$$D = \left[\frac{2^{7-3n}\rho Q^{2-n}}{m\pi^{2-n}N_{\text{Re,pl}}}\left(\frac{n}{3n+1}\right)^n\right]^{1/(4-3n)} \tag{6-84}$$

3. Bingham Plastic

The problem variables are

Given: DF, Q, $\mu_\infty \tau_o$, ρ, L *Find*: D

The combined group that is independent of D is equivalent to Eq. (6-78), i.e.,

$$fN_{\text{Re}}^5 = \left(\frac{\pi^2\ \text{DF}}{32LQ^2}\right)\left(\frac{4Q\rho}{\pi\mu_\infty}\right)^5 = \left(\frac{32\ \text{DF}\ Q^3\rho^5}{\pi^3 L\mu_\infty^5}\right) \tag{6-85}$$

The procedure is

1. Calculate fN_{Re}^5 from Eq. (6-85).
2. Assume $f = 0.01$.

3. Calculate N_{Re} from

$$N_{Re} = \left(\frac{fN_{Re}^5}{0.01}\right)^{1/5} \tag{6-86}$$

4. Calculate D from

$$D = \frac{4Q\rho}{\pi\mu_\infty N_{Re}} \tag{6-87}$$

5. Calculate N_{He} from

$$N_{He} = \frac{D^2\rho\tau_o}{\mu_\infty^2} \tag{6-88}$$

6. Calculate f from Eq. (6-65) using the values of N_{Re} and N_{He} from steps 3 and 5.
7. Compare the value of f from step 6 with the assumed value in step 2. If they do not agree, insert the result of step 6 for f into step 3 in place of 0.01, and repeat Steps 3–7 until they agree.

The resulting value of D is determined in step 4.

D. Use of Tables

The relationship between flow rate, pressure drop, and pipe diameter for water flowing at 60°F in Schedule 40 horizontal pipe is tabulated in Appendix G over a range of pipe velocities that cover the most likely conditions. For this special case, no iteration or other calculation procedures are required for any of the unknown driving force, unknown flow rate, or unknown diameter problems (although interpolation in the table is usually necessary). Note that the friction loss is tabulated in this table as pressure drop (in psi) per 100 ft of pipe, which is equivalent to $100\rho e_f/144L$ in Bernoulli's equation, where ρ is in lb_m/ft^3, e_f is in ft lb_f/lb_m, and L is in ft.

VII. TUBE FLOW (POISEUILLE) VISCOMETER

In Section II.B of Chapter 3, the tube flow viscometer was described in which the viscosity of any fluid with unknown viscous properties could be determined from measurements of the total pressure gradient $(-\Delta\Phi/L)$ and the volumetric flow rate (Q) in a tube of known dimensions. The viscosity is given by

$$\eta = \frac{\tau_w}{\dot{\gamma}_w} \tag{6-89}$$

where τ_w follows directly from the pressure gradient and Eq. (6-4), and the wall shear rate is given by

$$\dot{\gamma}_w = \Gamma\left(\frac{3n' + 1}{4n'}\right) \tag{6-90}$$

where $\Gamma = 4Q/\pi R^3 = 8V/D$ and

$$n' = \frac{d \log \tau_w}{d \log \Gamma} = \frac{d \log(-\Delta\Phi)}{d \log Q} \tag{6-91}$$

is the point slope of $-\Delta\Phi$ versus Q at each measured value of Q. Equation (6-90) is completely independent of the specific fluid viscous properties and can be dreived from Eq. (6-7) as follows. By using Eq. (6-4), the independent variable in Eq. (6-7) can be changed from r to τ_{rx}, i.e.,

$$Q = -\pi \int_0^R r^2 \dot{\gamma}\, dr = \frac{\pi R^3}{\tau_w^3} \int_0^{\tau_w} \tau_{rx}^2 \dot{\gamma}\, d\tau_{rx} \tag{6-92}$$

This can be solved for the shear rate at the tube wall ($\dot{\gamma}_w$) by first differentiating Eq. (6-92) with respect to the *parameter* τ_w by application of Leibnitz' rule* to give

$$\frac{d(\Gamma \tau_w^3)}{d\tau_w} = 4\tau_w^2 \dot{\gamma}_w \tag{6-93}$$

where $\Gamma = 4Q/\pi R^3$. Thus,

$$\dot{\gamma}_w = \frac{1}{4\tau_w^2}\frac{d(\Gamma\tau_w^3)}{d\tau_w} = \frac{\tau_w}{4}\left(\frac{d\Gamma}{d\tau_w} + 3\frac{\Gamma}{\tau_w}\right) = \Gamma\left(\frac{3n' + 1}{4n'}\right) \tag{6-94}$$

where $n' = d(\log \tau_w)/d(\log \Gamma)$ is the local slope of the log-log plot of τ_w versus Γ, (or $-\Delta\Phi$ versus Q), at each measured value of Q.

VIII. TURBULENT DRAG REDUCTION

A very remarkable effect was observed by Toms during World War II when pumping Napalm (a "jellied" solution of a polymer in gasoline). He found that the polymer solution could be pumped through pipes in turbulent flow with considerably lower friction loss than exhibited by the gasoline at the same flow rate in the same pipe without the polymer. This phenomenon,

---

* Leibnitz' rule:

$$\frac{\partial}{\partial x}\int_{A(x)}^{B(x)} I(x, y)\, dy = \int_{A(x)}^{B(x)} \frac{\partial I}{\partial x}\, dy + I(x, B)\frac{\partial B}{\partial x} - I(x, A)\frac{\partial A}{\partial x}$$

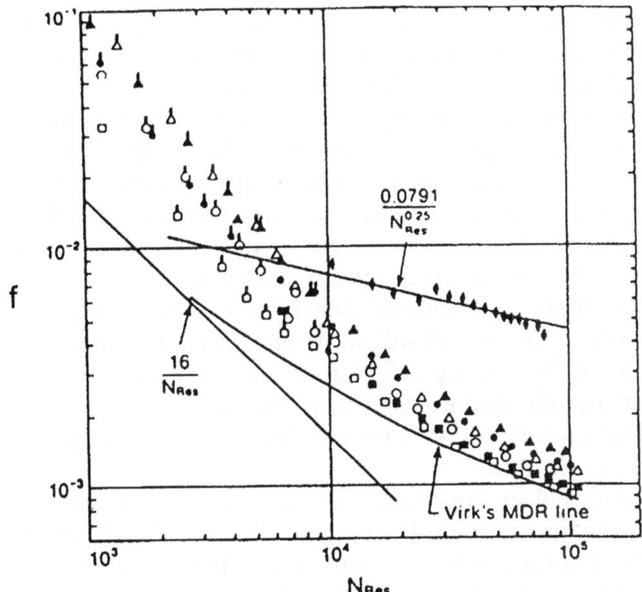

FIGURE 6-5 Drag reduction data for polyacrylamide solutions ($N_{Re,s}$ is the Reynolds number based on the solvent properties.) MDR is Virks' maximum drag reduction assymptote (Virk, 1985). (From Darby and Chang, 1984).

known as *turbulent drag reduction* (or the *Toms effect*), has been observed for solutions (mostly aqueous) of a variety of very high polymers (e.g., molecular weights on the order of 10^6) and has been the subject of a large amount of research. The effect is very significant, because as much as 85% *less* energy is required to pump solutions of some high polymers at concentrations of 100 ppm or less through pipes than is required to pump the solvent alone at the same flow rate through the same pipe. This is illustrated in Fig. 6-5, which shows some of Chang's data (Darby and Chang, 1984) for the Fanning friction factor versus (solvent) Reynolds number for fresh and "degraded" polyacrylamide solutions of concentrations from 100 to 500 ppm, in a 2 mm diameter tube. Note that the friction factor at low Reynolds numbers (laminar flow) is much larger than that for the (Newtonian) solvent, whereas it is much lower at high (turbulent) Reynolds numbers. The non-Newtonian viscosity of these solutions is shown in Fig. 3-7 in Chapter 3.

Although the exact mechanism is debatable, Darby and Chang (1984) and Darby and Pivsa-Art (1991) have presented a model for turbulent drag reduction based on the fact that solutions of very high polymers are viscoelastic and the concept that in any unsteady deformation (such as turbulent flow) elastic properties will store energy that would otherwise be dissipated in a purely viscous fluid. Since energy that is dissipated (i.e., the "friction loss") must be made up by adding energy (e.g., by a pump) to sustain the flow, that portion of the energy that is stored by elastic deformations remains in the flow and does not have to be made up by external energy sources. Thus, less energy must be supplied externally to sustain the flow, i.e., the drag is reduced. This concept is analogous to that of bouncing an elastic ball. If there is no viscosity (i.e., internal friction) to dissipate the energy, the ball will continue to bounce indefinitely with no external energy input needed. However, a viscous ball will not bounce at all, because all of the energy is dissipated by viscous deformation and is transformed to "heat." Thus, the greater the fluid elasticity in proportion to the viscosity, the less the energy that must be added to replace that which is dissipated by the turbulent motion of the flow.

The model for turbulent drag reduction developed by Darby and Chang (1984) and later modified by Darby and Pivsa-Art (1991) shows that for smooth tubes the friction factor versus Reynolds number relationship for Newtonian fluids (e.g., the Colebrook or Churchill equation) may also be used for drag-reducing flows, provided (1) the Reynolds number is defined with respect to the properties (e.g., viscosity) of the Newtonian solvent and (3) the Fanning friction factor is modified as follows:

$$f_p = \frac{f_s}{\sqrt{1 + N_{De}^2}} \tag{6-95}$$

Here, f_s is the solvent (Newtonian) Fanning friction factor, as predicted for a Newtonian fluid with the viscosity of the solvent using the (Newtonian) Reynolds number, f_p is a "generalized" Fanning friction factor that applies to (drag-reducing) polymer solutions as well as Newtonian fluids, and N_{De} is the dimensionless Deborah number, which depends upon the fluid viscoelastic properties and accounts for the storage of energy by the elastic deformations (for Newtonian fluids, $N_{De} = 0$ so that $f_p = f_s$). Figure 6-6 shows the data from Fig. 6-5 (and many other data sets, as well) replotted in terms of this generalized friction factor. The data are well represented by the classic Colebrook equation (for Newtonian fluids in smooth tubes) on this plot.

The complete expression for N_{De} is given by Darby and Pivsa-Art (1991) as a function of the viscoelastic fluid properties of the fluid (i.e., the Carreau parameters η_0, λ, and p). This expression is

$$N_{De} = \frac{0.0163 N_\zeta N_{Re,s}^{0.338} (\mu_s/\eta_o)^{0.5}}{[1/N_{Re,s}^{0.75} + 0.00476 N_\zeta^2 (\mu_s/\eta_o)^{0.75}]^{0.318}} \tag{6-96}$$

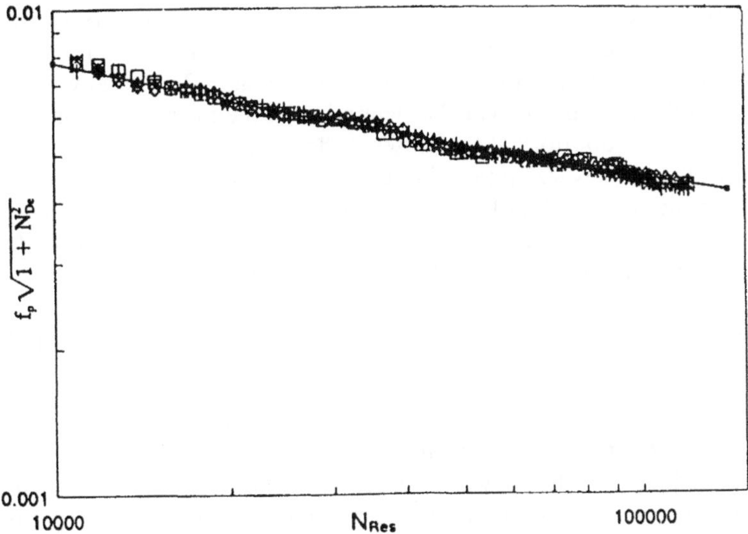

FIGURE 6-6 Drag reduction data replotted in terms of generalized friction factor. (From Darby and Pivsa-Art, 1991.)

where

$$N_{\zeta} = [(1 + N_{\lambda}^2)^p - 1]^{0.5} \tag{6-97}$$

and

$$N_{\lambda} = \frac{8V\lambda}{D} \tag{6-98}$$

$N_{\mathrm{Re,s}}$ is the Reynolds number based on the solvent properties, μ_s is the solvent viscosity, D is the pipe diameter, V is the velocity in the pipe, and λ is the fluid time constant (from the Carreau model fit of the viscosity curve).

Inasmuch as the rheological properties are very difficult to measure for very dilute solutions (100 ppm or less), a simplified expression was developed by Darby and Pivsa-Art (1991) in which these rheological parameters are contained within two "constants," k_1 and k_2:

$$N_{\mathrm{De}} = k_2\left(\frac{8\mu_s N_{\mathrm{Re,s}}}{\rho D^2}\right)^{k_1} N_{\mathrm{Re,s}}^{0.34} \tag{6-99}$$

where k_1 and k_2 depend only on the specific polymer solution and its concentration. Darby and Pivsa-Art (1991) examined a variety of drag-reducing data sets from the literature for various polymer solutions in various size pipes and determined the corresponding values of k_1 and k_2 that fit the model to the data. These values are given in Table 6-2. For any drag-reducing solution, k_1 and k_2 can be determined experimentally from two data points in the laboratory at two different flow rates (Reynolds numbers) in turbulent flow in any size pipe. The resulting values can be used with the model to predict friction loss for that solution at any Reynolds number in any size pipe. If the Colebrook equation for smooth tubes is used, the appropriate generalized expression for the friction factor is

$$f = \frac{0.41}{[\ln(N_{Re,s}/7)]^2} \frac{1}{(1 + N_{De}^2)^{1/2}} \tag{6-100}$$

Example 6-1: Friction Loss in Drag-Reducing Solutions. Determine the percentage reduction in the power required to pump water through a 3 in. ID smooth pipe at 300 gpm by adding 100 wppm of "degraded" Separan AP-30.

Solution. We first calculate the Reynolds number for the solvent (water) under the given flow conditions, using a viscosity of 0.01 poise and a density of 1 g/cm^3.

$$N_{Re,s} = \frac{4Q\rho}{\pi D\mu} = \frac{4(300\text{ gpm})[63.1\text{ cm}^3/(\text{s gpm})](1\text{ g/cm}^3)}{\pi(3\text{ in.})(2.54\text{ cm/in.})[0.01\text{ g/(cm s)}]} = 3.15 \times 10^5$$

Then calculate the Deborah number from Eq. (6-99), using $k_1 = 0.088$ and $k_2 = 0.0431$ from Table 6-2:

$$N_{De} = k_2 \left(\frac{8\mu_s N_{Re,s}}{\rho D^2} \right)^{k_1} N_{Re,s}^{0.34} = 5.45$$

These values can now be used to calculate the smooth pipe friction factor from Eq. (6-100). Excluding the N_{De} term gives the friction factor for the Newtonian solvent (f_s), and including the N_{De} term gives the friction factor for the polymer solution (f_p) under the same flow conditions:

$$f_s = \frac{0.41}{[\ln(N_{Re,s}/7)]^2} = 0.00357$$

$$f_p = \frac{0.41}{[\ln(N_{Re,s}/7)]^2} \left(\frac{1}{(1 + N_{De}^2)^{1/2}} \right) = 0.000645$$

The power (HP) required to pump the fluid is given by $-\Delta P\, Q$. Because $-\Delta P$ is proportional to fQ^2 and Q is the same with and without the

TABLE 6-2 Parameters for Eq. (6-93) for Various Polymer Solutions

Polymer	Conc. (mg/kg)	Dia. (cm)	k_1	k_2 (s^{k1})	Reference
Guar gum	20	1.27	0.05	0.009	Wang (1972)
(Jaguar A-20-D)	50		0.06	0.014	
	200		0.07	0.022	
	500		0.10	0.029	
	1000		0.16	0.028	
Guar gum	30		0.05	0.008	
	60		0.06	0.010	
	240		0.08	0.016	
	480		0.11	0.018	
Polyacrylamide	100	0.176–	0.093	0.0342	Darby and
Separan AP-30,	250	1.021	0.095	0.0293	Pivsa-Art (1991)
fresh	500		0.105	0.0244	
Separan AP-30,	100		0.088	0.0431	
degraded	250		0.095	0.0360	
	500		0.103	0.0280	
AP-273	10	1.090	0.12	0.0420	White and Gordon
PAM E198	10	0.945	0.21	0.0074	(1975)
	280			0.0078	Virk and Baher
PAA	300	2.0, 3.0	0.40	0.0050	(1970)
	700		0.53	0.0049	Hoffmann (1978)
ET-597	125	0.69,	0.47	0.00037	Astarita et al. (1969)
	250	1,1 2.05	0.39	0.0013	
	500		0.30	0.0061	
Hydroxyethyl	100	2.54	0.10	0.0074	Wang (1972)
cellulose	200		0.16	0.0072	
(OP-100M)	500		0.24	0.0068	
	1000		0.35	0.0063	
(HEC)	2860	4.8, 1.1, 2.05	0.02	0.0310	Savins (1969)
Polyethylene oxide					
WSR 301	10	5.08	0.22	0.017	Goren and Norbury
	20		0.21	0.016	(1967)
	50		0.19	0.014	
W205	10	0.945	0.31	0.0022	Virk and Baher
	105		0.26	0.0080	(1970)
Xanthan gum (Rhodopol 23)	1000	0.52	0.02	0.046	Bewersdorff (1988)

Source: Darby and Pivsa-Art (1991).

polymer, the fractional reduction in power is given by

$$DR = \frac{HP_s - HP_p}{HP_s} = \frac{f_s - f_p}{f_s} = 0.82$$

That is adding the polymer results in an 82% *reduction* in the power required to overcome drag!

PROBLEMS

Pipe Flows

1. Show how the Hagen–Poiseuille equation for the steady laminar flow of a Newtonian fluid in a uniform cylindrical tube can be derived starting from the general microscopic equations of motion (e.g., the continuity and momentum equations).
2. The Hagen–Poiseuille equation [Eq. (6-11)] describes the laminar flow of a Newtonian fluid in a tube. Since a Newtonian fluid is defined by the relation $\tau = \mu\dot{\gamma}$, rearrange the Hagen–Poiseuille equation to show that the shear rate at the tube wall for a Newtonian fluid is given by $\dot{\gamma}_w = 4Q/\pi R^3 = 8V/D$.
3. Derive the relation between the friction factor and Reynolds number in turbulent flow for smooth pipe [Eq. (6-34)], starting with the von Karman equation for the velocity distribution in the turbulent boundary layer [Eq. (6-26)].
4. Evaluate the kinetic energy correction factor α in Bernoulli's equation for turbulent flow assuming that the 1/7 power law velocity profile [Eq. (6-36)] is valid. Repeat this for laminar flow of a Newtonian fluid in a tube, for which the velocity profile is parabolic.
5. A Newtonian fluid with SG = 0.8 is forced through a capillary tube at a rate of $5\,cm^3/min$. The tube has a downward slope of 30° to the horizontal, and the pressure drop is measured between two taps located 40 cm apart on the tube using a mercury manometer, which reads 3 cm. When water is forced through the tube at a rate of $10\ cm^3/min$, the manometer reading is 2 cm.
   (a) What is the viscosity of the unknown Newtonian fluid?
   (b) What is the Reynolds number of the flow for each fluid?
   (c) If two separate pressure transducers, which read the total pressure directly in psig, were used to measure the pressure at each of the pressure taps directly instead of using the manometer, what would be the difference in the transducer readings?
6. A liquid is draining from a cylindrical vessel through a tube in the bottom of the vessel, as illustrated in Fig. 6-P6. If the liquid has a specific gravity of 0.85 and drains out at a rate of $1\,cm^3/s$, what is the viscosity of the liquid? The entrance loss coefficient from the tank to the tube is 0.4, and the system has the following dimensions:

   $D = 2$ in. $\quad d = 2$ mm

   $L = 10$ cm $\quad h = 5$ cm

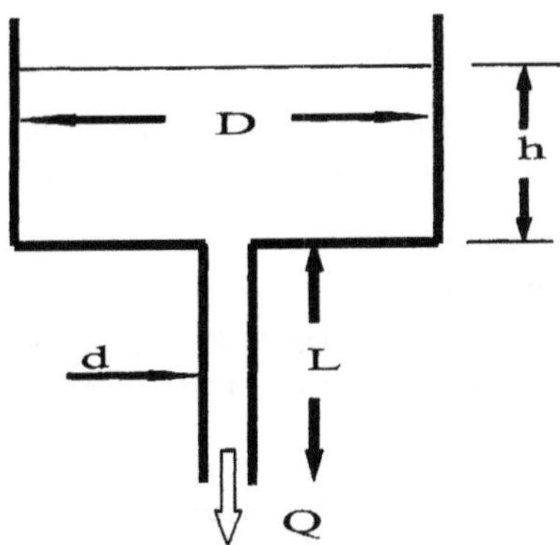

FIGURE 6-P6

7. You are given a liquid and are asked to find its viscosity. Its density is known to be 0.97 g/cm^3. You place the fluid in an open vessel to which a 20 cm long vertical tube with an inside diameter of 2 mm, is attached to the bottom. When the depth of the liquid in the container is 6 cm, you find that it drains out through the tube at a rate of 2.5 cm^3/s. If the diameter of the open vessel is much larger than that of the tube and friction loss from the vessel to the tube is negligible, what is the fluid viscosity?
8. Repeat Problem 7 accounting for the friction loss from the vessel to the tube, assuming a loss coefficient of 0.50 for the contraction.
9. You must measure the viscosity of an oil that has an SG of 0.92. To do this, you put the oil into a large container to the bottom of which a small vertical tube, 25 cm long, has been attached, through which the oil can drain by gravity. When the level of the oil in the container is 6 in. above the container bottom, you find that the flow rate through the tube is 50 cm^3/min. You run the same experiment with water instead of oil and find that under the same conditions the water drains out at a rate of 156 cm^3/min. If the loss coefficient for the energy dissipated in the contraction from the container to the tube is 0.5, what is the viscosity of the oil?
10. You want to transfer No. 3 fuel oil (30°API)from a storage tank to a power plant at a rate of 2000 bbl/day. The diameter of the pipeline is $1\frac{1}{2}$ in. sch 40, with a length of 1200 ft. The discharge of the line is 20 ft higher than the suction end, and both ends are at 1 atm pressure. The inlet temperature of the oil is 60°F, and the transfer pump is 60% efficient. If the specific heat of the oil is 0.5 Btu/(lb_m °F) and the pipeline is perfectly insulated, determine:
    (a) The horsepower of the motor required to drive the pump
    (b) The temperature of the oil leaving the pipeline
11. You must specify a pump to deliver 800 bbl/day of a 35°API distillate at 90°F from a distillation column to a storage tank in a refinery. If the level in the tank is 20 ft above that in the column, the total equivalent length of pipe is 900 ft, and both the column and tank are at atmospheric pressure, what horsepower would

be needed if you use $1\frac{1}{2}$ in. sch 40 pipe? What power would be needed if you use 1 in. sch 40 pipe?

12. Water is flowing at a rate of 700 gpm through a horizontal 6 in. sch 80 commercial steel pipe at 90°F. If the pressure drops by 2.23 psi over a 100 ft length of pipe:
    (a) What is the value of the Reynolds number?
    (b) What is the magnitude of the pipe wall roughness?
    (c) How much driving force (i.e., pressure difference) would be required to move the water at this flow rate through 10 mi of pipe if it were made of commercial steel?
    (d) What size commercial steel pipe would be required to transport the water at the same flow rate over the same distance if the driving force is the static head in a water tower 175 ft above the pipe?
13. A 35°API distillate at 60°F is to be pumped a distance of 2000 ft through a 4 in. sch 40 horizontal pipeline at a flow rate of 500 gpm. What power must the pump deliver to the fluid if the pipeline is made of: (a) drawn tubing; (b) commercial steel; (c) galvanized iron; (d) PVC plastic?
14. The Moody diagram illustrates the effect of roughness on the friction factor in turbulent flow but indicates no effect of roughness in laminar flow. Explain why this is so. Are there any restrictions or limitations that should be placed on this conclusion? Explain.
15. You have a large supply of very rusty 2 in. sch 40 steel pipe, which you want to use for a pipeline. Because rusty metal is rougher than clean metal, you want to know its effective roughness before laying the pipeline. To do this, you pump water at a rate of 100 gpm through a 100 ft long section of the pipe and find that the pressure drops by 15 psi over this length. What is the effective pipe roughness, in inches?
16. A 32 hp pump (100% efficient) is required to pump water through a 2 in. sch 40 pipeline, 6000 ft long, at a rate of 100 gpm.
    (a) What is the equivalent roughness of the pipe?
    (b) If the pipeline is replaced by new commercial steel 2 in. sch 40 pipe, what power would be required to pump water at a rate of 100 gpm through this pipe? What would be the percentage saving in power compared to the old pipe?
17. You have a piping system in your plant that has gotten old and rusty. The pipe is 2 in. sch 40 steel, 6000 ft long. You find that it takes 35 hp to pump water through the system at a rate of 100 gpm.
    (a) What is the equivalent roughness of the pipe?
    (b) If you replace the pipe with the same size new commercial steel pipe, what percentage savings in the required power would you expect at a flow rate of 100 gpm?
18. Water enters a horizontal tube through a flexible vertical rubber hose that can support no forces. If the tube is 1/8 in. sch. 40, 10 ft long, and the water flow rate is 2 gpm, what force (magnitude and direction) must be applied to the tube to keep it stationary? Neglect the weight of the tube and the water in it. The hose ID is the same as that of the tube.

19. A water tower that is 90 ft high provides water to a residential subdivision. The water main from the tower to the subdivision is 6 in. sch 40 steel, 3 miles long. If each house uses a maximum of 50 gal/hr (at peak demand) and the pressure in the water main is not to be less than 30 psig at any point, how many homes can be served by the water main?
20. A heavy oil ($\mu = 100\,\text{cP}$, SG = 0.85) is draining from a large tank through a 1/8 in. sch 40 tube into an open bucket. The level in the tank is 3 ft above the tube inlet, and the pressure in the tank is 10 psig. The tube is 30 ft long, and it is inclined downward at an angle of 45° to the horizontal. What is the flow rate of the oil, in gpm? What is the value of the Reynolds number in this problem?
21. SAE 10 lube oil (SG = 0.93) is being pumped upward through a straight 1/4 in. sch 80 pipe that is oriented at a 45° angle to the horizontal. The two legs of a manometer using water as the manometer fluid are attached to taps in the pipe wall that are 2 ft apart. If the manometer reads 15 in., what is the oil flow rate, in gal/hr?
22. Cooling water is fed by gravity from an open storage tank 20 ft above ground, through 100 ft of $1\frac{1}{2}$ in. ID steel pipe, to a heat exchanger at ground level. If the pressure entering the heat exchanger must be 5 psig for it to operate properly, what is the water flow rate through the pipe?
23. A water main is to be laid to supply water to a subdivision located 2 miles from a water tower. The water in the tower is 150 ft above ground, and the subdivision consumes a maximum of 10,000 gpm of water. What size pipe should be used for the water main? Assume Schedule 40 commercial steel pipe. The pressure above the water is 1 atm in the tank and is 30 psig at the subdivision.
24. A water main is to be laid from a water tower to a subdivision that is 2 mi away. The water level in the tower is 150 ft above the ground. The main must supply a maximum of 1000 gpm with a minimum of 5 psig at the discharge end, at a temperature of 65°F. What size commercial steel sch 40 pipe should be used for the water main? If plastic pipe (which is hydraulically smooth) were used instead, would this alter the result? If so, what diameter of plastic pipe should be used?
25. The water level in a water tower is 110 ft above ground level. The tower supplies water to a subdivision 3 mi away, through an 8 in. sch 40 steel water main. If the minimum water pressure entering the residential water lines at the houses must be 15 psig, what is the capacity of the water main (in gpm)? If there are 100 houses in the subdivision and each consumes water at a peak rate of 20 gpm, how big should the water main be?
26. A hydraulic press is powered by a remote high pressure pump. The gage pressure at the pump is 20 MPa, and the pressure required to operate the press is 19 MPa (gage) at a flow rate of 0.032 m^3/min. The press and pump are to be connected by 50 m of drawn stainless steel tubing. The fluid properties are those of SAE 10 lube oil at 40°C. What is the minimum tubing diameter that can be used?
27. Water is to be pumped at a rate of 100 gpm from a well that is 100 ft deep, through 2 mi of horizontal 4 in. sch 40 steel pipe, to a water tower that is 150 ft high.

(a) Neglecting fitting losses, what horsepower will the pump require if it is 60% efficient?
(b) If the elbow in the pipe at ground level below the tower breaks off, how fast will the water drain out of the tower?
(c) How fast would it drain out if the elbow at the top of the well gave way instead?
(d) What size pipe would you have to run from the water tower to the ground in order to drain it at a rate of 10 gpm?

28. A concrete pipe storm sewer, 4 ft in diameter, drops 3 ft in elevation per mile of length. What is the maximum capacity of the sewer (in gpm) when it is flowing full?
29. You want to siphon water from an open tank using a 1/4 in. diameter hose. The discharge end of the hose is 10 ft below the water level in the tank, and the siphon will not operate if the pressure falls below 1 psia anywhere in the hose. If you want to siphon the water at a rate of 1 gpm, what is the maximum height above the water level in the tank that the hose can extend and still operate?

Non-Newtonian Pipe Flows

30. Equation (6-43) describes the laminar flow of a power law fluid in a tube. Since a power law fluid is defined by the relation $\tau = m\dot{\gamma}^n$, rearrange Eq. (6-43) to show that the shear rate at the tube wall for a power law fluid is given by $\dot{\gamma}_w = (8V/D)(3n + 1)/4n$ where $8V/D$ is the wall shear rate for a Newtronian fluid.
31. A large tank contains SAE 10 lube oil at a temperature of 60°F and a pressure of 2 psig. The oil is 2 ft deep in the tank and drains out through a vertical tube in the bottom. The tube is 10 ft long and discharges the oil at atmospheric pressure. Assuming the oil to be Newtonian and neglecting the friction loss from the tank to the tube, how fast will it drain through the tube? If the oil is not Newtonian, but instead can be described as a power law fluid with a flow index of 0.4 and an apparent viscosity of 80 cP at a shear rate of $1\ s^{-1}$, how would this affect your answer? The tube diameter is 1/2 in.
32. A polymer solution is to be pumped at a rate of 3 gpm through a 1 in. diameter pipe. The solution behaves as a power law fluid with a flow index of 0.5, an apparent viscosity of 400 cP at a shear rate of $1\ s^{-1}$, and a density of 60 lb_m/ft^3.
(a) What is the pressure gradient in psi/ft?
(b) What is the shear rate at the pipe wall and the apparent viscosity of the fluid at this shear rate?
(c) If the fluid were Newtonian, with a viscosity equal to the apparent viscosity from (b) above, what would the pressure gradient be?
(d) Calculate the Reynolds numbers for the polymer solution and for the above Newtonian fluid.
33. A coal slurry that is characterized as a power law fluid has a flow index of 0.4 and an apparent viscosity of 200 cP at a shear rate of $1\ s^{-1}$. If the coal has a specific gravity of 2.5 and the slurry is 50% coal by weight in water, what pump horsepower will be required to transport 25 million tons of coal per year through a 36 in. ID, 1000 mi long pipeline? Assume that the entrance and exit

of the pipeline are at the same pressure and elevation and that the pumps are 60% efficient.

34. A coal slurry is found to behave as a power law fluid, with a flow index of 0.3, a specific gravity of 1.5. and an apparent viscosity of 70 cP at a shear rate of $100\,s^{-1}$. What volumetric flow rate of this fluid would be required to reach turbulent flow in a 1/2 in. ID smooth pipe that is 15 ft long? What is the pressure drop in the pipe (in psi) under these conditions?
35. A coal slurry is to be transported by pipeline. It has been determined that the slurry may be described by the power law model, with a flow index of 0.4, an apparent viscosity of 50 cP at a shear rate of $100\ s^{-1}$, and a density of 90 lb_m/ft^3. What horsepower would be required to pump the slurry at a rate of 900 gpm through an 8 in. sch 40 pipe that is 50 mi long?
36. A sewage sludge is to be transported a distance of 3 mi through a 12 in. ID pipeline at a rate of 2000 gpm. The sludge is a Bingham plastic with a yield stress of 35 dyn/cm^2, a limiting viscosity of 80 cP, and a specific gravity of 1.2. What size motor (in horsepower) would be required to drive the pump if it is 50% efficient?
37. A coal suspension is found to behave as a power law fluid, with a flow index of 0.4, a specific gravity of 1.5, and an apparent viscosity of 90 cP at a shear rate of $100\,s^{-1}$. What would the volumetric flow rate of this suspension be in a 15 ft long, 5/8 in. ID smooth tube, with a driving force of 60 psi across the tube? What is the Reynolds number for the flow under these conditions?
38. A coal–water slurry containing 65% (by weight) coal is pumped from a storage tank at a rate of 15 gpm through a 50 m long 1/2 in. sch 40 pipeline to a boiler, where it is burned. The storage tank is at 1 atm pressure and 80°F, and the slurry must be fed to the burner at 20 psig. The specific gravity of coal is 2.5, and it has a heat capacity of 0.5 Btu/(lb_m°F).
    (a) What power must the pump deliver to the slurry if it is assumed to be Newtonian with a viscosity of 200 cP?
    (b) In reality the slurry is non-Newtonian and can best be described as a Bingham plastic with a yield stress of 800 dyn/cm^2 and a limiting viscosity of 200 cP. Accounting for these properties, what would the required pumping power be?
    (c) If the pipeline is well insulated, what will the temperature of the slurry be when it enters the boiler, for both case (a) and case (b)?
39. A sludge is to be transported by pipeline. It has been determined that the sludge may be described by the power law model, with a flow index of 0.6, an apparent viscosity of 50 cP at a shear rate of $1\,s^{-1}$, and a density of 95 lb_m/ft^3. What hydraulic horsepower would be required to pump the slurry at a rate of 600 gpm through a 6 in. ID pipe that is 5 mi long?
40. You must design a transfer system to feed a coal slurry to a boiler. However, you don't know the slurry properties, so you measure them in the lab using a cup-and-bob (Couette) viscometer. The cup has a diameter of 10 cm and a bob diameter of 9.8 cm, and the length of the bob is 8 cm. When the cup is rotated at a rate of 2 rpm, the torque measured on the bob is 2.4×10^4 dyn cm, and at 20 rpm it is 6.5×10^4 dyn cm.

(a) If you use the Bingham plastic model to describe the slurry properties, what are the values of the yield stress and the limiting viscosity?
(b) If the power law model were used instead, what would be the values of the flow index and consistency?
(c) Using the Bingham plastic model for the slurry, with a value of the yield stress of 35 dyn/cm^2, a limiting viscosity of 35 cP, and a density of 1.2 g/cm^3, what horsepower would be required to pump the slurry through a 1000 ft long, 3 in. ID sch 40 pipe at a rate of 100 gpm?

41. A thick slurry with SG = 1.3 is to be pumped through a 1 in. ID pipe that is 200 ft long. You don't know the properties of the slurry, so you test it in the lab by pumping it through a 4 mm ID tube that is 1 m long. At a flow rate of 0.5 cm^3/s, the pressure drop in this tube is 1 psi, and at a flow rate of 5 cm^3/s it is 1.5 psi. Estimate the pressure drop that would be required to pump the slurry through the 1 in. pipe at a rate of 2 gpm and also at 30 gpm. Clearly explain the procedure you use, and state any assumptions that you make. Comment in detail about the possible accuracy of your predictions. Slurry SG = 1.3.
42. Drilling mud has to be pumped down into an oil well that is 8000 ft deep. The mud is to be pumped at a rate of 50 gpm to the bottom of the well and back to the surface, through a pipe having an effective ID of 4 in. The pressure at the bottom of the well is 4500 psi. What pump head is required to do this? The drilling mud has properties of a Bingham plastic, with a yield stress of 100 dyn/cm^2, a limiting (plastic) viscosity of 35 cP, and a density of 1.2 g/cm^3.
43. A straight vertical tube, 100 cm long and 2 mm ID, is attached to the bottom of a large vessel. The vessel is open to the atmosphere and contains a liquid with a density of 1 g/cm^3 to a depth of 20 cm above the bottom of the vessel.
    (a) If the liquid drains through the tube at a rate of 3 cm^3/s, what is it's viscosity?
    (b) What is the largest tube diameter that can be used in this system to measure the viscosity of liquids that are at least as viscous as water, for the same liquid level in the vessel? Assume that the density is the same as water.
    (c) A non-Newtonian fluid, represented by the power law model, is introduced into the vessel with the 2 mm diameter tube attached. If the fluid has a flow index of 0.65, an apparent viscosity of 5 cP at a shear rate of 10 s^{-1}, and a density of 1.2 g/cm^3, how fast will it drain through the tube, if the level is 20 cm above the bottom of the vessel?
44. A non-Newtonian fluid, described by the power law model, is flowing through a thin slit between two parallel planes of width W, separated by a distance H. The slit is inclined upward at an angle θ to the horizontal.
    (a) Derive an equation relating the volumetric flow rate of this fluid to the pressure gradient, slit dimensions, and fluid properties.
    (b) For a Newtonian fluid, this solution can be written in dimensionless form as

    $$f = 24/N_{\text{Re,h}}$$

    where the Reynolds number, $N_{\text{Re,h}}$, is based upon the hydraulic diameter of the channel. Arrange your solution for the power law fluid in dimensionless form, and solve for the friction factor, f.

(c) Set your result from (b) equal to $24/N_{Re,h}$ and determine an equivalent expression for the power law Reynolds number for slit flow.

45. You are drinking a milk shake through a straw that is 8 in. long and 0.3 in. in diameter. The milk shake has the properties of a Bingham plastic, with a yield stress of 300 dyn/cm^2, a limiting viscosity of 150 cP, and a density of 0.8 g/cm^3.
    (a) If the straw is inserted 5 in. below the surface of the milk shake, how hard must you suck to get the shake flowing through the entire straw (e.g., how much vacuum must you pull, in psi)?
    (b) If you pull a vacuum of 1 psi, how fast will the shake flow (in cm^3/s)?
46. Water is to be transferred at a rate of 500 gpm from a cooling lake through a 6 in. diameter sch 40 pipeline to an open tank in a plant that is 30 mi from the lake.
    (a) If the transfer pump is 70% efficient, what horsepower motor is required to drive the pump?
    (b) An injection station is installed at the lake that injects a high polymer into the pipeline, to give a solution of 50 ppm concentration with the following properties: a low shear limiting viscosity of 80 cP, a flow index of 0.5, and a transition point from low shear Newtonian to shear thinning behavior at a shear rate of 10 s^{-1}. What horsepower is now required to drive the same pump, to achieve the same flow rate?
47. You measure the viscosity of a sludge in the lab and conclude that it can be described as a power law fluid with a flow index of 0.45, a viscosity of 7 poise at a shear rate of 1 s^{-1}, and a density of 1.2 g/cm^3.
    (a) What horsepower would be required to pump the sludge through a 3 in. sch 40 pipeline, 1000 ft long, at a rate of 100 gpm?
    (b) The viscosity data show that the sludge could also be described by the Bingham plastic model, with a viscosity of 7 poise at a shear rate of 1 s^{-1} and a viscosity of 0.354 poise at a shear rate of 100 s^{-1}. Using this model, what required horsepower would you predict for the above pipeline?
    (c) Which answer do you think would be the most reliable, (a) or (b), and why?
48. An open drum, 3 ft in diameter, contains a mud that is known to be described by the Bingham plastic model, with a yield stress of 120 dyn/cm^2, a limiting viscosity of 85 cP, and a density of 98 lb$_m$/ft^3. A 1 in. ID hose, 10 ft long, is attached to a hole in the bottom of the drum to drain the mud out. How far below the surface of the mud should the end of the hose be lowered in order to drain the mud at a rate of 5 gpm?
49. You would like to determine the pressure drop–flow rate relation for a slurry in a pipeline. To do this, you must determine the rheological properties of the slurry, so you test it in the lab by pumping it through a 1/8 in. ID pipe that is 10 ft long. You find that it takes 5 psi pressure drop in the pipe to produce a flow rate of 100 cm^3/s and that 10 psi results in a flow rate of 300 cm^3/s.
    (a) What can you deduce about the rheological characteristics of the slurry from these data?
    (b) If it is assumed that the slurry can be adequately described by the power law model, what are the values of the fluid properties, as deduced from the data?

(c) If the Bingham plastic model is used instead of the power law model to describe the slurry, what are its properties?

50. A pipeline is installed to transport a red mud slurry from an open tank in an alumina plant to a disposal pond. The line is 5 in. sch 80 commercial steel, 12,000 ft long, and is designed to transport the slurry at a rate of 300 gpm. The slurry properties can be described by the Bingham plastic model, with a yield stress of 15 dyn/cm^2, a limiting viscosity of 20 cP, and an SG of 1.3. You may neglect any fittings in this pipeline.
   (a) What delivered pump head and hydraulic horsepower would be required to pump this mud?
   (b) What would be the required pump head and horsepower to pump water at the same rate through the same pipeline?
   (c) If 100 ppm of fresh Separan AP-30 polyacrylamide polymer were added to the water in case (b), what would the required pump head and horsepower be?
51. Determine the power required to pump water at a rate of 300 gpm through a 3 in. ID pipeline, 50 mi long, if:
   (a) The pipe is new commercial steel
   (b) The pipe wall is hydraulically smooth
   (c) The pipe wall is smooth, and "degraded" Separan AP-30 polyacrylamide is added to the water at a concentration of 100 wppm.

NOTATION

D	diameter, [L]
DF	driving force, Eq. (6-67), $[FL/M = L^2/t^2]$
e_f	energy dissipated per unit mass of fluid, $[FL/M = L^2/t^2]$
f	Fanning friction factor, [—]
F_x	force component in the x direction, $[F = ML/t^2]$
K_f	loss coefficient, [—]
L	length, [L]
m	power law consistency parameter, $[M/Lt^{2-n}]$
n	power law flow index, [—]
N_{De}	Deborah number, [—]
N_{He}	Hedstrom number, Eq. (6-62), [—]
N_{Re}	Reynolds number, [—]
$N_{Re,s}$	solvent Reynolds number, [—]
$N_{Re,pl}$	power law Reynolds number, [—]
P	pressure, $[F/L^2 = M/Lt^2]$
Q	volumetric flow rate, $[L^3/t]$
r	radial direction, [L]
R	tube radius, [L]
t	time, [t]
V	spatial average velocity, [L/t]
v_*	friction velocity, Eq. (6-28), [L/t]

v_x	local velocity in the x direction, [L/t]
v_x'	turbulent eddy velocity component in the x direction, [L/t]
v^+	dimensionless velocity, Eq. (6-27), [—]
w	external shaft work (e.g., – pump work) per unit mass of fluid $[FL/M = L^2/t^2]$
y^+	dimensionless distance from wall, Eq. (6-27), [—]
δ_L	laminar boundary layer thickness, [L]
δ_T	turbulent boundary layer thickness, [L]
ε	roughness, [L]
Φ	potential $(= P + \rho gz)$, $[F/L^2 = M/Lt^2]$
μ	viscosity (constant), [M/Lt]
ρ	density, $[M/L^3]$
τ_o	yield stress, $[F/L^2 = M/Lt^2]$
τ_{rx}	shear stress component, force in x direction on r surface, $[F/L^2 = M/Lt^2]$
τ_{rx}'	turbulent (Reynolds) stress component, $[F/L^2 = M/Lt^2]$
τ_w	stress exerted by the fluid on the wall, $[F/L^2 = M/Lt^2]$

Subscripts

x, y, z, θ	coordinate directions
w	wall location

REFERENCES

Astarita G, G Greco Jr, L Nicodemo. AIChE J. 15:564, 1969.

Bewersdorff HW, NS Berman. Rheol Acta 27:130, 1988.

Churchill SW, Chem Eng Nov 7, 1977, p 91.

Darby R, HD Chang. AIChE J 30:274, 1984.

Darby R, J Melson. Chem Eng Dec 28, 1981, p 59.

Darby R, S Pivsa-Art. Can J Chem Eng 69:1395, 1991.

Darby R, R Mun, DV Boger. Chem Eng September 1992,

Dodge DW, AB Metzner. AIChE J 5:189, 1959.

Goren Y, JF Norbury. ASME J Basic Eng 89:816, 1967.

Hoffmann L, P Schummer. Rheol Acta 17:98, 1978.

Savins JG. In CS Wells, ed. Viscous Drag Reduction. New York: Plenum Press, 1969, p 183.

Virk PS. AIChE J 21:625, 1975.

Virk PS, H Baher. Chem Eng Sci 25:1183, 1970.

Wang CB. Ind Eng Chem Fundam 11:566, 1972.

White A. J Mech Eng Sci, 8:452, 1966.

White D Jr, RJ Gordon. AIChE J 21:1027, 1975.

7

Internal Flow Applications

I. NONCIRCULAR CONDUITS

All the relationships presented in Chapter 6 apply directly to circular pipe. However, many of these results can also, with appropriate modification, be applied to conduits with noncircular cross sections. It should be recalled that the derivation of the momentum equation for uniform flow in a tube [e.g., Eq. (5-44)] involved no assumption about the shape of the tube cross section. The result is that the friction loss is a function of a geometric parameter called the "hydraulic diameter":

$$D_h = 4\frac{A}{W_\mathrm{p}} \tag{7-1}$$

where A is the area of the flow cross section and W_p is the wetted perimeter (i.e., the length of contact between the fluid and the solid boundary in the flow cross section). For a full circular pipe, $D_\mathrm{h} = D$ (the pipe diameter). The hydraulic diameter is the key characteristic geometric parameter for a conduit with any cross-sectional shape.

A. Laminar Flows

By either integrating the microscopic momentum equations (see Example 5-9) or applying a momentum balance to a "slug" of fluid in the center

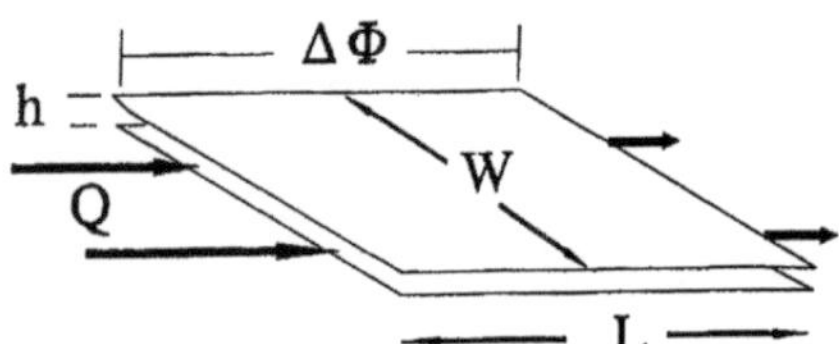

FIGURE 7-1 Flow in a slit.

of the conduit as was done for tube flow, a relationship can be determined between flow rate and driving force for laminar flow in a conduit with a noncircular cross section. This can also be done by application of the equivalent integral expressions analogous to Eqs. (6-6) to (6-10). The results for a few examples for Newtonian fluids will be given below. These results are the equivalent of the Hagen–Poiseuille equation for a circular tube and are given in both dimensional and dimensionless form.

1. Flow in a Slit

Flow between two flat parallel plates that are closely spaced ($h \ll W$) is shown in Fig. 7-1. The hydraulic diameter for this geometry is $D_h = 4A/W_p = 2h$, and the solution for a Newtonian fluid in laminar flow is

$$Q = -\frac{\Delta\Phi\, Wh^3}{12\mu L} \tag{7-2}$$

This can be rearranged into the equivalent dimensionless form

$$fN_{Re,h} = 24 \tag{7-3}$$

where

$$N_{Re,h} = \frac{D_h V\rho}{\mu} = \frac{D_h Q\rho}{\mu A} \tag{7-4}$$

Here, $A = Wh$, and the Fanning friction factor is, by definition,

$$f = \frac{e_f}{\left(\frac{V^2}{2}\right)\left(\frac{4L}{D_h}\right)} = \frac{-\Delta\Phi}{\left(\frac{\rho V^2}{2}\right)\left(\frac{4L}{D_h}\right)} \tag{7-5}$$

because the Bernoulli equation reduces to $e_f = -\Delta\Phi/\rho$ for this system.

2. Flow in a Film

The flow of a thin film down an inclined plane is illustrated in Fig. 7-2. The film thickness is h $\ll W$, and the plate is inclined at an angle θ to the vertical.

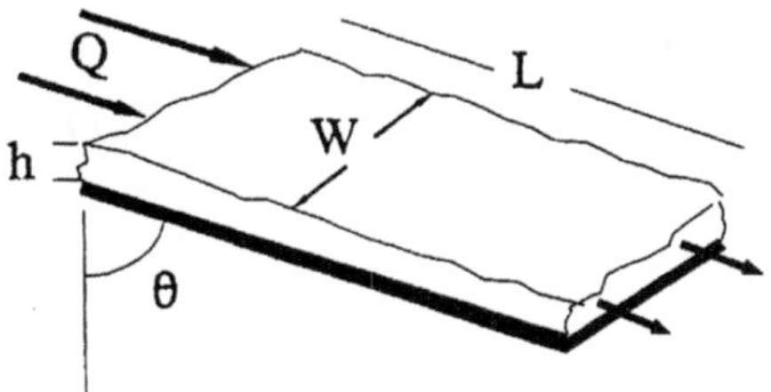

FIGURE 7-2 Flow in a film.

For this flow the hydraulic diameter is $D_h = 4h$ (since only one boundary in the cross section is a wetted surface). The laminar flow solution for a Newtonian fluid is

$$Q = -\frac{\Delta\Phi\, h^3 W}{3\mu L} = \frac{\rho g h^3 W \cos\theta}{3\mu} \tag{7-6}$$

The dimensionless form of this equation is

$$fN_{\mathrm{Re,h}} = 24 \tag{7-7}$$

where the Reynolds number and friction factor are given by Eqs. (7-4) and (7-5), respectively.

3. Annular Flow

Axial flow in the annulus between two concentric cylinders, as illustrated in Fig. 7-3, is frequently encountered in heat exchangers. For this geometry the hydraulic diameter is $D_h = (D_o - D_i$, and the Newtonian laminar flow solution is

$$Q = -\frac{\Delta\Phi\pi(D_o^2 - D_i^2)}{128\mu L}\left(D_o^2 + D_i^2 - \frac{D_o^2 - D_i^2}{\ln(D_o/D_i)}\right) \tag{7-8}$$

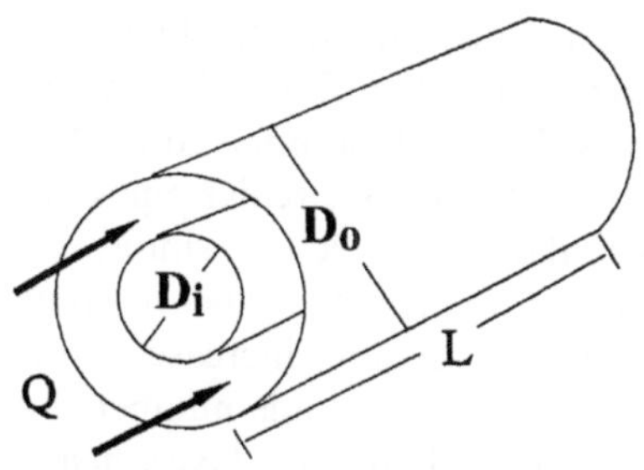

FIGURE 7-3 Flow in an annulus.

The dimensionless form of this expression is

$$fN_{\text{Re,h}} = 16\alpha \tag{7-9}$$

where

$$\alpha = \frac{(D_o - D_i)^2}{D_o^2 + D_i^2 - \dfrac{D_o^2 - D_i^2}{\ln(D_o/D_i)}} \tag{7-10}$$

It can be shown that as $D_i/D_o \to 0$, $\alpha \to 1$ and the flow approaches that for a circular tube. Likewise, as $D_i/D_o \to 1$, $\alpha \to 1.5$ and the flow approaches that for a slit.

The value of $fN_{\text{Re,h}}$ for laminar flow varies only by about a factor of 50% or so for a wide variety of geometries. This value has been determined for a Newtonian fluid in various geometries, and the results are summarized in Table 7-1. This table gives the expressions for the cross-sectional area and hydraulic diameter for six different conduit geometries, and the corresponding values of $fN_{\text{Re,h}}$, the dimensionless laminar flow solution. The total range of values for $fN_{\text{Re,h}}$ for all of these geometries is seen to be approximately 12–24. Thus, for any completely arbitrary geometry, the dimensionless expression $fN_{\text{Re,h}} \approx 18$ would provide an approximate solution for fully developed flow, with an error of about 30% or less.

B. Turbulent Flows

The effect of geometry on the flow field is much less pronounced for turbulent flows than for laminar flows. This is because the majority of the energy dissipation (e.g., flow resistance) occurs within the boundary layer, which, in typical turbulent flows, occupies a relatively small fraction of the total flow field near the boundary. In contrast, in laminar flow the "boundary layer" occupies the entire flow field. Thus, although the total solid surface contacted by the fluid in turbulent flow influences the flow resistance, the actual shape of the surface is not as important. Consequently, the hydraulic diameter provides an even better characterization of the effect of geometry for noncircular conduits with turbulent flows than with laminar flows. The result is that relationships developed for turbulent flows in circular pipes can be applied directly to conduits of noncircular cross section simply by replacing the tube diameter by the hydraulic diameter in the relevant dimensionless groups. The accuracy of this procedure increases with increasing Reynolds number, because the higher the Reynolds number the greater the turbulence intensity and the thinner the boundary layer; hence the less important the actual shape of the cross section.

TABLE 7-1 Laminar Flow Factors for Noncircular Conduits

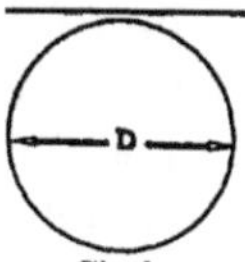

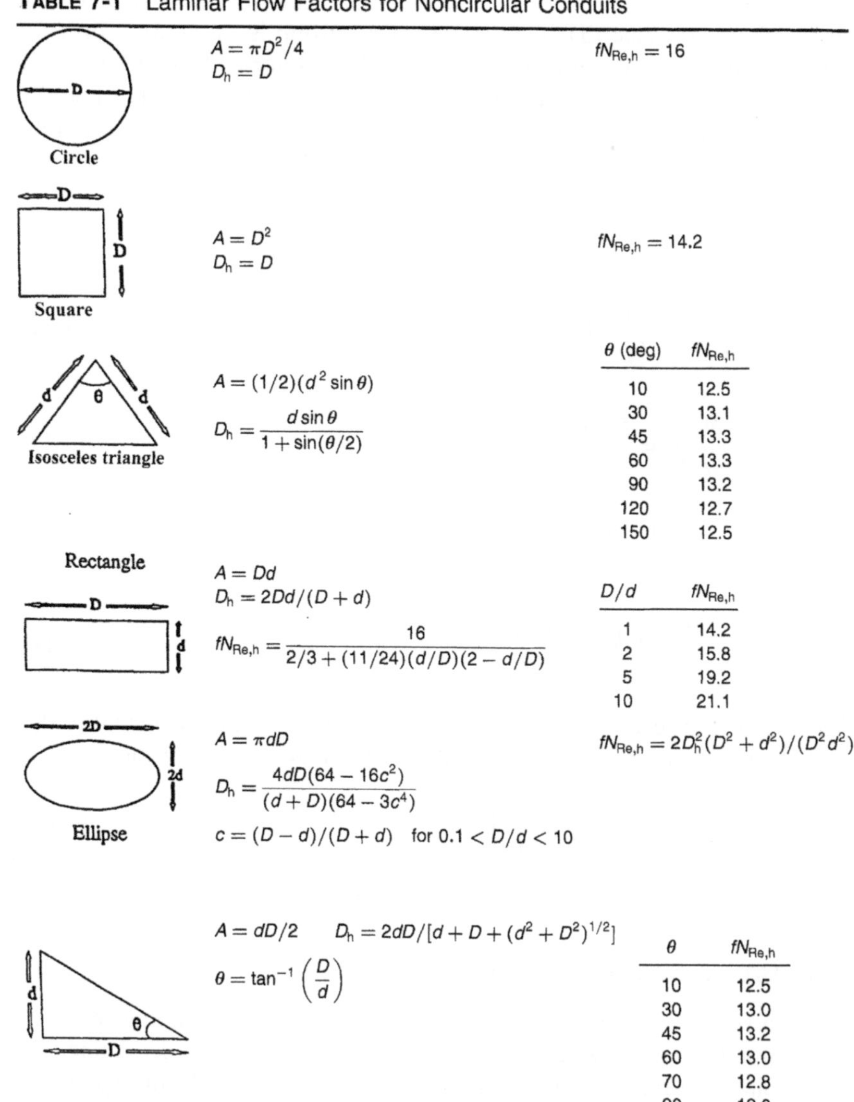

Circle

$A = \pi D^2/4$

$D_h = D$

$fN_{Re,h} = 16$

Square

$A = D^2$

$D_h = D$

$fN_{Re,h} = 14.2$

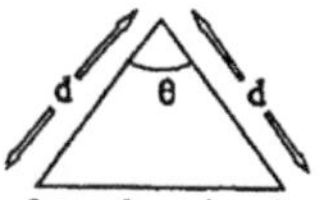

Isosceles triangle

$A = (1/2)(d^2 \sin\theta)$

$$D_h = \frac{d \sin\theta}{1 + \sin(\theta/2)}$$

θ (deg)	$fN_{Re,h}$
10	12.5
30	13.1
45	13.3
60	13.3
90	13.2
120	12.7
150	12.5

Rectangle

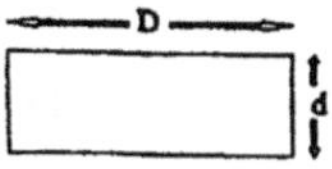

$A = Dd$

$D_h = 2Dd/(D + d)$

$$fN_{Re,h} = \frac{16}{2/3 + (11/24)(d/D)(2 - d/D)}$$

D/d	$fN_{Re,h}$
1	14.2
2	15.8
5	19.2
10	21.1

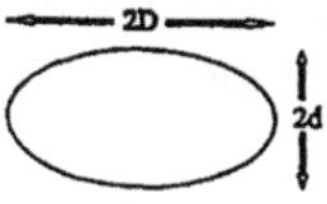

Ellipse

$A = \pi dD$

$$D_h = \frac{4dD(64 - 16c^2)}{(d + D)(64 - 3c^4)}$$

$c = (D - d)/(D + d)$ for $0.1 < D/d < 10$

$fN_{Re,h} = 2D_h^2(D^2 + d^2)/(D^2 d^2)$

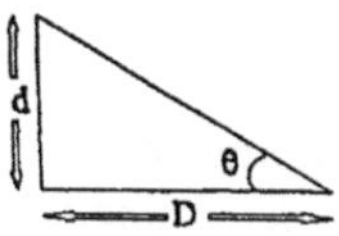

$A = dD/2$ $\quad D_h = 2dD/[d + D + (d^2 + D^2)^{1/2}]$

$$\theta = \tan^{-1}\left(\frac{D}{d}\right)$$

θ	$fN_{Re,h}$
10	12.5
30	13.0
45	13.2
60	13.0
70	12.8
90	12.0

It is important to use the hydraulic diameter substitution ($D = D_h$) in the appropriate (original) form of the dimensionless groups [e.g., $N_{Re} = DV\rho/\mu$, $f = e_f/(2LV^2/D)$] and not a form that has been adapted for circular tubes (e.g., $N_{Re} = 4Q\rho/\pi D\mu$). That is, the proper modification of the Reynolds number for a noncircular conduit is $[D_h V\rho/\mu$, not $4Q\rho/\pi D_h\mu$. One clue that the dimensionless group is the wrong form for a noncircular conduit is the presence of π, which is normally associated only with circular geometries (Remember: "pi are round, cornbread are square"). Thus, the appropriate dimensionless groups from the tube flow solutions can be modified for noncircular geometries as follows:

$$N_{Re,h} = \frac{D_h V\rho}{\mu} = \frac{4Q\rho}{W_p\mu} \tag{7-11}$$

$$f = \frac{e_f D_h}{2LV^2} = \frac{2e_f}{LQ^2}\left(\frac{A^3}{W_p}\right) \tag{7-12}$$

$$N_R = \frac{N_{Re,h}}{\varepsilon/D_h} = \frac{D_h^2 Q\rho}{\varepsilon A\mu} = \frac{16Q\rho}{\varepsilon\mu}\left(\frac{A}{W_p^2}\right) \tag{7-13}$$

$$fN_{Re,h}^2 = \frac{32e_f\rho^2}{L\mu^2}\left(\frac{A}{W_p}\right)^3 = \frac{e_f\rho^2 D_h^3}{2L\mu^2} \tag{7-14}$$

$$fN_{Re,h}^5 = \frac{2048e_f Q^3\rho^5}{L\mu^5}\left(\frac{A}{W_p^2}\right)^3 \tag{7-15}$$

The circular tube expressions for f and N_{Re} can also be transformed into the equivalent expressions for a noncircular conduit by the substitution

$$\pi \rightarrow \frac{W_p}{D_h} = 4\frac{A}{D_h^2} = \frac{1}{4}\left(\frac{W_p^2}{A}\right) \tag{7-16}$$

II. MOST ECONOMICAL DIAMETER

We have seen how to determine the driving force (e.g., pumping requirement) for a given pipe size and specified flow as well as how to determine the proper pipe size for a given driving force (e.g., pump head) and specified flow. However, when we install a pipeline or piping system we are usually free to select both the "best" pipe and the "best" pump. The term "best" in this case refers to that combination of pipe and pump that will minimize the total system cost.

The total cost of a pipeline or piping system includes the capital cost of both the pipe and pumps as well as operating costs, i.e. the cost of the energy required to drive the pumps:

Capital cost of pipe (CCP)
Capital cost of pump stations (CCPS)
Energy cost to power pumps (EC)

Although the energy cost is "continuous" and the capital costs are "one time," it is common to spread out (or amortize) the capital cost over a period of Y years i.e., over the "economic lifetime" of the pipeline. The reciprocal of this ($X = 1/Y$) is the fraction of the total capital cost written off per year. Taking 1 year as the time basis, we can combine the capital cost per year and the energy cost per year to get the total cost (there are other costs, such as maintenance, but these are minor and do not materially influence the result).

Data on the cost of typical pipeline installations of various sizes were reported by Darby and Melson (1982). They showed that these data can be represented by the equation:

$$\mathrm{CCP} = aD_{\mathrm{ft}}^{\mathrm{p}}L \tag{7-17}$$

where D_{ft} is the pipe ID in feet, and the parameters a and p depend upon the pipe wall thickness as shown in Table 7-2. Likewise, the capital cost of (installed) pump stations (for 500 hp and over) was shown to be a linear function of the pump power, as follows (see Fig. 7-4):

$$\mathrm{CCPS} = A + B\mathrm{HP}/\eta_e \tag{7-18}$$

where $A = 172,800$, $B = 450.8\,\mathrm{hp}^{-1}$ (in 1980 \$), and HP/η_e is the horsepower rating of the pump, (HP is the "hydraulic power," which is the power delivered directly to the fluid).

The energy cost is determined from the pumping energy requirement, which is in turn determined from the Bernoulli equation:

$$-w = \frac{\Delta\Phi}{\rho} + \frac{1}{2}\Delta V^2 + \sum e_{\mathrm{f}} \tag{7-19}$$

TABLE 7-2 Cost of Pipe (1980\$)[a]

	Pipe grade				
Parameter	ANSI 300#	ANSI 400#	ANSI 600#	ANSI 900#	ANSI 1500#
a	23.1	23.9	30.0	38.1	55.3
p	1.16	1.22	1.31	1.35	1.39

[a]Pipe cost (\$/ft) $= a(ID_{ft})^p$
(Note: The ANSI pipe grades correspond approximately to Sched 20, 30, 40, 80, and 120 for commercial steel pipe.

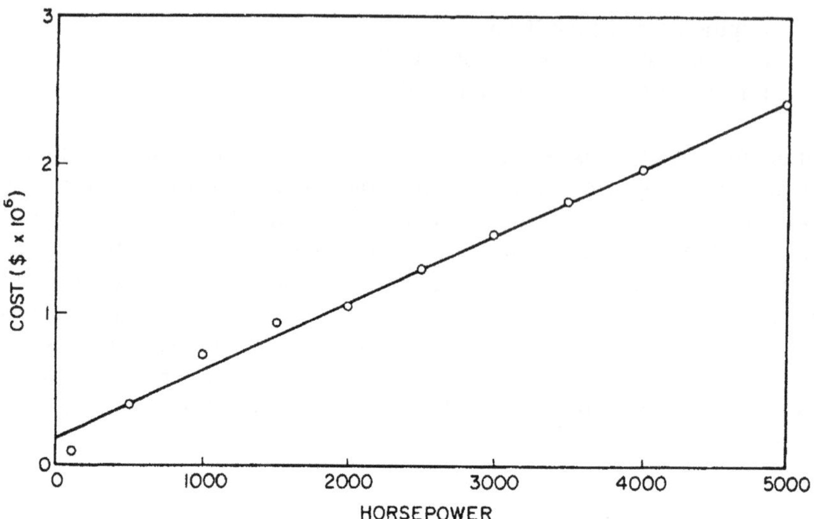

Figure 7-4 Cost of pump stations (1980 \$). Pump station cost (\$) = CCPS = $A + B\,\mathrm{hp}/\eta_e$ where $A = 172{,}800$ and $B = 451/\mathrm{hp}$ for stations of 500 hp or more.

where

$$\sum e_f = 4f\frac{V^2}{2}\sum\left(\frac{L}{D}\right)_{eq} \tag{7-20}$$

and $\sum (L/D)_{eq}$ is assumed to include the equivalent length of any fittings (which are usually a small portion of a long pipeline). The required hydraulic pumping power (HP) is thus

$$\mathrm{HP} = -w\dot{m} = \dot{m}\left(\frac{2fLV^2}{D} + \frac{\Delta\Phi}{\rho}\right) = \frac{32fL\dot{m}^3}{\pi^2\rho^2D^5} + \dot{m}\,\frac{\Delta\Phi}{\rho} \tag{7-21}$$

The total pumping energy cost per year is therefore

$$\mathrm{EC} = C\,\mathrm{HP}/\eta_e \tag{7-22}$$

where C is the unit energy cost (e.g., \$/(hp yr), ¢/kWh) and η_e is the pump efficiency. Note that the capital cost increases almost linearly with the pipe diameter, whereas the energy cost decreases in proportion to about the fifth power of the diameter.

The total annual cost of the pipeline is the sum of the capital and energy costs:

$$\mathrm{TC} = X(\mathrm{CCP} + \mathrm{CCPS}) + \mathrm{EC} \tag{7-23}$$

Substituting Eqs. (7-17), and (7-22) into Eq. (7-23) gives

$$\mathrm{TC} = XaD^pL + XA + \frac{BX + C}{\eta_e}\left[\frac{32fL\dot{m}^3}{\pi^2\rho^2D^5} + \dot{m}\,\frac{\Delta\Phi}{\rho}\right] \tag{7-24}$$

Now we wish to find the pipe diameter that minimizes this total cost. To do this, we differentiate Eq. (7-24) with respect to D, set the derivative equal to zero, and solve for D (i.e., D_{ec}, the most economical diameter):

$$D_{ec} = \left[\frac{B + CY}{ap\eta_e}\left(\frac{160f\dot{m}^3}{\pi^2\rho^2}\right)\right]^{1/(p+5)} \tag{7-25}$$

where $Y = 1/X$ is the "economic lifetime" of the pipeline.

One might question whether the cost information in Table 7-2 and Fig. 7-4 could be used today, because these data are based on 1980 information and prices have increased greatly since that time. However, as seen from Eq. (7-25), the cost parameters (i.e., B, C, and a) appear as a ratio. Since capital costs and energy costs tend to inflate at approximately the same rate (see, e.g., Durand et al., 1999), this ratio is essentially independent of inflation, and conclusions based on 1980 economic data should be valid today.

A. Newtonian Fluids

Equation (7-25) is implicit for D_{ec}, because the friction factor (f) depends upon D_{ec} through the Reynolds number and the relative roughness of the pipe. It can be solved by iteration in a straightforward manner, however, by the procedure used for the "unknown diameter" problem in Chapter 6. That is, first assume a value for f (say, 0.005), calculate D_{ec} from Eq. (7-25), and use this diameter to compute the Reynolds number and relative roughness; then use these values to find f (from the Moody diagram or Churchill equation). If this value is not the same as the originally assumed value, used it in place of the assumed value and repeat the process until the values of f agree.

Another approach is to regroup the characteristic dimensionless variables in the problem so that the unknown (D_{ec}) appears in only one group. After rearranging Eq. (7-25) for f, we see that the following group will be independent of D_{ec}:

$$fN_{\mathrm{Re}}^{p+5} = \left(\frac{4}{\pi}\right)^{p+3} \frac{\rho^2 ap\eta_e\dot{m}^{p+2}}{10(B + CY)\mu^{p+5}} = N_c \tag{7-26}$$

We can call this the "cost group" (N_c), because it contains all the cost parameters. We can also define a roughness group that does not include the diameter:

$$N_R = \frac{\varepsilon/D_{ec}}{N_{Re}} = \frac{\pi\mu\varepsilon}{4\dot{m}} \tag{7-27}$$

The remaining group is the Reynolds number, which is the dependent group because it contains D_{ec}:

$$N_{Re} = \frac{4\dot{m}}{\pi D_{ec}\mu} \tag{7-28}$$

The Moody diagram can be used to construct a plot of N_{Re} versus $N_c = fN^{p+5}$ for various values of p and N_R (a double parametric plot), which permits a direct solution to this problem (see Darby and Melson, 1982). The above equations can also be used directly to simplify the iterative solution. Since the value of N_c is known, assuming a value for f will give N_{Re} directly from Eq. (7-26). This, in turn, gives D_{ec} from Eq. (7-28), and hence ε/D_{ec}. These values of N_{Re} and ε/D_{ec} are used to find f (from the Moody diagram or Churchill equation), and the iteration is continued until successive values of f agree. The most difficult aspect of working with these groups is ensuring a consistent set of units for all the variables (with appropriate use of the conversion factor g_c, if working in engineering units). For this reason, it is easier to work with consistent units in a scientific system (e.g., SI or cgs), which avoids the need for g_c.

Example 7-1: Economic Pipe Diameter. What is the most economical diameter for a pipeline that is required to transport crude oil with a viscosity of 30 cP and an SG of 0.95, at a rate of 1 million barrels per day using ANSI 1500# pipe, if the cost of energy is 5¢ per kWh (in 1980 \$)? Assume that the economical life of the pipeline is 40 years and that the pumps are 50% efficient.

Solution. From Table 7-2, the pipe cost parameters are

$$p = 1.39, \qquad a = 55.3\,\frac{\$}{\text{ft}^{2.39}} \times \left(\frac{3.28\,\text{ft}}{\text{m}}\right)^{2.39} = 945.5\,\$/\text{m}^{2.39}$$

Using SI units will simplify the problem. After converting, we have

$$\dot{m} = \rho Q = 1748\,\text{kg/s}, \qquad \mu = 0.03\,\text{Pa s}, \qquad CY = \$17.52/\text{W}$$

From Fig. 7-4 we get the pump station cost factor B:

$$B = 451\,\$/\text{hp} = 0.604\,\$/\text{W}$$

and the "cost group" is [Eq. (7-26)]

$$N_c = \left(\frac{4}{\pi}\right)^{p+3} \frac{\rho^2 ap\eta_e \dot{m}^{p+2}}{10(B+CY)\mu^{p+5}} = 5.07 \times 10^{27} = fN_{Re}^{6.39}$$

Assuming a roughness of 0.0018 in., we can solve for D_{ec} by iteration as follows.

First, assume $f = 0.005$ and use this to get N_{Re} from $N_c = fN_{Re}^{6.39}$. From N_{Re} we find D_{ec}, and thus ε/D_{ec}. Then, using the Churchill equation or Moody diagram, we find a valyue for f and compare it with the assumed value. This is repeated until convergence is achieved:

Assumed f	N_{Re}	D_{ec}(m)	ε/D_{ec}	f (Churchill)
0.005	4.96×10^4	1.49	3.07×10^{-5}	0.00523
0.00523	4.93×10^4	1.50	3.05×10^{-5}	0.00524

This agreement is close enough. The most economical diameter is 1.5 m, or 59.2 in. The "standard pipe size" closest to this value on the high side (or the closest size that can readily be manufactured) would be used.

B. Non-Newtonian Fluids

A procedure analogous to the one followed can be used for non-Newtonian fluids that follow the power law or Bingham plastic models (Darby and Melson, 1982).

1. Power Law

For power law fluids, the basic dimensionless variables are the Reynolds number, the friction factor, and the flow index (n). If the Reynolds number is expressed in terms of the mass flow rate, then

$$N_{Re,pl} = \left(\frac{4}{\pi}\right)^{2-n} \left(\frac{4n}{3n+1}\right)^n \left(\frac{\dot{m}^{2-n}\rho^{n-1}}{D_{ec}^{4-3n} 8^{n-1} m}\right) \tag{7-29}$$

Eliminating D_{ec} from Eqs. (7-25) and (7-29), the equivalent cost group becomes

$$f^{4-3n} N_{Re,pl}^{5+p} = \frac{(52.4)(10^{3n})(2^{7p-3n(1+p)})}{\pi^{(2+n)(1+p)} m^{5+p}} \left(\frac{ap\eta_e}{B+CY}\right)^{4-3n} \times \left(\frac{\rho^{3-p+n(p-1)} \dot{m}^{2(p-1)+n(4-p)}}{[(3n+1)/n]^{n(5+p)}}\right) \tag{7-30}$$

Since all values on the right hand side of Eq. (7-30) are known, assuming a value of f allows a corresponding value of $N_{\mathrm{Re\,pl}}$ to be determined. This value can then be used to check the assumed value of f using the general expression for the power law friction factor [Eq. (6-44)] and iterating until agreement is attained.

2. Bingham Plastic

The basic dimensionless variables for the Bingham plastic are the Reynolds number, the Hedstrom number, and the friction factor. Eliminating D_{ec} from the Reynolds number and Eq. (7-25) (as above), the cost group is:

$$fN_{\mathrm{Re}}^{p+5} = \left(\frac{4}{\pi}\right)^{p+3}\left(\frac{\rho^2 ap\eta_e \dot{m}^{p+2}}{10(B+CY)\mu_\infty^{p+5}}\right) \tag{7-31}$$

D_{ec} can also be eliminated from the Hedstrom number by combining it with the Reynolds number:

$$N_{\mathrm{He}}N_{\mathrm{Re}}^2 = \left(\frac{4}{\pi}\right)^2\left(\frac{\tau_0 \rho \dot{m}^2}{\mu_\infty^4}\right) \tag{7-32}$$

These equations can readily be solved by iteration, as follows. Assuming a value of f allows N_{Re} to be determined from Eq. (7-31). This is then used with Eq. (7-32) to find N_{He}. The friction factor is then calculated using these values of N_{Re} and N_{He} and the Bingham plastic pipe friction factor equation [Eq. (6.65)]. The result is compared with the assumed value, and the process is repeated until agreement is attained.

Graphs have been presented by Darby and Melson (1982) that can be used to solve these problems directly without iteration. However, interpolation on double-parametric logarithmic scales is required, so only approximate results can be expected from the precision of reading these plots. As mentioned before, the greatest difficulty in using these equations is that of ensuring consistent units. In many cases it is most convenient to use cgs units in problems such as these, because fluid properties (density and viscosity) are frequently found in these units, and the scientific system (e.g., cgs) does not require the conversion factor g_c. In addition, the energy cost is frequently given in cents per kilowatt-hour, which is readily converted to cgs units (e.g., \$/erg).

III. FRICTION LOSS IN VALVES AND FITTINGS

Evaluation of the friction loss in valves and fittings involves the determination of the appropriate loss coefficient (K_f), which in turn defines the energy loss per unit mass of fluid:

$$e_f = K_f V^2/2 \tag{7-33}$$

where V is (usually) the velocity in the pipe upstream of the fitting or valve (however, this is not always true, and care must be taken to ensure that the value of V that is used is the one that is used in the defining equation for K_f). The actual evaluation of K_f is done by determining the friction loss e_f from measurements of the pressure drop across the fitting (valve, etc.). This is not straightforward, however, because the pressure in the pipe is influenced by the presence of the fitting for a considerable distance both upstream and downstream of the fitting. It is not possible, therefore, to obtain accurate values from measurements taken at pressure taps immediately adjacent to the fitting. The most reliable method is to measure the total pressure drop through a long run of pipe both with and without the fitting, at the same flow rate, and determine the fitting loss by difference.

There are several "correlation" expressions for K_f, which are described below in the order of increasing accuracy. The "3-K" method is recommended, because it accounts directly for the effect of both Reynolds number and fitting size on the loss coefficient and more accurately reflects the scale effect of fitting size than the 2-K method. For highly turbulent flow, the Crane method agrees well with the 3-K method but is less accurate at low Reynolds numbers and is not recommended for laminar flow. The loss coefficient and $(L/D)_{eq}$ methods are more approximate but give acceptable results at high Reynolds numbers and when losses in valves and fittings are "minor losses" compared to the pipe friction. They are also appropriate for first estimates in problems that require iterative solutions.

A. Loss Coefficient

Values of K_f for various types of valves, fittings, etc. are tabulated in various textbooks and handbooks. The assumption that these values are constant for a given type of valve or fitting is not accurate, however, because in reality the value of K_f varies with both the size (scale) of the fitting and the level of turbulence (Reynolds number). One reason that K_f is not the same for all fittings of the same type (e.g., all 90° elbows) is that all the dimensions of a fitting, such as the diameter and radius of curvature, do not scale by the same factor for large and small fittings. Most tabulated values for K_f are close to the values of K_∞ from the 3-K method.

B. Equivalent L/D Method

The basis for the equivalent L/D method is the assumption that there is some length of pipe (L_{eq}) that has the same friction loss as that which occurs in the fitting, at a given (pipe) Reynolds number. Thus, the fittings are conceptually replaced by the equivalent additional length of pipe that has the same friction loss as the fitting:

$$e_f = \frac{4fV^2}{2} \sum \left(\frac{L}{D}\right)_{eq} \tag{7-34}$$

where f is the friction factor in the pipe at the given pipe Reynolds number and relative roughness. This is a convenient concept, because it allows the solution of pipe flow problems with fittings to be carried out in a manner identical to that without fittings if L_{eq} is known. Values of $(L/D)_{eq}$ are tabulated in various textbooks and handbooks for a variety of fittings and valves and are also listed in Table 7-3. The method assumes that (1) sizes of all fittings of a given type can be scaled by the corresponding pipe diameter (D) and (2) the influence of turbulence level (i.e., Reynolds number) on the friction loss in the fitting is identical to that in the pipe (because the pipe f values is used to determine the fitting loss). Neither of these assumptions is accurate (as pointed out above), although the approximation provided by this method gives reasonable results at high turbulence levels (high Reynolds numbers), especially if fitting losses are minor.

C. Crane Method

The method given in Crane Technical Paper 410 (Crane Co., 1991) is a modification of the preceding methods. It is equivalent to the $(L/D)_{eq}$ method except that it recognizes that there is generally a higher degree of turbulence in the fitting than in the pipe at a given (pipe) Reynolds number. This is accounted for by always using the "fully turbulent" value for f (e.g., f_T) in the expression for the friction loss in the fitting, regardless of the actual Reynolds number in the pipe, i.e.,

$$e_f = \frac{K_f V^2}{2}, \qquad \text{where } K_f = 4f_T(L/D)_{eq} \tag{7-35}$$

The value of f_T can be calculated from the Colebrook equation,

$$f_T = \frac{0.0625}{[\log(3.7D/\varepsilon)]^2} \tag{7-36}$$

in which ε is the pipe roughness (0.0018 in. for commercial steel). This is a two-constant model [f_T and $(L/D)_{eq}$], and values of these constants are tabulated in the Crane paper for a wide variety of fittings, valves, etc. This method gives satisfactory results for high turbulence levels (high Reynolds numbers) but is less accurate at low Reynolds numbers.

D. 2-K (Hooper) Method

The 2-K method was published by Hooper (1981, 1988) and is based on experimental data in a variety of valves and fittings, over a wide range of Reynolds numbers. The effect of both the Reynolds number and scale (fitting size) is reflected in the expression for the loss coefficient:

$$e_f = \frac{K_f V^2}{2}, \qquad \text{where } K_f = \frac{K_1}{N_{Re}} + K_\infty\left(1 + \frac{1}{ID_{in}}\right) \tag{7-37}$$

Here, ID_{in} is the internal diameter (in inches) of the pipe that contains the fitting. This method is valid over a much wider range of Reynolds numbers than the other methods. However. the effect of pipe size (e.g., $1/ID_{in}$) in Eq. (7-37) does not accurately refect observations, as discussed below.

E. 3-K (Darby) Method

Although the 2-K method applies over a wide range of Reynolds numbers, the scaling term (1/ID) does not accurately reflect data over a wide range of sizes for valves and fittings, as reported in a variety of sources (Crane, 1988, Darby, 2001, Perry and Green, 1998, CCPS, 1998 and references therein). Specifically, all preceding methods tend to underpredict the friction loss for pipes of larger diameters. Darby (2001) evaluated data from the literature for various valves, tees, and elbows and found that they can be represented more accurately by the following "3-K" equation:

$$K_f = \frac{K_1}{N_{Re}} + K_i\left(1 + \frac{K_d}{D_{n,in.}^{0.3}}\right) \tag{7-38}$$

Values of the 3 K's (K_1, K_i, and K_d) are given in Table 7-3 (along with represesentative values of $[L/D]_{eq}$) for various valves and fittings. These values were determined from combinations of literature values from the references listed above, and were all found to accurately follows the scaling law given in Eq. (7-38). The values of K_1 are mostly those of the Hooper 2-K method, and the values of K_i were mostly determined from the Crane data. However, since there is no single comprehensive data set set for many fittings over a wide range of sizes and Reynolds numbers, some estimation was necessary for some values. Note that the values of K_d are all very close to 4.0, and this can be used to scale known values of K_f for a given pipe size to apply to other sizes. This method is the most accurate of the methods described for all Reynolds numbers and fitting sizes. Tables 7-4 and 7-5 list values for K_f for Expansions and Contractions, and Entrance and Exit conditions, respectively, from Hooper (1988).

Table 7-3 3-K Constants for Loss Coefficients for Valves and Fittings

$$K_f = K_1/N_{Re} + K_i(1 + K_d/D_n^{0.3})$$

where D_n is the nominal diameter in inches

Fitting				(L/D)eq	K_1	K_i	K_d
Elbows	90°	Threaded, standard	$(r/D = 1)$	30	800	0.14	4.0
		Threaded, long radius	$(r/D = 1.5)$	16	800	0.071	4.2
		Flanged, welded, bends	$(r/D = 1)$	20	800	0.091	4.0
			$(r/D = 2)$	12	800	0.056	3.9
			$(r/D = 4)$	14	800	0.066	3.9
			$(r/D = 6)$	17	800	0.075	4.2
		Mitered 1 weld	(90°)	60	1000	0.27	4.0
		2 welds	(45°)	15	800	0.068	4.1
		3 welds	(30°)	8	800	0.035	4.2
	45°	Threaded standard	$(r/D = 1)$	16	500	0.071	4.2
		Long radius	$(r/D = 1.5)$		500	0.052	4.0
		Mitered 1 weld	(45°)	15	500	0.086	4.0
		2 welds	(22.5°)	6	500	0.052	4.0
	180°	Threaded, close return bend	$(r/D = 1)$	50	1000	0.23	4.0
		Flanged	$(r/D = 1)$		1000	0.12	4.0
		All	$(r/D = 1.5)$		1000	0.10	4.0

Tees	Through-Branch (as elbow) Threaded	$(r/D = 1)$	60	500	0.274	4.0
		$(r/D = 1.5)$		800	0.14	4.0
	Flanged	$(r/D = 1)$	20	800	0.28	4.0
	Stub-in branch			1000	0.34	4.0
	Run Through Threaded	$(r/D = 1)$	20	200	0.091	4.0
	Flanged	$(r/D = 1)$		150	0.017	4.0
	Stub-in branch			100	0	0
Valves	angle valve – 45° full line size, $\beta = 1$		55	950	0.25	4.0
	–90° full line size, $\beta = 1$		150	1000	0.69	4.0
	Globe valve Standard, $\beta = 1$		340	1500	1.70	3.6
	Plug valve Branch flow		90	500	0.41	4.0
	Straight through		18	300	0.084	3.9
	Three-way (flow through)		30	300	0.14	4.0
	Gate valve Standard, $\beta = 1$		8	300	0.037	3.9
	Ball valve Standard, $\beta = 1$		3	300	0.017	4.0
	Diaphragm Dam-type			1000	0.69	4.9
	Swing check $V_{min} = 35[\rho(lb_m/ft^3)]^{-1/2}$		100	1500	0.46	4.0
	Lift check $V_{min}40[\rho lb_m/ft^3)]^{1/2}$		600	2000	2.85	3.8

TABLE 7-4 Loss Coefficients for Expansions and Contractions

K_f to be used with upstream velocity head, $V_1^2/2$. $\beta = d/D$

Contraction

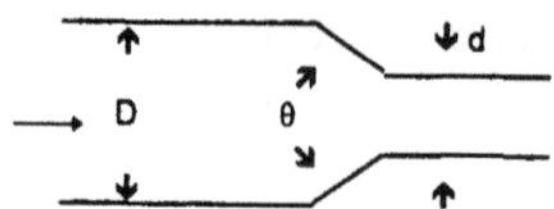

$\theta < 45^\circ$

$N_{Re,1} < 2500$:

$$K_f = 1.6\left[1.2 + \frac{160}{N_{Re,1}}\right]\left[\frac{1}{\beta^4} - 1\right] \sin\frac{\theta}{2}$$

$N_{Re,1} > 2500$:

$$K_f = 1.6[0.6 + 1.92 f_1]\left[\frac{1-\beta^2}{\beta^4}\right] \sin\frac{\theta}{2}$$

$N_{Re,1} < 2500$:

$$K_f = \left[1.2 + \frac{160}{N_{Re,1}}\right]\left[\frac{1}{\beta^4} - 1\right]\left[\sin\frac{\theta}{2}\right]^{1/2}$$

$N_{Re,1} > 2500$:

$$K_f = [0.6 + 0.48 f_1]\left[\frac{1-\beta^2}{\beta^4}\right]\left[\sin\frac{\theta}{2}\right]^{1/2}$$

Expansion

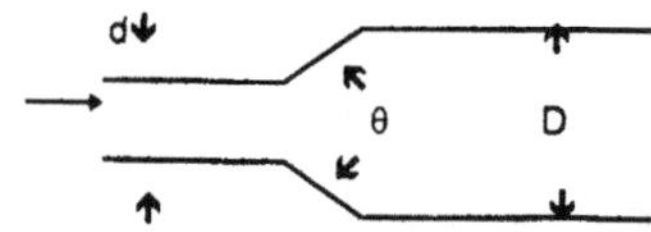

$\theta < 45^\circ$

$N_{Re,1}$, 4000:

$$K_f = 5.2(1 - \beta^4) \sin\frac{\theta}{2}$$

$N_{Re,1} > 4000$:

$$K_f = 2.6(1 + 3.2 f_1)(1 - \beta^2)^2 \sin\frac{\theta}{2}$$

$\theta > 45^\circ$

$N_{Re,1} < 4000$:

$$K_f = 2(1 - \beta^4)$$

$N_{Re,1} > 4000$:

$$K_{f1} = (1 + 3.2 f_1)(1 - \beta^2)^2$$

$N_{Re,1}$ is the upstream Reynolds number, and f_1 is the pipe friction factor at this Reynolds number.
Source: Hooper (1988).

TABLE 7-5 Loss Coefficients for Pipe Entrance and Exit

$$K_f = K_1/N_{Re} + K_\infty$$

Entrance

Inward projecting (Borda)

$K_1 = 160$, $K_\infty = 1.0$

Flush (rounded)

$K_1 = 160$

r/d	K_∞
0.0 (sharp)	0.5
0.02	0.28
0.04	0.24
0.06	0.15
0.10	0.09
0.15 & up	0.04

For pipe exit: $K_\infty = 1.0$ for all geometries
$K_1 = 0.0$

Orifice: $K_\infty = \dfrac{2.91}{\beta^4}(1 - \beta^2)(1 - \beta^4) = \dfrac{(1 - \beta^2)(1 - \beta^4)}{C_o^2\beta^4}$

$\beta = D_o/D_p$

$K_1 = 0.0$

Source: William B. Hooper, Chemical Engineering, p. 97, 1981.

The definition of K_f (i.e., $K_f = 2e_f/V^2$) involves the kinetic energy of the fluid, $V^2/2$. For sections that undergo area changes (e.g., pipe entrance, exit, expansion, or contraction), the entering and leaving velocities will be different. Because the value of the velocity used with the definition of K_f is arbitrary, it is very important to know which velocity is the reference value for a given loss coefficient. Values of K_f are usually based on the larger velocity entering or leaving the fitting (through the smaller area), but this should be verified if any doubt exists.

A note is in order relative to the exit loss coefficient, which is listed in Table 7-5 as equal to 1.0. Actually, if the fluid exits the pipe into unconfined space, the loss coefficient is zero, because the velocity of a fluid exiting the pipe (in a free jet) is the same as that of the fluid inside the pipe (and the kinetic energy change is also zero). However, when the fluid exits into a confined space the kinetic energy is dissipated as friction in the mixing process as the velocity goes to zero, so the loss coefficient is 1.0. In this case the change in the kinetic energy and the friction loss at the exit cancel out.

IV. NON-NEWTONIAN FLUIDS

There are insufficient data in the literature to enable reliable correlation or prediction of friction loss in valves and fittings for non-Newtonian fluids. As a first approximation, however, it can be assumed that a correlation similar to the 3-K method should apply to non-Newtonian fluids if the (Newtonian) Reynolds number in Eq. (7-38) could be replaced by a single corresponding dimensionless group that adequately characterizes the influence of the non-Newtonian properties. For the power law and Bingham plastic fluid models, two rheological parameters are required to describe the viscous properties, which generally results in two corresponding dimensionless groups ($N_{Re,pl}$ and n for the power law, and N_{Re} and N_{He} for the Bingham plastic). However, it is possible to define an "effective viscosity" for a non-Newtonian fluid model that has the same significance for the Reynolds number as the viscosity for a Newtonian fluid and incorporates all the appropriate parameters for that model and that can be used to define an equivalent non-Newtonian Reynolds number (see Darby and Forsyth, 1992). For a Newtonian fluid, the Reynolds number can be rearranged as follows:

$$N_{Re} = \frac{DV\rho}{\mu} = \frac{\rho v^2}{\mu V/D} = \frac{\rho V^2}{\tau_w/8} \tag{7-39}$$

Introducing $\tau_w = m[(8V/D)(3n+1)/4n]^n$ for the power law model, the result is

$$N_{Re,pl} = \frac{2^{7-3n}\rho Q^{2-n}}{m\pi^{2-n}D^{4-3n}}\left(\frac{n}{3n+1}\right)^n \tag{7-40}$$

which is identical to the expression derived in Chapter 6. For the Bingham plastic, the corresponding expression for the Reynolds number is:

$$N_{Re,BP} = \frac{4Q\rho}{\pi D\mu_\infty(1+\pi D^3\tau_0/32Q\mu_\infty)} = \frac{N_{Re}}{1+N_{He}/8N_{Re}} \tag{7-41}$$

This is determined by replacing τ_w for the Newtonian fluid in Eq. (7-39) with $\tau_0 + \mu_\infty\dot{\gamma}_w$ and using the approximation $\dot{\gamma}_w = 8V/D$. The ratio $N_{He}/N_{Re} = D\tau_0/V\mu_\infty$ is also called the Bingham number (N_{Bi}). Darby and Forsyth (1992) showed experimentally that mass transfer in Newtonian and non-

Newtonian fluids can be correlated by this method; that is, the same correlation applies to both Newtonian and non-Newtonian fluids when the Newtonian Reynolds number is replaced by either Eq. (7-40) for the power law fluid model or Eq. (7-41) for the Bingham plastic fluid model. As a first approximation, therefore, we may assume that the same result would apply to friction loss in valves and fittings as described by the 2-K or 3-K models [Eq. 7-38)].

V. PIPE FLOW PROBLEMS WITH FITTINGS

The inclusion of significant fitting friction loss in piping systems requires a somewhat different procedure for the solution of flow problems than that which was used in the absence of fitting losses in Chapter 6. We will consider the same classes of problems as before, i.e. unknown driving force, unknown flow rate, and unknown diameter for Newtonian, power law, and Bingham plastics. The governing equation, as before, is the Bernoulli equation, written in the form

$$\mathrm{DF} = -\left(\frac{\Delta\Phi}{\rho} + w\right) = \sum e_{\mathrm{f}} + \frac{1}{2}\,\Delta(\alpha V^2) \tag{7-42}$$

where

$$\sum e_{\mathrm{f}} = \frac{1}{2}\sum (V^2 K_{\mathrm{f}}) = \frac{8Q^2}{\pi^2}\sum \frac{K_{\mathrm{f}}}{D^4} \tag{7-43}$$

$$K_{\mathrm{pipe}} = \frac{4fL}{D}, \qquad K_{\mathrm{fit}} = \frac{K_1}{N_{\mathrm{Re}}} + K_\infty\left(1 + \frac{K_{\mathrm{o}}}{D_{\mathrm{n}}^{0.3}}\right) \tag{7-44}$$

and the summation is over each fitting and segment of pipe (of diameter D) in the system. The loss coefficients for the pipe and fittings are given by the Fanning friction factor and 3-K formula, as before. Substituting Eq. (7-43) into Eq. (7-42) gives the following form of the Bernoulli equation:

$$\mathrm{DF} = \frac{8Q^2}{\pi^2}\left(\sum_i \frac{K_i}{D_i^4} + \frac{\alpha_2}{D_2^4} - \frac{\alpha_1}{D_1^4}\right) \tag{7-45}$$

The α's are the kinetic energy correction factors at the upstream and downstream points (recall that $\alpha = 2$ for laminar flow and $\alpha = 1$ for turbulent flow for a Newtonian fluid).

A. Unknown Driving Force

Here we wish to find the net driving force required to transport a given fluid at a given rate through a given pipeline containing a specified array of valves and fittings.

1. Newtonian Fluid

The knowns and unknowns for this case are

Given: Q, μ, ρ, D_i, L_i, ε_i, fittings *Find*: DF

The driving force (DF) is given by Eq. (7-45), in which the K_i's are related to the other variables by the Moody diagram (or Churchill equation) for each pipe segment (K_{pipe}), and by the 3-K method for each valve and fitting (K_{fit}), as a function of the Reynolds number:

$$N_{Re\,i} = \frac{4Q\rho}{\pi D_i \mu} \tag{7-46}$$

The solution procedure is as follows:

1. Calculate $N_{Re\,i}$ from Eq. (7-46) for each pipe segment, valve, and fitting (i).
2. For each pipe segment of diameter D_i, get f_i from the Churchill equation or Moody diagram using $N_{Re\,i}$ and ε_i/D_i, and calculate $K_{pipe} = 4(fL/D)_i$.
3. For each valve and fitting, calculate K from $N_{Re\,i}$, and D_i, using the 3-K method.
4. Calculate the driving force, DF, from Eq. (7-45).

2. Power Law Fluid

The knowns and unknowns for this case are:

Given: Q, D_i, L_i, ε_i, m, n, fittings *Find*: DF

The appropriate expressions that apply are the Bernoulli equation [Eq. (7-45)], the power law Reynolds number [Eq. (7-40)], the pipe friction factor as a function of $N_{Re,pl}$ and n [Eq. (6-44)], and the 3-K equation for fitting losses [Eq. (7-38)] with the Reynolds number replaced by $N_{Re,pl}$. The procedure is

1. From the given values, calculate $N_{Re,pl}$ from Eq. [7-40].
2. Using $N_{Re\,pl}$ and n, calculate f (and the corresponding K_{pipe}) for each pipe section from the power law friction factor equation [Eq. (6-44)], and calculate K_f for each valve and fitting using the 3-K method [Eq. (7-38)].

3. Calculate the driving force (DF) from the Bernoulli equation, Eq. (7-45).

3. Bingham Plastic

The procedure for the Bingham plastic is identical to that for the power law fluid, except that Eq. (7-41) is used for the Reynolds number in the 3-K equation for fittings instead of Eq. (7-39), and the expression for the Bingham pipe friction factor is given by Eq. (6-62).

B. Unknown Flow Rate

The Bernoulli equation [Eq. (7-45)] can be rearranged for the flow rate, Q, as follows:

$$Q = \frac{\pi}{2\sqrt{2}} \left[\frac{\text{DF}}{\sum_i (K_i/D_i^4) + \alpha_2/D_2^4 - \alpha_1/D_1^4} \right]^{1/2} \tag{7-47}$$

The flow rate can be readily calculated if the loss coefficients can be determined. The procedure involves an iteration, starting with estimated values for the loss coefficients. These are used in Eq. (7-47) to find Q, which is used to calculate the Reynolds number(s), which are then used to determine revised values for the K_i's, as follows.

1. Newtonian Fluid

The knowns and unknowns are

Given: DF, D, L, ε, μ, ρ, fittings *Find*: Q

1. A first estimate for the pipe friction factor and the K_i's can be made by assuming that the flow is fully turbulent (and the α's = 1). Thus,

$$f_1 = \frac{0.0625}{[\log(3.7D/\varepsilon)]^2} \tag{7-48}$$

and

$$K_{\text{fit}} = K_{\text{i}} \left(1 + \frac{K_{\text{d}}}{D_{\text{n, in.}}^{0.3}} \right) \tag{7-49}$$

2. Using these values, calculate Q from Eq. (7-47), from which the Reynolds number is determined: $N_{\text{Re}} = 4Q\rho/\pi D\mu$).

3. Using this Reynolds number, determine the revised pipe friction factor (and hence $K_{pipe} = 4fL/D$) from the Moody diagram (or Churchill equation), and the K_{fit} values from the 3-K equation.
4. Repeat steps 2 and 3 until Q does not change.

The solution is the last value of Q calculated from step 2.

2. Power Law Fluid

The knowns and unknowns are

Given: DF, D, L, m, n, ρ *Find*: Q

The procedure is essentially identical to the one followed for the Newtonian fluid, except that Eq. (7-40) is used for the Reynolds number in step 2 and Eq. (6-44) is used for the pipe friction factor in step 3.

3. Bingham Plastic

The knowns and unknowns are

Given: DF, D, L, μ_∞, τ_0 *Find*: Q

The procedure is again similar to the one for the Newtonian fluid, except that the pipe friction factor in step 3 (thus K_{pipe}) is determined from Eq. (6-62) using $N_{Re} = 4Q\rho/\pi D\mu_\infty$ and $N_{He} = D^2\rho\tau_0/\mu_\infty^2$. The values of K_{fit} are determined from the 3-K equation using Eq. (7-41) for the Reynolds number.

C. Unknown Diameter

It is assumed that the system contains only one size (diameter) of pipe. The Bernoulli equation can be rearranged to give D:

$$D = \left[\frac{8Q^2(\sum_i K_i + \alpha_2 D^4/D_2^4 - \alpha_1 D^4/D_1^4)}{\pi^2 \mathrm{DF}}\right]^{1/4} \tag{7-50}$$

This is obviously implicit in D (the terms involving the α's can be neglected for the initial estimate). If the K_i values can be estimated, then the diameter can be determined from Eq. (7-50). However, since D is unknown, so is ε/D, so a "cruder" first estimate for f and for the K_{fit} values is required. Also, since $K_{pipe} = 4fL/D$, an estimated value for f still does not allow determination of K_{pipe}. Therefore, the initial estimate for K_{pipe} can be made by neglecting the fittings altogether, as outlined in Chapter 6.

1. Newtonian Fluids

The knowns and unknowns are

Given: Q, DF, $L, \varepsilon, \mu, \rho$ *Find*: D

If fittings are neglected, the following group can be evaluated from known values:

$$fN_{\text{Re}}^5 = \frac{32\,\text{DF}\,\rho^5 Q^3}{\pi^3 L\mu^5} \tag{7-51}$$

The procedure is as follows.

1. For a first estimate, assume $f = 0.005$.
2. Use this in Eq. (7-51) to estimate the Reynolds number:

$$N_{\text{Re}} = \left(\frac{fN_{\text{Re}}^5}{0.005}\right)^{1/5} = \left(\frac{32\,DF\,\rho^5 Q^3}{0.005\pi^3 L\mu^5}\right)^{1/5} \tag{7-52}$$

3. Get a first estimate for D from this Reynolds number:

$$D = \frac{4Q\rho}{\pi\mu N_{\text{Re}}} \tag{7-53}$$

   Now the complete equations for f and K_{fit} can be used for further iteration.
4. Using the estimates of D and N_{Re} obtained in steps 2 and 3, determine f and K_{pipe} from the Moody diagram (or Churchill equation) and the K_{fit} from the 3-K formula.
5. Calculate D from Eq. (7-50), using the previous value of D (from step 3) in the α terms.
6. If the values of D from steps 3 and 5 do not agree, calculate the value of N_{Re} using the D from step 5, and use these N_{Re} and D values in step 4.
7. Repeat steps 4–6 until D does not change.

2. Power Law Fluid

The knowns and unknowns are

Given: Q, DF, L, m, n, ρ *Find*: D

The basic procedure for the power law fluid is the same as above for the Newtonian fluid. We get a first estimate for the Reynolds number by ignoring fittings and assuming turbulent flow. This is used to estimate the value of f (hence K_{pipe}) using Eq. (6-44) and the K_{fit} values from the equivalent 3-K equation. Inserting these into Eq. (7-50) then gives a first estimate for the diameter, which is then used to revise the Reynolds number. The iteration continues until successive values agree, as follows:

1. Assume $f = 0.005$.

2. Ignoring fittings, the first estimate for $N_{Re,pl}$ is

$$N_{Re,pl} = \left(\frac{fN_{Re,pl}^{5/(4-3n)}}{0.005}\right)^{(4-3n)/5}$$

$$= \left(\frac{\pi^2 DF}{0.16LQ^2}\right)^{(4-3n)/5} \left[\frac{2^{7-3n}Q^{2-n}}{m\pi^{2-n}} \left(\frac{n}{3n+1}\right)^n\right] \tag{7-54}$$

3. Get a first estimate for D from this value and the definition of the Reynolds number:

$$D = \left[\frac{2^{7-3n}\rho Q^{2-n}}{m\pi^{2-n}N_{Re,pl}} \left(\frac{n}{3n+1}\right)^n\right]^{1/(4-3n)} \tag{7-55}$$

4. Using the values of $N_{Re,pl}$ from step 2 and D from step 3, calculate the value of f and K_{pipe}) from Eq. (6-44), and the K_{fit} values from the 3-K equation.
5. Insert the K values into Eq. (7-50) to find a new value of D.
6. If the value of D from step 5 does not agree with that from step 3, use the value from step 5 to calculate a revised $N_{Re,pl}$, and repeat steps 4–6 until agreement is attained.

3. Bingham Plastic

The knowns and unknowns are

Given: Q, DF, L, τ_0, μ_∞, ρ *Find*: D

The procedure for a Bingham plastic is similar to the foregoing, using Eq. (6-62) for the pipe friction factor:

1. Assume $f = 0.02$.
2. Calculate

$$fN_{Re}^5 = \frac{32\,DF\,Q^3\rho^5}{\pi^3 L\mu_\infty^5} \tag{7-56}$$

3. Get a first estimate of Reynolds number from

$$N_{Re} = \left(\frac{fN_{Re}^5}{0.02}\right)^{1/5} \tag{7-57}$$

4. Use this to estimate D:

$$D = \frac{4Q\rho}{\pi N_{Re}\mu_\infty} \tag{7-58}$$

5. Using this D and N_{Re}, calculate $N_{He} = D^2 \rho \tau_0 / \mu_\infty^2$, the pipe friction factor from Eq. (6-62), $K_{pipe} = 4fL/D$, and the K_{fit}'s from the 3-K equation using Eq. (7-41) for the Bingham plastic Reynolds number.
6. Insert the K_f values into Eq. (7-50) to get a revised value of D.
7. Using this value of D, revise the values of N_{Re} and N_{He}, and repeat steps 5–7 until successive values agree.

VI. SLACK FLOW

A special condition called "slack flow" can occur when the gravitational driving force exceeds the full pipe friction loss, such as when a liquid is being pumped up and down over hilly terrain. Consider the situation shown in Fig. 7-5, in which the pump upstream provides the driving force to move the liquid up the hill at a flow rate of Q. Since gravity works against the flow on the uphill side and aids the flow on the downhill side, the job of the pump is to get the fluid to the top of the hill. The minimum pressure is at point 2 at the top of the hill, and the flow rate (Q) is determined by the balance between the pump head ($H_p = -w/g$) and the frictional and gravitational resistance to flow on the uphill side (i.e., the Bernoulli equation applied from point 1 to point 2):

$$H_p = h_{f_{12}} + \frac{\Phi_2 - \Phi_1}{\rho g} \tag{7-59}$$

where

$$h_{f_{12}} = \frac{4fL_{12}}{gD}\left(\frac{V^2}{2}\right) \tag{7-60}$$

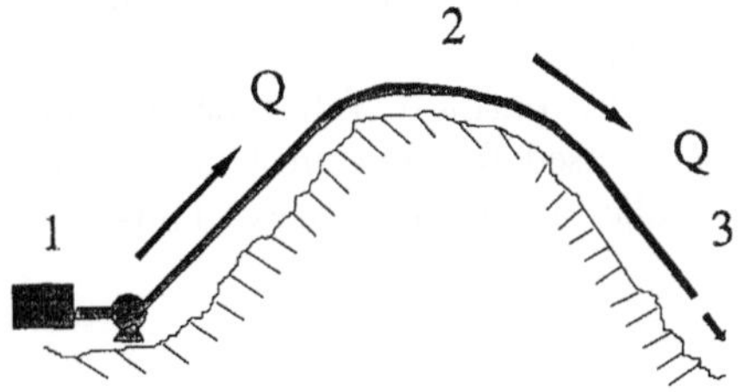

FIGURE 7-5 Conditions for slack flow.

and H_p is the required pump head (e.g., $-w/g$). Now the driving force in the pipe on the downhill side (from point 2 to point 3) is determined only by the potential (pressure and gravity) difference between these two points, which is independent of the flow rate. However, this driving force must be balanced by the friction loss (resistance) in the pipe:

$$\frac{\Phi_2 - \Phi_3}{\rho g} = h_{f_{23}} \tag{7-61}$$

The friction loss is determined by the fluid properties, the fluid velocity, and the pipe size. If the pipe is full of liquid, the velocity is determined by the pipe diameter and flow rate (Q), both of which are the same on the downhill side as on the uphill side for a constant area pipe. Since the downhill driving force is mainly gravity, the higher the hill the greater is the driving force relative to the "full pipe" flow resistance. Thus it is quite feasible that, for a full pipe, the downhill conditions will be such that

$$\frac{\Phi_2 - \Phi_3}{\rho g} > (h_{f_{23}})_{\text{full}} \tag{7-62}$$

Because the energy balance [Eq. (7-61)] must be satisfied, we see that the friction loss on the downhill side must increase to balance this driving force. The only way this can happen is for the velocity to increase, and the only way this can occur is for the flow cross-sectional area to decrease (because Q is fixed). The only way the flow area can change is for the liquid to fill only part of the pipe, i.e., it must flow partly full (with the remaining space filled with vapor). This condition is known as *slack flow*; the pipe is full on the upstream side of the hill but only partly full on the downstream side, with a correspondingly higher velocity in the downhill pipe such that the friction loss on the downhill side balances the driving force. Since the pressure in the vapor space is uniform, there will be no "pressure drop" in the slack flow downhill pipe, and the driving force in this section is due only to gravity.

The cross section of the fluid in the partially full pipe will not be circular (see Fig. 7-6), so the methods used for flow in a noncircular conduit are applicable, i.e., the hydraulic diameter applies. Thus, Eq. (7-61) becomes

$$z_2 - z_3 = h_{f_{23}} = \frac{2fLQ^2}{gD_h A^2} \tag{7-63}$$

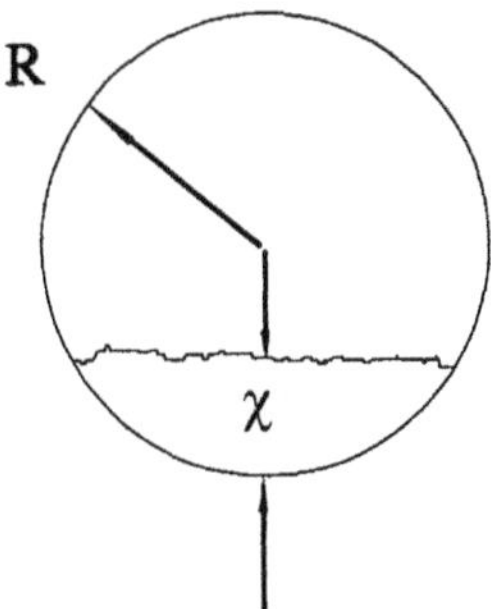

FIGURE 7-6 Pipe flowing less than full.

where $D_h = 4A/W_p$. If the depth of the liquid in the pipe is χ (which can be either larger or smaller than R; see Fig. 7-6), the expressions for the flow cross section and wetted perimeter are

$$A = R^2\left\{\cos^{-1}\left(1-\frac{\chi}{R}\right) - \left(1-\frac{\chi}{R}\right)\left[1-\left(\frac{\chi}{R}\right)^2\right]^{1/2}\right\} \tag{7-64}$$

and

$$W_p = 2R\cos^{-1}\left(1-\frac{\chi}{R}\right) \tag{7-65}$$

In order to find χ for a given pipe, fluid, and flow rate, a trial-and-error (iterative) procedure is required:

1. Assume a value of χ/R and calculate A, W_p, and D_h.
2. Calculate $N_{Re} = (D_h Q\rho)/A\mu$, and determine f from the Moody diagram (or Churchill equation).
3. Calculate the right-hand side (RHS) of Eq. (7-63). If $z_2 - z_3 <$ RHS, then increase the assumed value of χ/R and repeat the process. If $z_2 - z_3 >$ RHS, then decrease the assumed value of χ/R and repeat. The solution is obtained when $z_2 - z_3 =$ RHS of Eq. (7-63).

Example 7-2: Slack Flow. A commercial steel pipeline with a 10 in ID carries water over a 300 ft high hill. The actual length of the pipe is 500 ft on the upstream side, and 500 ft on the downstream side of the hill. Find (a) the minimum flow rate at which slack flow will not occur in the pipe and (b) the position of the interface in the pipe when the flow rate is 80% of this value.

Solution. Slack flow will not occur until the driving force (due to gravity) on the downstream side of the hill (from 2 to 3 in Fig. 7-5) exceeds the friction loss in this part of the pipeline; that is, when Eq. (7-63) is no longer satisfied with $A = \pi D^2/4$ and $D_h = D$.

(a) Since this an "unknown flow rate" problem, the flow rate can most easily be determined by first computing the value:

$$fN_{Re}^2 = \frac{g\,\Delta z\,\rho^2 D^3}{2L_{23}\mu^2} = 4.82 \times 10^{10}$$

where the fluid density has been taken to be 1 g/cm^3 and the viscosity 1 cp. This is solved iteratively with the Churchill equation for f and N_{Re}, by first assuming $f = 0.005$, then using this to get N_{Re} from the above equation. Using this N_{Re} value, with $\varepsilon/D = 0.0018/10 = 0.00018$, f is found from the Churchill equation. This process is repeated until successive values of f agree. This process gives $f = 0.0035$ and $N_{Re} = 3.73 \times 10^6$. The flow rate is then

$$Q = \frac{\pi D \mu N_{Re}}{4\rho} = 7.44 \times 10^5\,\text{cm}^3/\text{s} = 11,800\,\text{gpm}$$

(b) For a flow rate of 80% of the value found in (a), slack flow will occur, and Eq. (7-63) must be satisfied for the resulting noncircular flow section (partly full pipe). In this case, we cannot calculate either f, A, or $D_h = 4A/W_p$ a priori. Collecting the known quantities together on one side of the Eq. (7-63), we get

$$\frac{f}{D_h A^2} = \frac{g\Delta z}{2LQ^2} = 2.13 \times 10^{-9}\,\text{cm}^{-3}$$

This value is used to determine f, D_h, A, W_p, and R by iteration using Eqs. (7-64) (7-65) and the Churchill equation, as follows. Assuming a value of χ/R permits calculation of A and W_p from Eqs. (7-64) and (7-65), which also gives $D_h = 4A/W_p$. The Reynolds number is then determined from $N_{Re} = D_h Q\rho/A\mu$, which is used to determine f from the Churchill equation. These values are combined to calculate the value of $f/D_h A^2$, and the process is repeated until this value equals $2.13 \times 10^{-9}\,\text{cm}^{-3}$. The results are

$$\chi/R = 1.37, \qquad A = 57.2\,\text{in}^2, \qquad W_p = 19.5\,\text{in},$$

$$N_{Re} = 3.01 \times 10^6, \qquad f = 0.00337$$

That is, the water interface in the pipe is a little more than two-thirds of the pipe diameter above the bottom of the pipe.

VII. PIPE NETWORKS

Piping systems often involve interconnected segments in various combinations of series and/or parallel arrangements. The principles required to analyze such systems are the same as those have used for other systems, e.g., the conservation of mass (continuity) and energy (Bernoulli) equations. For each pipe junction or "node" in the network, continuity tells us that the sum of all the flow rates into the node must equal the sum of all the flow rates out of the node. Also, the total driving force (pressure drop plus gravity head loss, plus pump head) between any two nodes is related to the flow rate and friction loss by the Bernoulli equation applied between the two nodes.

If we number each of the nodes in the network (including the entrance and exit points), then the continuity equation as applied at node i relates the flow rates into and out of the node:

$$\sum_{n=1}^{n} Q_{ni} = \sum_{m=1}^{m} Q_{im} \tag{7-66}$$

where Q_{ni} represents the flow rate from any upstream node n into node i, and Q_{im} is the flow rate from node i out to any downstream node m.

Also, the total driving force in a branch between any two nodes i and j is determined by Bernoulli's equation [Eq. (7-45)] as applied to this branch. If the driving force is expressed as the total head loss between nodes (where $h_i = \Phi_i/\rho g$), then

$$h_i - h_j - \frac{w_{ij}}{g} = \frac{8Q_{ij}^2}{g\pi^2 D_{ij}^4} \sum_i^j K_{fij} \tag{7-67}$$

where $-w_{ij}/g$ is the pump head (if any) in the branch between nodes i and j, D_{ij} is the pipe diameter, Q_{ij} is the flow rate, and $\sum_i^j K_{fij}$ represents the sum of the loss coefficients for all of the fittings, valves, and pipe segments in the branch between nodes i and j. The latter are determined by the 3-K equation for all valves and fittings and the Churchill equation for all pipe segments and are functions of the flow rates and pipe sizes (Q_{ij} and D_{ij}) in the branch between nodes i and j. The total number of equations is thus equal to the number of branches plus the number of (internal) nodes, which then equals the number of unknowns that can be determined in the network.

These network equations can be solved for the unknown driving force (across each branch) or the unknown flow rate (in each branch of the network) or an unknown diameter for any one or more of the branches, subject to constraints on the pressure (driving force) and flow rates. Since the solution involves simultaneous coupled nonlinear equations, the process is best done by iteration on a computer and can usually be done by using a spreadsheet. The simplest procedure is usually to assume values for the total head h_i at one or more intermediate nodes, because these values are bounded by upstream and downstream values that are usually known, and then iterate on these internal head values.

A typical procedure for determining the flow rates in each branch of a network given the pipe sizes and pressures entering and leaving the network is illustrated by the following example.

Example 7-3: Flow in a Manifold. A manifold, or "header," distributes fluid from a common source into various branch lines, as shown in the Fig. 7-7. The manifold diameter is chosen to be much larger than that of the branches, so the pressure drop in the manifold is much smaller than that in the branch lines, ensuring that the pressure is essentially the same entering each branch. However, these conditions cannot always be satisfied in practice, especially if the total flow rate is large and/or the manifold is not sufficiently larger than the branch lines, so the assumptions should be verified.

The header illustrated is 0.5 in. in diameter and feeds three branch lines, each 0.25 in. in diameter. The five nodes are labeled in the diagram. The fluid exits each of the branches at atmospheric pressure and the same elevation, so each of the branch exit points is labeled "5" because the exit conditions are the same for all three branches. The distance between the branches on the header is 60 ft, and each branch is 200 ft long. Water enters the header (node 1) at a pressure of 100 psi and exits the branches (nodes 5) at atmospheric pressure. The entire network is assumed to be horizontal. Each branch contains two globe valves in addition to the 200 ft of pipe and the entrance fitting from the header to the branch. We must determine the

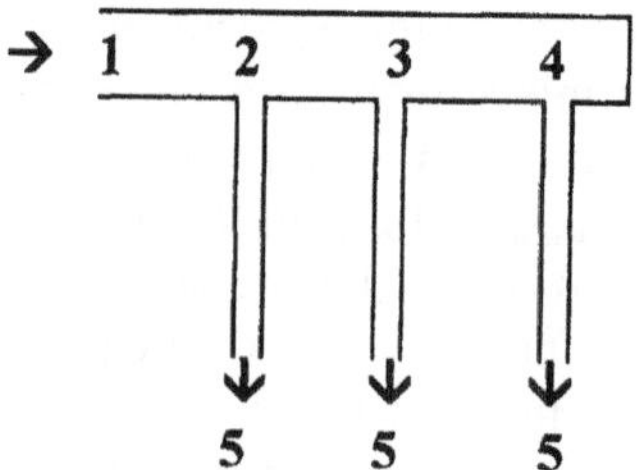

FIGURE 7-7 Flow in a header.

pressure at the entrance to each branch, the flow rate through each branch, and the total flow rate entering the system. The equations for this problem are set up in a spreadsheet, the output of which is shown in Table 7-6. The procedure followed is outlined below.

The head at both the entrance to the header ($h_1 = 230.8$ ft) and the exit from the branches ($h_5 = 0$) is known. If the head at node 2 were known, Bernoulli's equation [Eq. (7-67)] could be used to calculate the flow rate from 1 to 2 (Q_{12}) and the flow rate from 2 to 5 (Q_{25}). By continuity, the flow rate from 2 to 3 must be the difference between these ($Q_{23} = Q_{12} - Q_{25}$). This flow rate is then used in Eq. (7-67) to determine the total head at node 3

TABLE 7-6 Spreadsheet Output for Example 7-3

Example - Manifold		Fluid: Water				
	Rho =	1	g/cc	Mu =	0.01	Poise
	P1 =	100	psig	P5 =	0	psig
Branch:	D12 =	0.5	in	L12 =	60	ft.
	D23 =	0.5	in.	L23 =	60	ft.
	D34 =	0.5	in.	L34 =	60	ft.
	D25 =	0.25	in.	L25 =	200	ft.
	D35 =	0.25	in.	L35 =	200	ft.
	D45 =	0.25	in.	L45 =	200	ft.
Loss Cof	K12 =	28.8				
(First Est.	K23 =	28.8		e =	0.0018	in.
	K34 =	28.8				
	K25 =	232.5		hi is in feet		
	K35 =	232.5				
	K45 =	232.5		Qij is in ft^3/s		

	Node I:	1		2		3		4		5	
				Assume							
ft(gage)	hi	230.769		216		208.757		206.408		0	
ft^3/s	Qij		0.00783		0.00411		0.00221		0.00217		
	Qi5			0.00264		0.00259		0.00258			
	NRe		22237	14965	11680	14712	6267	14629	12342		
	B		4.3E+03	2.4E+06	1.3E+08	3.2E+06	2.7E+12	3.5E+06	5.3E+07		
	A		1.4E+19	3.1E+18	6.6E+18	3.1E+18	2.5E+18	8.7E+18	2.6E+18		
	f		0.00811	0.00976	0.00889	0.00978	0.01001	0.00858	0.00998		
	SumK		46.7	415.6	51.2	416.3	57.7	370.1	423.8		
ft^3/s	RevQij		0.00615		0.00411		0.00221		0.00217		
ft^3/s	RevQi5			0.00197		0.00194		0.00204			
	RevNRe		17458	11194	11680	10995	6267	11595	12342		
	B		2.1E+05	2.5E+08	1.3E+08	3.4E+08	2.7E+12	1.5E+08	5.3E+07		
	A		1.1E+19	2.4E+18	6.6E+18	2.3E+18	2.5E+18	2.7E+19	2.6E+18		
	f		0.00836	0.0101	0.00889	0.01012	0.01001	0.00744	0.00998		
	RevSumK		48.2	428.4	51.2	429.3	57.7	326.6	423.8		
ft^3/s	RevQij		0.00606		0.00411		0.00221		0.00217		
ft^3/s	RevQi5			0.00194		0.00191		0.00217	0.00191		CHECK
	SumQin=0			6.4E-05		2.2E-19		3.4E-05			
ft(gage)	Calc h5			-1.9E-14		1.2E-13		3.8E-14			
		Total Flow Rate = Q25 + Q35 + Q45 =				0.00602	=?Q12 =	0.00606	ft^3/s		CHECK
					=	2.70	gpm				

(h_3). With h_3 known, Q_{35} can be determined from Eq. (7-67), as above. Q_{34} is then determined from continuity ($Q_{34} = Q_{23} - Q_{35}$). Knowing Q_{34} and h_3 allows h_4 to be determined from Eq. (7-67). Q_{45} can then be determined from the known values of h_4 and h_5, as well as by continuity, since $Q_{34} = Q_{45}$. Comparison of these two values of Q_{45} provides a check on the converged solution, as does the overall continuity equation $Q_{12} = Q_{25} + Q_{35} + Q_{45}$.

The spreadsheet calculations are done by first assuming a value for h_2 and checking the continuity of the flow rates for agreement. The value of h_2 is adjusted until these checks are in reasonable agreement. The calculation of flow rate from Bernoulli's equation [Eq. (7-67)] is also iterative, because the equation involves the loss coefficients, which depend upon the flow rate through the Reynolds number. Thus, initial estimates for these loss coefficients must be made, as shown on the spreadsheet output (these are based on $f = 0.005$ for the pipe segments, and $K_f = K_\infty$ for each fitting). The iteration revises these values using the Churchill equation for $K_{pipe} = 4fL/D$ and the 3-K method for the fittings. Only two iteration steps are needed for each unknown Q calculation, as shown on the output. The result shows Spreadsheet Output for Example 7-3 that the head in the manifold drops about 10%, but this does not result in the flow rates in he branch lines varying significantly. The spreadsheet is also very convenient for "what if?" analyses, because it is easy to change any of the known conditions, pipe sizes, fluid properties, etc. and immediately observe the results.

PROBLEMS

1. You must design a pipeline to carry crude oil at a rate of 1 million barrels per day. If the viscosity of the oil is 25 cP and its SG is 0.9, what is the most economic diameter for the pipeline if the pipe cost is \$3 per foot of length and per inch of diameter, if the power cost \$0.05/kWh, and the cost of the pipeline is to be written off over a 3 year period? The oil enters and leaves the pipeline at atmospheric pressure. What would the answer be if the economic lifetime of the pipeline were 30 years?
2. A crude oil pipeline is to be built to carry oil at the rate of 1 million barrels per day (1 bbl = 42 gal). If the pipe costs \$12 per foot of length per inch of diameter, power to run the pumps costs \$0.07/kWh, and the economic lifetime of the pipeline is 30 yr, what is the most economic diameter for the pipeline? What total pump horsepower would be required if the line is 800 mi long, assuming 100% efficient pumps? (Oil: $\mu = 35\,\text{cP}$, $\rho = 0.85\,\text{g/cm}^3$).
3. A coal slurry pipeline is to be built to transport 45 million tons/yr of coal slurry a distance of 1500 mi. The slurry can be approximately described as Newtonian, with a viscosity of 35 cP and SG of 1.25. The pipeline is to be built from ANSI 600# commercial steel pipe, the pumps are 50% efficient, energy costs are \$0.06/

kWh, and the economic lifetime of the pipeline is 25 years. What would be the most economical pipe diameter, and what would be the corresponding velocity in the line?

4. The Alaskan pipeline was designed to carry crude oil at a rate of 1.2 million bbl/day (1 bbl = 42 gal). If the oil is assumed to be Newtonian, with a viscosity of 25 cP and SG = 0.85, the cost of energy is \$0.1/kWh, and the pipe grade is 600# ANSI, what would be the most economical diameter for the pipeline? Assume that the economic lifetime of the pipeline is 30 years.
5. What is the most economical diameter of a pipeline that is required to transport crude oil ($\mu = 30$ cP, SG = 0.95) at a rate of 1 million bbl/day using ANSI 1500# pipe if the cost of energy is \$0.05/kWh (in 1980 dollars), the economic lifetime of the pipeline is 40 yr, and the pumps are 50% efficient.
6. Find the most economical diameter of sch 40 commercial steel pipe that would be needed to transport a petroleum fraction with a viscosity of 60 cP and SG of 1.3 at a rate of 1500 gpm. The economic life of the pipeline is 30 yr, the cost of energy is \$0.08/kWh, and the pump efficiency is 60%. The cost of pipe is \$20 per ft length per inch ID. What would be the most economical diameter to use if the pipe is stainless steel, at a cost of \$85 per (ft in. ID), all other things being equal?
7. You must design and specify equipment for transporting 100% acetic acid (SG = 1.0, $\mu = 1$ cP) at a rate of 50 gpm from a large vessel at ground level into a storage tank that is 20 ft above the vessel. The line includes 500 ft of pipe and eight flanged elbows. It is necessary to use stainless steel for the system (pipe is hydraulically smooth), and you must determine the most economical size pipe to use. You have 1.5 in. and 2 in. nominal sch 40 pipe available for the job. Cost may be estimated from the following approximate formulas:

   Pump cost: Cost (\$) = 75.2 (gpm)$^{0.3}$ (ft of head)$^{0.25}$
   Motor cost: Cost (\$) = 75 (hp)$^{0.85}$
   Pipe cost: Cost (\$)/ft = 2.5 (nom. dia, in.)$^{3/2}$
   90° elbow: Cost (\$) = 5 (nom. dia, in)$^{1.5}$
   Power: Cost = 0.03 \$/kWh

   (a) Calculate the total pump head (i.e., pressure drop) required for each size pipe, in ft of head.
   (b) Calculate the motor hp required for each size pipe assuming 80% pump efficiency (motors available only in multiples of 1/4 hp)
   (c) Calculate the total capital cost for pump, motor, pipe, and fittings for each size pipe.
   (d) Assuming that the useful life of the installation is 5 yr, calculate the total operating cost over this period for each size pipe.
   (e) Which size pipe results in the lowest total cost over the 5 yr period?
8. A large building has a roof with dimensions 50 ft × 200 ft, that drains into a gutter system. The gutter contains three drawn aluminum downspouts that have a square cross section, 3 in. on a side. The length of the downspouts from the roof to the ground is 20 ft. What is the heaviest rainfall (in in./hr) that the downspouts can handle before the gutter will overflow?

9. A roof drains into a gutter, which feeds the water into a downspout with a square cross section (4 in. × 4 in.). The discharge end of the downspout is 12 ft below the entrance, and terminates in a 90° mitered (one weld) elbow. The downspout is made of smooth sheet metal.
   (a) What is the capacity of the downspout, in gpm?
   (b) What would the capacity be if there were no elbow on the end?
10. An open concrete flume is to be constructed to carry water from a plant unit to a cooling lake by gravity flow. The flume has a square cross section and is 1500 ft long. The elevation at the upstream end is 10 ft higher than that of the lower end. If the flume is to be designed to carry 10,000 gpm of water when full, what should its size (i.e., width) be? Assume rough cast concrete.
11. An open drainage canal with a rectangular cross section is 10 ft wide and 5 ft deep. If the canal slopes 5 ft in 1 mi, what is the capacity of the canal in gpm when running full of water?
12. A concrete lined drainage ditch has a triangular cross section that is an equilateral triangle 8 ft on each side. The ditch has a slope of 3 ft/mi. What is the flow capacity of the ditch, in gpm?
13. An open drainage canal is to be constructed to carry water at a maximum rate of 10^6 gpm. The canal is concrete-lined and has a rectangular cross section with a width that is twice its depth. The elevation of the canal drops 3 ft per mile of length. What should the dimensions of the canal be?
14. A drainage ditch is to be built to carry rainfall runoff from a subdivision. The maximum design capacity is to be 1 million gph (gal/hr), and it will be concrete lined. If the ditch has a cross section that is an equilateral triangle (open at the top) and if it has a slope of 2 ft/mi, what should the width at the top be?
15. A drainage canal is to be dug to keep a low-lying area from flooding during heavy rains. The canal would carry the water to a river that is 1 mi away and 6 ft lower in elevation. The canal will be lined with cast concrete and will have a semicircular cross section. If it is sized to drain all of the water falling on a 1 mi^2 area during a rainfall of 4 in./hr, what should the diameter of the semicircle be?
16. An open drainage canal with a rectangular cross section and a width of 20 ft is lined with concrete. The canal has a slope of 1 ft/1000 yd. What is the depth of water in the canal when the water is flowing through it at a rate of 500,000 gpm?
17. An air ventilating system must be designed to deliver air at 20°F and atmospheric pressure at a rate of 150 ft^3/s, through 4000 ft of square duct. If the air blower is 60% efficient and is driven by a 30 hp motor, what size duct is required if it is made of sheet metal?
18. Oil with a viscosity of 25 cP and SG of 0.78 is contained in a large open tank. A vertical tube made of commercial steel, with a 1 in. ID and a length of 6 ft, is attached to the bottom of the tank. You want the oil to drain through the tube at a rate of 30 gpm.
   (a) How deep should the oil in the tank be for it to drain at this rate?
   (b) If a globe valve is installed in the tube, how deep must the oil be to drain at the same rate, with the valve wide open?

19. A vertical tube is attached to the bottom of a vessel that is open to the atmosphere. A liquid with SG = 1.2 is draining from the vessel through the tube, which is 10 cm long and has an ID of 3 mm. When the depth of the liquid in the vessel is 4 cm, the flow rate through the tube is 5 cm^3/s.
    (a) What is the viscosity of the liquid (assuming it is Newtonian)?
    (b) What would your answer be if you neglected the entrance loss from the tank to the tube?
20. Heat is to be transferred from one process stream to another by means of a double pipe heat exchanger. The hot fluid flows in a 1 in. sch 40 tube, which is inside (concentric with) a 2 in. sch 40 tube, with the cold fluid flowing in the annulus between the tubes. If both fluids are to flow at a velocity of 8 ft/s and the total equivalent length of the tubes is 1300 ft, what pump power is required to circulate the colder fluid? Properties at average temperature: $\rho = 55\,\text{lb}_m/\text{ft}^3$, $\mu = 8$ cP.
21. A commercial steel ($e = 0.0018$ in.) pipeline is $1\frac{1}{4}$ in. sch 40 diameter, 50 ft long, and includes one globe valve. If the pressure drop across the entire line is 22.1 psi when it is carrying water at a rate of 65 gpm, what is the loss coefficient for the globe valve? The friction factor for the pipe is given by the equation

    $$f = 0.0625/[\log(3.7D/e)]^2$$

22. Water at 68°F is flowing through a 45° pipe bend at a rate of 2000 gpm. The inlet to the bend is 3 in. ID, and the outlet is 4 in. ID. The pressure at the inlet is 100 psig, and the pressure drop in the bend is equal to half of what it would be in a 3 in. 90° elbow. Calculate the net force (magnitude and direction) that the water exerts on the pipe bend.
23. What size pump (horsepower) is required to pump oil (SG = 0.85, $\mu = 60$ cP) from tank A to tank B at a rate of 2000 gpm through a 10 in. sch 40 pipeline, 500 ft long, containing 20 90° elbows, one open globe valve, and two open gate valves? The oil level in tank A is 20 ft below that in tank B, and both are open to the atmosphere.
24. A plant piping system takes a process stream ($\mu = 15$ cP, $\rho = 0.9\,\text{g/cm}^3$) from one vessel at 20 psig, and delivers it to another vessel at 80 psig. The system contains 900 ft of 2 in. sch 40 pipe, 24 standard elbows, and five globe valves. If the downstream vessel is 10 ft higher than the upstream vessel, what horsepower pump would be required to transport the fluid at a rate of 100 gpm, assuming a pump efficiency of 100%?
25. Crude oil ($\mu = 40$ cP, SG = 0.87) is to be pumped from a storage tank to a refinery through a 10 in. sch 20 commercial steel pipeline at a flow rate of 2000 gpm. The pipeline is 50 mi long and contains 35 90° elbows and 10 open gate valves. The pipeline exit is 150 ft higher than the entrance, and the exit pressure is 25 psig. What horsepower is required to drive the pumps in the system if they are 70% efficient?
26. The Alaskan pipeline is 48 in. ID, 800 mi long, and carries crude oil at a rate of 1.2 million bbl/day (1 bbl = 42 gal). Assuming the crude oil to be a Newtonian fluid with a viscosity of 25 cP and SG of 0.87, what is the total pumping horse-

power required to operate the pipeline? The oil enters and leaves the pipeline at sea level, and the line contains the equivalent of 150 90° elbows and 100 open gate valves. Assume that inlet and discharge pressures are 1 atm.

27. A 6 in. sch 40 pipeline carries oil ($\mu = 15$ cP, SG $= 0.85$) at a velocity of 7.5 ft/s from a storage tank at 1 atm pressure to a plant site. The line contains 1500 ft of straight pipe, 25 90° elbows, and four open globe valves. The oil level in the storage tank is 15 ft above ground, and the pipeline discharges at a point 10 ft above ground at a pressure of 10 psig. What is the required flow capacity in gpm and the pressure head to be specified for the pump needed for this job? If the pump is 65% efficient, what horsepower motor is required to drive the pump?
28. An open tank contains 5 ft of water. The tank drains through a piping system containing ten 90° elbows, ten branched tees, six gate valves, and 40 ft of horizontal sch 40 pipe. The top surface of the water and the pipe discharge are both at atmospheric pressure. An entrance loss factor of 1.5 accounts for the tank-to-pipe friction loss and kinetic energy change. Calculate the flow rate (in gpm) and Reynolds number for a fluid with a viscosity of 10 cp draining through sch 40 pipe with nominal diameters of 1/8, 1/4, 1/2, 1, 1.5, 2, 4, 6, 8, 10, and 12 in., including all of the above fittings, using (a) constant K_f values, (b) $(L/D)_{eq}$ values, (c) the 2-K method, and (d) the 3-K method. Constant K_f and $(L/D)_{eq}$ values from the literature are given below for these fittings:

Fitting	Constant K_f	$(L/D)_{eq}$
90° elbow	0.75	30
Branch tee	1.0	60
Gate valve	0.17	8

29. A pump takes water from a reservoir and delivers it to a water tower. The water in the tower is at atmospheric pressure and is 120 ft above the reservoir. The pipeline is composed of 1000 ft of straight 2 in. sch 40 pipe containing 32 gate valves, two globe valves, and 14 standard elbows. If the water is to be pumped at a rate of 100 gpm using a pump that is 70% efficient, what horsepower motor would be required to drive the pump?
30. You must determine the pump pressure and power required to transport a petroleum fraction ($\mu = 60$ cP, $\rho = 55\ \text{lbm/ft}^3$) at a rate of 500 gpm from a storage tank to the feed plate of a distillation column. The pressure in the tank is 2 psig, and that in the column is 20 psig. The liquid level in the tank is 15 ft above ground, and the column inlet is 60 ft high. If the piping system contains 400 ft of 6 in. sch 80 steel pipe, 18 standard elbows, and four globe valves, calculate the required pump head (i.e., pressure rise) and the horsepower required if the pump is 70% efficient.
31. What horsepower pump would be required to transfer water at a flow rate of 100 gpm from tank A to tank B if the liquid surface in tank A is 8 ft above ground and that in tank B is 45 ft above ground? The piping between tanks

consists of 150 ft of $1\frac{1}{2}$ in. sch 40 pipe and 450 ft of 2 in. sch 40 pipe, including 16 90° standard elbows and four open globe valves.

32. A roof drains into a gutter, which feeds the water into a downspout that has a square cross section (4 in. × 4 in.). The discharge end of the downspout is 12 ft below the entrance and terminates in a 90° mitered (one weld) elbow. The downspout is made of smooth sheet metal.
    (a) What is the capacity of the downspout, in gpm?
    (b) What would the capacity be if there were no elbow on the end?
33. An additive having a viscosity of 2 cP and a density of 50 lb_m/ft^3 is fed from a reservoir into a mixing tank. The pressure in the reservoir and in the tank is 1 atm, and the level in the reservoir is 2 ft above the end of the feed line in the tank. The feed line contains 10 ft of 1/4 in. sch 40 pipe, four elbows, two plug valves, and one globe valve. What will the flow rate of the additive be, in gpm, if the valves are fully open?
34. The pressure in the water main serving your house is 90 psig. The plumbing between the main and your outside faucet contains 250 ft of galvanized 3/4 in. sch 40 pipe, 16 elbows, and the faucet which is an angle valve. When the faucet is wide open, what is the flow rate, in gpm?
35. You are filling your beer mug from a keg. The pressure in the keg is 5 psig, the filling tube from the keg is 3 ft long with 1/4 in. ID, and the valve is a diaphragm dam type. The tube is attached to the keg by a (threaded) tee, used as an elbow. If the beer leaving the tube is ft above the level of the beer inside the keg and there is a 2 ft long, 1/4 in. ID stainless steel tube inside the keg, how long will it take to fill your mug if it holds 500 cm^3? (Beer: $\mu = 8$ cP, $\rho = 64\ lb_m/ft^3$)
36. You must install a piping system to drain SAE 10 lube oil at 70°F (SG = 0.928) from tank A to tank B by gravity flow. The level in tank A is 10 ft above that in tank B, and the pressure in A is 5 psi greater than that in tank B. The system will contain 200 ft of sch 40 pipe, eight std elbows, two gate valves, and a globe valve. What size pipe should be used if the oil is to be drained at a rate of 100 gpm?
37. A new industrial plant requires a supply of water at a rate of 5.7 m^3/min. The gauge pressure in the water main, which is located 50 m from the plant, is 800 kPa. The supply line from the main to the plant will require a total length of 65 m of galvanized iron pipe, four standard elbows, and two gate valves. If the water pressure at the plant must be no less than 500 kPa, what size pipe should be used?
38. A pump is used to transport water at 72°F from tank A to tank B at a rate of 200 gpm. Both tanks are vented to the atmosphere. Tank A is 6 ft above the ground with a water depth of 4 ft in the tank, and tank B is 40 ft above ground with a water depth of 5 ft . The water enters the top of tank B, at a point 10 ft above the bottom of the tank. The pipeline joining the tanks contains 185 ft of 2 in. sch 40 galvanized iron pipe, three standard elbows, and one gate valve.
    (a) If the pump is 70% efficient, what horsepower motor would be required to drive the pump?
    (b) If the pump is driven by a 5 hp motor, what is the maximum flow rate that can be achieved, in gpm?

39. A pipeline carrying gasoline (SG = 0.72, $\mu = 0.7$ cP) is 5 mi long and is made of 6 in. sch 40 commercial steel pipe. The line contains 24 of 90° elbows, eight open gate valves, two open globe valves, and a pump capable of producing a maximum head of 400 ft . The line inlet pressure is 10 psig and the exit pressure is 20 psig. The discharge end is 30 ft higher than the inlet end.
    (a) What is the maximum flow rate possible in the line, in gpm?
    (b) What is the horsepower of the motor required to drive the pump if it is 60% efficient?
40. A water tower supplies water to a small community of 800 houses. The level of the water in the tank is 120 ft above ground level, and the water main from the tower to the housing area is 1 mi of sch 40 commercial steel pipe. The water system is designed to provide a minimum pressure of 15 psig at peak demand, which is estimated to be 2 gpm per house.
    (a) What nominal size pipe should be used for the water main?
    (b) If this pipe is installed, what would be the actual flow rate through the water main, in gpm?
41. A 12 in. sch 40 pipe, 60 ft long, discharges water at 1 atm pressure from a reservoir. The pipe is horizontal, and the inlet is 12 ft below the surface of the water in the reservoir.
    (a) What is the flow rate in gpm?
    (b) In order to limit the flow rate to 3,500 gpm, an orifice is installed at the end of the pipe. What should the orifice diameter be?
    (c) What size pipe would have to be used to limit the flow rate to 3500 gpm without using an orifice?
42. Crude oil with a viscosity of 12.5 cP and SG = 0.88 is to be pumped through a 12 in. sch 30 commercial steel pipe at a rate of 1900 bbl/hr. The pipeline is 15 mi long, with a discharge that is 250 ft above the inlet, and contains 10 standard elbows and four gate
    (a) What pump horsepower is required if the pump is 67% efficient?
    (b) If the cost of energy is $0.08/kWh and the pipe is 600# ANSI steel, is the 12 in. pipe the most economical one to use (assume a 30 yr economic life of the pipeline)? If not, what is the most economical diameter?
43. A pipeline to carry crude oil at a rate of 1 million bbl/day is constructed with 50 in. ID pipe and is 700 mi long with the equivalent of 70 gate valves installed but no other fittings:
    (a) What is the total power required to drive the pumps if they are 70% efficient?
    (b) How many pump stations will be required if the pumps develop a discharge pressure of 100 psi?
    (c) If the pipeline must go over hilly terrain, what is the steepest downslope grade that can be tolerated without creating slack flow in the pipe line? (Crude oil viscosity is 25 cP, SG = 0.9.)
44. You are building a pipeline to transport crude oil (SG = 0.8, $\mu = 30$ cP) from a seaport over a mountain to a tank farm. The top of the mountain is 3000 ft above the seaport and 1000 ft above the tank farm. The distance from the port

to the mountain top is 200 mi, and from the mountain top to the tank farm it is 75 mi. The oil enters the pumping station at the port at 1 atm pressure and is to be discharged at the tank farm at 20 psig. The pipeline is 20 in. sch 40, and the oil flow rate is 2000 gpm.

(a) Will slack flow occur in the line? If so, you must install a restriction (orifice) in the line to ensure that the pipe is always full. What should the pressure loss across the orifice be, in psi?

(b) How much pumping power will be required, if the pumps are 70% efficient? What pump head is required?

45. You want to siphon water from an open tank using a hose. The discharge end of the hose is 10 ft below the water level in the tank. The minimum allowable pressure in the hose for proper operation is 1 psia. If you wish the water velocity in the hose to be 10 ft/s, what is the maximum height that the siphon hose can extend above the water level in the tank for proper operation?

46. A liquid is draining from a cylindrical vessel through a tube in the bottom of the vessel, as illustrated in Fig. 7-46. The liquid has a specific gravity of 1.2 and a viscosity of 2 cP. The entrance loss coefficient from the tank to the tube is 0.4, and the system has the following dimensions:

$$D = 2\,\text{in.}, \qquad d = 3\,\text{mm}, \qquad L = 20\,\text{cm}, \qquad h = 5\,\text{cm}, \qquad e = 0.0004\,\text{in.}$$

(a) What is the volumetric flow rate of the liquid in cm^3/s?

(b) What would the answer to (a) be if the entrance loss were neglected?

(c) Repeat part (a) for a value of $h = 75$ cm.

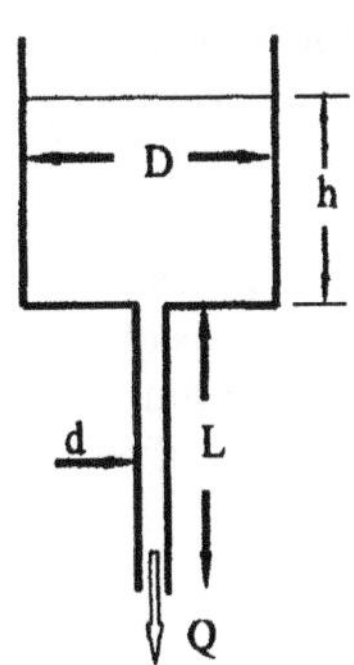

FIGURE 7-P46.

47. Water from a lake is flowing over a concrete spillway at a rate of 100,000 gpm. The spillway is 100 ft wide and is inclined at a 30° angle to the vertical. If the effective roughness of the concrete is 0.03 in., what is the depth of water in the stream flowing down the spillway?

48. A pipeline consisting of 1500 ft of 6 in. sch 40 pipe containing 25 90° elbows and four open gate valves carries oil with a viscosity of 35 cP and a specific gravity of 0.85 at a velocity of 7.5 ft/s from a storage tank to a plant site. The storage

tank is at atmospheric pressure and the level in the tank is 15 ft above ground. The pipeline discharge is 10 ft above ground, and the discharge pressure is 10 psig.

(a) What is the required pump capacity (in gpm) and pump head (in ft) needed in this pipeline?
(b) If the pump has an efficiency of 65%, what horsepower motor would be required to drive it?

49. A centrifugal pump is located 4 ft above the surface of water in a tank. The suction line to the pump is 6 in. sch 40 pipe and extends 6 ft below the surface of the water. If the water temperature is 50°F, what is the pressure (in psia) at the pump inlet when the flow rate is 500 gpm?
50. Water is pumped at a rate of 500 gpm through a 10 in. ID pipeline, 50 ft long, that contain two standard elbows and a swing check valve. The pressure is 1 atm entering and leaving the pipeline. Calculate the pressure drop (in psi) through the pipeline due to friction using (a) the 2-K method; (b) the $(L/D)_{eq}$ method; (c) the 3-K method.
51. Water at 70°F is flowing in a film down the outer surface of a 4 in. OD vertical tube at a rate of 1 gpm. What is the thickness of the film?
52. What diameter of pipe would be required to transport a liquid with a viscosity of 1 cP and a density of 1 g/cm^3 at a rate of 1500 gpm, if the length of the pipe is 213 ft, the wall roughness is 0.006 in., and the total driving force is 100 ft lb$_f$/lb$_m$?
53. The ETSI pipeline was designed to carry a coal slurry from Wyoming to Texas at a rate of 30×10^6 tons/yr. The slurry behaves like a Bingham plastic, with a yield stress of 100 dyn/cm^2, a limiting viscosity of 40 cP, and a density of 1.4 g/cm^3. Using the cost of ANSI 1500# pipe and 7¢/kWh for electricity, determine the most economical diameter for the pipeline if its economical lifetime is 25 yr and the pumps are 50% efficient.
54. A mud slurry is drained from a tank through a 50 ft long plastic hose. The hose has an elliptical cross section with a major axis of 4 in. and a minor axis of 2 in. The open end of the hose is 10 ft below the level in the tank. The mud is a Bingham plastic, with a yield stress of 100 dyn/cm^2, a limiting viscosity of 50 cP, and a density of 1.4 g/cm^3.

(a) At what rate will the mud drain through the hose (in gpm)?
(b) At what rate would water drain through the hose?

55. A 90° threaded elbow is attached to the end of a 3 in. sch 40 pipe, and a reducer with an inside diameter of 1 in. is threaded into the elbow. If water is pumped through the pipe and out the reducer into the atmosphere at a rate of 500 gpm, calculate the forces exerted on the pipe at the point where the elbow is attached.
56. A continuous flow reactor vessel contains a liquid reacting mixture with a density of 0.85 g/cm^3 and a viscosity of 7 cP at 1 atm pressure. Near the bottom of the vessel is a $1\frac{1}{2}$ in. outlet line containing a safety relief valve. There is 4 ft of pipe with two 90° elbows between the tank and the valve. The relief valve is a spring-loaded lift check valve, which opens when the pressure on the upstream

side of the valve reaches 5 psig. Downstream of the valve is 30 ft of horizontal pipe containing four elbows and two gate valves that empties into a vented catch tank. The check valve essentially serves as a level control for the liquid in the reactor because the static head in the reactor is the only source of pressure on the valve. Determine

(a) The fluid level in the reactor at the point where the valve opens.
(b) When the valve opens, the rate (in gpm) at which the liquid will drain from the reactor into the catch tank.

NOTATION

A	cross sectional area, [L^2]
A	pump station cost parameter, Fig. 7-4, [\$]
a	pipe cost parameter, [$4/L^{p+1}$]
B	pump station cost parameter, Fig. 7-4, [$\$t/FL = \$t^3/ML^2$]
C	energy cost, [$\$/FL = \$t^2/ML^2$]
D	diameter, [L]
D	hydraulic diameter, [L]
DF	driving force, Eq. (7-42), [$FL/M = L^2/t^2$]
e_f	energy dissipated per unit mass of fluid, [$FL/M = L^2/t^2$]
f	Fanning friction factor, [L]
f_T	fully turbulent friction factor, Eq. (7-36), [—]
h	fluid layer thickness, [L], or total/head potenbtial (m/pg), [L]
H_p	required pump head, [L]
h_f	friction loss head, [2]
HP	power, [$FL/t = ML^2/t^3$]
ID_{in}	pipe inside diameter in inches, [L]
K_f	loss coefficient, [—]
K_1, K_∞	2-K loss coefficient parameters, [—]
K_1, K_i, K_d	3-K loss coefficient parameters, [—]
L	length, [L]
$\dot{m}$	mass flow rate, [M/t]
N_c	cost group, Eq. (7-26), [—]
N_{He}	Hedstrom number, [—]
$N_{Re,h}$	Reynolds number based on hydraulic diameter, [—]
$N_{Re,pl}$	power law Reynolds number, [—]
Q	volumetric flow rate, [L^3/t]
R	pipe radius, [L]
V	spatial average velocity, [L/t]
W_p	wetted perimeter, [L]
χ	position of influence in partially full pipe [L]
X	fraction of capital cost charged per unit time, [1/1t]
Y	economic lifetime $= 1/X$, [t]

α	kinetic energy correction factor, [—]
$\Delta(\ )$	$(\ )_1 - (\ )_2$
ε	roughness, [L]
η_e	efficiency, [—]
ρ	density, $[M/L^3]$
Φ	potential $(P + \rho g z)$, $[F/L^2 = M/Lt^2]$

Subscripts

1	reference point 1
2	reference point 2
ij	difference in values between points *i* and *j*
n	nominal pipe size in inches

REFERENCES

CCPS (Center for Chemical Process Safety), Guidelines for Pressure Relief and Effluent Handling Systems. New York: AIChE, 1998.

Crane Co. Flow of Fluids Through Valves, Fittings, and Pipe. Tech Paper 410. New York: Crane Co, 1991.

Darby, R. Friction loss in valves and fittings—Part II. *Chem Eng* 2001 (in press).

Darby R, J Forsyth. *Can J Chem Eng* **70**:97–103, 1992.

Darby R, JD Melson. *J Pipelines* **2**:11–21, 1982.

Durand AA, JA Boy, JL Corral, LO Barra, JS Trueba, PV Brena. Update rules for pipe sizing. *Chem Eng* May 1999, pp 153–156.

Hooper, WB. *Chem Eng* Aug 24, 1981, p. 97.

Hooper, WB. *Chem Eng* November 1988, p 89.

8

Pumps and Compressors

I. PUMPS

There exist a wide variety of pumps that are designed for various specific applications. However, most of them can be broadly classified into two categories: positive displacement and centrifugal. The most significant characteristics of each of these are described below.

A. Positive Displacement Pumps

The term *positive displacement pump* is quite descriptive, because such pumps are designed to displace a more or less fixed volume of fluid during each cycle of operation. They include piston, diaphragm, screw, gear, progressing cavity, and other pumps. The volumetric flow rate is determined by the displacement per cycle of the moving member (either rotating or reciprocating) times the cycle rate (e.g., rpm). The flow capacity is thus fixed by the design, size, and operating speed of the pump. The pressure (or head) that the pump develops depends upon the flow resistance of the system in which the pump is installed and is limited only by the size of the driving motor and the strength of the parts. Consequently, the discharge line from the pump should never be closed off without allowing for recycle around the pump or damage to the pump could result.

In general positive displacement pumps have limited flow capacity but are capable of relatively high pressures. Thus these pumps operate at essentially constant flow rate, with variable head. They are appropriate for high pressure requirements, very viscous fluids, and applications that require a precisely controlled or metered flow rate.

B. Centrifugal Pumps

The term "centrifugal pumps" is also very descriptive, because these pumps operate by the transfer of energy (or angular momentum) from a rotating impeller to the fluid, which is normally inside a casing. A sectional view of a typical centrifugal pump is shown in Fig. 8-1. The fluid enters at the axis or "eye" of the impeller (which may be open or closed and usually contains radial curved vanes) and is discharged from the impeller periphery. The kinetic energy and momentum of the fluid are increased by the angular momentum imparted by the high-speed impeller. This kinetic energy is then converted to pressure energy ("head") in a diverging area (the "volute") between the impeller discharge and the casing before the fluid exits the pump. The head that these pumps can develop depends upon the pump design and the size, shape, and speed of the impeller and the flow capacity is determined by the flow resistance of the system in which the pump is installed. Thus, as will be shown, these pumps operate at approxi-

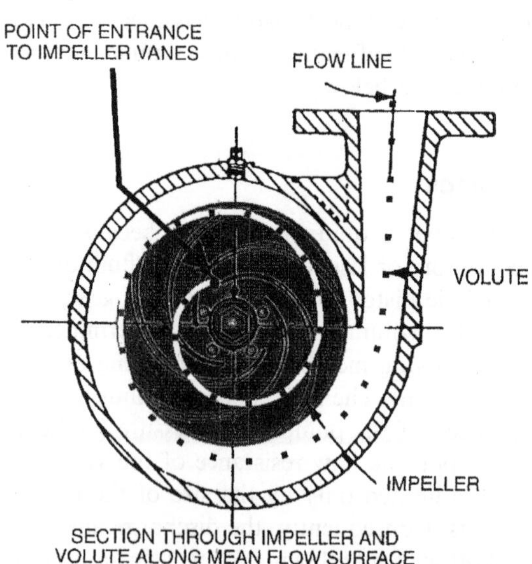

FIGURE 8-1 Sectional view of a typical centrifugal pump.

mately constant head and variable flow rate, within limits, of course, determined by the size and design of the pump and the size of the driving motor.

Centrifugal pumps can be operated in a "closed off" condition (i.e., closed discharge line), bceause the liquid will recirculate within the pump without causing damage. However, such conditions should be avoided, because energy dissipation within the pump could result in excessive heating of the fluid and/or the pump or unstable operation, with adverse consequences. Centrifugal pumps are most appropriate for "ordinary" (i.e., low to moderate viscosity) liquids under a wide variety of flow conditions and are thus the most common type of pump. The following discussion applies primarily to centrifugal pumps.

II. PUMP CHARACTERISTICS

Bernoulli's equation applied between the suction and the discharge of a pump gives

$$-w = \frac{\Delta P}{\rho} = gH_p \tag{8-1}$$

That is, the *net* energy or work put into the fluid by the pump goes to increasing the fluid pressure or the equivalent pump head, H_p. However, because pumps are not 100% efficient, some of the energy delivered from the motor to the pump is dissipated or "lost" due to friction. It is very difficult to separately characterize this friction loss, so it is accounted for by the pump efficiency, η_e, which is the ratio of the useful work (or hydraulic work) done by the pump on the fluid ($-w$) to the work put into the pump by the motor ($-w_m$):

$$\eta_e = \frac{-w}{-w_m} \tag{8-2}$$

The efficiency of a pump depends upon the pump and impeller design, the size and speed of the impeller, and the conditions under which it is operating and is determined by tests carried out by the pump manufacturer. This will be discussed in more detail later.

When selecting a pump for a particular application, it is first necessary to specify the flow capacity and head required of the pump. Although many pumps might be able to meet these specifications, the "best" pump is normally the one that has the highest efficiency at the specified operating conditions. The required operating conditions, along with a knowledge of the pump efficiency, then allow us to determine the required size (e.g., brake horsepower, HP) of the driving motor for the pump:

$$HP = -w_m \dot{m} = \frac{\Delta P Q}{\eta_e} = \frac{\rho g H_p Q}{\eta_e} \tag{8-3}$$

Now the power delivered from the motor to the pump is also the product of the torque on the shaft driving the pump (Γ) and the angular velocity of the shaft (ω):

$$HP = \Gamma \omega = \frac{\rho g H_p Q}{\eta_e} \tag{8-4}$$

If it is assumed that the fluid leaves the impeller tangentially at the same speed as the impeller (an approximation), then an angular momentum balance on the fluid in contact with the impeller gives:

$$\Gamma = \dot{m} \omega R_i^2 = \rho Q \omega R_i^2 \tag{8-5}$$

where R_i is the radius of the impeller and the angular momentum of the fluid entering the eye of the impeller has been neglected (a good assumption). By eliminating Γ from Eqs. (8-4) and (8-5) and solving for H_p, we obtain

$$H_p \cong \frac{\eta_e \omega^2 R_i^2}{g} \tag{8-6}$$

This shows that the pump head is determined primarily by the size and speed of the impeller and the pump efficiency, independent of the flow rate of the fluid. This is approximately correct for most centrifugal pumps over a wide range of flow rates. However, there is a limitation to the flow that a given pump can handle, and as the flow rate approaches this limit the developed head will start to drop off. The maximum efficiency for most pumps occurs near the flow rate where the head starts to drop significantly.

Figure 8-2 shows a typical set of pump characteristic curves as determined by the pump manufacturer. "Size 2 × 3" means that the pump has a 2 in. discharge and a 3 in. suction port. "R&C" and "$1\frac{7}{8}$ pedestal" are the manufacturer's designations, and 3500 rpm is the speed of the impeller. Performance curves for impellers with diameters from $6\frac{1}{4}$ to $8\frac{3}{4}$ in. are shown, and the efficiency is shown as contour lines of constant efficiency. The maximum efficiency for this pump is somewhat above 50%, although some pumps may operate at efficiencies as high as 80% or 90%. Operation at conditions on the right-hand branch of the efficiency contours (i.e., beyond the "maximum normal capacity" line in Fig. 8-2) should be avoided, because this could result in unstable operation. The pump with the characteristics in Fig. 8-2 is a slurry pump, with a semiopen impeller, designed to pump solid suspensions (this pump can pass solid particles as

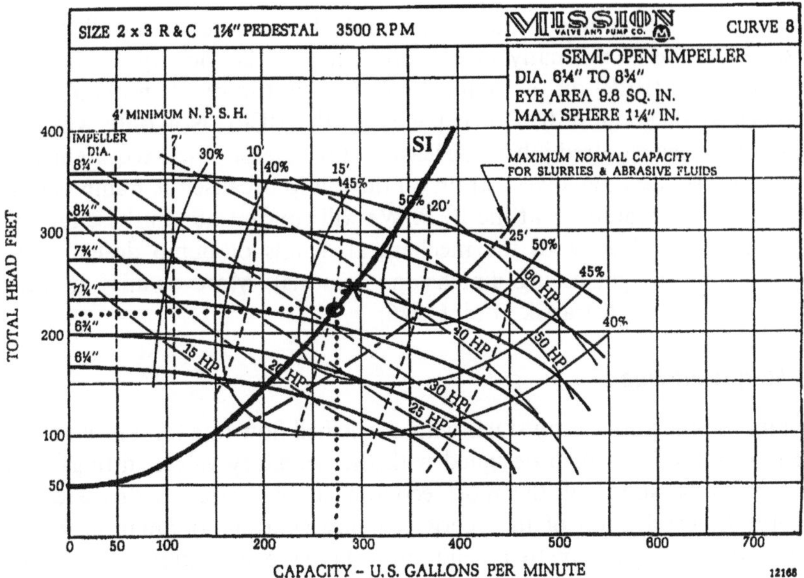

Figure 8-2 Typical pump characteristic curves. (From TRW Mission Pump Brochure.)

large as $1\frac{1}{4}$ in. in diameter). Pump characteristic curves for a variety of other pumps are shown in Appendix H.

Such performance curves are normally determined by the manufacturer from operating data using water at 60°F. Note from Eq. (8-6) that the head is independent of fluid properties, although from Eq. (8-4) the power is proportional to the fluid density (as is the developed pressure). The horsepower curves in Fig. 8-2 indicate the motor horsepower required to pump water at 60°F and must be corrected for density when operating with other fluids and/or at other temperatures. Actually, it is better to use Eq. (8-4) to calculate the required motor horsepower from the values of the head, flow rate, and efficiency at the operating point. The curves on Fig. 8-2 labeled "minimum NPSH" refer to the cavitation characteristics of the pump, which will be discussed later.

III. PUMPING REQUIREMENTS AND PUMP SELECTION

When selecting a pump for a given application (e.g., a required flow capacity and head), we must specify the appropriate pump type, size and type of impeller, and size (power) and speed (rpm) of the motor that will do the "best" job. "Best" normally means operating in the vicinity of the *best efficiency point* (BEP) on the pump curve (i.e., not lower than about 75% or higher than about 110% of capacity at the BEP). Not only will this condition do the required job at the least cost (i.e., least power requirement), but it also provides the lowest strain on the pump because the pump design is optimum for conditions at the BEP. We will concentrate on these factors and not get involved with the mechanical details of pump design (e.g., impeller vane design, casing dimensions or seals). More details on these topics are given by Karassik et al. (1976).

A. Required Head

A typical piping application starts with a specified flow rate for a given fluid. The piping system is then designed with the necessary valves, fittings, etc. and should be sized for the most economical pipe size, as discussed in Chapter 7. Application of the energy balance (Bernoulli) equation to the entire system, from the upstream end (point 1) to the downstream end (point 2) determines the overall net driving force (DF) in the system required to overcome the frictional resistance:

$$\mathrm{DF} = \sum e_{\mathrm{f}} \tag{8-7}$$

(where the kinetic energy change is assumed to be negligible).

The total head (driving force) is the net sum of the pump head, the total pressure drop, and the elevation drop:

$$\frac{\mathrm{DF}}{g} = H_{\mathrm{p}} + \frac{P_1 - P_2}{\rho g} + (z_1 - z_2) \tag{8-8}$$

The friction loss ($\sum e_{\mathrm{f}}$) is the sum of all of the losses from point 1 (upstream) to point 2 (downstream):

$$\sum e_{\mathrm{f}} = \sum_i \left(\frac{V^2}{2} K_{\mathrm{f}} \right)_i = \frac{8Q^2}{\pi^2} \sum_i \left(\frac{K_{\mathrm{f}}}{D^4} \right)_i \tag{8-9}$$

where the loss coefficients (K_{f}'s) include all pipe, valves, fittings, contractions, expansions, etc. in the system. Eliminating DF and $\sum e_{\mathrm{f}}$ from Eqs. (8-7), (8-8), and (8-9) and solving for the pump head, H_{p}, gives

$$H_{\mathrm{p}} = \frac{P_2 - P_1}{\rho g} + (z_2 - z_1) + \frac{8Q^2}{g\pi^2} \sum_i \left(\frac{K_{\mathrm{f}}}{D^4} \right)_i \tag{8-10}$$

This relates the system pump head requirement to the specified flow rate and the system loss parameters (e.g., the K_{f} values). Note that H_{p} is a quadratic function of Q for highly turbulent flow (i.e., constant K_f). For laminar flow, the K_f values are inversely proportional to the Reynolds number, which results in a linear relationship between H_p and Q. A plot of H_p versus Q from Eq. (8-10), illustrated in Fig. 8-2 as line S1, is called the *operating line* for the system. Thus the required pump head and flow capacity are determined by the system requirements, and we must select the best pump to meet this requirement.

B. Composite Curves

Most pump manufacturers provide composite curves, such as those shown in Fig. 8-3, that show the operating range of various pumps. For each pump that provides the required flow rate and head, the individual pump characteristics (such as those shown in Fig. 8-2 and Appendix H) are then consulted. The intersection of the system curve with the pump characteristic curve for a given impeller determines the pump operating point. The impeller diameter is selected that will produce the required head (or greater at the specified flow rate). This is repeated for all possible pump, impeller, and speed combinations to determine the combination that results in the highest efficiency (i.e., least power requirement). Note that if the operating point (H_p, Q) does not fall exactly on one of the (impeller) curves, then the

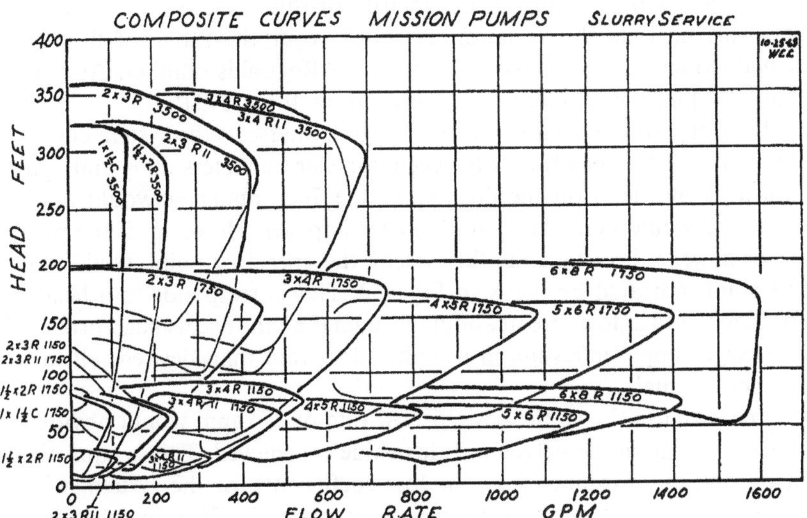

FIGURE 8-3 Typical pump composite curve. (From TRW Mission Pump Brochure [manufacturer's catalog.])

actual impeller diameter that produces the higher head at the required flow rate Q is chosen. However, when this pump is installed in the system, the actual operating point will correspond to the intersection of the system curve [Eq. (8-10)] and the actual pump impeller curve at this point, as indicated by the X in Fig. 8-2.

Example 8-1: Pump Selection. Consider a piping system that must deliver water at a rate of 275 gpm from one storage tank to another, both of which are at atmospheric pressure, with the level in the downstream tank being 50 ft higher than in the upstream tank. The piping system contains 65 ft of 2 in. sch 40 pipe, one globe valve, and six elbows. If the pump to be used has the characteristics shown in Fig. 8-2, what diameter impeller should be used with this pump, and what motor horsepower would be required?

Solution. The head requirement for the piping system is given by Eq. (8-10). Here, $z_2 - z_1 = 50\,\text{ft}$ and, since both upstream and downstream pressures are 1 atm, $\Delta P = 0$. The Reynolds number at 275 gpm for water at 60°F is 4.21×10^5, which gives a friction factor of 0.00497 in commercial steel pipe ($\varepsilon/D = 0.0018/2.067$). The corresponding loss coefficient for the pipe is $K_{\text{pipe}} = 4fL/D = 7.51$, and the loss coefficients for the fittings from Table 7-3 are (assuming flanged connections) elbow, $K_1 = 800$, $K_i = 0.091$, $K_d = 4.0$; globe valve, $K_1 = 1500$, $K_i = 1.7$, $K_d = 3.6$. At the pipe Reynolds number, this gives $\sum(K_f) = (K_{\text{pipe}} + K_{\text{Glbv}} + 6K_{\text{el}} = 16.4$. The curve labeled S1 in Fig. 8-2 is H_p vs. Q from Eq. (8-10), for this value of the loss coefficients. This neglects the variation of the K_f over the range of flow rate indicated, which is a good assumption at this Reynolds number. At a flow rate of 275 gpm, the required head from Eq. (8-10) is 219 ft.

The point where the flow rate of 275 gpm intersects the system curve in Fig. 8-2 (at 219 ft of head) falls between impeller diameters of $7\frac{1}{4}$ and $7\frac{3}{4}$ in., as indicated by the O on the line. Thus, the $7\frac{1}{4}$ in. diameter would be too small, so we would need the $7\frac{1}{4}$ in. diameter impeller. However, if the pump with this impeller is installed in the system, the operating point would move to the point indicated by the X in Fig. 8-2. This corresponds to a head of almost 250 ft and a flow rate of about 290 gpm (i.e., the excess head provided by the larger impeller results in a higher flow rate than desired, all other things being equal).

One way to achieve the desired flow rate of 275 gpm would obviously be to close down on the valve until this value is achieved. This is equivalent to increasing the resistance (i.e., the loss coefficient) for the system, which will shift the system curve upward until it intersects the $7\frac{3}{4}$ in. impeller curve at the desired flow rate of 275 gpm. The pump will still provide 250 ft of head, but about 30 ft of this head is "lost" (dissipated) due to the additional resistance in the partly closed valve. The pump efficiency at this operating point is about 47%, and the motor power (H_p) required to pump water at 60°F at this point is $HP = \rho g H_p Q/\eta_e = 37\,hp$.

A control valve operates in this mode automatically (as discussed in Chapter 10), but this is obviously not an efficient use of the available energy. A more efficient way of controlling the flow rate, instead of closing the valve, might be to adjust the speed of the impeller by using a variable speed drive. This would save energy because it would not increase the friction loss as does closing down on the valve, but it would require greater capital cost because variable speed drives are more expensive than fixed speed motors.

IV. CAVITATION AND NET POSITIVE SUCTION HEAD (NPSH)

A. Vapor Lock and Cavitation

As previously mentioned, a centrifugal pump increases the fluid pressure by first imparting angular momentum (or kinetic energy) to the fluid, which is converted to pressure in the diffuser or volute section. Hence, the fluid velocity in and around the impeller is much higher than that either entering or leaving the pump, and the pressure is the lowest where the velocity is highest. The minimum pressure at which a pump will operate properly must be above the vapor pressure of the fluid; otherwise the fluid will vaporize (or "boil"), a condition known as *cavitation*. Obviously, the higher the temperature the higher the vapor pressure and the more likely that this condition will occur. When a centrifugal pump contains a gas or vapor it will still develop the same head, but because the pressure is proportional to the fluid density it will be several orders of magnitude lower than the pressure for a liquid at the same head. This condition (when the pump is filled with a gas or vapor) is known as *vapor lock*, and the pump will not function when this occurs.

However, cavitation may result in an even more serious condition than vapor lock. When the pressure at any point within the pump drops below the vapor pressure of the liquid, vapor bubbles will form at that point (this generally occurs on or near the impeller). These bubbles will then be transported to another region in the fluid where the pressure is greater than the vapor pressure, at which point they will collapse. This formation and collapse of bubbles occurs very rapidly and can create local "shock waves," which can cause erosion and serious damage to the impeller or pump. (It is often obvious when a pump is cavitating, because it may sound as though there are rocks in the pump!)

B. NPSH

To prevent cavitation, it is necessary that the pressure at the pump suction be sufficiently high that the minimum pressure anywhere in the pump will be above the vapor pressure. This required minimum suction pressure (in excess of the vapor pressure) depends upon the pump design, impeller size and speed, and flow rate and is called the *minimum required net positive suction head* (NPSH). Values of the minimum required NPSH for the pump in Fig. 8-2 are shown as dashed lines. The NPSH is almost independent of impeller diameter at low flow rates and increases with flow rate as well as with impeller diameter at higher flow rates. A distinction is sometimes made between the minimum NPSH "required" to prevent cavitation (sometimes termed the NPSHR) and the actual head (e.g., pressure) "available" at the pump suction (NPSHA). A pump will not cavitate if NPSHA > (NPSHR + vapor pressure head).

The NPSH at the operating point for the pump determines where the pump can be installed in a piping system to ensure that cavitation will not occur. The criterion is that the pressure head at the suction (entrance) of the pump (e.g., the NPSHA) must exceed the vapor pressure head by at least the value of the NPSH (or NPSHR) to avoid cavitation. Thus, if the pressure at the pump suction is P_s and the fluid vapor pressure is P_v at the operating temperature, cavitation will be prevented if

$$\text{NPSHA} = \frac{P_s}{\rho g} \geq \text{NPSH} + \frac{P_v}{\rho g} \tag{8-11}$$

The suction pressure P_s is determined by applying the Bernoulli equation to the suction line upstream of the pump. For example, if the pressure at the entrance to the upstream suction line is P_1, the maximum distance above this point that the pump can be located without cavitating (i.e., the maximum suction lift) is determined by Bernoulli's equation from P_1 to P_s:

$$h_{max} = \frac{P_1 - P_v}{\rho g} - \text{NPSH} + \frac{V_1^2 - V_s^2}{2g} - \frac{\sum (e_f)_s}{g} \tag{8-12}$$

where Eq. (8-11) has been used for P_s. V_1 is the velocity entering the suction line, V_s is the velocity at the pump inlet (suction), and $\sum (e_f)_s$ is the total friction loss in the suction line from the upstream entrance (point 1) to the pump inlet, including all pipe, fittings, etc. The diameter of the pump suction port is usually bigger than the discharge or exit diameter in order to minimize the kinetic energy head entering the pump, because this kinetic energy decreases the maximum suction lift and enhances cavitation. Note that if the maximum suction lift (h_{max}) is negative, the pump must be located below the upstream entrance to the suction line to prevent cavitation. It is best to be conservative when interpreting the NPSH requirements to prevent cavitation.

The minimum required NPSH on the pump curves is normally determined using water at 60°F with the discharge line fully open. However, even though a pump will run with a closed discharge line with no bypass, there will be much more recirculation within the pump if this occurs, which increases local turbulence and local velocities as well as dissipative heating, both of which increase the minimum required NPSH. This is especially true with high efficiency pumps, which have close clearances between the impeller and pump casing.

Example 8-2: Maximum Suction Lift. A centrifugal pump with the characteristics shown in Fig. 8-2 is to be used to pump an organic liquid from a reboiler to a storage tank, through a 2 in. sch 40 line, at a rate of 200 gpm. The pressure in the reboiler is 2.5 atm, and the liquid has a vapor pressure of 230 mmHg, an SG of 0.85, and a viscosity of 0.5 cP at the working temperature. If the suction line upstream of the pump is also 2 in. sch 40 and has elbows and one globe valve, and the pump has a $7\frac{3}{4}$ in. impeller, what is the maximum height above the reboiler that the pump can be located without cavitating?

Solution. The maximum suction lift is given by Eq. (8-12). From Fig. 8-2, the NPSH required for the pump at 200 gpm is about 11 ft. The velocity in the reboiler (V_1) can be neglected, and the velocity in the pipe (see Appendix E-1) is $V_s = 200/10.45 = 19.1$ ft/s.

The friction loss in the suction line

$$e_f = \frac{V_s^2}{2} \sum (K_{pipe} + K_{GlbV} + 2K_{el})$$

where $K_{pipe} = 4fh/D$ and the fitting losses are given by the 3-K formula and Table 7-3 (elbow: $K_1 = 800$, $K_\infty = 0.091$, $K_d = 4.0$; globe valve: $K_1 = 1500$, $K_\infty = 1.7$, $K_d = 3.6$). The value of the Reynolds number for this flow is 5.23×10^5, which, for commercial steel pipe ($\varepsilon/D = 0.0018/2.067$), gives $f = 0.00493$. Note that the pipe length is h in K_{pipe}, which is the same as the maximum suction length (h_{max}) on the left of Eq. (8-12), assuming that the suction line is vertical. The unknown (h) thus appears on both sides of the equation. Solving Eq. (8-12) for h gives 17.7 ft.

C. Specific Speed

The flow rate, head, and impeller speed at the maximum or "best" efficiency point (BEP) of the pump characteristic can be used to define a dimensionless group called the *specific speed*:

$$N_s = \frac{N\sqrt{Q}}{H^{3/4}} \quad \left(\text{in } \frac{\text{rpm}\sqrt{\text{gpm}}}{\text{ft}^{3/4}}\right) \tag{8-13}$$

Although this group is dimensionless (and hence unitless), it is common practice to use selected mixed (inconsistent) units when quoting the value of N_s, i.e., N in rpm, Q in gpm, and H in feet. The value of the specific speed represents the ratio of the pump flow rate to the head at the speed corresponding to the maximum efficiency point (BEP) and depends primarily on the design of the pump and impeller. As previously stated, most centrifugal pumps operate at relatively low heads and high flow rates, e.g., high values of N_s. However, this value depends strongly on the impeller design, which can vary widely from almost pure radial flow to almost pure axial flow (like a fan). Some examples of various types of impeller design are shown in Fig. 8-4. Radial flow impellers have the highest head and lowest flow capacity (low N_s), whereas axial flow impellers have a high flow rate and low head characteristic (high N_s). Thus the magnitude of the specific speed is a direct indication of the impeller design and performance, as shown in Fig. 8-5. Figure 8-5 also indicates the range of flow rates and efficiencies of the various impeller designs, as a function of the specific speed. As indicated in Fig. 8-5, the maximum efficiency corresponds roughly to a specific speed of about 3000.

D. Suction Specific Speed

Another "dimensionless" group, analogous to the specific speed, that relates directly to the cavitation characteristics of the pump is the *suction specific speed*, N_{ss}:

$$N_{ss} = \frac{NQ^{1/2}}{(\text{NPSH})^{3/4}} \tag{8-14}$$

The units used in this group are also rpm, gpm and ft. This identifies the *inlet conditions* that produce similar flow behavior in the inlet for geometrically similar pump inlet passages. Note that the suction specific speed (N_{ss}) relates only to the pump cavitation characteristics as related to the inlet conditions, whereas the specific speed (N_s) relates to the entire pump at the BEP. The suction specific speed can be used, for example, to characterize the conditions under which excessive recirculation may occur at the inlet to the impeller vanes. Recirculation involves flow reversal and reentry resulting from undesirable pressure gradients at the inlet or discharge of the impeller vanes, and its occurrence generally defines the stable operating limits of the pump. For example, Fig. 8-6 shows the effect of the suction specific speed on the stable "recirculation-free" operating window,

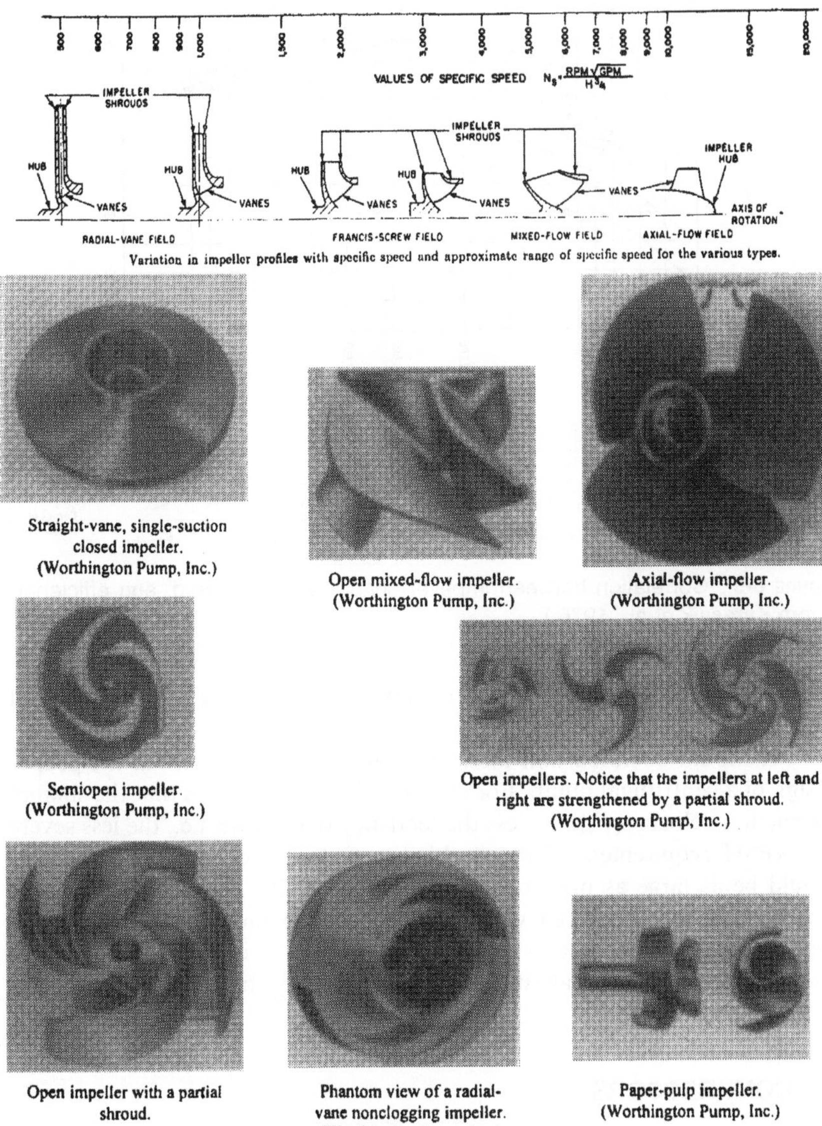

Variation in impeller profiles with specific speed and approximate range of specific speed for the various types.

Straight-vane, single-suction closed impeller. (Worthington Pump, Inc.)

Open mixed-flow impeller. (Worthington Pump, Inc.)

Axial-flow impeller. (Worthington Pump, Inc.)

Semiopen impeller. (Worthington Pump, Inc.)

Open impellers. Notice that the impellers at left and right are strengthened by a partial shroud. (Worthington Pump, Inc.)

Open impeller with a partial shroud.

Phantom view of a radial-vane nonclogging impeller. (Worthington Pump, Inc.)

Paper-pulp impeller. (Worthington Pump, Inc.)

Figure 8-4 Impeller designs and specific speed characteristics. (From Karassik et al., 1976.)

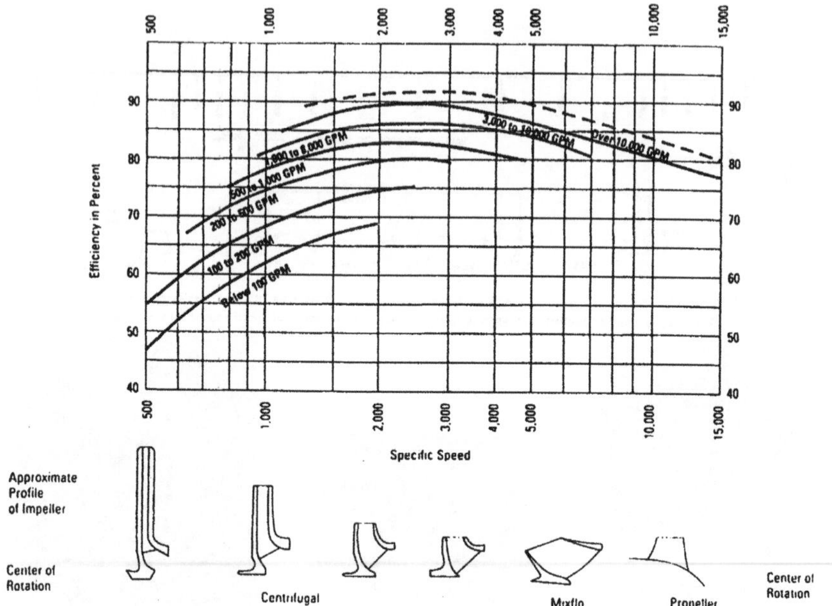

FIGURE 8-5 Correlation between impeller shape, specific speed, and efficiency. (From Karassik et al., 1976.)

expressed as NPSH versus percent of capacity at BEP, for various values of N_{SS}.

It should be noted that there are conflicting parameters in the proper design of a centrifugal pump. For example, Eq. (8-12) shows that the smaller the suction velocity (V_s), the less the tendency to cavitate, i.e., the less severe the NPSH requirement. This would dictate that the eye of the impeller should be as large as practical in order to minimize Vs. However, a large impeller eye means a high vane tip speed at the impeller inlet, which is destabilizing with respect to recirculation. Hence, it is advisable to design the impeller with the smallest eye diameter that is practicable.

V. COMPRESSORS

A compressor may be thought of as a high pressure pump for a compressible fluid. By "high pressure" is meant conditions under which the compressible properties of the fluid (gas) must be considered, which normally occur when the pressure changes by as much as 30% or more. For "low pressures" (i.e., smaller pressure changes), a fan or blower may be an appropriate "pump"

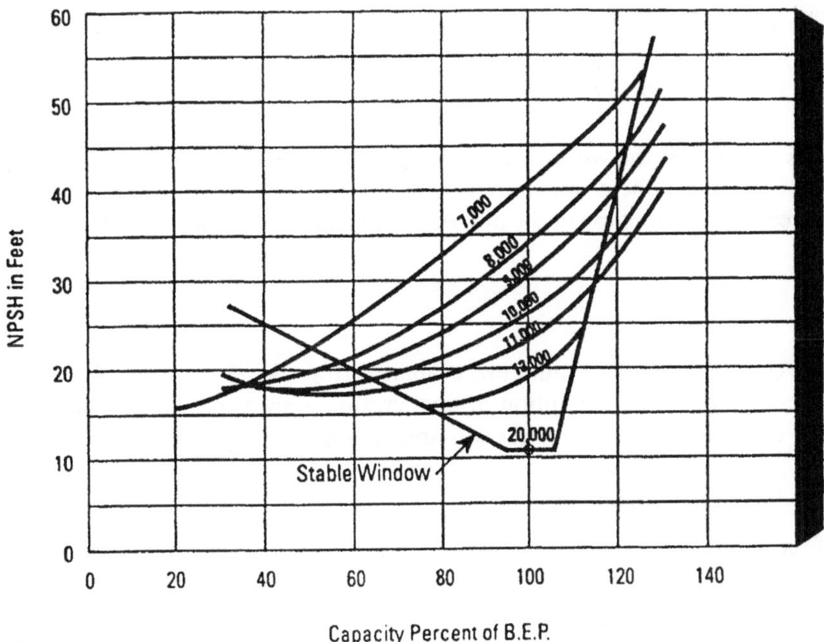

Figure 8-6 Effect of suction speed on stable operating window due to recirculation. (Numbers on the curves are the values of the suction specific speed, N_{ss}.) (From Raymer, 1993.)

for a gas. Fan operation can be analyzed by using the incompressible flow equations, because the relative pressure difference and hence the relative density change are normally small. Like pumps, compressors may be either positive displacement or centrifugal, the former being suitable for relatively high pressures and low flow rates whereas the latter are designed for higher flow rates but lower pressures. The major distinction in the governing equations, however, depends upon the conditions of operation, i.e., whether the system is isothermal or adiabatic. The following analyses assume that the gas is adequately described by the ideal gas law. This assumption can be modified, however, by an appropriate compressibility correction factor, as necessary. For an ideal (frictionless) compression, the work of compression is given by the Bernoulli equation, which reduces to

$$-w = \int_{P_1}^{P_2} \frac{dP}{\rho} \tag{8-15}$$

The energy balance equation for the gas can be written

$$\Delta h = q + e_f + \int_{P_1}^{P_2} \frac{dP}{\rho} \tag{8-16}$$

which says that the work of compression plus the energy dissipated due to friction and any heat transferred into the gas during compression all go to increasing the enthalpy of the gas. Assuming ideal gas properties, the density is

$$\rho = \frac{PM}{RT} \tag{8-17}$$

The compression work cannot be evaluated from Eq. (8-15) using Eq. (8-17) unless the operating condition or temperature is specified. We will consider two cases: isothermal compression and adiabatic compression.

A. Isothermal Compression

If the temperature is constant, eliminating ρ from Eqs. (8-17) and (8-15) and evaluating the integral gives

$$-w = \frac{RT}{M} \ln \frac{P_2}{P_1} \tag{8-18}$$

where the ratio P_2/P_1 is the *compression ratio* (r).

B. Isentropic Compression

For an ideal gas under adiabatic frictionless (i.e., isentropic) conditions,

$$\frac{P}{\rho^k} = \text{constant}, \qquad \text{where } k = \frac{c_p}{c_v};\ c_p = c_v + \frac{R}{M} \tag{8-19}$$

The specific heat ratio k is approximately 1.4 for diatomic gases (O_2, N_2, etc.) and 1.3 for triatomic and higher gases (NH_3, H_2O, CO_2, etc.). The corresponding expression for isothermal conditions follows from Eq. (8-17):

$$P/\rho = \text{constant} \tag{8-20}$$

Note that the isothermal condition can be considered a special case of the isentropic condition for $k = 1$. The "constant" in Eq. (8-19) or (8-20) can be evaluated from known conditions at some point in the system (e.g., P_1 and T_1). Using Eq. (8-19) to eliminate the density from Eq. (8-15) and evaluating the integral leads to

$$-w = \frac{RT_1 k}{M(k-1)} \left[\left(\frac{P_2}{P_1} \right)^{(k-1)/k} - 1 \right] \tag{8-21}$$

Although it is not obvious by inspection, setting $k = 1$ in Eq. (8-21) reduces that equation to Eq. (8-18) (this follows by application of l'Hospital's rule).

If we compare the work required to compress a given gas to a given compression ratio by isothermal and isentropic processes, we see that the isothermal work is always less than the isentropic work. That is, less energy would be required if compressors could be made to operate under isothermal conditions. However, in most cases a compressor operates under more nearly adiabatic conditions (isentropic, if frictionless) because of the relatively short residence time of the gas in the compressor, which allows very little time for heat generated by compression to be transferred away. The temperature rise during an isentropic compression is determined by eliminating ρ from Eqs. (8-17) and (8-19):

$$\frac{T_2}{T_1} = \left(\frac{P_2}{P_1}\right)^{(k-1)/k} = r^{(k-1)/k} \tag{8-22}$$

In reality, most compressor conditions are neither purely isothermal nor purely isentropic but somewhere in between. This can be accounted for in calculating the compression work by using the isentropic equation [Eq. (8-21)], but replacing the specific heat ratio k by a "polytropic" constant, γ, where $1 < \gamma < k$. The value of γ is a function of the compressor design as well as the properties of the gas.

C. Staged Operation

It is often impossible to reach a desired compression ratio with a single compressor, especially a centrifugal compressor. In such cases multiple compressor "stages" can be arranged in series to increase the overall compression ratio. Furthermore, to increase the overall efficiency it is common to cool the gas between stages by using "interstage coolers." With interstage cooling to the initial temperature (T_1), it can be shown that as the number of stages increases, the total compression work for isentropic compression approaches that of isothermal compression at T_1.

For multistage operation, there will be an optimum compression ratio for each stage that will minimize the total compression work. This can be easily seen by considering a two stage compressor with interstage cooling. The gas enters stage 1 at (P_1, T_1), leaves stage 1 at (P_2, T_2) and is then cooled to T_1. It then enters stage 2 at (P_2, T_1), and leaves at P_3. By computing the total isentropic work for both stages [using Eq. (8-21)] and setting the derivative of this with respect to the interstage pressure (P_2) equal to zero, the value of P_2 that results in the minimum total work can be found.

The result is that the optimum interstage pressure that minimizes the total work for a two stage compression with intercooling to T_1 is

$$P_2 = (P_1 P_3)^{1/2} \qquad \text{or } \frac{P_2}{P_1} = \frac{P_3}{P_2} = r = \left(\frac{P_3}{P_1}\right)^{1/2} \tag{8-23}$$

That is, the total work is minimized if the compression ratio for each stage is the same. This result can easily be generalized to any number (n) of stages (with interstage cooling to the initial temperature), as follows:

$$r = \frac{P_2}{P_1} = \frac{P_3}{P_2} = \cdots = \frac{P_{n+1}}{P_n} = \left(\frac{P_{n+1}}{P_1}\right)^{1/n} \tag{8-24}$$

If there is no interstage cooling or if there is interstage cooling to a temperature other than T_1, it can be shown that the optimum compression ratio for each stage (i) is related to the temperature entering that stage (T_i) by

$$T_i\left(\frac{P_{i+1}}{P_i}\right)^{(k-1)/k} = T_i r_i^{(k-1)/k} = \text{const} \tag{8-25}$$

D. Efficiency

The foregoing equations apply to ideal (frictionless) compressors. To account for friction losses, the ideal computed work is divided by the compressor efficiency, η_e, to get the total work that must be supplied to the compressor:

$$(-w)_{\text{total}} = \frac{(-w)_{\text{ideal}}}{\eta_e} \tag{8-26}$$

The energy "lost" due to friction is actually dissipated into thermal energy, which raises the temperature of the gas. This temperature rise is in addition to that due to the isentropic compression, so that the total temperature rise across an adiabatic compressor stage is given by

$$T_2 = T_1 r^{(k-1)/k} + \frac{1-\eta_e}{\eta_e}\left(\frac{-w_{\text{ideal}}}{c_v}\right) \tag{8-27}$$

PROBLEMS

Pumps

1. The pressure developed by a centrifugal pump for Newtonian liquids that are not highly viscous depends upon the liquid density, the impeller diameter, the rotational speed, and the volumetric flow rate.

(a) Determine a suitable set of dimensionless groups that should be adequate to relate all of these variables.

You want to know what pressure a pump will develop with a liquid having an SG of 1.4 at a flow rate of 300 gpm using an impeller with a diameter of 12 in. driven by a motor running at 1100 rpm. You have a similar test pump in the lab with a 6 in. impeller driven by an 1800 rpm motor. You want to run a test with the lab pump under conditions that will allow you to determine the pressure developed by the larger pump.

(b) Should you use the same liquid in the lab as in the larger pump, or can you use a different liquid? Why?
(c) If you use the same liquid, at what flow rate will the operation of the lab pump simulate that of the larger pump?
(d) If the lab pump develops a pressure of 150 psi at the proper flow rate, what pressure will the field pump develop at 300 gpm?
(e) What pressure will the field pump develop with water at 300 gpm?

2. The propeller of a speed boat is 1 ft in diameter and is 1 ft below the surface of the water. At what speed (rpm) will cavitation occur at the propeller? Water density = $64\ \text{lb}_m/\text{ft}^3$, P_v of water = 18.65 mm Hg.
3. You must specify a pump to be used to transport water at a rate of 5000 gpm through 10 mi of 18 in. sch 40 pipe. The friction loss in valves and fittings is equivalent to 10% of the pipe length, and the pump is 70% efficient. If a 1200 rpm motor is used to drive the pump, determine:
   (a) The required horsepower and torque rating of the motor.
   (b) The diameter of the impeller that should be used in the pump.
4. You must select a centrifugal pump that will develop a pressure of 40 psi when pumping a liquid with an SG of 0.88 at a rate of 300 gpm. From all the pump characteristic curves in Appendix H, select the best pump for this job. Specify pump head, impeller diameter, motor speed, efficiency, and motor horsepower.
5. An oil with a 32.6° API gravity at 60°F is to be transferred from a storage tank to a process unit that is 10 ft above the tank, at a rate of 200 gpm. The piping system contains 200 ft of 3 in. sch 40 pipe, 25 90° screwed elbows, six stub-in tees used as elbows, two lift check valves, and four standard globe valves. From the pump performance curves in Appendix H, select the best pump to do this job. Specify the pump size, motor speed, impeller diameter, operating head and efficiency and the horsepower of the motor required to drive the pump.
6. You must purchase a centrifugal pump to circulate cooling water that will deliver 5000 gpm at a pressure of 150 psi. If the pump is driven by an 1800 rpm motor, what should the horsepower and torque rating of the motor be, and how large (diameter) should the pump impeller be, assuming an efficiency of 60%?
7. In order to pump a fluid of SG = 0.9 at a rate of 1000 gpm through a piping system, a hydraulic power of 60 hp is required. Determine the required pump head, the torque of the driving motor, and the estimated impeller diameter, if an 1800 rpm motor is used.
8. From your prior analysis of pumping requirements for a water circulating system, you have determined that a pump capable of delivering 500 gpm at a

pressure of 60 psi is required. If a motor operating at 1800 rpm is chosen to drive the pump, which is 70% efficient, determine:

(a) The required horsepower rating of the motor.
(b) The required torque rating of the motor.
(c) The diameter of the impeller that should be used in the pump.
(d) What color the pump should be painted.

9. You want to pump water at 70°F from an open well, 200 ft deep, at a rate of 30 gpm through a 1 in. sch 40 pipe, using a centrifugal pump having an NPSH of 8 ft. What is the maximum distance above the water level in the well that the pump can be located without cavitating? (Vapor pressure of water at 60°F = 18.7 mmHg.)
10. Steam condensate at 1 atm and 95°C (P_v = 526 mmHg) is returned to a boiler from the condenser by a centrifugal boiler feed pump. The flow rate is 100 gpm through a 2.5 in. sch 40 pipe. If the equivalent length of the pipe between the condenser and the pump is 50 ft, and the pump has an NPSH of 6 ft, what is the maximum height above the condenser that the pump can be located?
11. Water at 160°F is to be pumped at a rate of 100 gpm through a 2 in. sch 80 steel pipe from one tank to another located 100 ft directly above the first. The pressure in the lower tank is 1 atm. If the pump to be used has a required NPSH of 6 ft of head, what is the maximum distance above the lower tank that the pump may be located?
12. A pump with a 1 in. diameter suction line is used to pump water from an open hot water well at a rate of 15 gpm. The water temperature is 90°C, with a vapor pressure of 526 mmHg and density of 60 lb_m/ft^3. If the pump NPSH is 4 ft, what is the maximum distance above the level of the water in the well that the pump can be located and still operate properly?
13. Hot water is to be pumped out of an underground geothermally heated aquifer located 500 ft below ground level. The temperature and pressure in the aquifer are 325°F and 150 psig. The water is to be pumped out at a rate of 100 gpm through 2.5 in. pipe using a pump that has a required NPSH of 6 ft. The suction line to the pump contains four 90° elbows and one gate valve. How far below ground level must the pump be located in order to operate properly?
14. You must install a centrifugal pump to transfer a volatile liquid from a remote tank to a point in the plant 500 ft from the tank. To minimize the distance that the power line to the pump must be strung, it is desirable to locate the pump as close to the plant as possible. If the liquid has a vapor pressure of 20 psia, the pressure in the tank is 30 psia, the level in the tank is 30 ft above the pump inlet, and the required pump NPSH is 15 ft, what is the closest that the pump can be located to the plant without the possibility of cavitation? The line is 2 in. sch 40, the flow rate is 100 gpm and the fluid properties are $\rho = 45\,lb_m/ft^3$ and $\mu = 5$ cP.
15. It is necessary to pump water at 70°F (P_v = 0.35 psia) from a well that is 150 ft deep, at a flow rate of 25 gpm. You do not have a submersible pump, but you do have a centrifugal pump with the required capacity that cannot be submerged. If a 1 in. sch 40 pipe is used, and the NPSH of the pump is 15 ft,

how close to the surface of the water must the pump be lowered for it to operate properly?

16. You must select a pump to transfer an organic liquid with a viscosity of 5 cP and SG of 0.87 at a rate of 1000 gpm through a piping system that contains 1000 ft of 8 in. sch 40 pipe, four globe valves, 16 gate valves, and 43 standard 90° elbows. The discharge end of the piping system is 30 ft above the entrance, and the pressure at both ends is 10 psia.
    (a) What pump head is required?
    (b) What is the hydraulic horsepower to be delivered to the fluid?
    (c) Which combination of pump size, motor speed, and impeller diameter from the pump charts in Appendix H would you choose for this application?
    (d) For the pump selected, what size motor would you specify to drive it?
    (e) If the vapor pressure of the liquid is 5 psia, how far directly above the liquid level in the upstream tank could the pump be located without cavitating?
17. You need a pump that will develop at least 40 psi at a flow rate of 300 gpm of water. What combination of pump size, motor speed, and impeller diameter from the pump characteristics in Appendix H would be the best for this application? State your reasons for the choice you make. What are the pump efficiency, motor horsepower and torque requirement, and NPSH for the pump you choose at these operating conditions?
18. A centrifugal pump takes water from a well at 120°F ($P_v = 87.8$ mmHg) and delivers it at a rate of 50 gpm through a piping system to a storage tank. The pressure in the storage tank is 20 psig, and the water level is 40 ft above that in the well. The piping system contains 300 ft of 1.5 in. sch 40 pipe, 10 standard 90° elbows, six gate valves, and an orifice meter with a diameter of 1 in.
    (a) What are the specifications required for the pump?
    (b) Would any of the pumps represented by the characteristic curves in Appendix H be satisfactory for this application? If more than one of them would work, which would be the best? What would be the pump head, impeller diameter, efficiency, NPSH, and required horsepower for this pump at the operating point?
    (c) If the pump you select is driven by an 1800 rpm motor, what impeller diameter should be used?
    (d) What should be the minimum torque and horsepower rating of the motor, if the pump is 50% efficient?
    (e) If the NPSH rating of the pump is 6 ft at the operating conditions, where should it be located in order to prevent cavitation?
    (f) What is the reading of the orifice meter, in psi?
19. Water at 20°C is pumped at a rate of 300 gpm from an open well in which the water level is 100 ft below ground level into a storage tank that is 80 ft above ground. The piping system contains 700 ft of $3\frac{1}{2}$ in. sch 40 pipe, eight threaded elbows, two globe valves, and two gate valves. The vapor pressure of the water is 17.5 mmHg.
    (a) What pump head and hydraulic horsepower are required?

(b) Would a pump whose characteristics are similar to those shown in Fig. 8-2 be suitable for this job? If so, what impeller diameter, motor speed, and motor horsepower should be used?

(c) What is the maximum distance above the surface of the water in the well at which the pump can be located and still operate properly?

20. An organic fluid is to be pumped at a rate of 300 gpm, from a distillation column reboiler to a storage tank. The liquid in the reboiler is 3 ft above ground level, the storage tank is 20 ft above ground, and the pump will be at ground level. The piping system contains 14 standard elbows, four gate valves, and 500 ft of 3 in. sch 40 pipe. The liquid has an SG of 0.85, a viscosity of 8 cP, and a vapor pressure of 600 mmHg. If the pump to be used has characteristics similar to those given in Appendix H, and the pressure in the reboiler is 5 psig, determine

(a) The motor speed to be used.
(b) The impeller diameter.
(c) The motor horsepower and required torque.
(d) Where the pump must be located to prevent cavitation.

21. A liquid with a viscosity of 5 cP, density of 45 lb_m/ft^3, and vapor pressure of 20 psia is transported from a storage tank in which the pressure is 30 psia to an open tank 500 ft downstream, at a rate of 100 gpm. The liquid level in the storage tank is 30 ft above the pump, and the pipeline is 2 in. sch 40 commercial steel. If the transfer pump has a required NPSH of 15 ft, how far downstream from the storage tank can the pump be located without danger of cavitation?

22. You must determine the specifications for a pump to transport water at 60°C from one tank to another at a rate of 200 gpm. The pressure in the upstream tank is 1 atm, and the water level in this tank is 2 ft above the level of the pump. The pressure in the downstream tank is 10 psig, and the water level in this tank is 32 ft above the pump. The pipeline contains 250 ft of 2 in. sch 40 pipe, with 10 standard 90° flanged elbows and six gate valves.

(a) Determine the pump head required for this job.

(b) Assuming your pump has the same characteristics as the one shown in Fig. 8-2, what size impeller should be used, and what power would be required to drive the pump with this impeller at the specified flow rate?

(c) If the water temperature is raised, the vapor pressure will increase accordingly. Determine the maximum water temperature that can be tolerated before the pump will start to cavitate, assuming that it is installed as close to the upstream tank as possible.

23. A piping system for transporting a liquid ($\mu = 50$ cP, $\rho = 0.85\ g/cm^3$) from vessel A to vessel B consists of 650 ft of 3 in. sch 40 commercial steel pipe containing four globe valves and 10 elbows. The pressure is atmospheric in A and 5 psig in B, and the liquid level in B is 10 ft higher than that in A. You want to transfer the liquid at a rate of 250 gpm at 80°F using a pump with the characteristics shown in Fig. 8-2. Determine

(a) The diameter of the impeller that you would use with this pump.

(b) The head developed by the pump and the power (in horsepower) required to pump the liquid.
(c) The power of the motor required to drive the pump
(d) The torque that the motor must develop.
(e) The NPSH of the pump at the operating conditions.

24. You must chose a centrifugal pump to pump a coal slurry. You have determined that the pump must deliver 200 gpm at a pressure of at least 35 psi. Given the pump characteristic curves in Appendix H, tell which pump you would specify (give pump size, speed, and impeller diameter) and why? What is the efficiency of this pump at its operating point, what horsepower motor would be required to drive the pump, and what is the required NPSH of the pump? The specific gravity of the slurry is 1.35.
25. You must specify a pump to take an organic stream from a distillation reboiler to a storage tank. The liquid has a viscosity of 5 cP, an SG of 0.78, and a vapor pressure of 150 mmHg. The pressure in the storage tank is 35 psig, and the inlet to the tank is located 75 ft above the reboiler, which is at a pressure of 25 psig. The pipeline in which the pump is to be located is $2\frac{1}{2}$ in. sch 40, 175 ft long, and there will be two flanged elbows and a globe valve in each of the pump suction and discharge lines. The pump must deliver a flow rate of 200 gpm. If the pump you use has the same characteristics as that illustrated in Fig. 8-2, determine
    (a) The proper impeller diameter to use with this pump.
    (b) The required head that the pump must deliver.
    (c) The actual head that the pump will develop.
    (d) The horsepower rating of the motor required to drive the pump.
    (e) The maximum distance above the reboiler that the pump can be located without cavitating.
26. You have to select a pump to transfer benzene from the reboiler of a distillation column to a storage tank at a rate of 250 gpm. The reboiler pressure is 15 psig and the temperature is 60°C. The tank is 5 ft higher than the reboiler and is at a pressure of 25 psig. The total length of piping is 140 ft of 2 in. sch 40 pipe. The discharge line from the pump containsthree gate valves and 10 elbows, and the suction line has two gate valves and six elbows. The vapor pressure of benzene at 60°C is 400 mmHg.
    (a) Using the pump curves shown in Fig. 8-2, determine the impeller diameter to use in the pump, the head that the pump would develop, the power of the motor required to drive the pump, and the NPSH required for the pump.
    (b) If the pump is on the same level as the reboiler, how far from the reboiler could it be located without cavitating?
27. [Deleted in second printing.]

28. A circulating pump takes hot water at 85°C from a storage tank, circulates it through a piping system at a rate of 150 gpm, and discharges it to the atmosphere. The tank is at atmospheric pressure, and the water level in the tank is 20 ft above the pump. The piping consists of 500 ft of 2 in. sch 40 pipe, with one globe valve upstream of the pump and three globe valves and eight threaded elbows downstream of the pump. If the pump has the characteristics shown in Fig. 8-2, determine
    (a) The head that the pump must deliver, the best impeller diameter to use with the pump, the pump efficiency and NPSH at the operating point, and the motor horsepower required to drive the pump.
    (b) How far the pump can be located from the tank without cavitating. Properties of water at 85°C: Viscosity 0.334 cP, density 0.970 g/cm^3, vapor pressure 433.6 mmHg.
29. A slurry pump operating at 1 atm must be selected to transport a coal slurry from an open storage tank to a rotary drum filter, at a rate of 250 gpm. The slurry is 40% solids by volume and has an SG of 1.2. The level in the filter is 10 ft above that in the tank, and the line contains 400 ft of 3 in. sch 40 pipe, two gate valves, and six 90° elbows. A lab test shows that the slurry can be described as a Bingham plastic with $\nu = 50$ cP and $\tau_0 = 80$ dyn/cm^2.
    (a) What pump head is required?
    (b) Using the pump curves in Appendix H, choose the pump that would be the best for this job. Specify the pump size, motor speed, impeller diameter, efficiency, and NPSH. Tell what criteria you used to make your decision.
    (c) What horsepower motor would you need to drive the pump?
    (d) Assuming the pump you choose has an NPSH of 6 ft at the operating conditions, what is the maximum elevation above the tank that the pump could be located, if the maximum temperature is 80°C? (P_v of water is 0.4736 bar at this temperature.)
30. A red mud slurry residue from a bauxite processing plant is to be pumped from the plant to a disposal pond, at a rate of 1000 gpm, through a 6 in. ID pipeline that is 2500 ft long. The pipeline is horizontal, and the inlet and discharge of the line are both at atmospheric pressure. The mud has properties of a Bingham plastic, with a yield stress of 250 dyn/cm^2, a limiting viscosity of 50 cP, and a density of 1.4 g/cm^3. The vapor pressure of the slurry at the operating temperature is 50 mmHg. You have available several pumps with the characteristics given in Appendix H.

(a) Which pump, impeller diameter, motor speed and motor horsepower would you use for this application?
(b) How close to the disposal pond could the pump be located without cavitating?
(c) It is likely that none of these pumps would be adequate to pump this slurry. Explain why, and explain what type of pump might be better.

31. A pipeline is installed to transport a red mud slurry from an open tank in an alumina plant to a disposal pond. The line is 5 in. sch 80 commercial steel, 12,000 ft long, and is designed to transport the slurry at a rate of 300 gpm. The slurry properties can be described by the Bingham plastic model, with a yield stress of 15 dyn/cm^2, a limiting viscosity of 20 cP, and an SG of 1.3. You may neglect any fittings in this pipeline.
(a) What delivered pump head and hydraulic horsepower would be required to pump this mud?
(b) What would be the required pump head and horsepower to pump water at the same rate through the same pipeline?
(c) If 100 ppm of fresh Separan AP-30 polyacrylamide polymer were added to the water in case (b), above, what would the required pump head and horsepower be?
(d) If a pump with the same characteristics as those illustrated in Fig. 8-2 could be used to pump these fluids, what would be the proper size impeller and motor horsepower to use for each of cases (a), (b), and (c), above. Explain your choices.

32. An organic liquid is to be pumped at a rate of 300 gpm from a distillation column reboiler at 5 psig to a storage tank at atmospheric pressure. The liquid in the reboiler is 3 ft above ground level, the storage tank is 20 ft above ground, and the pump will be at ground level. The piping system contains 14 standard elbows, four gate valves, and 500 ft of 3 in. sch 40 pipe. The liquid has an SG of 0.85, a viscosity of 8 cP, and a vapor pressure of 600 mmHg. Select the best pump for this job from those for which the characteristics are given in Appendix H, and determine
(a) The motor speed
(b) The impeller diameter
(c) The motor horsepower and required torque
(d) Where the pump must be located to prevent cavitation.

Compressors

33. Calculate the work per pound of gas required to compress air from 70°F and 1 atm to 2000 psi with an 80% efficient compressor under the following conditions:
(a) Single stage isothermal compression.
(b) Single stage adiabatic compression.
(c) Five stage adiabatic compression with intercooling to 70°F and optimum interstage pressures.

(d) Three stage adiabatic compression with interstage cooling to 120°F and optimum interstage pressures.

Calculate the outlet temperature of the air for cases (b), (c), and (d), above. For air: $c_p = 0.24$ Btu (lb_m °F), $k = 1.4$.

34. It is desired to compress ethylene gas [MW = 28, $k = 1.3$, $c_p = 0.357$ Btu/(lb_m °F)] from 1 atm and 80°F to 10,000 psia. Assuming ideal gas behavior, calculate the compression work required per pound of ethylene under the following conditions:
    (a) A single stage isothermal compressor.
    (b) A four stage adiabatic compressor with interstage cooling to 80°F and optimum interstage pressures.
    (c) A four stage adiabatic compressor with no intercooling, assuming the same interstage pressures as in (b) and 100% efficiency.
35. You have a requirement to compress natural gas ($k = 1.3$, MW = 18) from 1 atm and 70°F to 5000 psig. Calculate the work required to do this per pound of gas in a 100% efficient compressor under the following conditions:
    (a) Isothermal single stage compressor.
    (b) Adiabatic three stage compressor with interstage cooling to 70°F.
    (c) Adiabatic two stage compressor with interstage cooling to 100°F.
36. Air is to be compressed from 1 atm and 70°F to 2000 psia. Calculate the work required to do this per pound of air using the following methods:
    (a) A single stage 80% efficient isothermal compressor.
    (b) A single stage 80% efficient adiabatic compressor.
    (c) A five stage 80% efficient adiabatic compressor with interstage cooling to 70°F.
    (d) A three stage 80% efficient adiabatic compressor with interstage cooling to 120°F. Determine the expression relating the pressure ratio and inlet temperature for each stage for this case by induction from the corresponding expression for optimum operation of the corresponding two stage case.
    (e) Calculate the final temperature of the gas for cases (b), (c), and (d).
37. It is desired to compress 1000 scfm of air from 1 atm and 70°F to 10 atm. Calculate the total horsepower required if the compressor efficiency is 80% for
    (a) Isothermal compression.
    (b) Adiabatic single stage compression.
    (c) adiabatic three stage compression with interstage cooling to 70°F and optimum interstage pressures.
    (d) Calculate the gas exit temperature for cases (b) and (c).

    Note: $c_p = 7$ Btu/(lb mol°F); assume ideal gas.
38. You want to compress air from 1 atm, 70°F, to 2000 psig, using a staged compressor with interstage cooling to 70°F. The maximum compression ratio per stage you can use is about 6, and the compressor efficiency is 70%.
    (a) How many stages should you use?
    (b) Determine the corresponding interstage pressures.
    (c) What power would be required to compress the air at a rate of 105 scfm?
    (d) Determine the temperature leaving the last stage.

(e) How much heat (in Btu/hr) must be removed by the interstage coolers?

39. A natural gas (methane) pipeline is to be designed to transport the gas at a rate of 50,000 scfm. The pipe is to be 6 in. ID, and the maximum pressure that the compressors can develop is 10,000 psig. The compressor stations are to be located in the pipeline at the point at which the pressure drops to 100 psi above that at which choked flow would occur (this is the suction pressure for the compressors). If the design temperature for the pipeline is 60°F, the compressors are 60% efficient, and the compressor stations each operate with three stages and interstage cooling to 60°F, determine
    (a) The proper distance between compressor stations, in miles.
    (b) The optimum interstage pressure and compression ratio for each compressor stage.
    (c) The total horsepower required for each compressor station.
40. A compressor feeds ethylene to a pipeline, that is 500 ft long and 6 in. in diameter. The compressor suction pressure in 50 psig at 70°F, the discharge pressure is 800 psig, and the downstream pressure at the end of the pipeline is 300 psig. For each of the two following cases, determine (1) the flow rate in the pipeline in scfm and (2) the power delivered from the compressor to the gas, in horsepower:
    (a) The compressor operates with a single stage;
    (b) The compressor has three stages, with interstage cooling to the entering temperature.

NOTATION

D	diameter, [L]
DF	driving force, Eq. (8-8), $[L^2/t^2]$
e_f	energy dissipated per unit mass of fluid, $[FL/M = L^2/t^2]$
g	acceleration due to gravity, $[L/t^2]$
H_p	pump head, [L]
HP	power, $[FL/t = ML^2/t^3]$
h_{max}	maxmaximum suction lift, [L]
k	isentropic exponent ($= c_v/c_p$ for ideal gas), [—]
K_f	loss coefficient, [—]
M	molecular weight, [M/mol]
N_s	specific speed, Eq. (8-13)
N_{ss}	suction specific speed, Eq. (8-14)
NPSH	net positive suction head, [L]
$\dot{m}$	mass flow rate, [M/t]
P	pressure, $[F/L^2 = M/(Lt^2]$
P_v	vapor pressure, $[F/L^2 = M/Lt^2]$
Q	volumetric flow rate, $[L^3/t]$
R	radius, [L]
r	compression ratio, [—]
T	temperature, [T]

w	work done by fluid system per unit mass of fluid, $[FL/M = L^2/t^2]$
Γ	moment or torque, $[FL = ML^2/t^2]$
$\Delta(\,)$	$(\,)_2 - (\,)_1$
η_e	efficiency, [—]
ρ	density, $[M/L^3]$
ω	angular velocity, $[1/t]$

Subscripts

1	reference point 1
2	reference point 2
Glbv	Globe value
i	impeller, ideal (frictionless)
m	motor
s	suction line

REFERENCES

Karassik IJ, WC Krutzsch, WH Fraser, JP Messina. Pump Handbook. New York: McGraw-Hill, 1976.

Raymer, RE. Watch suction specific speed. CEP 89(3), 79–84, 1993.

9

Compressible Flows

I. GAS PROPERTIES

The main difference between the flow behavior of incompressible and compressible fluids, and between the equations that govern them, is the effect of variable density, e.g., the dependence of density upon pressure and temperature. At low velocities (relative to the speed of sound), relative changes in pressure and associated effects are often small and the assumption of incompressible flow with a constant (average) density may be reasonable. It is when the gas velocity approaches the speed at which a pressure change propagates (i.e., the speed of sound) that the effects of compressibility become the most significant. It is this condition of high-speed gas flow (e.g., "fast gas") that is of greatest concern to us here.

A. Ideal Gas

All gases are "non-ideal" in that there are conditions under which the density of the gas may not be accurately represented by the ideal gas law,

$$\rho = PM/R \tag{9-1}$$

However, there are also conditions under which this law provides a very good representation of the density for virtually any gas. In general, the higher the temperature and the lower the pressure relative to the critical temperature and pressure of the gas, the better the ideal gas law represents gas properties. For example, the critical conditions for CO_2 are 304 K, 72.9 atm, whereas for N_2 they are 126 K, 33.5 atm. Thus, at normal atmospheric conditions (300 K, 1 atm) N_2 can be described very accurately by the ideal gas law, whereas CO_2 deviates significantly from this law under such conditions. This is readily discernible from the P–H diagrams for the compound (see e.g., Appendix D), because ideal gas behavior can be identified with the conditions under which the enthalpy is independent of pressure, i.e., the constant temperature lines on the P–H diagram are vertical (see Section 5.III.B of Chapter 5). For the most common gases (e.g., air) at conditions that are not extreme, the ideal gas law provides a quite acceptable representation for most engineering purposes.

We will consider gases under two possible conditions: isothermal and isentropic (or adiabatic). The *isothermal* (constant temperature) condition may be approximated, for example, in a long pipeline in which the residence time of the gas is long enough that there is plenty of time to reach thermal equilibrium with the surroundings. Under these conditions, for an ideal gas,

$$\frac{P}{\rho} = \text{constant} = \frac{P_1}{\rho_1} = \frac{P_2}{\rho_2}, \text{etc} \tag{9-2}$$

The *adiabatic* condition occurs, for example, when the residence time of the fluid is short as for flow through a short pipe, valve, orifice, etc. and/or for well-insulated boundaries. When friction loss is small, the system can also be described as *locally isentropic*. It can readily be shown that an ideal gas under isentropic conditions obeys the relationship

$$\frac{P}{\rho^k} = \text{constant} \frac{P_1}{\rho_1^k} = \frac{P_2}{\rho_2^k}, \text{etc} \tag{9-3}$$

where $k = c_p/c_v$ is the "isentropic exponent" and, for an ideal gas, $c_p = c_v + R/M$. For diatomic gases $k \approx 1.4$, whereas for triatomic and higher gases $k \approx 1.3$. Equation (9-3) is also often used for non-ideal gases, for which k is called the "isentropic exponent." A table of properties of various gases, including the isentropic exponent, is given in Appendix C, which also includes a plot of k as a function of temperature and pressure for steam.

B. The Speed of Sound

Sound is a small-amplitude compression pressure wave, and the speed of sound is the velocity at which this wave will travel through a medium. An expression for the speed of sound can be derived as follows. With reference

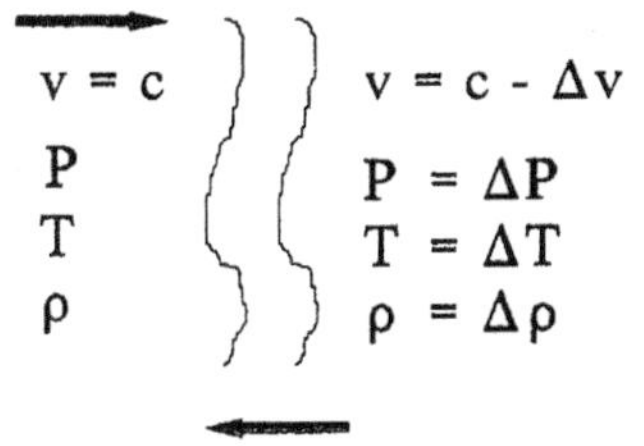

v = c
P
T
ρ

v = c - Δv
P = ΔP
T = ΔT
ρ = Δρ

FIGURE 9-1 Sound wave moving at velocity c.

to Fig. 9-1, we consider a sound wave moving from left to right with velocity c. If we take the wave as our reference, this is equivalent to considering a standing wave with the medium moving from right to left with velocity c. Since the conditions are different upstream and downstream of the wave, we represent these differences by ΔV, ΔT, ΔP, and $\Delta\rho$. The conservation of mass principle applied to the flow through the wave reduces to

$$\dot{m} = \rho A c = (\rho + \Delta\rho)A(c - \Delta V) \tag{9-4}$$

or

$$\Delta V = c\frac{\Delta\rho}{\rho + \Delta\rho} \tag{9-5}$$

Likewise, a momentum balance on the fluid "passing through" the wave is

$$\sum F = \dot{m}(V_2 - V_1) \tag{9-6}$$

which becomes, in terms of the parameters in Fig. 9-1,

$$PA - (P + \Delta P)A = \rho A c(c - \Delta V - c) \tag{9-7}$$

or

$$\Delta P = \rho c \Delta V \tag{9-8}$$

Eliminating ΔV from Eqs. (9-5) and (9-8) and solving for c^2 gives

$$c^2 = \frac{\Delta P}{\Delta\rho}\left(1 + \frac{\Delta\rho}{\rho}\right) \tag{9-9}$$

For an infinitesimal wave under isentropic conditions, this becomes

$$c = \left[\left(\frac{\partial P}{\partial\rho}\right)_s\right]^{1/2} = \left[k\left(\frac{\partial P}{\partial\rho}\right)_T\right]^{1/2} \tag{9-10}$$

where the equivalence of the terms in the two radicals follows from Eqs. (9-2) and (9-3).

For an ideal gas, Eq. (9-10) reduces to

$$c = \left(\frac{kP}{\rho}\right)^{1/2} = \left(\frac{kRT}{M}\right)^{1/2} \tag{9-11}$$

For solids and liquids,

$$\left(\frac{\partial P}{\partial \rho}\right)_s = \frac{K}{\rho} \tag{9-12}$$

where K is the *bulk modulus* (or "compressive stiffness") of the material. It is evident that the speed of sound in a completely incompressible medium would be infinite. From Eq. (9-11) we see that the speed of sound in an ideal gas is determined entirely by the nature of the gas (M and k) and the temperature (T).

II. PIPE FLOW

Consider a gas flowing in a uniform (constant cross section) pipe. The mass flow rate and mass flux ($G = \dot{m}/A$) are the same at all locations along the pipe:

$$G = \dot{m}/A = \rho V = \text{constant} \tag{9-13}$$

Now the pressure drops along the pipe because of energy dissipation (e.g., friction), just as for an incompressible fluid. However, because the density decreases with decreasing pressure and the product of the density and velocity must be constant, the velocity must increase as the gas moves through the pipe. This increase in velocity corresponds to an increase in kinetic energy per unit mass of gas, which also results in a drop in temperature. There is a limit as to how high the velocity can get in a straight pipe, however, which we will discuss shortly.

Because the fluid velocity and properties change from point to point along the pipe, in order to analyze the flow we apply the differential form of the Bernoulli equation to a differential length of pipe (dL):

$$\frac{dP}{\rho} + g\,dz + d\left(\frac{V^2}{2}\right) + \delta e_f = -\delta w = 0 \tag{9-14}$$

If there is no shaft work done on the fluid in this system and the elevation (potential energy) change can be neglected, Eq. (9-14) can be rewritten using Eq. (9-13) as follows:

$$\frac{dP}{\rho} + \frac{G^2}{\rho}\,d\left(\frac{1}{\rho}\right) = -\delta e_f = -\frac{2fV^2\,dL}{D} = -\frac{2f}{D}\left(\frac{G}{\rho}\right)^2 dL \tag{9-15}$$

where the friction factor f is a function of the Reynolds number:

$$f = \text{fn}\left(N_{\text{Re}} = \frac{DG}{\mu} \cong \text{constant}\right) \tag{9-16}$$

Because the gas viscosity is not highly sensitive to pressure, for isothermal flow the Reynolds number and hence the friction factor will be very nearly constant along the pipe. For adiabatic flow, the viscosity may change as the temperature changes, but these changes are usually small. Equation (9-15) is valid for any prescribed conditions, and we will apply it to an ideal gas in both isothermal and adiabatic (isentropic) flow.

A. Isothermal Flow

Substituting Eq. (9-1) for the density into Eq. (9-15), rearranging, integrating from the inlet of the pipe (point 1) to the outlet (point 2), and solving the result for G gives

$$G = \left(\frac{M(P_1^2 - P_2^2)/2RT}{2fL/D + \ln(P_1/P_2)}\right)^{1/2}$$

$$= \sqrt{P_1\rho_1}\left(\frac{1 - P_2^2/P_1^2}{4fL/D - 2\ln(P_2/P_1)}\right)^{1/2} \tag{9-17}$$

If the logarithmic term in the denominator (which comes from the change in kinetic energy of the gas) is neglected, the resulting equation is called the *Weymouth equation*. Furthermore, if the average density of the gas is used in the Weymouth equation, i.e.,

$$\bar{\rho} = \frac{(P_1 + P_2)M}{2RT}, \qquad \text{or} \qquad \frac{M}{2RT} = \frac{\bar{\rho}}{P_1 + P_2} \tag{9-18}$$

Eq. (9-17) reduces identically to the Bernoulli equation for an incompressible fluid in a straight, uniform pipe, which can be written in the form

$$G = \left(\frac{\bar{\rho}(P_1 - P_2)}{2fL/D}\right)^{1/2} = \sqrt{(P_1\bar{\rho}}\left(\frac{1 - P_1/P_2}{2fL/D}\right)^{1/2} \tag{9-19}$$

Inspection of Eq. (9-17) shows that as P_2 decreases, both the numerator and denominator increase, with opposing effects. By setting the derivative of Eq. (9-17) with respect to P_2 equal to zero, the value of P_2 that maximizes G and the corresponding expression for the maximum G

can be found. If the conditions at this state (maximum mass flux) are denoted by an asterisk, e.g., P_2^*, G^*, the result is

$$G^* = P_2^* \sqrt{\frac{M}{RT}} = P_2^* \sqrt{\frac{\rho_1}{P_1}} \tag{9-20}$$

or:

$$V_2^* = \sqrt{\frac{RT}{M}} = \sqrt{\frac{P_1}{\rho_1}} = c \tag{9-21}$$

That is, as P_2 decreases, the mass velocity will increase up to a maximum value of G^*, at which point the velocity at the end of the pipe reaches the speed of sound. Any further reduction in the downstream pressure can have no effect on the flow in the pipe, because the speed at which pressure information can be transmitted is the speed of sound. That is, since pressure changes are transmitted at the speed of sound, they cannot propagate upstream in a gas that is already traveling at the speed of sound. Therefore, the pressure inside the downstream end of the pipe will remain at P_2^*, regardless of how low the pressure outside the end of the pipe (P_2) may fall. This condition is called *choked flow* and is a very important concept, because it establishes the conditions under which maximum gas flow can occur in a conduit. When the flow becomes choked, the mass flow rate in the pipe will be insensitive to the exit pressure but will still be dependent upon the upstream conditions.

Although Eq. (9-17) appears to be explicit for G, it is actually implicit because the friction factor depends on the Reynolds number, which depends on G. However, the Reynolds number under choked flow conditions is often high enough that fully turbulent flow prevails, in which case the friction factor depends only on the relative pipe roughness:

$$\frac{1}{\sqrt{f}} = -4 \log\left[\frac{\varepsilon/D}{3.7}\right] \tag{9-22}$$

If the upstream pressure and flow rate are known, the downstream pressure (P_2) can be found by rearranging Eq. (9-17), as follows

$$\frac{P_2}{P_1} = \left\{1 - \frac{G^2}{P_1 \rho_1}\left[\frac{4fL}{D} - 2\ln\left(\frac{P_2}{P_1}\right)\right]\right\}^{1/2} \tag{9-23}$$

which is implicit in P_2. A first estimate for P_2 can be obtained by neglecting the last term on the right (corresponding to the Weymouth approximation). This first estimate can then be inserted into the last term in Eq. (9-23) to provide a second estimate for P_2, and the process can be repeated as necessary.

B. Adiabatic Flow

In the case of adiabatic flow we use Eqs. (9-1) and (9-3) to eliminate density and temperature from Eq. (9-15). This can be called the *locally isentropic* approach, because the friction loss is still included in the energy balance. Actual flow conditions are often somewhere between isothermal and adiabatic, in which case the flow behavior can be described by the isentropic equations, with the isentropic constant k replaced by a "polytropic" constant (or "isentropic exponent") γ, where $1 < \gamma < k$, as is done for compressors. (The isothermal condition corresponds to $\gamma = 1$, whereas truly isentropic flow corresponds to $\gamma = k$.) This same approach can be used for some non-ideal gases by using a variable isentropic exponent for k (e.g., for steam, see Fig. C-1).

Combining Eqs. (9-1) and (9-3) leads to the following expressions for density and temperature as a function of pressure:

$$\rho = \rho_1 \left(\frac{P}{P_1}\right)^{1/k}, \qquad T = T_1 \left(\frac{P}{P_1}\right)^{(k-1)/k} \tag{9-24}$$

Using these expressions to eliminate ρ and T from Eq. (9-15) and solving for G gives

$$G = \sqrt{P_1 \rho_1} \left[\frac{2\left(\dfrac{k}{k+1}\right)\left(1 - \left(\dfrac{P_2}{P_1}\right)^{(k+1)/k)}\right)}{\dfrac{4fL}{D} - \dfrac{2}{k} \ln\left(\dfrac{P_2}{P_1}\right)} \right]^{1/2} \tag{9-25}$$

If the system contains fittings as well as straight pipe, the term $4fL/D$ $(= K_{f,pipe})$ can be replaced by $\sum K_f$, i.e., the sum of all loss coefficients in the system.

C. Choked Flow

In isentropic flow (just as in isothermal flow), the mass velocity reaches a maximum when the downstream pressure drops to the point where the velocity becomes sonic at the end of the pipe (e.g., the flow is choked). This can be shown by differentiating Eq. (9-25) with respect to P_2 (as before) or, alternatively, as follows

$$G = \frac{\dot{m}}{A} = \rho V \tag{9-26a}$$

$$\frac{\partial G}{\partial P} = \frac{\partial(\rho V)}{\partial P} = \rho \frac{\partial V}{\partial P} + V \frac{\partial \rho}{\partial P} = 0 \qquad \text{(for max } G\text{)} \tag{9-26b}$$

For isentropic conditions, the differential form of the Bernoulli equation is

$$\frac{dP}{\rho} + VdV = 0 \qquad \text{or} \qquad \frac{\partial V}{\partial P} = -\frac{1}{\rho V} \tag{9-27}$$

Substituting this into Eq. (9-26b) gives

$$-\frac{1}{V} + V\frac{\partial \rho}{\partial P} = 0 \tag{9-28}$$

However, since

$$c^2 = \left(\frac{\partial P}{\partial \rho}\right)_s \tag{9-29}$$

Eq. (9-28) can be written

$$-\frac{1}{V} + \frac{V}{c^2} = 0, \qquad \text{or} \qquad V = c \tag{9-30}$$

This shows that when the mass velocity reaches a maximum (e.g., the flow is choked), the velocity is sonic.

1. Isothermal

Under isothermal conditions, choked flow occurs when

$$V_2 = c = V_2^* = \sqrt{\frac{RT}{M}} = \sqrt{\frac{P_1}{\rho_1}} \tag{9-31}$$

where the asterisk denotes the sonic state. Thus,

$$G^* = \rho_2 V_2^* = \frac{P_2^* M}{RT}\sqrt{\frac{RT}{M}} = \sqrt{P_1 \rho_1}\,\frac{P_2^*}{P_1} \tag{9-32}$$

If G^* is eliminated from Eqs. (9-17) and (9-32) and the result is solved for $\sum K_f$, the result is

$$\sum K_f = \left(\frac{P_1}{P_2^*}\right)^2 - 2\ln\left(\frac{P_1}{P_2^*}\right) - 1 \tag{9-33}$$

where $4fL/D$ in Eq. (9-17) has been replaced by $\sum K_f$. Equation (9-33) shows that the pressure at the (inside of the) end of the pipe at which the flow becomes sonic (P_2^*) is a unique function of the upstream pressure (P_1) and the sum of the loss coefficients in the system ($\sum K_f$). Since Eq. (9-33) is implicit in P_2^*, it can be solved for P_2^* by iteration for given values of $\sum K_f$ and P_1. Equation (9-33) thus enables the determination of the "choke pressure" P_2^* for given values of $\sum K_f$ and P_1.

2. Adiabatic

For adiabatic (or locally isentropic) conditions, the corresponding expressions are

$$V_2 = c = V_2^* = \left(\frac{kRT_2}{M}\right)^{1/2}, \qquad \frac{T_2}{T_1} = \left(\frac{P_2}{P_1}\right)^{(k-1)/k} \tag{9-34}$$

and

$$G^* = \frac{P_2^* M}{RT_2^*}\left(\frac{kRT_2^*}{M}\right)^{1/2} = \sqrt{P_1\rho_1}\left[k\left(\frac{P_2^*}{P_1}\right)^{(k+1)/k}\right]^{1/2} \tag{9-35}$$

Eliminating G^* from Eqs. (9-25) and (9-35) and solving for $\sum K_f$ gives

$$\sum K_f = \frac{2}{k+1}\left[\left(\frac{P_1}{P_2^*}\right)^{(k+1)/k} - 1\right] - \frac{2}{k}\ln\left(\frac{P_1}{P_2^*}\right) \tag{9-36}$$

Just as for isothermal flow, this is an implicit expression for the "choke pressure" (P_2^*) as a function of the upstream pressure (P_1), the loss coefficients ($\sum K_f$), and the isentropic exponent (k), which is most easily solved by iteration. It is very important to realize that once the pressure at the end of the pipe falls to P_2^* and choked flow occurs, all of the conditions within the pipe ($G = G^*$, $P_2 = P_2^*$, etc.) will remain the same regardless of how low the pressure outside the end of the pipe falls. The pressure drop *within the pipe* (which determines the flow rate) is always $P_1 - P_2^*$ when the flow is choked.

D. The Expansion Factor

The adiabatic flow equation [Eq. (9-25)] can be represented in a more convenient form as

$$G = Y\left(\frac{2\rho_1 \Delta P}{\sum K_f}\right)^{1/2} = Y\sqrt{P_1\rho_1}\left(\frac{2(1 - P_2/P_1)}{\sum K_f}\right)^{1/2} \tag{9-37}$$

where $\rho_1 = P_1 M/RT_1$, $\Delta P = P_1 - P_2$, and Y is the *expansion factor.* Note that Eq. (9-37) without the Y term is the Bernoulli equation for an incompressible fluid of density ρ_1. Thus, the expansion factor $Y = G_{\text{adiabatic}}/G_{\text{incompressible}}$ is simply the ratio of the adiabatic mass flux [Eq. (9-25)] to the corresponding incompressible mass flux and is a unique function of P_2/P_1, k, and K_f. For convenience, values of Y are shown in Fig. 9-2a for $k = 1.3$ and Fig. 9-2b for $k = 1.4$ as a function of $\Delta P/P_1$ and $\sum K_f$ (which is denoted simply K on these plots). The conditions corresponding to the lower ends of the lines on the plots (i.e., the "button") represent the sonic

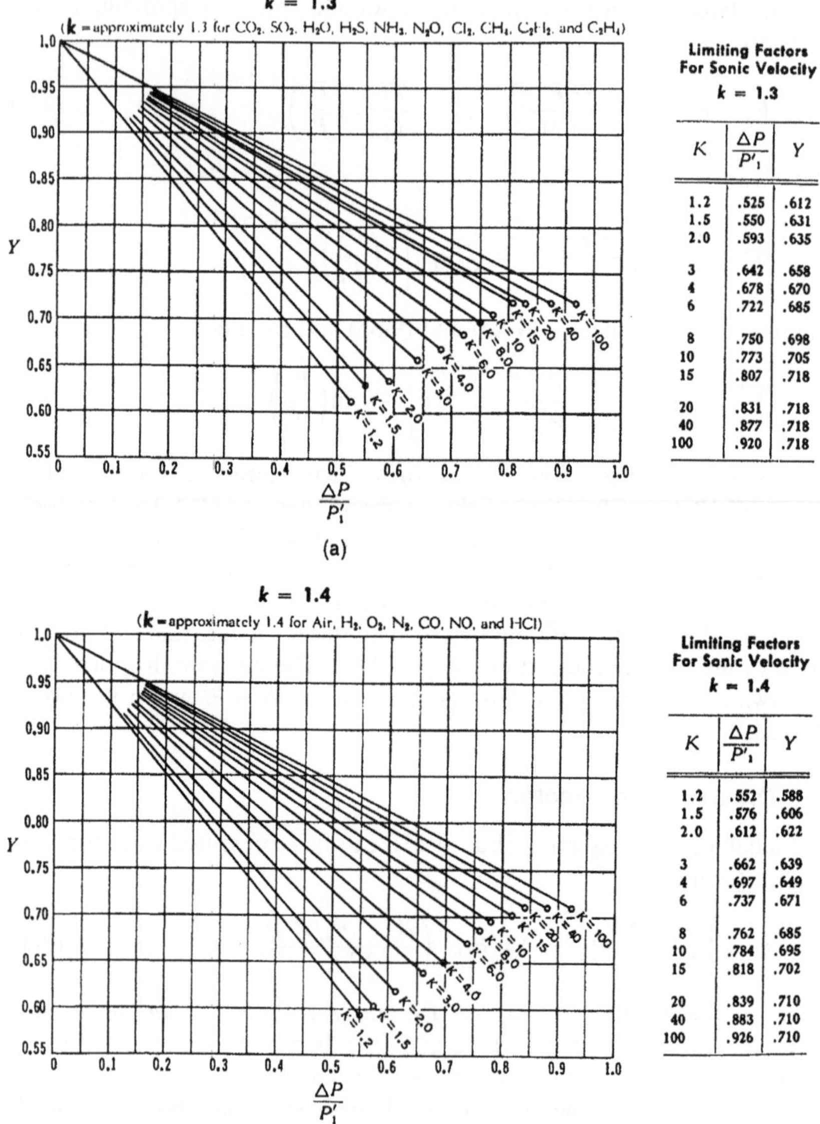

Limiting Factors For Sonic Velocity $k = 1.3$

K	$\frac{\Delta P}{P'_1}$	Y
1.2	.525	.612
1.5	.550	.631
2.0	.593	.635
3	.642	.658
4	.678	.670
6	.722	.685
8	.750	.698
10	.773	.705
15	.807	.718
20	.831	.718
40	.877	.718
100	.920	.718

Limiting Factors For Sonic Velocity $k = 1.4$

K	$\frac{\Delta P}{P'_1}$	Y
1.2	.552	.588
1.5	.576	.606
2.0	.612	.622
3	.662	.639
4	.697	.649
6	.737	.671
8	.762	.685
10	.784	.695
15	.818	.702
20	.839	.710
40	.883	.710
100	.926	.710

Figure 9-2 Expansion factor for adiabatic flow in piping systems. (a) $k = 1.3$; (b) $k = 1.4$. (From Crane Co., 1991.)

(choked flow) state where $P_2 = P_2^*$. These same conditions are given in the tables accompanying the plots, which enables the relationships for choked flow to be determined more precisely than is possible from reading the plots. Note that it is not possible to extrapolate beyond the "button" at the end of the lines in Figs. 9-2a and 9-2b because this represents the choked flow state, in which $P_2 = P_2^*$ (inside the pipe) and is independent of the external exit pressure.

Figure 9-2 provides a convenient way of solving compressible adiabatic flow problems for piping systems. Some iteration is normally required, because the value of K_f depends on the Reynolds number, which cannot be determined until G is found. An example of the procedure for solving a typical problem follows.

Given: P_1, D, L, ε, k, M *Find*: P_2^* and G^*

1. Estimate $\sum K_f$ by assuming fully turbulent flow. This requires a knowledge of ε/D to get $K_f = 4fL/D$ for the pipe and K_i and K_d for each fitting.
2. From Fig. 9-2a (for $k = 1.3$) or Fig. 9-2b (for $k = 1.4$), at the end of the line corresponding to the value of $K = \sum K_f$ (or from the table beside the plot) read the values of Y and $\Delta P^*/P_1 = (P_1 - P_2^*)/P_1$.
3. Calculate $G = G^*$ from Eq. (9-37).
4. Calculate $N_{Re} = DG/\mu$, and use this to revise the value of $K = \sum K_f$ for the pipe ($K_f = 4fL/D$) and fittings (3-K) accordingly.
5. Repeat steps 2–4 until there is no change in G.

The value of the downstream pressure (P_2) at which the flow becomes sonic ($P_2 = P_2^*$) is given by $P_2^* = P_1(1 - \Delta P^*/P_1)$. If the exit pressure is equal to or less than this value, the flow will be choked and G is calculated using P_2^*. Otherwise, the flow will be subsonic, and the flow rate will be determined using the pressure P_2.

E. Ideal Adiabatic Flow

The adiabatic flow of an ideal gas flowing through a frictionless conduit or a constriction (such as an orifice nozzle, or valve) can be analyzed as follows. The total energy balance is

$$\Delta h + g\Delta z + \tfrac{1}{2}\Delta V^2 = q + w \tag{9-38}$$

For horizontal adiabatic flow with no external work, this becomes

$$\Delta h + \tfrac{1}{2}\Delta V^2 = 0 \tag{9-39}$$

where

$$\Delta h = \Delta(c_p T) = \frac{k}{k-1}\,\Delta\left(\frac{P}{\rho}\right) \tag{9-40}$$

which follows from the ideal gas relation $c_p - c_v = R/M$ and the definition of k ($k = c_p/c_v$). Equation (9-39) thus becomes

$$\frac{k}{k-1}\left(\frac{P_2}{\rho_2} - \frac{P_1}{\rho_1}\right) + \frac{V_2^2 - V_1^2}{2} = 0 \tag{9-41}$$

Using the isentropic condition (P/ρ^k = constant) to eliminate ρ_2, this can be written

$$V_2^2 - V_1^2 = \frac{2k}{k-1}\left(\frac{P_1}{\rho_1}\right)\left[1 - \left(\frac{P_2}{P_1}\right)^{(k-1)/k}\right] \tag{9-42}$$

If V_1 is eliminated by using the continuity equation, $(\rho VA)_1 = (\rho VA)_2$, this becomes

$$V_2 = \left[\frac{2k}{k-1}\left(\frac{P_1}{\rho_1}\right)\frac{[1 - (P_2/P_1)^{(k-1)/k}]}{[1 - (A_2/A_1)^2(P_2/P_1)^{2/k}]}\right]^{1/2} \tag{9-43}$$

Because

$$G = V_2\rho_2 = V_2\rho_1(P_2/P_1)^{1/k} \tag{9-44}$$

and assuming that the flow is from a larger conduit through a small constriction, such that $A_1 \gg A_2$ (i.e., $V_1 \ll V_2$), Eq. (9-44) becomes

$$G = \left\{\frac{2k}{k-1}\left(\frac{P_1^2 M}{RT_1}\right)\left(\frac{P_2}{P_1}\right)^{2/k}\left[1 - \left(\frac{P_2}{P_1}\right)^{(k-1)/k}\right]\right\}^{1/2}$$

or

$$G = \sqrt{P_1\rho_1}\left\{\frac{2k}{k-1}\left(\frac{P_2}{P_1}\right)^{2/k}\left[1 - \left(\frac{P_2}{P_1}\right)^{(k-1)/k}\right]\right\}^{1/2} \tag{9-45}$$

Equation (9-45) represents flow through an "ideal nozzle," i.e., an isentropic constriction.

From the derivative of Eq. (9-45) (setting $\partial G/\partial r = 0$ where $r = P_2/P_1$), it can be shown that the mass flow is a maximum when

$$\frac{P_2^*}{P_1} = \left(\frac{2}{k+1}\right)^{k/(k-1)} \tag{9-46}$$

which, for $k = 1.4$ (e.g., air), has a value of 0.528. That is, if the downstream pressure is approximately one half or less of the upstream pressure, the flow will be choked. In such a case, the mass velocity can be determined by using Eq. (9-35) with P_2^* from Eq. (9-46):

$$G^* = P_1\left(\frac{2}{k+1}\right)^{(k+1)/2(k-1)}\sqrt{\frac{kM}{RT_1}} = \sqrt{kP_1\rho_1}\left(\frac{2}{k+1}\right)^{(k+1)/2(k-1)} \tag{9-47}$$

For $k = 1.4$, this reduces to

$$G^* = 0.684P_1\sqrt{\frac{M}{RT_1}} = 0.684\sqrt{P_1\rho_1} \tag{9-48}$$

The mass flow rate under adiabatic conditions is always somewhat greater than that under isothermal conditions, but the difference is normally <20%. In fact, for long piping systems ($L/D > 1000$), the difference is usually less than 5% (see, e.g., Holland, 1973). The flow of compressible (as well as incompressible) fluids through nozzles and orifices will be considered in the following chapter on flow-measuring devices.

III. GENERALIZED EXPRESSIONS

For adiabatic flow in a constant area duct, the governing equations can be formulated in a more generalized dimensionless form that is useful for the solution of both subsonic and supersonic flows. We will present the resulting expressions and illustrate how to apply them here, but we will not show the derivation of all of them. For this, the reader is referred to publications such as that of Shapiro (1953) and Hall (1951).

A. Governing Equations

For steady flow of a gas (at a constant mass flow rate) in a uniform pipe, the pressure, temperature, velocity, density, etc. all vary from point to point along the pipe. The governing equations are the conservation of mass (continuity), conservation of energy, and conservation of momentum, all applied to a differential length of the pipe, as follows.

1. *Continuity*

$$\frac{\dot{m}}{A} = G = \rho V = \text{constant} \tag{9-49a}$$

or

$$\frac{d\rho}{\rho} + \frac{dV}{V} = 0 \tag{9-49b}$$

2. *Energy*

$$h + \tfrac{1}{2}V^2 = \text{constant} = h_0 = c_p T_0 = c_p T + \tfrac{1}{2}V^2 \tag{9-50a}$$

or

$$dh + VdV = 0 \tag{9-50b}$$

Since the fluid properties are defined by the entropy and enthalpy, Eqs. (9-50) represent a curve on an h–s diagram, which is called a *Fanno line.*

3. *Momentum*

$$\frac{dP}{\rho} + VdV = -\frac{4\tau_w}{\rho D_h}\,dL = -\frac{2fV^2}{D}\,dL \tag{9-51}$$

By making use of the isentropic condition (i.e. P/ρ^k = constant), the following relations can be shown

$$h = c_p T = \frac{kRT}{(k-1)M} = \frac{P}{\rho}\left(\frac{k}{k-1}\right) \tag{9-52}$$

$$\frac{P}{\rho} = \frac{RT}{M} = \frac{c^2}{k} \tag{9-53}$$

$$N_{Ma} = \frac{V}{c} \tag{9-54}$$

where N_{MA} is the Mach number.

An "impulse function" (F) is also useful in some problems where the force exerted on bounding surfaces is desired:

$$F = PA + \rho AV^2 = PA(1 + kN_{Ma}^2) \tag{9-55}$$

These equations can be combined to yield the dimensionless forms

$$\frac{dP}{P} = -\frac{kN_{Ma}^2[1 + (k-1)N_{Ma}^2]}{2(1 - N_{Ma}^2)}\left(4f\,\frac{dL}{D}\right) \tag{9-56}$$

$$\frac{dN_{\text{Ma}}^2}{N_{\text{Ma}}^2} = \frac{kN_{\text{Ma}}^2[1 + (k-1)N_{\text{Ma}}^2/2]}{1 - N_{\text{Ma}}^2}\left(4f\,\frac{dL}{D}\right) \tag{9-57}$$

$$\frac{dV}{V} = \frac{kN_{\text{Ma}}^2}{2(1 - N_{\text{Ma}}^2)}\left(4f\,\frac{dL}{D}\right) = -\frac{d\rho}{\rho} \tag{9-58}$$

$$\frac{dT}{T} = -\frac{k(k-1)N_{\text{Ma}}^4}{2(1 - N_{\text{Ma}}^2)}\left(4f\,\frac{dL}{D}\right) \tag{9-59}$$

$$\frac{dP_0}{P_0} = -\frac{kN_{\text{Ma}}^2}{2}\left(4f\,\frac{dL}{D}\right) \tag{9-60}$$

and

$$\frac{dF}{F} = \frac{dP}{P} + \frac{kN_{\text{Ma}}^2}{1 + kN_{\text{Ma}}^2}\,\frac{dN_{\text{Ma}}^2}{N_{\text{Ma}}^2} \tag{9-61}$$

The subscript 0 represents the "stagnation" state, i.e., the conditions that would prevail if the gas were to be slowed to a stop and all kinetic energy converted reversibly to internal energy. For a given gas, these equations show that all conditions in the pipe depend uniquely on the Mach number and dimensionless pipe length. In fact, if $N_{\text{Ma}} < 1$, an inspection of these equations shows that as the distance down the pipe (dL) increases, V will increase but P, ρ, and T will decrease. However, if $N_{\text{Ma}} > 1$, just the opposite is true, i.e., V decreases while P, ρ, and T increase with distance down the pipe. That is, a flow that is initially subsonic will approach (as a limit) sonic flow as L increases, whereas an initially supersonic flow will also approach sonic flow as L increases. Thus all flows, regardless of their starting conditions, will tend toward the speed of sound as the gas progresses down a uniform pipe. Therefore, the only way a subsonic flow can be transformed into a supersonic flow is through a converging–diverging nozzle, where the speed of sound is reached at the nozzle throat. We will not be concerned here with supersonic flows, but the interested reader can find this subject treated in many fluid mechanics books (such as Hall (1951) and Shapiro (1953).

B. Applications

It is convenient to take the sonic state ($N_{\text{Ma}} = 1$) as the reference state for application of these equations. Thus, if the upstream Mach number is N_{Ma}, the length of pipe through which this gas must flow to reach the speed of sound ($N_{\text{Ma}} = 1$) will be L^*. This can be found by integrating Eq. (9-57) from ($L = 0$, N_{Ma}) to ($L = L^*$, $N_{Ma} = 1$). The result is

$$\frac{4\bar{f}L^*}{D} = \frac{1 - N_{Ma}^2}{kN_{Ma}^2} + \frac{k+1}{2k} \ln\left[\frac{(k+1)N_{Ma}^2}{2 + (k-1)N_{Ma}^2}\right] \tag{9-62}$$

where $\bar{f}$ is the average friction factor over the pipe length L^*. Because the mass velocity is constant along the pipe, the Reynolds number (and hence f) will vary only as a result of variation in the viscosity, which is usually small. If $\Delta L = L = L_1^* - L_2^*$ is the pipe length over which the Mach number changes from N_{Ma1} to N_{Ma2}, then

$$\frac{4\bar{f}\Delta L}{D} = \left(\frac{4\bar{f}L^*}{D}\right)_1 - \left(\frac{4\bar{f}L^*}{D}\right)_2 \tag{9-63}$$

Likewise, the following relationships between the problem variables and their values at the sonic (reference) state can be obtained by integrating Eqs. (9-56)–(9-60).

$$\frac{P}{P^*} = \frac{1}{N_{Ma}}\left[\frac{k+1}{2 + (k-1)N_{Ma}^2}\right]^{1/2} \tag{9-64}$$

$$\frac{T}{T^*} = \left(\frac{c}{c^*}\right)^2 = \frac{k+1}{2 + (k-1)N_{Ma}^2} \tag{9-65}$$

$$\frac{\rho}{\rho^*} = \frac{V^*}{V} = \frac{1}{N_{Ma}}\left[\frac{2 + (k-1)N_{Ma}^2}{k+1}\right]^{1/2} \tag{9-66}$$

$$\frac{P_0}{P_0^*} = \frac{1}{N_{Ma}}\left[\frac{2 + (k-1)N_{Ma}^2}{k+1}\right]^{(k+1)/2(k-1)} \tag{9-67}$$

With these relationships in mind, the conditions at any two points (1 and 2) in the pipe are related by

$$\frac{T_2}{T_1} = \frac{T_2/T^*}{T_1/T^*}, \qquad \frac{P_2}{P_1} = \frac{P_2/P^*}{P_1/P^*} \tag{9-68}$$

and

$$\frac{4\bar{f}\Delta L}{D} = \frac{4\bar{f}}{D}(L_1^* - L_2^*) = \left(\frac{4\bar{f}L^*}{D}\right)_1 - \left(\frac{4\bar{f}L^*}{D}\right)_2 \tag{9-69}$$

Also, the mass velocity at N_{Ma} and at the sonic state are given by

$$G = N_{Ma}P\sqrt{\frac{kM}{RT}}, \qquad G^* = P^*\sqrt{\frac{kM}{RT^*}}. \tag{9-70}$$

For pipe containing fittings, the term $4fL/D$ would be replaced by the sum of the loss coefficients ($\sum K_f$) for all pipe sections and fittings. These equations apply to adiabatic flow in a constant area duct, for which the sum of the enthalpy and kinetic energy is constant [e.g., Eq. (9-50)], which also defines the Fanno line. It is evident that each of the dependent variables at any point in the system is a unique function of the nature of the gas (k) and the Mach number of the flow (N_{Ma}) at that point. Note that although the dimensionless variables are expressed relative to their values at sonic conditions, it is not always necessary to determine the actual sonic conditions to apply these relationships. Because the Mach number is often the unknown quantity, an iterative or trial-and-error procedure for solving the foregoing set of equations is required. However, these relationships may be presented in tabular form (Appendix I) or in graphical form (Fig. 9-3), which can be used directly for solving various types of problems without iteration, as shown below.

C. Solution of High-Speed Gas Problems

We will illustrate the procedure for solving the three types of pipe flow problems for high-speed gas flows: unknown driving force, unknown flow rate, and unknown diameter.

1. Unknown Driving Force

The unknown driving force could be either the upstream pressure, P_1, or the downstream pressure, P_2. However, one of these must be known, and the other can be determined as follows.

Given: P_1, T_1, G, D, L *Find*: P_2

1. Calculate $N_{Re} = DG/\mu_1$ and use this to find f_1 from the Moody diagram or the Churchill equation.
2. Calculate $N_{Ma1} = (G/P_1)(RT_1/kM)^{1/2}$. Use this with Eqs. (9-62), (9-64), and (9-65) or Fig. 9-3 or Appendix I to find $(4fL_1^*/D)_1$, P_1/P^*, and T_1/T^*. From these values and the given quantities, calculate L_1^*, P^*, and T^*.
3. Calculate $L_2^* = L_1^* - L$, and use this to calculate $(4f_1L_2^*/D)_2$. Use this with Fig. 9-3 or Appendix I or Eqs. (9-62), (9-64), and (9-65) to get N_{Ma2}, P_2/P^*, and T_2/T^*. [Note that Eq. (9-62) is implicit for N_{Ma2}]. From these values, determine P_2 and T_2.
4. Revise μ by evaluating it at an average temperature, $(T_1 + T_2)/2$, and pressure, $(P_1 + P_2)/2$. Use this to revise N_{Re} and thus f, and repeat steps 3 and 4 until no change occurs.

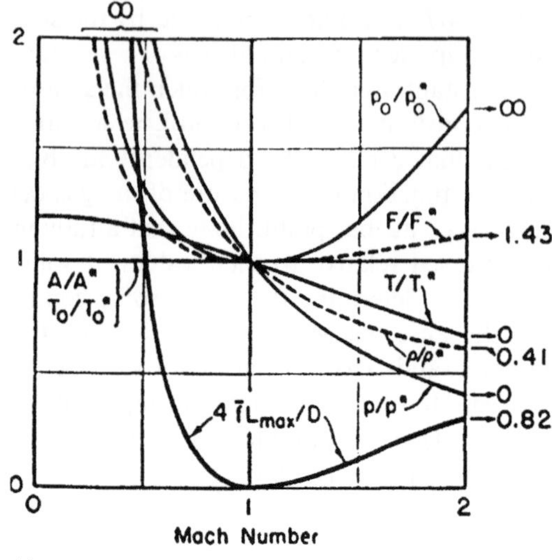

(a)

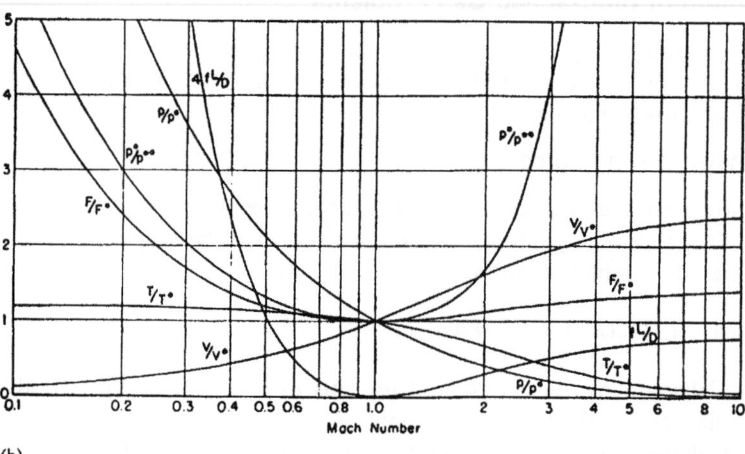

(b)

Figure 9-3 Fanno line functions for $k = 1.4$. ([a] From Hall, 1951 and [b] from Shapiro, 1953.)

2. Unknown Flow Rate

The mass velocity (G) is the unknown, which is equivalent to the mass flow rate because the pipe diameter is known. This requires a trial and error procedure, because neither the Reynolds nor Mach numbers can be calculated a priori.

Given: P_1, P_2, T_1, L, and D *Find*: G

1. Assume a value for N_{Ma1}. Use Eqs. (9-62), (9-64), and (9-65) or Fig. 9-3 or Appendix I with this value to find P_1/P^*, T_1/T^*, and $(4fL_1^*/D)_1$. From these and known quantities, determine P^* and T^*.
2. Calculate $G_1 = N_{Ma1}P_1(kM/RT_1)^{1/2}$ and $N_{Re1} = DG/\mu$. From the latter, find f_1 from the Moody diagram or Churchill equation.
3. Calculate $(4fL_2^*/D)_2 = (4fL_1^*/D)_1 - 4f_1L/D$. Use this with Eq. (9-62) (implicit) and Eqs. (9-64) and (9-65) or Fig. 9-3 or Appendix I to find N_{Ma2}, P_2/P^*, and T_2/T^* at point 2.
4. Calculate $P_2 = (P_2/P^*)P^*$, $T_2 = (T_2/T^*)T^*$, $G_2 = N_{Ma2}P_2(kM/RT_2)^{1/2}$, and $N_{Re} = DG_2/\mu$. Use the latter to determine a revised value of $f = f_2$.
5. Using $f = (f_1 + f_2)/2$ for the revised friction factor, repeat steps 3 and 4 until there is no change.
6. Compare the given value of P_2 with the calculated value from step 4. If they agree, the answer is the calculated value of G_2 from step 4. If they do not agree, return to step 1 with a new guess for N_{Ma1}, and repeat the procedure until agreement is achieved.

3. Unknown Diameter

The procedure for an unknown diameter involves a trial-and-error procedure similar to the one for the unknown flow rate.

Given: $P_1, T_1, L, P_2, \dot{m}$ *Find*: D

1. Assume a value for N_{Ma1} and use Eqs. (9-62), (9-64), and (9-65) or Fig. 9-3, or Appendix I to find P_1/P^*, T_1/T^* and $(4fL_1^*/D)_1$. Also, calculate $G = N_{Ma1}P_1(kM/RT_1)^{1/2}$, $D = (4\dot{m}/\pi G)^{1/2}$, and $N_{Re1} = DG/\mu$. Use N_{Re1} to find f_1 from the Moody diagram or Churchill equation.
2. Calculate $P_2/P^* = (P_1/P^*)(P_2/P_1)$, and use this with Fig. 9-3 or Appendix I or Eqs. (9-64) (implicitly), (9-62), and (9-65) to find N_{Ma2}, $(4fL_2^*/D)_2$, and T_2/T^*. Calculate $T_2 = (T_2/T^*)(T^*/T_1)T_1$, and use P_2 and T_2 to determine μ_2. Then use μ_2 to determine $N_{Re2} = DG/\mu_2$, which determines f_2 from the Moody diagram or Churchill equation.
3. Using $f = (f_1 + f_2)/2$, calculate $L = L_1^* - L_2^* = [(4fL_1^*/D)_1 - (4fL_2^*/D)_2](D/4f)$.
4. Compare the value of L calculated in step 3 with the given value. If they agree, the value of D determined in step 1 is correct. If they

do not agree, return to step 1, revise the assumed value of N_{Ma1}, and repeat the entire procedure until agreement is achieved.

PROBLEMS

Compressible Flow

1. A 12 in. ID gas pipeline carries methane (MW = 16) at a rate of 20,000 scfm. The gas enters the line at a pressure of 500 psia, and a compressor station is located every 100 mi to boost the pressure back up to 500 psia. The pipeline is isothermal at 70°F, and the compressors are adiabatic with an efficiency of 65%. What is the required horsepower for each compressor? Assume ideal gas.
2. Natural gas (CH_4) is transported through a 6 in. ID pipeline at a rate of 10,000 scfm. The compressor stations are 150 mi apart, and the compressor suction pressure is to be maintained at 10 psig above that at which choked flow would occur in the pipeline. The compressors are each two stage, operate adiabatically with interstage cooling to 70°F, and have an efficiency of 60%. If the pipeline temperature is 70°F, calculate:
   (a) The discharge pressure, interstage pressure, and compression ratio for the compressor stations.
   (b) The horsepower required at each compressor station.
3. Natural gas (methane) is transported through a 20 in. sch 40 commercial steel pipeline at a rate of 30,000 scfm. The gas enters the line from a compressor at 100 psi and 70°F. Identical compressor stations are located every 10 mi along the line, and at each station the gas is recompressed to 100 psia and cooled to 70°F.
   (a) Determine the suction pressure at each compressor station.
   (b) Determine the horsepower required at each station if the compressors are 80% efficient.
   (c) How far apart could the compressor stations be located before the flow in the pipeline becomes choked?
4. Natural gas (methane) is transported through an uninsulated 6 in. ID commercial steel pipeline, 1 mi long. The inlet pressure is 100 psi and the outlet pressure is 1 atm. What are the mass flow rate of the gas and the compressor power required to pump it? $T_1 = 70°F$, $\mu_{gas} = 0.02$ cP.
5. It is desired to transfer natural gas (CH_4) at a pressure of 200 psia and a flow rate of 1000 scfs through a 1 mi long uninsulated commercial steel pipeline into a storage tank at 20 psia. Can this be done using either a 6 in. or 12 in. ID pipe? What diameter pipe would you recommend? $T_1 = 70°F$, $\mu = 0.02$ cP.
6. A natural gas (methane) pipeline is to be designed to transport gas at a rate of 5000 scfm. The pipe is to be 6 in ID, and the maximum pressure that the compressors can develop is 1500 psig. The compressor stations are to be located in the pipeline at the point at which the pressure drops to 100 psi above that at which choked flow would occur (i.e., the suction pressure for the compressor stations). If the design temperature for the pipeline is 60°F, the compressors are

60% efficient, and the compressor stations each operate with three stages and interstage cooling to 60°F, determine

(a) The proper distance between compressor stations, in miles.
(b) The optimum interstage pressure and compression ratio for each compressor stage.
(c) The total horsepower required for each compressor station.

7. Ethylene leaves a compressor at 3500 psig and is carried in a 2 in, sch 40 pipeline, 100 ft long, to a unit where the pressure is 500 psig. The line contains two plug valves, one swing check valve, and eight flanged elbows. If the temperature is 100°F, what is the flow rate (in scfm)?
8. A 12 in. ID natural gas (methane) pipeline carries gas at a rate of 20,000 scfm. The compressor stations are 100 mi apart, and the discharge pressure of the compressors is 500 psia. If the temperature of the surroundings is 70°F, what is the required horsepower of each compressor station, assuming 65% efficiency? If the pipeline breaks 10 mi downstream from a compressor station, what will be the flow rate through the break?
9. The pressure in a reactor fluctuates between 10 and 30 psig. It is necessary to feed air to the reactor at a constant rate of 20 lb_m/hr, from an air supply at 100 psig, 70°F. To do this, you insert an orifice into the air line that will provide the required constant flow rate. What should the diameter of the orifice be?
10. Oxygen is to be fed to a reactor at a constant rate of 10 lb_m/s from a storage tank in which the pressure is constant at 100 psig and the temperature is 70°F. The pressure in the reactor fluctuates between 2 and 10 psig, so you want to insert a choke in the line to maintain the flow rate constant. If the choke is a 2 ft length of tubing, what should the diameter of the tubing be?
11. Methane is to be fed to a reactor at a rate of 10 lb_m/min. The methane is available in a pipeline at 20 psia, 70°F, but the pressure in the reactor fluctuates between 2 and 10 psia. To control the flow rate, you want to install an orifice plate that will choke the flow at the desired flow rate. What should the diameter of the orifice be?
12. Ethylene gas ($MW = 28$, $k = 1.3$, $\mu = 0.1$ cP) at 100°F is fed to a reaction vessel from a compressor through 100 ft of 2 in. sch 40 pipe containing two plug valves, one swing check valve, and eight flanged elbows. If the compressor discharge pressure is 3500 psig and the pressure in the vessel is 500 psig, what is the flow rate of the gas in scfm (1 atm, 60°F)?
13. Nitrogen is fed from a high pressure cylinder, through 1/4 in. ID stainless steel tubing, to an experimental unit. The line ruptures at a point 10 ft from the cylinder. If the pressure of the nitrogen in the cylinder is 3000 psig and the temperature is 70°F, what are the mass flow rate of the gas through the line and the pressure in the tubing at the point of the break?
14. A storage tank contains ethylene at 200 psig and 70°F. If a 1 in. line that is 6 ft long and has a globe valve on the end is attached to the tank, what would be the rate of leakage of the ethylene (in scfm) if

(a) The valve is fully open?
(b) The line breaks off right at the tank?

15. A 2 in. sch 40 pipeline is connected to a storage tank containing ethylene at 100 psig and 80°C.
    (a) If the pipe breaks at a distance of 50 ft from the tank, determine the rate at which the ethylene will leak out of the pipe (in lb_m/s). There is one globe valve in the line between the tank and the break.
    (b) If the pipe breaks off right at the tank, what would the leak rate be?
16. Saturated steam at 200 psig (388°F, 2.13 ft^3/lb_m, $\mu = 0.015$ cP) is fed from a header to a direct contact evaporator that operates at 10 psig. If the steam line is 2 in. sch. 40 pipe, 50 ft long, and includes four flanged elbows and one globe valve, what is the steam flow rate in lb_m/hr?
17. Air is flowing from a tank at a pressure of 200 psia and $T = 70°F$ through a venturi meter into another tank at a pressure of 50 psia. The meter is mounted in a 6 in. ID pipe section (that is quite short) and has a throat diameter of 3 in. What is the mass flow rate of air?
18. A tank containing air at 100 psia and 70°F is punctured with a hole 1/4 in. in diameter. What is the mass flow rate of the air out of the tank?
19. A pressurized tank containing nitrogen at 800 psig is fitted with a globe valve, to which is attached a line with 10 ft of 1/4 in. ID stainless steel tubing and three standard elbows. The temperature of the system is 70°F. If the valve is left wide open, what is the flow rate of nitrogen, in lb_m/s and also in scfm?
20. Gaseous chlorine (MW = 71) is transferred from a high pressure storage tank at 500 psia and 60°F, through an insulated 2 in. sch 40 pipe 200 ft long, into another vessel where the pressure is 200 psia. What are the mass flow rate of the gas and its temperature at the point where it leaves the pipe?
21. A storage tank containing ethylene at a pressure of 200 psig and a temperature of 70°F springs a leak. If the hole through which the gas is leaking is 1/2 in. in diameter, what is the leakage rate of the ethylene, in scfm?
22. A high pressure cylinder containing N_2 at 200 psig and 70°F is connected by 1/4 in. ID stainless steel tubing, 20 ft long to a reactor in which the pressure is 15 psig. A pressure regulator at the upstream end of the tubing is used to control the pressure, and hence the flow rate, of the N_2 in the tubing.
    (a) If the regulator controls the pressure entering the tubing at 25 psig what is the flow rate of the N_2 (in scfm)?
    (b) If the regulator fails so that the full cylinder pressure is applied at the tubing entrance, what will the flow rate of the N_2 into the reactor be (in scfm)?
23. Oxygen is supplied to an astronaut through an umbilical hose that is 7 m long. The pressure in the oxygen tank is 200 kPa at a temperature of 10°C, and the pressure in the space suit is 20 kPa. If the umbilical hose has an equivalent roughness of 0.01 mm, what should the hose diameter be to supply oxygen at a rate of 0.05 kg/s? If the suit springs a leak and the pressure drops to zero, at what rate will the oxygen escape?
24. Ethylene (MW = 28) is transported from a storage tank, at 250 psig and 70°F, to a compressor station where the suction pressure is 100 psig. The transfer line is 1 in. sch 80, 500 ft long, and contains two ball valves and eight threaded

elbows. An orifice meter with a diameter of 0.75 in. is installed near the entrance to the pipeline.

(a) What is the flow rate of the ethylene through the pipeline (in scfh)?
(b) If the pipeline breaks at a point 200 ft from the storage tank and there are 4 elbows and one valve in the line between the tank and the break, what is the flow rate of the ethylene (in scfh)?
(c) What is the differential pressure across the orifice for both cases (a) and (b), in inches of water?

25. Air passes from a large reservoir at 70°F through an isentropic converging–diverging nozzle into the atmosphere. The area of the throat is 1 cm^2, and that of the exit is 2 cm^2. What is the reservoir pressure at which the flow in the nozzle just reaches sonic velocity, and what are the mass flow rate and exit Mach number under these conditions?
26. Air is fed from a reservoir through a converging nozzle into a 1/2 in. ID drawn steel tube that is 15 ft long. The flow in the tube is adiabatic, and the reservoir temperature and pressure are 70°F and 100 psia.
    (a) What is the maximum flow rate (in lb_m/s) that can be achieved in the tube?
    (b) What is the maximum pressure at the tube exit at which this flow rate will be reached?
    (c) What is the temperature at this point under these conditions?
27. A gas storage cylinder contains nitrogen at 250 psig and 70°F. Attached to the cylinder is a 3 in. long, 1/4 in. sch 40 stainless steel pipe nipple, and attached to that is a globe valve followed by a diaphragm valve. Attached to the diaphragm valve is a 1/4 in. (ID) copper tubing line. Determine the mass flow rate of nitrogen (in lb_m/s) if
    (a) The copper tubing breaks off at a distance of 30 ft downstream of the diaphragm valve.
    (b) The pipe breaks off right at the cylinder.
28. A natural gas pipeline (primarily CH_4) is supplied by a compressor. The compressor suction pressure is 20 psig, and the discharge pressure is 1000 psig. The pipe is 5 in. sch 40, and the ambient temperature is 80°F.
    (a) If the pipeline breaks at a point 2 mi downstream from the compressor station, determine the rate at which the gas will escape (in scfm).
    (b) If the compressor efficiency is 80%, what power is required to drive it?
29. You have to feed a gaseous reactant to a reactor at a constant rate of 1000 scfm. The gas is contained at 80°F and a pressure of 500 psigin a tank that is located 20 ft from the reactor, and the pressure in the reactor fluctuates between 10 and 20 psig. You know that if the flow is choked in the feed line to the reactor, then the flow rate will be independent of the pressure in the reactor, which is what you require. If the feed line has a roughness of 0.0018 in., what should its diameter be in order to satisfy your requirements? The gas has an MW of 35, an isentropic exponent of 1.25, and a viscosity of 0.01 cP at 80°F.
30. A pressure vessel containing nitrogen at 300°F has a relief valve installed on its top. The valve is set to open at a pressure of 125 psig and exhausts the contents

to atmospheric pressure. The valve has a nozzle that is 1.5 in. in diameter and 4 in. long, which limits the flow through the valve when it opens.

(a) If the flow resistance in the piping between the tank and the valve, and from the valve discharge to the atmosphere are neglected, determine the mass flow through the valve when it opens in lb_m/s.

(b) In reality, there is a 3 ft length of 3 in. pipe between the tank and the valve, and a 6 ft length of 4 in. pipe downstream of the valve discharge. What is the effect on the calculated flow rate of including this piping?

31. A storage tank contains ethylene at 80°F and has a relief valve that is set to open at a pressure of 250 psig. The valve must be sized to relieve the gas at a rate of 85 lb_m/s when it opens. The valve has a discharge coefficient (the ratio of the actual to the theoretical mass flux) of 0.975.

(a) What should be the diameter of the nozzle in the valve, in inches?

What horsepower would be required to compress the gas from 1 atm to the maximum tank pressure at a rate equal to the valve flow rate for:

(b) A single stage compressor;

(c) A two-stage compressor with intercooling. Assume 100% efficiency for the compressor.

NOTATION

A	cross sectional area, $[L^2]$
c	speed of sound, $[L/t]$
D	diameter, $[L]$
F	force $[F = ML/t^2]$
f	Fanning friction factor, [—]
G	mass flux, $[M/tL^2]$
h	enthalpy per unit mass, $[FL/M = L^2/t^2]$
k	isentropic exponent ($= c_v/c_p$ for ideal gas), [—]
K_f	loss coefficient, [—]
L	length, $[L]$
M	molecular weight, $[M/mol]$
$\dot{m}$	mass flow rate, $[M/t]$
N_{Ma}	Mach number, [—]
N_{Re}	Reynolds number, [—]
P	pressure, $[F/L_2 = M/(Lt^2)]$
q	heat transferred to the fluid per unit mass, $[FL/M = L^2/t^2]$
R	gas constant, $[FL/(T\ mole) = ML^2/(t^2T\ mol)]$
T	temperature, $[T]$
V	spatial averaged velocity, $[L/t]$
w	work done by the fluid system per unit mass, $[FL/M = L^2/t^2]$
Y	expansion factor, [—]
z	vertical distance measured upward, $[L]$
ρ	density, $[M/L^3]$

Subscripts

1	reference point 1
2	reference point 2
s	constant entropy
T	constant temperature

Superscripts

*	sonic state

REFERENCES

Crane C. Flow of Fluids Through Valves, Fittings, and Pipe. Tech Manual 410. New York: Crane Co, 1991.

Hall NA. Thermodynamics of Fluid Flow. Englewood Cliffs, NJ: Prentice-Hall, 1951.

Holland FA, R. Bragg. Fluid Flow for Chemical Engineers, 2nd Ed., London, Edward Arnold, 1995.

Shapiro AH. The Dynamics and Thermodynamics of Compressible Fluid Flow, Vol. I. New York: Ronald Press, 1953.

10

Flow Measurement and Control

I. SCOPE

In this chapter we will illustrate and analyze some of the more common methods for measuring flow rate in conduits, including the pitot tube, venturi, nozzle, and orifice meters. This is by no means intended to be a comprehensive or exhaustive treatment, however, as there are a great many other devices in use for measuring flow rate, such as turbine, vane, Coriolis, ultrasonic, and magnetic flow meters, just to name a few. The examples considered here demonstrate the application of the fundamental conservation principles to the analysis of several of the most common devices. We also consider control valves in this chapter, because they are frequently employed in conjunction with the measurement of flow rate to provide a means of controlling flow.

II. THE PITOT TUBE

As previously discussed, the volumetric flow rate of a fluid through a conduit can be determined by integrating the local ("point") velocity over the cross section of the conduit:

$$Q = \int_A \mathrm{v}\, dA \tag{10-1}$$

If the conduit cross section is circular, this becomes

$$Q = \int_0^{\pi R^2} \mathrm{v}(r)\, d(\pi r^2) = 2\pi \int_0^R \mathrm{v}(r) r\, dr \tag{10-2}$$

The pitot tube is a device for measuring $\mathrm{v}(r)$, the local velocity at a given position in the conduit, as illustrated in Fig. 10-1. The measured velocity is then used in Eq. (10-2) to determine the flow rate. It consists of a differential pressure measuring device (e.g., a manometer, transducer, or DP cell) that measures the pressure difference between two tubes. One tube is attached to a hollow probe that can be positioned at any radial location in the conduit, and the other is attached to the wall of the conduit in the same axial plane as the end of the probe. The local velocity of the streamline that impinges on the end of the probe is $\mathrm{v}(r)$. The fluid element that impacts the open end of the probe must come to rest at that point, because there is no flow through the probe or the DP cell; this is known as the *stagnation point*. The Bernoulli equation can be applied to the fluid streamline that impacts the probe tip:

$$\frac{P_2 - P_1}{\rho} + \frac{1}{2}(\mathrm{v}_2^2 - \mathrm{v}_1^2) = 0 \tag{10-3}$$

where point 1 is in the free stream just upstream of the probe and point 2 is just inside the open end of the probe (the stagnation point). Since the friction loss is negligible in the free stream from 1 to 2, and $\mathrm{v}_2 = 0$ because the fluid in the probe is stagnant, Eq. (10-3) can be solved for v_1 to give

$$\mathrm{v}_1 = \left(\frac{2(P_2 - P_1)}{\rho}\right)^{1/2} \tag{10-4}$$

The measured pressure difference ΔP is the difference between the "stagnation" pressure in the velocity probe at the point where it connects to the DP cell and the "static" pressure at the corresponding point in the tube connected to the wall. Since there is no flow in the vertical direction, the difference in pressure between any two vertical elevations is strictly

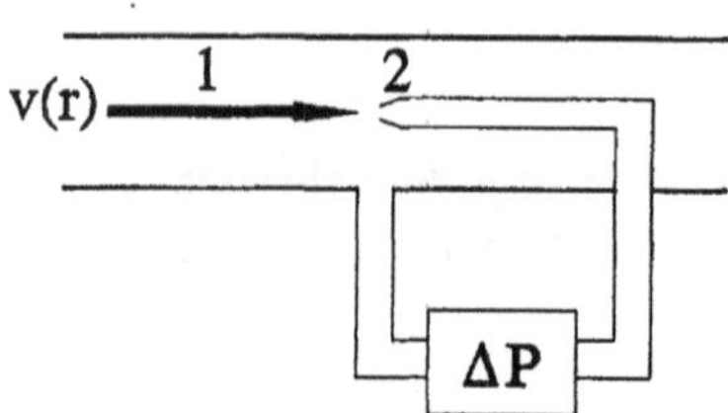

Figure 10-1 Pitot tube.

hydrostatic. Thus, the pressure difference measured at the DP cell is the same as that at the elevation of the probe, because the static head between point 1 and the pressure device is the same as that between point 2 and the pressure device, so that $\Delta P = P_2 - P_1$.

We usually want to determine the total flow rate (Q) through the conduit rather than the velocity at a point. This can be done by using Eq. (10-1) or Eq. (10-2) if the local velocity is measured at a sufficient number of radial points across the conduit to enable accurate evaluation of the integral. For example, the integral in Eq. (10-2) could be evaluated by plotting the measured v(r) values as v(r) vs. r^2, or as rv(r) vs. r [in accordance with either the first or second form of Eq. (10-2), respectively], and the area under the curve from $r = 0$ to $r = R$ could be determined numerically.

The pitot tube is a relatively complex device and requires considerable effort and time to obtain an adequate number of velocity data points and to integrate these over the cross section to determine the total flow rate. On the other hand the probe offers minimal resistance to the flow and hence is very efficient from the standpoint that it results in negligible friction loss in the conduit. It is also the only practical means for determining the flow rate in very large conduits such as smokestacks. There are standardized methods for applying this method to determine the total amount of material emitted through a stack, for example.

III. THE VENTURI AND NOZZLE

There are other devices, however, that can be used to determine the flow rate from a single measurement. These are sometimes referred to as *obstruction meters*, because the basic principle involves introducing an "obstruction" (e.g., a constriction) into the flow channel and then measuring the pressure drop across the obstruction that is related to the flow rate. Two such devices are the venturi meter and the nozzle, illustrated in Figs. 10-2 and 10-3 respectively. In both cases the fluid flows through a reduced area, which results in an increase in the velocity at that point. The corresponding change in pressure between point 1 upstream of the constriction and point 2 at the position of the minimum area (maximum velocity) is measured and is then related to the flow rate through the energy balance. The velocities are related by the continuity equation, and the Bernoulli equation relates the velocity change to the pressure change:

$$\rho_1 V_1 A_1 = \rho_2 V_2 A_2 \tag{10-5}$$

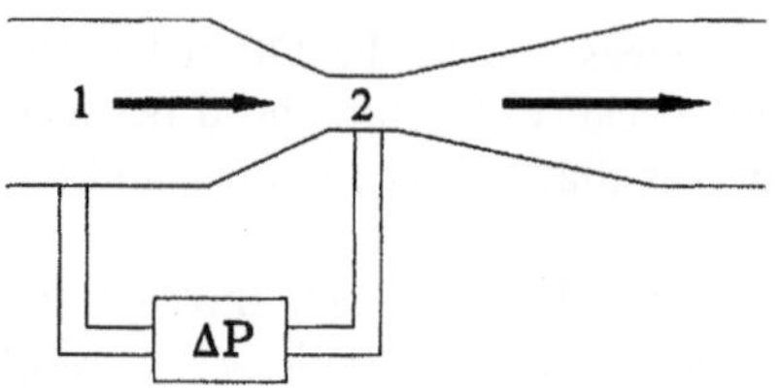

FIGURE 10-2 Venturi meter.

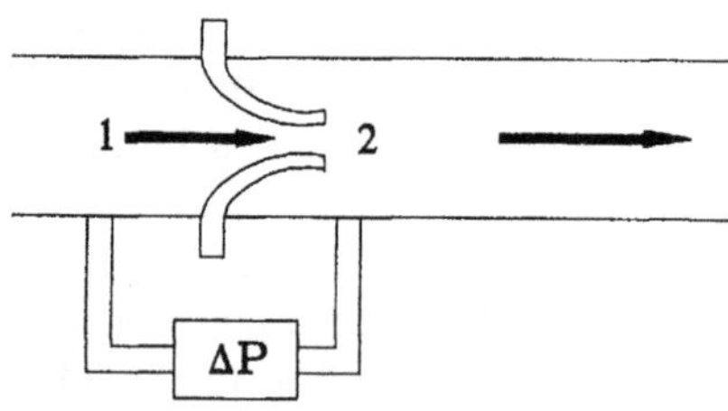

FIGURE 10-3 Nozzle.

For constant density,

$$V_1 = V_2 \frac{A_2}{A_1} \tag{10-6}$$

and the Bernoulli equation is

$$\frac{P_2 - P_1}{\rho} + \frac{1}{2}(V_2^2 - V_1^2) + e_f = 0 \tag{10-7}$$

where plug flow has been assumed. Using Eq. (10-6) to eliminate V_1 and neglecting the friction loss, Eq. (10-7) can be solved for V_2:

$$V_2 = \left(\frac{-2\Delta P}{\rho(1 - \beta^4)}\right)^{1/2} \tag{10-8}$$

where $\Delta P = P_2 - P_1$ and $\beta = d_2/D_1$ (where d_2 is the minimum diameter at the throat of the venturi or nozzle). To account for the inaccuracies introduced by assuming plug flow and neglecting friction, Eq. (10-8) is written

$$V_2 = C_d\left(\frac{-2\Delta P}{\rho(1 - \beta^4)}\right)^{1/2} \tag{10-9}$$

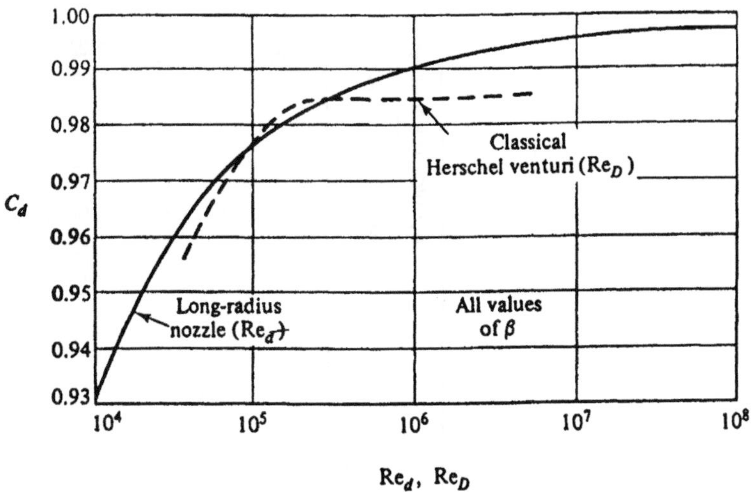

FIGURE 10-4 Venturi and nozzle discharge coefficient versus Reynolds number. (From White, 1994.)

where C_d is the "discharge" or venturi (or nozzle) coefficient and is determined by calibration as a function of the Reynolds number in the conduit. Typical values are shown in Fig. 10-4, where

$$N_{Re_D} = \frac{D_1 V_1 \rho}{\mu} \quad \text{and} \quad N_{Re_d} = \frac{d_2 V_2 \rho}{\mu} = N_{Re_D}/\beta.$$

Because the discharge coefficient accounts for the non-idealities in the system (such as the friction loss), one would expect it to decrease with increasing Reynolds number, which is contrary to the trend in Fig. 10-4. However, the coefficient also accounts for deviation from plug flow, which is greater at lower Reynolds numbers. In any event, the coefficient is not greatly different from 1.0, having a value of about 0.985 for (pipe) Reynolds numbers above about 2×10^5, which indicates that these non-idealities are small.

According to Miller (1983), for $N_{Re_D} > 4000$ the discharge coefficient for the venturi, as well as for the nozzle and orifice, can be described as a function of N_{Re_D} and β by the general equation

$$C_d = C_\infty + \frac{b}{N_{Re_D}^n} \tag{10-10}$$

where the parameters C_∞, b, and n are given in Table 10-1 as a function of β. The range over which Eq. (10-10) applies and its approximate accuracy are given in Table 10-2 (Miller, 1983). Because of the gradual expansion

TABLE 10-1 Values for Discharge Coefficient Parameters[a] in Eq. (10-10)

Primary device	Discharge coefficient C_∞ at infinite Reynolds number	Reynolds number term: Coefficient b	Reynolds number term: Exponent n
Venturi			
Machined inlet	0.995	0	0
Rough cast inlet	0.984	0	0
Rough welded sheet-iron inlet	0.985	0	0
Universal venturi tube[b]	0.9797	0	0
Lo-Loss tube[c]	$1.05 - 0.471\beta + 0.564\beta^2 - 0.514\beta^3$	0	0
Nozzle			
ASME long radius	0.9975	$-6.53\beta^{0.5}$	0.5
ISA	$0.9900–0.2262\beta^{4.1}$	$1708–8936\beta + 19,779\beta^{4.7}$	1.15
Venturi nozzle (ISA inlet)	$0.9858–0.195\beta^{4.5}$	0	0

Orifice			
Corner taps	$0.5959 + 0.0312\beta^{2.1} - 0.184\beta^{8}$	$91.71\beta^{2.5}$	0.75
Flange taps (D in in.)			
$D \geq 2.3$	$0.5959 + 0.0312\beta^{2.1} - 0.184\beta^{8} + 0.09\dfrac{\beta^4}{D(1-\beta^4)}; -0.0337\dfrac{\beta^3}{D}$	$91.71\beta^{2.5}$	0.75
$2 \leq D \leq 2.3$[d]	$0.5959 + 0.0312\beta^{2.1} - 0.184\beta^{8} + 0.039\dfrac{\beta^4}{1-\beta^4} - 0.0337\dfrac{\beta^3}{D}$	$91.71\beta^{2.5}$	0.75
Flange taps (D^* in mm)			
$D^* \geq 58.4$	$0.5959 + 0.0312\beta^{2.1} - 0.184\beta^{8} + 2.286\dfrac{\beta^4}{D^*(1-\beta^4)} - 0.856\dfrac{\beta^3}{D^*}$	$91.71\beta^{2.5}$	0.75
$50.8 \leq D^* \leq 58.4$	$0.5959 + 0.0312\beta^{2.1} - 0.184\beta^{8} + 0.039\dfrac{\beta^4}{1-\beta^4} - 0.856\dfrac{\beta^3}{D^*}$	$91.71\beta^{2.5}$	0.75
D and $D/2$ taps	$0.5959 + 0.0312\beta^{2.1} - 0.184\beta^{8} + 0.039\dfrac{\beta^4}{1-\beta^4} - 0.0158\beta^3$	$91.71\beta^{2.5}$	0.75
$2\frac{1}{2}D$ and $8D$ taps[d]	$0.5959 + 0.461\beta^{2.1} + 0.48\beta^{8} + 0.039\dfrac{\beta^4}{1-\beta^4}$	$91.71\beta^{2.5}$	0.75

[a] Detailed Reynolds number, line size, beta ratio, and other limitations are given in Table 10-2.
[b] From BIF CALC-440/441; the manufacturer should be consulted for exact coefficient information.
[c] Derived from the Badger meter, Inc. Lo-Loss tube coefficient curve; the manufacturer should be consulted for exact coefficient information.
[d] From Stolz (1978).
Source: Miller (1983).

TABLE 10-2 Applicable Range and Accuracy of Eqn (10-10), with Parameters from Table 10-1

Primary device	Nominal pipe diameter D, in (mm)	Beta ratio β	Pipe Reynolds number N_{ReD} range	Coefficient accuracy, %[a]
Venturi				
Machine inlet	2–10 (50–250)	0.4–0.75	2×10^5 to 10^6	±1
Rough cast	4–32 (100–800)	0.3–0.75	2×10^5 to 10^6	±0.7
Rough-welded sheet-iron inlet	8–48 (100–1500)	0.4–0.7	2×10^5 to 10^6	±1.5
Universal venturi tube[b]	≥ 3 (≥ 75)	0.2–0.75	$> 7.5 \times 10^4$	±0.5
Lo-Loss[b]	3–120 (75—3000)	0.35–0.85	1.25×10^5 to 3.5×10^6	±1
Nozzle				
ASME	2–16 (50–400)	0.25–0.75	10^4 to 10^7	±2.0
ISA	2–20 (50–500)	0.3–0.6	10^5 to 10^6	±0.8
		0.6–0.75	2×10^5 to 10^7	$2\beta - 0.4$
Venturi nozzle	3–20 (75–500)	0.3–0.75	2×10^5 to 2×10^6	$\pm 1.2 \pm 1.54\,\beta^4$

Orifice				
Corner, flange, D and $D/2$	2–36 (50–900)	0.2–0.6	10^4 to 10^7	±0.6
		0.6–0.75	10^4 to 10^7	±β
		0.2–0.75	2×10^3 to 10^4	±0.6 ±β
$2\frac{1}{2}D$ and $8D$ (pipe taps)	2–36 (50–900)	0.2–0.5	10^4 to 10^7	±0.8
		0.51–0.7		±1.6
Eccentric				
Flange and vena contracta	4 (100)	0.3–0.75	10^4 to 10^6	±2
	6–14 (150–350)	0.3–0.75	10^4 to 10^6	±1.5
Segmental				
Flange and vena contracta	4–14 (150–350)	0.35–0.75	10^4 to 10^6	±2
Quadrant-edged				
Flange and corner	1–4 (25–100)	0.25–0.6	250 to 6×10^4	±2–±2.5
Conical entrance				
Corner		0.1–0.3	25 to 2×10^4	±2–±2.5

[a] ISO 5167 (1980) and ASME *Fluid Meters* (1971) show slightly different values for some devices.
[b] The manufacturer should be consulted for recommendations.
Source: Miller (1983).

designed into the venturi meter, the pressure recovery is relatively large, so the net friction loss across the entire meter is a relatively small fraction of the measured (maximum) pressure drop, as indicated in Fig. 10-5. However, because the flow area changes abruptly downstream of the orifice and nozzle, the expansion is uncontrolled, and considerable eddying occurs down-

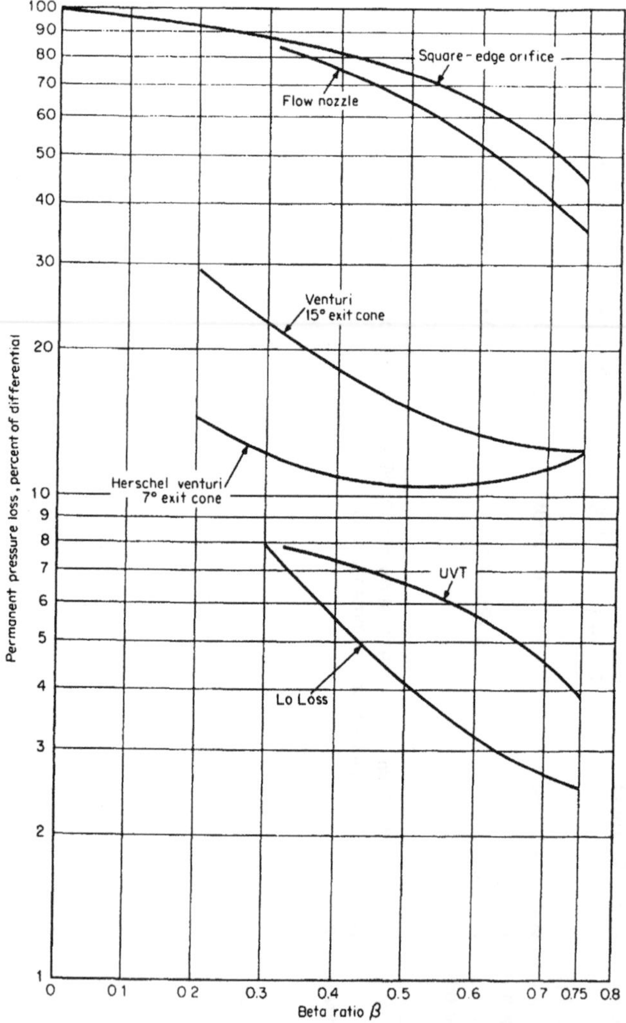

FIGURE 10-5 Unrecovered (friction) loss in various meters as a percentage of measured pressure drop. (From Cheremisinoff and Cheremisinoff, 1987.)

stream. This dissipates more energy, resulting in a significantly higher net friction loss and lower pressure recovery.

The foregoing equations assume that the device is horizontal, i.e., that the pressure taps on the pipe are located in the same horizontal plane. If such is not the case, the equations can be easily modified to account for changes in elevation by replacing the pressure P at each point by the total potential $\Phi = P + \rho g z$.

The flow nozzle, illustrated in Fig.10-3, is similar to the venturi meter except that it does not include the diffuser (gradually expanding) section. In fact, one standard design for the venturi meter is basically a flow nozzle with an attached diffuser (see Fig.10-6). The equations that relate the flow rate and measured pressure drop in the nozzle are the same as for the venturi

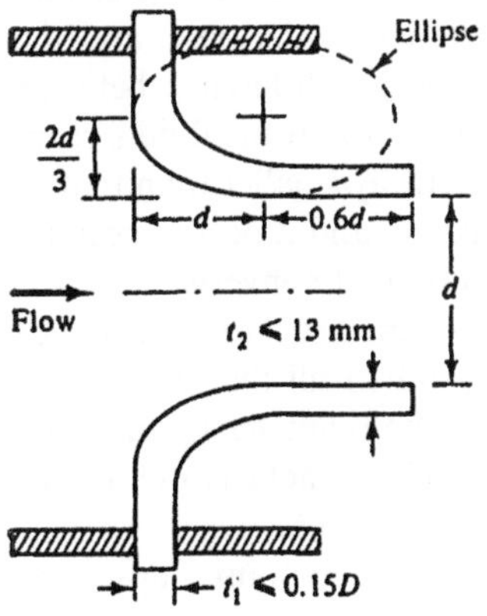

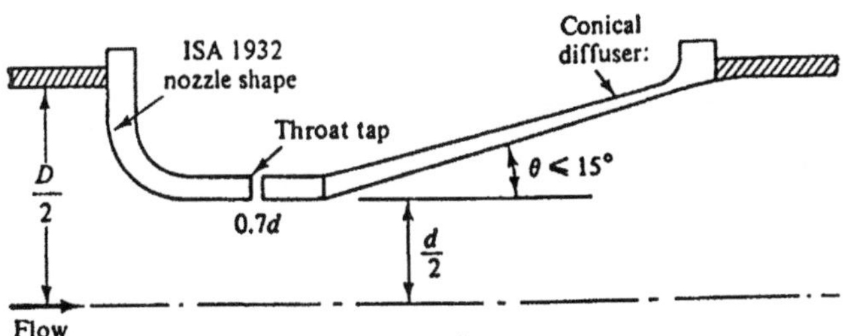

FIGURE 10-6 International standard shapes for nozzle and venturi meter. (From White, 1994.)

[e.g., Eq. (10-9)], and the nozzle (discharge) coefficient is also shown in Fig. 10-4. It should be noted that the Reynolds number that is used for the venturi coefficient in Fig. 10-4 is based on the pipe diameter (D), whereas the Reynolds number used for the nozzle coefficient is based on the nozzle diameter (d) (note that $N_{\mathrm{Re}_D} = \beta N_{\mathrm{Re}_d}$). There are various "standard" designs for the nozzle, and the reader should consult the literature for details (e.g., Miller, 1983). The discharge coefficient for these nozzles can also be described by Eq. (10-10), with the appropriate parameters given in Table 10-1.

IV. THE ORIFICE METER

The simplest and most common device for measuring flow rate in a pipe is the orifice meter, illustrated in Fig. 10-7. This is an "obstruction" meter that consists of a plate with a hole in it that is inserted into the pipe, and the pressure drop across the plate is measured. The major difference between this device and the venturi and nozzle meters is the fact that the fluid stream leaving the orifice hole contracts to an area considerably smaller than that of the orifice hole itself. This is called the *vena contracta*, and it occurs because the fluid has considerable inward radial momentum as it converges into the orifice hole, which causes it to continue to flow "inward" for a distance downstream of the orifice before it starts to expand to fill the pipe. If the pipe diameter is D, the orifice diameter is d, and the diameter of the vena contracta is d_2, the contraction ratio for the vena contracta is defined as $C_c = A_2/A_o = (d_2/d)^2$. For highly turbulent flow, $C_c \approx 0.6$.

The complete Bernoulli equation, as applied between point 1 upstream of the orifice where the diameter is D and point 2 in the vena contracta where the diameter is d_2, is

$$\int_{P_2}^{P_1} \frac{dP}{\rho} = \frac{1}{2}(\alpha_2 V_2^2 - \alpha_1 V_1^2) + \frac{K_f}{2} V_1^2 \tag{10-11}$$

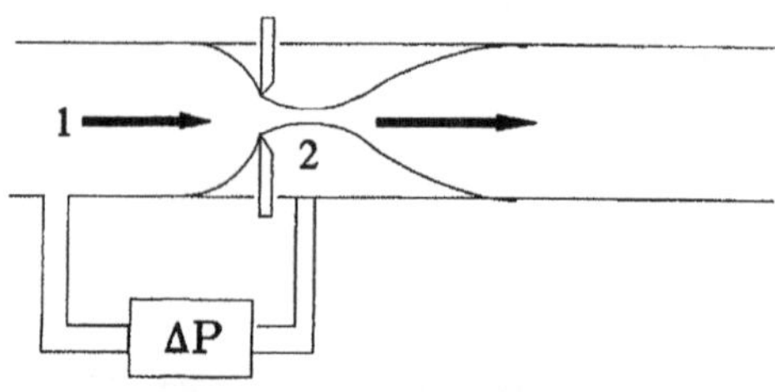

FIGURE 10-7 Orifice meter.

As for the other obstruction meters, when the continuity equation is used to eliminate the upstream velocity from Eq. (10-11), the resulting expression for the mass flow rate through the orifice is

$$\dot{m} = \frac{C_o A_o \rho_2}{(1-\beta^4)^{1/2}} \left[2 \int_{P_2}^{P_1} \frac{dP}{\rho} \right]^{1/2} \tag{10-12}$$

where $\beta = d/D$ and C_o is the orifice coefficient:

$$C_o = \frac{C_c}{\sqrt{\alpha_2}} \left[\frac{1-\beta^4}{1-\beta^4 [C_c(\rho_2/\rho_1)]^2 [(\alpha_1 - K_f)/\alpha_2]} \right]^{1/2} \tag{10-13}$$

C_o is obviously a function of β and the loss coefficient K_f (which depends on N_{Re}).

A. Incompressible Flow

For incompressible flow, Eq. (10-12) becomes

$$\dot{m} = C_o A_o \left(\frac{2\rho \Delta P}{1-\beta^4} \right)^{1/2} \tag{10-14}$$

It is evident that the orifice coefficient incorporates the effects of both friction loss and velocity changes and must therefore depend upon the Reynolds number and beta ratio. This is reflected in Fig. 10-8, in which the orifice (discharge) coefficient is shown as a function of the orifice Reynolds number (N_{Re_d}) and β.

Actually, there are a variety of "standard" orifice plate and pressure tap designs (Miller, 1983). Figure 10-9 shows the ASME specifications for the most common concentric square edged orifice. The various pressure tap locations, illustrated in Fig. 10-10, are radius taps ($1D$ upstream and $D/2$ downstream); flange taps (1 in. upstream and downstream); pipe taps ($2\frac{1}{2}D$ upstream and $8D$ downstream); and corner taps. Radius taps, for which the location is scaled to the pipe diameter, are the most common. Corner taps and flange taps are the most convenient, because they can be installed in the flange that holds the orifice plate and so do not require additional taps through the pipe wall. Pipe taps are less commonly used and essentially measure the total unrecovered pressure drop, or friction loss, through the entire orifice (which is usually quite a bit lower than the maximum pressure drop across the orifice plate). Vena contracta taps are sometimes specified, with the upstream tap $1D$ from the plate and the downstream tap at the vena contracta location, although the latter varies with the Reynolds number and beta ratio and thus is not a fixed position.

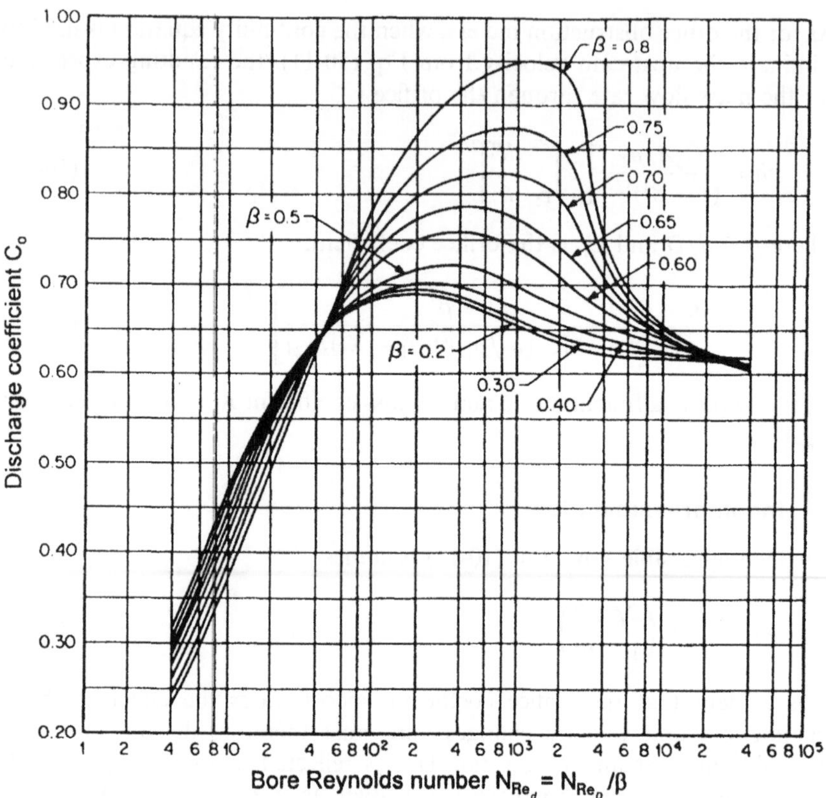

FIGURE 10-8 Orifice discharge coefficient for square-edged orifice and flange, corner, or radius type (From Miller, 1983.)

The orifice coefficient shown in Fig. 10-8 is valid to within about 2–5% (depending upon the Reynolds number) for all pressure tap locations except pipe and vena contracta taps. More accurate values can be calculated from Eq. (10-10), with the parameter expressions given in Table 10-1 for the specific orifice and pressure tap arrangement.

B. Compressible Flow

Equation (10-14) applies to incompressible fluids, such as liquids. For an ideal gas under adiabatic conditions, Eq. (10-12) gives

$$\dot{m} = C_o A_o \sqrt{\frac{P_1 \rho_1}{1-\beta^4}} \left\{ \frac{2k}{k-1} \left(\frac{P_2}{P_1}\right)^{2/k} \left[\left(\frac{P_1}{P_2}\right)^{(k-1)/k} - 1 \right] \right\}^{1/2} \tag{10-15}$$

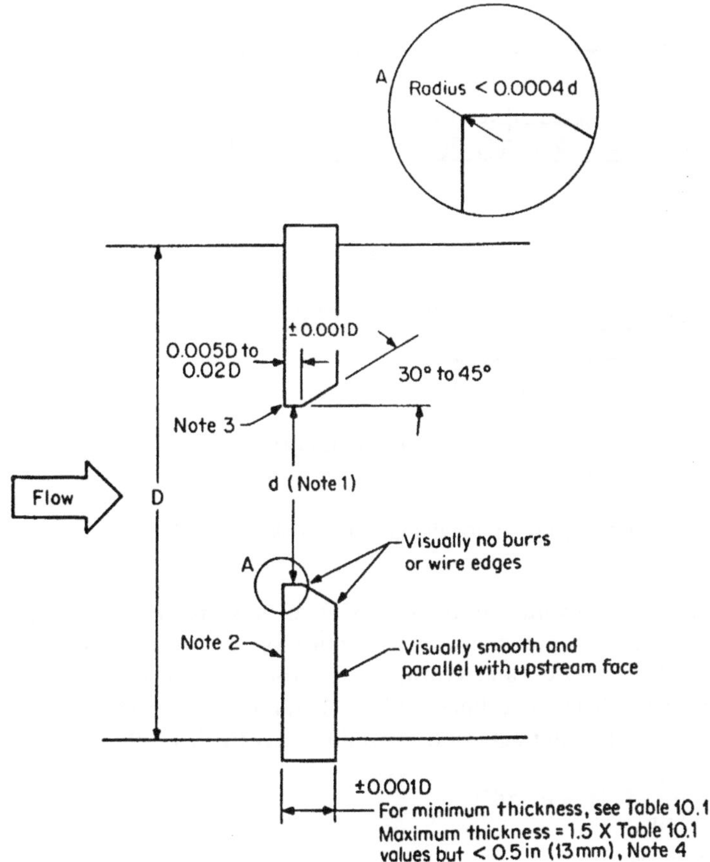

Notes: (1) Mean of four diameters, no diameter > 0.05% of mean diameter. (2) Maximum slope less than 1% from perpendicular; relative roughness < 10^{-4}d over a circle not less than 1.5d. (3) Visually does not reflect a beam of light, finish with a fine radial cut from center outward. (4) ANSI/ASME MFC Draft 2 (July 1982), ASME, N.Y. 1982

FIGURE 10-9 Concentric square-edged orifice specifications. (From Miller, 1983.)

It is more convenient to express this result in terms of the ratio of Eq. (10-15) to the corresponding incompressible equation, Eq. (10-14), which is called the *expansion factor Y*:

$$\dot{m} = C_o A_o Y \sqrt{\frac{P_1 \rho_1}{1 - \beta^4}} \left[2 \left(1 - \frac{P_2}{P_1} \right) \right]^{1/2} \tag{10-16}$$

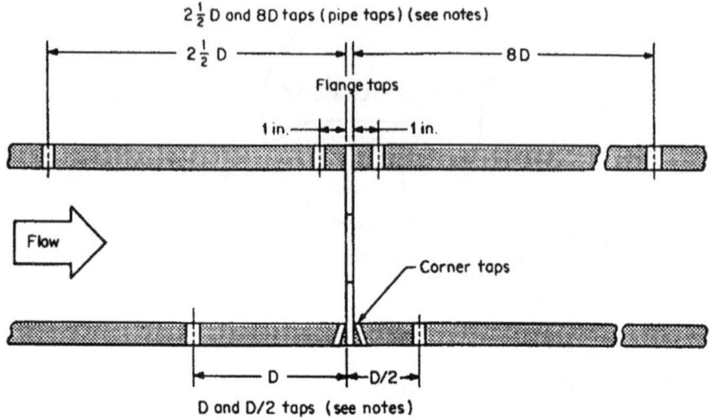

FIGURE 10-10 Orifice pressure tap locations. (From Miller, 1983.)

where the density ρ_1 is evaluated at the upstream pressure (P_1). For convenience, the values of Y are shown as a function of $\Delta P/P_1$ and β for the square-edged orifice, nozzles, and venturi meters for values of $k = c_p/c_v$ of 1.3 and 1.4 in Fig. 10-11. The lines on Fig. 10-11 for the orifice can be represented by the following equation for radius taps (Miller, 1983):

$$Y = 1 - \frac{\Delta P}{kP_1}(0.41 + 0.35\beta^4) \tag{10-17}$$

and for pipe taps by

$$Y = 1 - \frac{\Delta P}{kP_1}[0.333 + 1.145(\beta^2 + 0.7\beta^5 + 12\beta^{13})] \tag{10-18}$$

V. LOSS COEFFICIENT

The total friction loss in an orifice meter, after all pressure recovery has occurred, can be expressed in terms of a loss coefficient, K_f, as follows. With reference to Fig. 10-12, the total friction loss is $P_1 - P_3$. By taking the system to be the fluid in the region from a point just upstream of the orifice plate (P_1) to a downstream position where the stream has filled the pipe (P_3), the momentum balance becomes

$$\sum_{\text{on system}} F = \dot{m}(V_3 - V_1) = 0 = P_1 A_o + P_2(A_1 - A_o) - P_3 A_1 \tag{10-19}$$

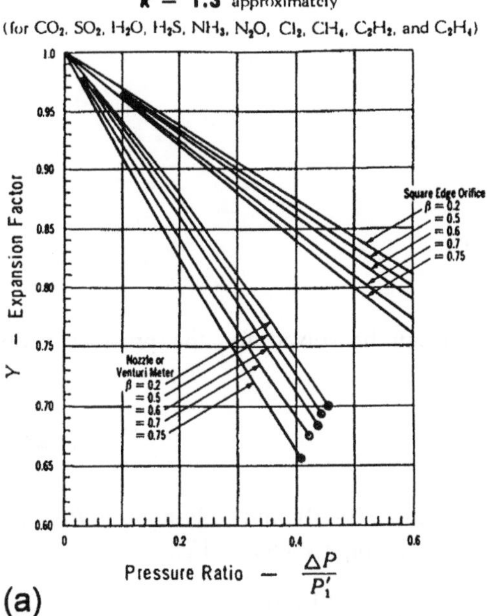

(a)

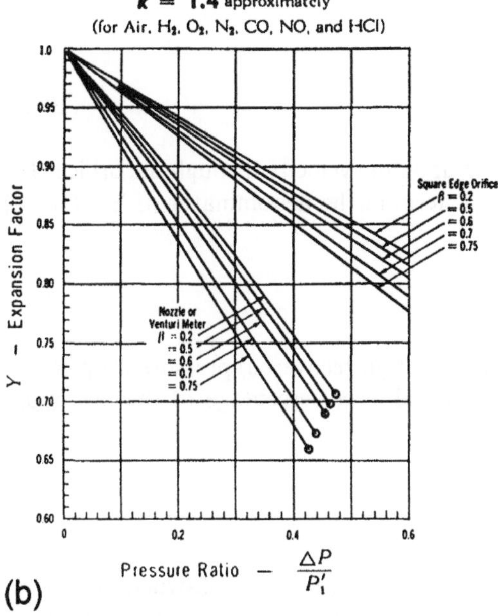

(b)

Figure 10-11 Expansion factor for orifice, nozzle, and venturi meter. (a) $k = 1.3$; (b) $k = 1.4$. (From Crane Co., 1978.)

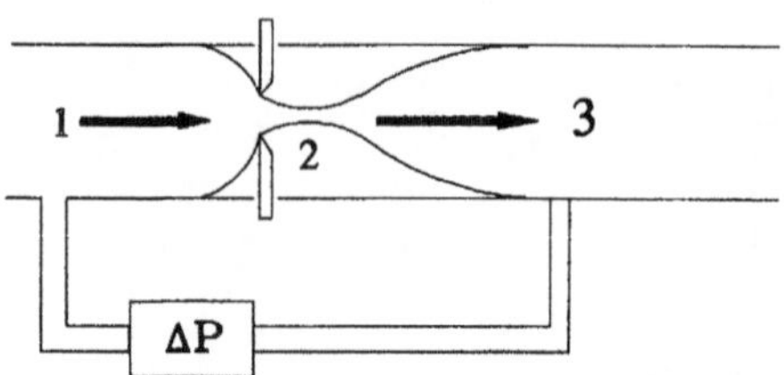

Figure 10-12 Pressure loss in orifice.

The orifice equation [Eq. (10-14)] can be solved for the pressure drop $P_1 - P_2$ to give

$$P_1 - P_2 = \frac{\rho V_o^2}{2}\left(\frac{1-\beta^4}{C_o^2}\right) \tag{10-20}$$

Eliminating P_2 from Eqs. (10-19) and (10-20) and solving for $P_1 - P_3$ provides a definition for K_f based on the pipe velocity (V_1):

$$P_1 - P_3 = \frac{\rho V_o^2(1-\beta^4)(1-\beta^2)}{2C_o^2} = \rho e_f = \frac{\rho V_1^2 K_f}{2} \tag{10-21}$$

Thus the loss coefficient is

$$K_f \approx \frac{(1-\beta^4)(1-\beta^2)}{C_o^2\beta^4} \tag{10-22}$$

If the loss coefficient is based upon the velocity through the orifice (V_o) instead of the pipe velocity, the β^4 term in the denominator of Eq. (10-22) does not appear:

$$K_f = \frac{(1-\beta^4)(1-\beta^2)}{C_o^2} \tag{10-23}$$

Equation (10-21) represents the net total (unrecovered) pressure drop due to friction in the orifice. This is expressed as a percentage of the maximum (orifice) pressure drop in Fig. 10-5.

VI. ORIFICE PROBLEMS

Three classes of problems involving orifices (or other obstruction meters) that the engineer might encounter are similar to the types of problems encountered in pipe flows. These are the "unknown pressure drop," "unknown flow rate," and "unknown orifice diameter" problems. Each involves relationships between the same five basic dimensionless variables: C_d, N_{Re_D}, β, $\Delta P/P_1$, and Y, where C_d represents the discharge coefficient for the meter. For liquids, this list reduces to four variables, because $Y = 1$ by definition. The basic orifice equation relates these variables:

$$\dot{m} = \frac{\pi D^2 \beta^2 Y C_d}{4} \left(\frac{P_1 \rho_1}{1 - \beta^4} \right)^{1/2} \left[2\left(1 - \frac{P_2}{P_1}\right) \right]^{1/2} \tag{10-24}$$

$$N_{Re_D} = \frac{4\dot{m}}{\pi D \mu}, \qquad \beta = \frac{d}{D} \tag{10-25}$$

and $Y = \text{fn}(\beta, \Delta P/P_1)$ [as given by Eq. (10-17) or (10-18) or Fig. 10-11], and $C_d = \text{fn}(\beta, N_{Re_D})$ [as given by Eq. (10-10) or Fig. 10-8]. The procedure for solving each of these problems is as follows.

A. Unknown Pressure Drop

In the case of an unknown pressure drop we want to determine the pressure drop to be expected when a given fluid flows at a given rate through a given orifice.

Given: $\dot{m}$, μ, ρ_1, D, d ($\beta = d/D$), P_1 *Find*: ΔP

The procedure is as follows.

1. Calculate N_{Re_D} and $\beta = d/D$ from Eq. (10-25).
2. Get $C_d = C_o$ from Fig. 10-8 or Eq. (10-10).
3. Assume $Y = 1$, and solve Eq. (10-24) for $(\Delta P)_1$:

$$(\Delta P)_1 = \left(\frac{4\dot{m}}{\pi D^2 \beta^2 C_o} \right)^2 \left(\frac{1 - \beta^4}{2\rho_1} \right) \tag{10-26}$$

4. Using $(\Delta P)_1/P_1$ and β, get Y from Eq. (10-17) or (10-18) or Fig. 10-11.
5. Calculate $\Delta P = (\Delta P)_1/Y^2$.
6. Use the value of ΔP from step 5 in step 4, and repeat steps 4–6 until there is no change.

B. Unknown Flow Rate

In the case of an unknown flow rate, the pressure drop across a given orifice is measured for a fluid with known properties, and the flow rate is to be determined.

Given: ΔP, P_1, D, d ($\beta = d/D$), μ, ρ_1 *Find*: $\dot{m}$

1. Using $\Delta P/P_1$ and β, get Y from Eq. (10-17) or (10-18) or Fig. 10-11.
2. Assume $C_o = 0.61$.
3. Calculate $\dot{m}$ from Eq. (10-24).
4. Calculate N_{Re_D} from Eq. (10-25).
5. Using N_{Re_D} and β, get C_o from Fig. 10-8 or Eq. (10-10).
6. If $C_o \neq 0.61$, use the value from step 5 in step 3, and repeat steps 3–6 until there is no change.

C. Unknown Diameter

For design purposes, the proper size orifice (d or β) must be determined for a specified (maximum) flow rate of a given fluid in a given pipe with a ΔP device that has a given (maximum) range.

Given: $\Delta P, P_1, \mu, \rho, D, \dot{m}$ *Find*: d (i.e., β)

1. Solve Eq. (10-24) for β, i.e.,

$$\beta = \left(\frac{X}{1+X}\right)^{1/4}, \qquad X = \frac{8}{\rho_1 \Delta P}\left(\frac{\dot{m}}{\pi D^2 Y C_o}\right)^2 \tag{10-27}$$

2. Assume that $Y = 1$ and $C_o = 0.61$.
3. Calculate $N_{Re_d} = N_{Re_D}/\beta$, and get C_o from Fig. 10-8 or Eq. (10-10) and Y from Fig. 10-11 or Eq. (10-17) or (10-18).
4. Use the results of step 3 in step 1, and repeat steps 1–4 until there is no change. The required orifice diameter is $d = \beta D$.

VII. CONTROL VALVES

Flow control is achieved by a control valve, which is automatically adjusted (opened or closed) continuously to achieve a desired flow rate. The valve is controlled by a computer that senses the output signal from a flow meter and adjusts the control valve by pneumatic or electrical signals in response to deviations of the measured flow rate from a desired set point. The control valve acts as a variable resistance in the flow line, because closing down on the valve is equivalent to increasing the flow resistance (i.e., the K_f) in the line. The nature of the relationship between the valve stem or plug position (which is the manipulated variable) and the flow rate through the valve (which is the desired variable) is a nonlinear function of the pressure–flow characteristics of the piping system, the driver (i.e., pump) characteristic, and the valve trim characteristic, which is determined by the design of the valve plug. This will be illustrated shortly.

A. Valve Characteristics

Different valve plugs (or "cages" that surround the plug) are usually available for a given valve, each providing a different flow response (or "trim") characteristic when the valve setting (i.e., the stem position) is changed. A specific valve characteristic must be chosen to match the response of the flow system and pump characteristic to give the desired response, as will be demonstrated later.

Figure 10-13 illustrates the flow versus valve stem travel characteristic for various typical valve trim functions (Fisher Controls, 1987). The "quick opening" characteristic provides the maximum change in flow rate at low opening or stem travel, with a fairly linear relationship. As the valve approaches the wide open position, the change in flow with travel approaches zero. This is best suited for on–off control but is

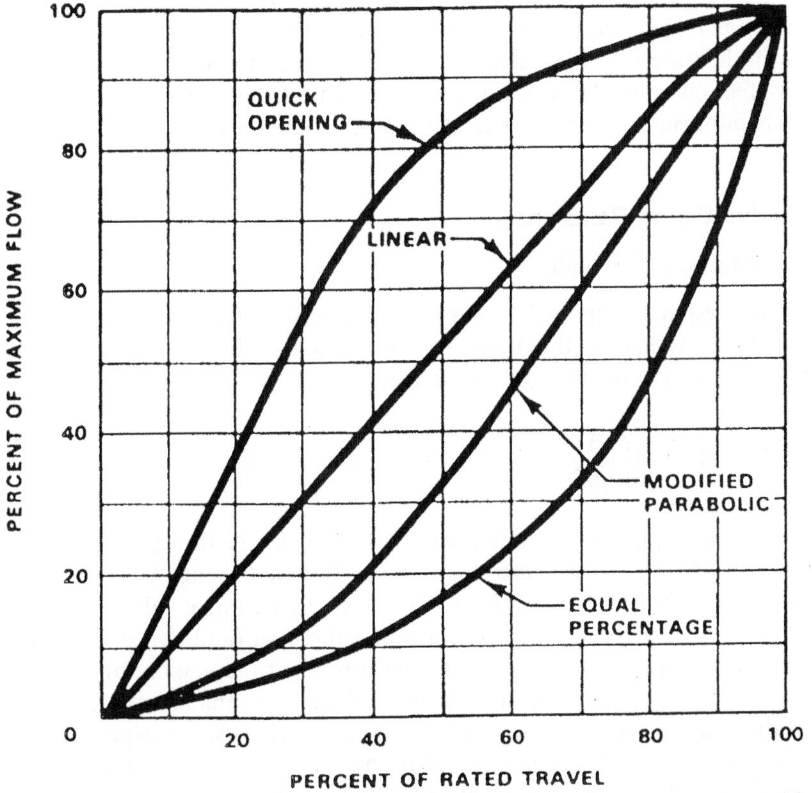

FIGURE 10-13 Control valve flow characteristics. (From Fisher Controls, 1987.)

also appropriate for some applications where a linear response is desired. The "linear" flow characteristic has a constant "valve gain," that is, the incremental change in flow rate with the change in valve plug position is the same at all flow rates. The "equal percentage" trim provides the same percentage change in flow for equal increments of valve plug position. With this characteristic the change in flow is always proportional to the value of the flow rate just before the change is made. This characteristic is used in pressure control applications and where a relatively small pressure drop across the valve is required relative to that in the rest of the system. The modified parabolic characteristic is intermediate to the linear and equal percentage characteristics and can be substituted for equal percentage valve plugs in many applications with some loss in performance.

Some general guidelines for the application of the proper valve characteristic are shown in Fig. 10-14. These are rules of thumb and the proper valve can be determined only by a complete analysis of the system in which the valve is to be used [see also Baumann (1991) for simplified guidelines]. We will illustrate how the valve trim characteristic interacts with the pump and system characteristics to affect the flow rate in the system and how to use this information to select the most appropriate valve trim.

B. Valve Sizing Relations

1. Incompressible Fluids

Bernoulli's equation applied across the valve relates the pressure drop and flow rate in terms of the valve loss coefficient. This equation can be rearranged to give the flow rate as follows:

$$Q = AV = A\left(\frac{2\Delta P_v}{\rho K_f}\right)^{1/2} \tag{10-28}$$

where A is an appropriate flow area, V is the velocity through that area, $\Delta P_v = P_1 - P_2$ is the pressure drop across the valve, and K_f is the loss coefficient referred to the velocity V. However, in a control valve the internal flow geometry is relatively complex and the area (and hence V) varies throughout the valve. Also, the pressure drop is not the maximum value in the valve (which would occur if P_2 is at the discharge vena contracta, as for the orifice meter) but is the *net unrecovered* pressure loss, corresponding to P_2 which is far enough downstream that any possible pressure recovery has occurred. The flow area and geometrical factors are thus combined, along with the density of the reference fluid and the friction loss coefficient, into a

Liquid Level Systems

Control Valve Pressure Drop	Best Inherent Characteristic
Constant ΔP	Linear
Decreasing ΔP with Increasing Load, ΔP at Maximum Load > 20% of Minimum Load ΔP	Linear
Decreasing ΔP with Increasing Load, ΔP at Maximum Load < 20% of Minimum Load ΔP	Equal-Percentage
Increasing ΔP with Increasing Load, ΔP at Maximum Load < 200% of Minimum Load ΔP	Linear
Increasing ΔP with Increasing Load, ΔP at Maximum Load > 200% of Minimum Load ΔP	Quick Opening

Pressure Control Systems

Application	Best Inherent Characteristic
Liquid Process	Equal-Percentage
Gas Process, Small Volume, Less Than 10 ft. of Pipe Between Control Valve and Load Valve	Equal-Percentage
Gas Process, Large Volume (Process has a Receiver, Distribution System or Transmission Line Exceeding 100 ft. of Nominal Pipe Volume) Decreasing ΔP with Increasing Load, ΔP at Maximum Load > 20% of Minimum Load ΔP	Linear
Gas Process, Large Volume, Decreasing ΔP with Increasing Load, ΔP at Maximum Load < 20% of Minimum Load ΔP	Equal-Percentage

Flow Control Processes

FLOW MEASUREMENT SIGNAL TO CONTROLLER	LOCATION OF CONTROL VALVE IN RELATION TO MEASURING ELEMENT	BEST INHERENT CHARACTERISTIC	
		Wide Range of Flow Set Point	Small Range of Flow but Large ΔP Change at Valve with Increasing Load
Proportional To Flow	In Series	Linear	Equal-Percentage
	In Bypass*	Linear	Equal-Percentage
Proportional To Flow Squared	In Series	Linear	Equal-Percentage
	In Bypass*	Equal-Percentage	Equal-Percentage

*When control valve closes, flow rate increases in measuring element.

FIGURE 10-14 Guidelines for control valve applications. (From Fisher Controls, 1987.)

single coefficient, resulting in the following equation for incompressible fluids:

$$Q = C_v\sqrt{\frac{\Delta P_v}{SG}} = C_v\sqrt{\rho_w g h_v} = 0.658 C_v \sqrt{h_v} \tag{10-29}$$

This equation defines the *flow coefficient*, C_v. Here, SG is the fluid specific gravity (relative to water), ρ_w is the density of water, and h_v is the "head loss" across the valve. The last form of Eq. (10-29) applies only for units of Q in gpm and h_v in ft. Although Eq. (10-29) is similar to the flow equation for flow meters, the flow coefficient C_v is not dimensionless, as are the flow meter discharge coefficient and the loss coefficient (K_f), but has dimensions of $[L^3][L/M]^{1/2}$. The value of C_v is thus different for each valve and also varies with the valve opening (or stem travel) for a given valve. Values for the valve C_v are determined by the manufacturer from measurements on each valve type. Because they are not dimensionless, the values will depend upon the specific units used for the quantities in Eq. (10-29). More specifically, the "normal engineering" (inconsistent) units of C_v are gpm/(psi)$^{1/2}$. [If the fluid density were included in Eq. (10-29) instead of SG, the dimensions of C_v would be L_2, which follows from the inclusion of the effective valve flow area in the definition of C_v]. The reference fluid for the density is water for liquids and air for gases.

The units normally used in the United States are the typical "engineering" units, as follows:

Q = volumetric flow rate (gpm for liquids or scfh for gas or steam)
SG = specific gravity [relative to water for liquids (62.3 lb_m/ft^3) or air at 60°F and 1 atm for gases (0.0764 lb_m/ft^3)]
ρ_1 = density at upstream conditions (lb_m/ft^3)
P_1 = upstream pressure (psia)
ΔP_v = total (net unrecovered) pressure drop across valve (psi)

Typical manufacturer's values of C_v to be used with Eq. (10-29) require the variables to be expressed in the above units, with h_v in ft. [For liquids, the value of 0.658 includes the value of the density of water, $\rho_w = 62.3\,lb_m/ft^3$, the ratio g/g_c (which has a magnitude of 1), and 144 $(in./ft)^2$]. For each valve design, tables for the values of the flow coefficients as a function of valve size and percent of valve opening are provided by the manufacturer (see Table 10-3, pages 318–319). In Table 10-3, K_m applies to cavitating and flashing liquids and C_1 applies to critical (choked) compressible flow, as discussed later.

a. Valve–System Interaction. In normal operation, a linear relation between the manipulated variable (valve stem position) and the desired variable (flow rate) is desired. However, the valve is normally a component of a flow system that includes a pump or other driver, pipe and fittings characterized by loss coefficients, etc. In such a system the flow rate is a nonlinear function of the component loss coefficients. Thus the control valve must have a nonlinear response (i.e., trim) to compensate for the nonlinear system characteristics if a linear response is to result. Selection of the proper size and trim of the valve to be used for a given application requires matching the valve, piping system, and pump characteristics, all of which interact (Darby, 1997). The operating point for a piping system depends upon the pressure-flow behavior of both the system and the pump, as described in Chapter 8 and illustrated in Fig. 8-2 (see also Example 8-1). The control valve acts like a variable resistance in the piping system, that is, the valve loss coefficient K_f increases (and the discharge coefficient C_v decreases) as the valve is closed. The operating point for the system is where the pump head (H_p) characteristic intersects the system head requirement (H_s):

$$H_s = \frac{\Delta P}{\rho g} + \Delta z + \frac{Q^2}{g}\left[\frac{8}{\pi^2}\sum_i \left(\frac{K_f}{D^4}\right)_i + \frac{1}{\rho_w C_v^2}\right] = H_p \tag{10-30}$$

where the last term in brackets is the head loss through the control valve, h_v, from Eq. (10-29), and C_v depends upon the valve stem travel, X (see, e.g., Fig. 10-13):

$$C_v = C_{v,\max} f(X) \tag{10-31}$$

A typical situation is illustrated in Fig. 10-15, which shows the pump curve and a system curve with no control valve and the same system curve with a valve that is partially closed. Closing down on the valve (i.e., reducing X) decreases the valve C_v and increases the head loss (h_v) through the valve. The result is to shift the system curve upward by an amount h_v at a given flow rate (note that h_v also depends on flow rate). The range of possible flow rates for a given valve (also known as the "turndown" ratio) lies between the intersection on the pump curve of the system curve with a "fully open" valve (Q_{max}, corresponding to $C_{v,max}$) and the intersection of the system curve with the (partly) closed valve. Of course the minimum flow rate is zero when the valve is fully closed. The desired operating point should be as close as practical to Q_{max}, because this corresponds to an open valve with minimum flow resistance. The flow is then controlled by closing down on the valve (i.e., reducing X and C_v, and thus raising h_v). The minimum operating flow rate (Q_{min}) is established by the turndown ratio (i.e., the operating range) required for proper control. These limits set the size of the valve (e.g., the required $C_{v,max}$), and the head flow rate characteristic of the system

TABLE 10-3 Example Flow Coefficient Values for a Control Valve with Various Trim Characteristics

Linear

Linear Characteristic

Coefficients	Body Size, Inch	Port Diameter, Inch	Total Travel, Inch	Valve Opening, Percent of Total Travel										K_m[1] and C_1
				10	20	30	40	50	60	70	80	90	100	
C_v (Liquid)	2 & 3 x 2	1-7/8	1-1/2	1.69	9.45	21.9	33.4	42.7	50.0	55.6	59.6	61.9	63.6	.72
	3 & 4 x 3	2-7/8	2	3.41	25.4	52.6	76.0	96.4	114	127	133	135	136	.91
	4 & 6 x 4	3-5/8	2	6.89	25.1	50.1	77.9	106	134	157	175	185	188	.86
	6 & 8 x 6	5-3/8	3	9.40	63.8	138	212	282	339	373	389	398	405	.81
C_g (Gas)	2 & 3 x 2	1-7/8	1-1/2	60.8	326	729	1110	1400	1600	1710	1780	1810	1840	28.9
	3 & 4 x 3	2-7/8	2	142	839	1760	2540	3240	3880	4320	4490	4540	4570	33.6
	4 & 6 x 4	3-5/8	2	229	791	1530	2350	3250	4190	5090	5850	6360	6580	35.0
	6 & 8 x 6	5-3/8	3	287	1910	4060	6160	8400	10,600	12,300	13,300	13,800	14,100	34.8
C_s (Steam)	2 & 3 x 2	1-7/8	1-1/2	3.04	16.3	36.5	55.5	70.0	80.0	85.5	89.0	90.5	92.0	28.9
	3 & 4 x 3	2-7/8	2	7.10	42.0	88.0	127	162	194	216	225	227	229	33.6
	4 & 6 x 4	3-5/8	2	11.5	39.6	76.5	118	163	210	255	293	318	329	35.0
	6 & 8 x 6	5-3/8	3	14.4	95.5	203	308	420	530	615	665	690	705	34.8

Equal Percentage

Equal Percentage Characteristic

Coefficients	Body Size, Inch	Port Diameter, Inch	Total Travel, Inch	10	20	30	40	50	60	70	80	90	100	K_m[1] and C_1
C_v (Liquid)	2 & 3 x 2	1-7/8	1-1/8	1.04	1.59	3.52	6.99	12.1	19.7	30.5	40.9	44.6	50.7	.79
	3 & 4 x 3	2-7/8	1-1/2	2.56	51.7	10.80	18.2	28.9	44.9	62.6	82.9	104	117	.91
	4 & 6 x 4	3-5/8	1-1/2	3.44	7.12	13.1	21.8	34.8	54.0	80.4	109	132	154	.71
	6 & 8 x 6	5-3/8	2-1/2	5.27	13.0	22.1	35.3	57	93	141	194	246	308	.64
C_g (Gas)	2 & 3 x 2	1-7/8	1-1/8	41.5	61.2	123	233	401	653	996	1320	1460	1590	31.4
	3 & 4 x 3	2-7/8	1-1/2	88.9	175	381	638	985	1530	2190	2890	3610	4000	34.2
	4 & 6 x 4	3-5/8	1-1/2	134	240	430	700	1080	1650	2480	3440	4210	5140	33.4
	6 & 8 x 6	5-3/8	2-1/2	152	422	673	1020	1710	2730	3990	5490	7350	9220	29.9
C_s (Steam)	2 & 3 x 2	1-7/8	1-1/8	2.08	3.06	6.15	11.7	20.1	32.7	49.8	66.0	73.0	79.5	31.4
	3 & 4 x 3	2-7/8	1-1/2	4.45	8.75	19.1	31.9	49.3	76.5	110	145	181	200	34.2
	4 & 6 x 4	3-5/8	1-1/2	6.70	12.0	21.5	35.0	54.0	82.5	124	172	211	257	33.4
	6 & 8 x 6	5-3/8	2-1/2	7.60	21.1	33.7	51.0	85.5	137	200	275	368	461	29.9

Modified Equal Percentage														Equal Percentage Characteristic
C_v (Liquid)	2 & 3 x 2	1-7/8	1-1/2	1.07	2.65	6.87	15.1	26.6	38.3	47.6	53.7	57.3	60.4	.73
	3 & 4 x 3	2-7/8	2	3.08	8.63	18.5	34.3	57.8	84.5	108	123	131	135	.88
	4 & 6 x 4	3-5/8	2	4.49	10.7	21.8	41.2	71.0	107	141	166	183	193	.85
	6 & 8 x 6	5-3/8	3	6.67	16.4	29.3	52.0	92.5	151	217	280	346	380	.75
C_g (Gas)	2 & 3 x 2	1-7/8	1-1/2	43.0	95.9	230	493	874	1260	1530	1660	1720	1800	29.8
	3 & 4 x 3	2-7/8	2	105	295	635	1140	1930	2890	3720	4250	4470	4540	33.6
	4 & 6 x 4	3-5/8	2	172	337	663	1280	2240	3380	4470	5480	6460	6670	34.6
	6 & 8 x 6	5-3/8	3	200	298	894	1520	2620	4330	6270	8210	10,700	12,500	32.9
C_s (Steam)	2 & 3 x 2	1-7/8	1-1/2	2.15	4.80	11.5	24.7	43.7	63.0	76.5	83.0	86.0	90.0	29.8
	3 & 4 x 3	2-7/8	2	5.25	14.8	31.8	57.0	96.5	145	186	213	224	227	33.6
	4 & 6 x 4	3-5/8	2	8.60	16.9	33.2	64.0	112	169	224	274	323	334	34.6
	6 & 8 x 6	5-3/8	3	10.0	24.9	44.7	76.0	131	217	314	411	535	625	32.9

1. This column lists the K_m values for the C_v coefficients and the C_1 values for the C_g and C_s coefficients at 100% travel

Source: Fisher Controls (1987).

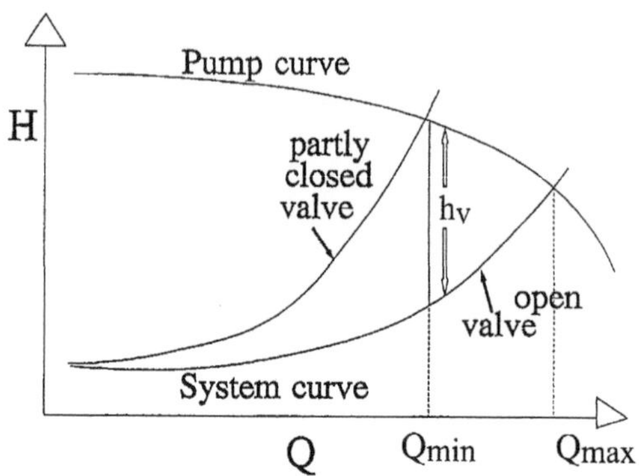

Figure 10-15 Effect of control valve on system operating point.

(including pump and valve) over the desired flow range determines the proper trim for the valve, as follows. As the valve is closed (reducing X) the system curve is shifted up by an amount h_v:

$$h_v = \frac{Q^2}{\rho_w g C_{v,max}^2 f^2(X)} \tag{10-32}$$

where $f(X)$ represents the valve trim characteristic function. Equation (10-32) follows directly from Eqs. (10-29) and (10-31). Thus, as X (the relative valve stem travel) is reduced, $f(X)$ and C_v are also reduced. This increases h_v, with the result that the system system curve now intersects the pump curve further to the left, at a lower value of Q. Substituting Eq. (10-32) into Eq. (10-30) gives

$$H_s = \frac{\Delta P}{\rho g} + \Delta z + Q^2 \left[\frac{8}{g\pi^2} \sum_i \left(\frac{K_f}{D^4} \right)_i + \frac{1}{[0.658 C_{v,max} f(X)]^2} \right] = H_p \tag{10-33}$$

which shows how the system required head (H_s) depends upon the valve stem position (X).

b. Matching Valve Trim to the System. The valve trim function is chosen to provide the desired relationship between valve stem travel (X) and flow rate (Q). This is usually a linear relation and/or a desired sensitivity. For example, if the operating point were far to the left on the diagram where the pump curve is fairly flat, then h_v would be nearly independent of flow rate. In this case, Q would be proportional to $C_v = C_{v,max} f(X)$, and a linear valve characteristic [$f(X)$ vs. X] would be desired. However, the operating point usually occurs where both curves are nonlinear, so that h_v depends strongly on Q which in turn is a nonlinear function of the valve stem position, X. In this case, the most appropriate valve trim can be determined by evaluating Q as a function of X for various trim characteristics (e.g., Fig. 10-13) and choosing the trim that provides the most linear response over the operating range. For a given valve (e.g., $C_{v,max}$), and a given trim response [e.g., $f(x)$ this can be done by calculating the system curve [e.g., Eq. (10-33)] for various valve settings (X) and determining the corresponding values of Q from the intersection of these curves with the pump curve (e.g., the operating point). The trim that gives the most linear (or most sensitive) relation between X and Q is then chosen. This process can be aided by fitting the trim function (e.g., Fig. 10-13) by an empirical equation such as:

Linear trim:

$$f(X) = X \tag{10-34}$$

Parabolic trim:

$$f(X) = X^2 \text{ or } X^n \tag{10-35}$$

Equal percentage trim:

$$f(X) = \frac{\exp(aX^n) - 1}{\exp(a) - 1} \tag{10-36}$$

Quick opening trim:

$$f(X) = 1 - [a(1 - X) - (a - 1)(1 - X)^n] \tag{10-37}$$

where a and n are parameters that can be adjusted to give the best fit to the trim curves.

Likewise, the pump characteristic can usually be described by a quadratic equation of the form

$$H_p = H_o - cQ - bQ^2 \tag{10-38}$$

where H_o, c, and b are curve-fit parameters ($H_o = H_p$ at $Q = 0$).

The operating point is where the pump head [Eq. 10-38] matches the system head requirement [Eq. 10-33]. Thus if Eq. (10-33) for the system head is set equal to Eq. (10-38) for the pump head, the result can be solved for $1/f(X)^2$ to give

$$\left(\frac{1}{f(X)}\right)^2 = 0.433C_{max}^2 \left[\frac{H_o - cQ - bQ^2 - \Delta z - \Delta P/\rho g}{Q^2} - \frac{0.00259 \sum K_f}{D_{in.}^4}\right] \tag{10-39}$$

where conversion factors have been included for units of Q in gpm, H_o, Δz, and $\Delta P/\rho g$ in ft; $D_{in.}$ in inches, and $C_{v,max}$ in (gpm/psi$^{1/2}$). The valve position X corresponding to a given flow rate Q is determined by equating the value of $f(X)$ obtained from Eq. (10-39) to that corresponding to a specific valve trim, e.g., Eqs. (10-34)–(10-37). This procedure is illustrated by the following example.

Example 10-1: Control Valve Trim Selection. It is desired to find the trim for a control valve that gives the most linear relation between stem position (X) and flow rate (Q) when used to control the flow rate in the fluid transfer system shown in Fig. 10-16. The fluid is water at 60°F, flowing through 100 ft of 3 in. sch 40 pipe containing 12 standard threaded elbows in addition to the control valve. The fluid is pumped from tank 1 upstream at atmospheric pressure to tank 2 downstream, which is also at atmospheric pressure and at an elevation $Z_2 = 20$ ft higher than tank 1. The pump is a 2×3 centrifugal pump with an $8\frac{3}{4}$ in. diameter impeller, for which the head curve can be represented by Eq. (10-38) with $H_o = 360$ ft, $a = 0.0006$, and $b = 0.0005$. A 3 in. control valve with the following possible trim plug characteristics will be considered.

Equal percentage (EP):

$$C_v = 51 \frac{\exp(0.5X^{2.5}) - 1}{\exp(0.5) - 1}$$

Modified parabolic (MP):

$$C_v = 60X^{1.6}$$

Linear (L):

$$C_v = 64X$$

Quick opening (QO):

$$C_v = 70\{1 - [0.1(1 - X) - (0.1 - 1)(1 - X)^{2.5}]\}$$

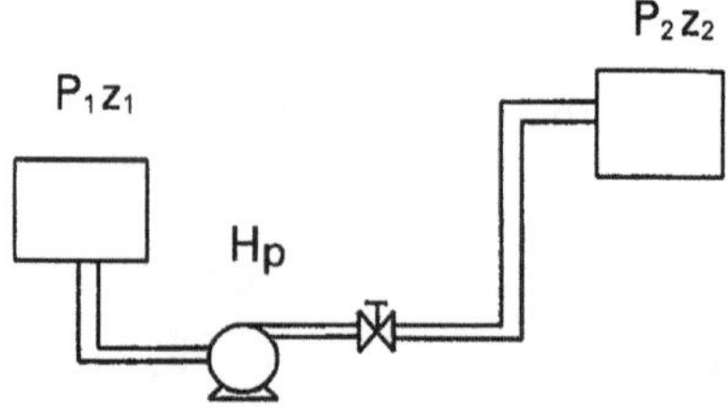

FIGURE 10-16 Fluid transfer system with control valve.

The range of flow rates possible with the control valve can be estimated by inserting the linear valve trim [i.e., $C_{v,\max} f(X) = 64X$] into Eq. (10-33) and calculating the system curves for the valve open, half open, and one-fourth open ($X = 1$, 0.5, 0.25). The intersection of these system curves with the pump curve shows that the operating range with this valve is approximately 150–450 gpm, as shown in Fig. 10-17.

The Q vs. X relation for various valve trim functions can be determined as follows. First, a flow rate (Q) is assumed, which is used to calculate the Reynolds number and thence the pipe friction factor and the loss coefficients for the pipe and fittings. Then a valve trim characteristic function is assumed and, using the pump head function parameters, the right-hand side of Eq. (10-39) is evaluated. This gives the value for $f(X)$ for that trim which corresponds to the assumed flow rate. This is then equated to the appropriate trim function for the valve given above (e.g., EP, MP, L, QO) and the resulting equation is solved for X (this may require an iteration procedure or the use of a nonlinear equation solver). The procedure is repeated over a range of assumed Q values for each of the given trim functions, giving the Q vs. X response for each trim as shown in the Fig. 10-18. It is evident that the equal percentage (EP) trim results in the most linear response and also results in the greatest rangeability or "turndown" (which is inversely proportional to the slope of the line).

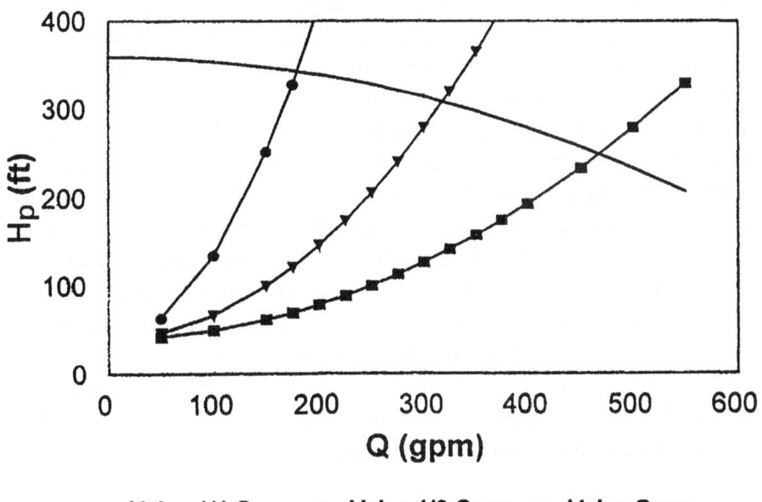

Figure 10-17 Flow range with linear trim.

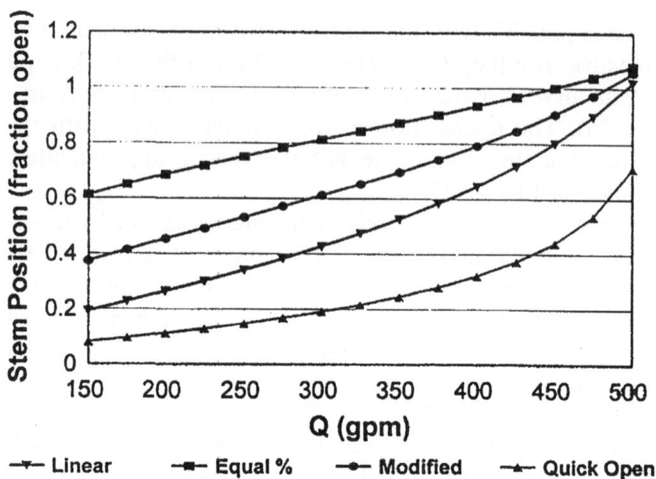

FIGURE 10-18 Flow rate versus stem position for various trim functions.

2. Cavitating and Flashing Liquids

The minimum pressure in the valve (P_{vc}) generally occurs at the vena contracta, just downstream of the flow orifice. The pressure then rises downstream to P_2, with the amount of pressure recovery depending upon the valve design. If P_{vc} is less than the fluid vapor pressure (P_v), the liquid will partially vaporize forming bubbles. If the pressure recovers to a value greater than P_v, these bubbles may collapse suddenly, setting up local shock waves, which can result in considerable damage. This situation is referred to as *cavitation*, as opposed to *flashing*, which occurs if the recovered pressure remains below P_v so that the vapor does not condense. After the first vapor cavities form, the flow rate will no longer be proportional to the square root of the pressure difference across the valve because of the decreasing density of the mixture. If sufficient vapor forms the flow can become choked, at which point the flow rate will be independent of the downstream pressure as long as P_1 remains constant. The critical pressure ratio ($r_c = P_{2c}/P_v$) at which choking will occur is shown in Fig. 10-19 for water and Fig. 10-20 for other liquids, as a function of the liquid vapor pressure (P_v) relative to the fluid critical pressure (P_c). Table 10-4 lists the critical pressure values for some common fluids. An equation that represents the critical pressure ratio, r_c, with acceptable accuracy is (Fisher Controls, 1977)

$$r_c = 0.96 - 0.28\sqrt{\frac{P_v}{P_c}} \tag{10-40}$$

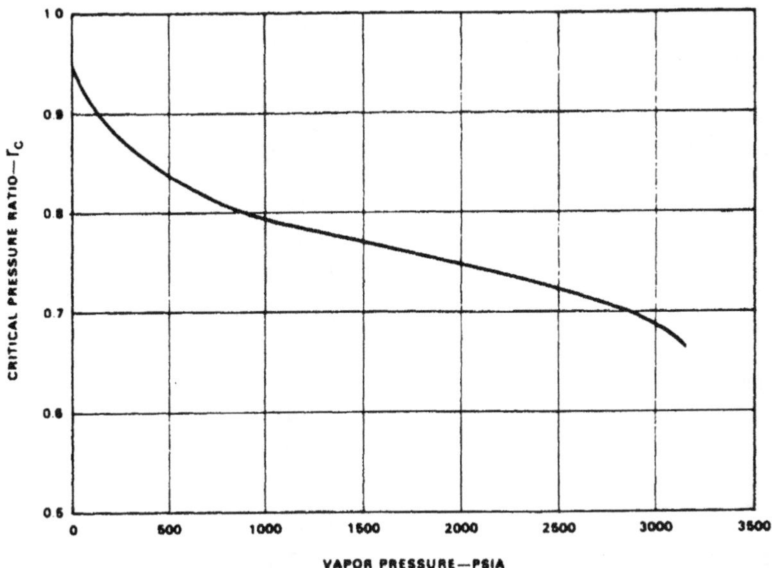

FIGURE 10-19 Critical pressure ratios for water. The abscissa is the water vapor pressure at the valve inlet. The ordinate is the corresponding critical pressure ratio, r_c. (From Fisher Controls, 1987.)

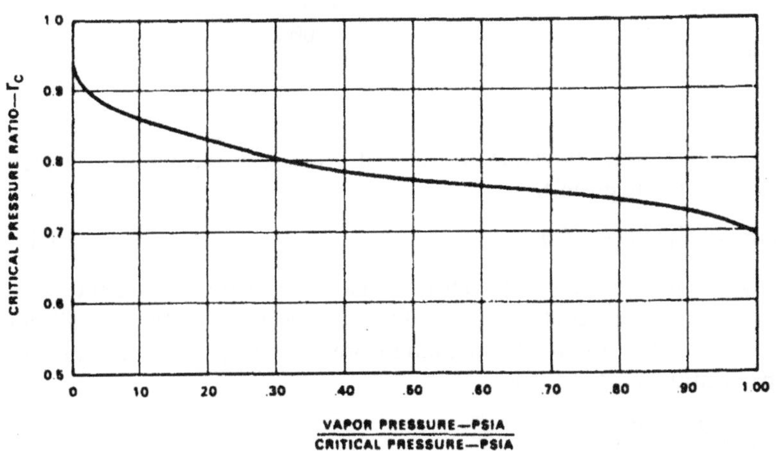

FIGURE 10-20 Critical pressure ratio for cavitating and flashing liquids other than water. The abscissa is the ratio of the liquid vapor pressure at the valve inlet divided by the thermodynamic critical pressure of the liquid. The ordinate is the corresponding critical pressure ratio, r_e. (From Fisher Controls, 1987.)

TABLE 10-4 Critical Pressure of Various Fluids, psia

Fluid	Critical pressure
Ammonia	1636
Argon	705.6
Butane	550.4
Carbon dioxide	1071.6
Carbon monoxide	507.5
Chlorine	1118.7
Dowtherm A	465
Ethane	708
Ethylene	735
Fluorine	808.5
Helium	33.2
Hydrogen	188.2
Hydrogen chloride	1198
Isobutane	529.2
Isobutylene	580
Methane	673.3
Nitrogen	492.4
Nitrous oxide	1047.6
Oxygen	736.5
Phosgene	823.2
Propane	617.4
Propylene	670.3
Refrigerant 11	635
Refrigerant 12	596.9
Refrigerant 22	716
Water	3206.2

With r_c known, the allowable pressure drop across the valve at which cavitation occurs is given by

$$\Delta P_c = K_m(P_1 - r_c P_v) \tag{10-41}$$

where K_m is the valve recovery coefficient (which is a function of the valve design). The recovery coefficient is defined as the ratio of the overall net pressure drop $(P_1 - P_2)$ to the maximum pressure drop from upstream to the vena contracta $(P_1 - P_{vc})$:

$$K_m = \frac{P_1 - P_2}{P_1 - P_{vc}} \tag{10-42}$$

Values of K_m for the Fisher Controls design EB valve are given in the last column of Table 10-3, and representative values for other valves at the full-open condition are given in Table 10-5.

TABLE 10-5 Representative Full-Open K_m Values for Various Valves

Body type	K_m
Globe: single port, flow opens	0.70–0.80
Globe: double port	0.70–0.80
Angle: flow closes	
Venturi outlet liner	0.20–0.25
Standard seat ring	0.50–0.60
Angle: flow opens	
Maximum orifice	0.70
Minimum orifice	0.90
Ball valve	
V-notch	0.40
Butterfly valve	
60° open	0.55
90° open	0.30

Source: Hutchison (1971).

If the pressure drop across the valve is $\Delta P > \Delta P_c$, the value of ΔP_c is used as the pressure drop in the standard liquid sizing equation to determine Q:

$$Q = C_v\sqrt{\Delta P_c/\mathrm{SG}} \tag{10-43}$$

Otherwise the value of $P_1 - P_2$ is used.

The notation used here is that of the Fisher Controls literature (e.g., Fisher Controls, 1990). The ANSI/ISAS75.01 standard for control valves (e.g., Baumann, 1991; Hutchison, 1971) uses the same equations, except that it uses the notation $F_L = (K_m)^{1/2}$ and $F_F = r_c$ in place of the factors K_m and r_c.

C. Compressible Fluids

1. Subsonic Flow

For relatively low pressure drops, the effect of compressibility is negligible, and the general flow equation [Eq. (10-29)] applies. Introducing the conversion factors to give the flow rate in standard cubic feet per hour (scfh) and the density of air at standard conditions (1 atm, 520°R), this equation becomes

$$Q_{\mathrm{scfh}} = 1362 C_v P_1 \left(\frac{\Delta P}{P_1\,\mathrm{SG}\,T_1}\right)^{1/2} \tag{10-44}$$

The effect of variable density can be accounted for by an expansion factor Y as has been done for flow in pipes and meters, in which case Eq. (10-44) can be written

$$Q_{\mathrm{scfh}} = 1362 C_v P_1 Y \left(\frac{X}{\mathrm{SG}\, T_1} \right)^{1/2} \tag{10-45}$$

where

$$X = \frac{\Delta P}{P_1} = \frac{P_1 - P_2}{P_1} = 1 - \frac{P_2}{P_1} \tag{10-46}$$

The expansion factor Y depends on the pressure drop X, the dimensions (clearance) in the valve, the gas specific heat ratio k, and the Reynolds number (the effect of which is often negligible). It has been found from measurements (Hutchison, 1971) that the expansion factor for a given valve can be represented, to within about ±2%, by the expression

$$Y = 1 - \frac{X}{3X_T} \tag{10-47}$$

where X_T is specific to the valve, as illustrated in Fig. 10-21. Deviations from the ideal gas law can be incorporated by multiplying T_1 in Eq. (10-44) or (10-45) by the compressibility factor, Z, for the gas.

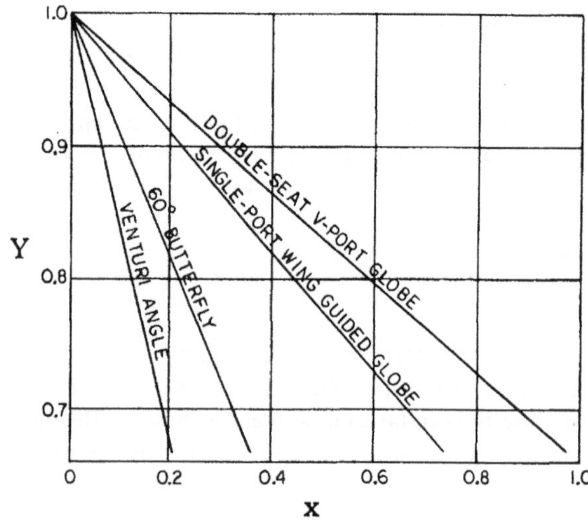

Figure 10-21 Expansion factor (Y) as a function of pressure drop ratio (X) for four different types of control valves. (From Hutchison, 1971.)

2. Choked Flow

When the gas velocity reaches the speed of sound, choked flow occurs and the mass flow rate reaches a maximum. It can be shown from Eq. (10-45) that this is equivalent to a maximum in $YX^{1/2}$, which occurs at $Y = 0.667$, and corresponds to the terminus of the lines in Fig 10-21. That is, X_T is the pressure ratio across the valve at which choking occurs, and any further increase in X (e.g., ΔP) due to lowering P_2 can have no effect on the flow rate.

The flow coefficient C_v is determined by calibration with water, and it is not entirely satisfactory for predicting the flow rate of compressible fluids under choked flow conditions. This has to do with the fact that different valves exhibit different pressure recovery characteristics with gases and hence will choke at different pressure ratios, which does not apply to liquids. For this reason, another flow coefficient, C_g, is often used for gases. C_g is determined by calibration with air under critical flow conditions (Fisher Controls, 1977). The corresponding flow equation for gas flow is

$$Q_{\text{critical}} = C_g P_1 \left(\frac{520}{\text{SG}\, T} \right)^{1/2} \tag{10-48}$$

3. Universal Gas Sizing Equation

Equation (10-44), which applies at low pressure drops, and Eq. (10-48), which applies to critical (choked) flow, have been combined into one general "universal" empirical equation by Fisher Controls (1977), by using a sine function to represent the transition between the limits of these two states:

$$Q_{\text{scfh}} = C_g \left(\frac{520}{\text{SG}\, T_1} \right)^{1/2} P_1 \sin \left[\frac{3417}{C_1} \sqrt{\frac{\Delta P}{P_1}} \right]_{\text{deg}} \tag{10-49}$$

Here, $C_1 = C_g/C_v$ and is determined by measurements on air. For the valve in Table 10-3, values of C_1 are listed in the last column. C_1 is also approximately equal to $40X_T^{1/2}$ (Hutchison, 1971). For steam or vapor at any pressure, the corresponding equation is

$$Q_{\text{lb/hr}} = 1.06 C_g \sqrt{\rho_1 P_1} \sin \left[\frac{3417}{C_1} \sqrt{\frac{\Delta P}{P_1}} \right]_{\text{deg}} \tag{10-50}$$

where ρ_1 is the density of the gas at P_1, in lb_m/ft^3. When the argument of the sine term (in brackets) in Eq. (10-49) or (10-50) is equal to 90° or more, the flow has reached critical flow conditions (choked) and cannot increase above this value without increasing P_1. Under these conditions, the sine term is set equal to unity for this and all larger values of ΔP.

The foregoing equations are based on flow coefficients determined by calibration with air. For application with other gases, the difference between the properties of air and those of the other gas must be considered. The gas density is incorporated into the equations, but a correction must be made for the specific heat ratio ($k = c_p/c_v$) as well. This can be done by considering the expression for the ideal (isentropic) flow of a gas through a nozzle, which can be written (in "engineering units") as follows:

$$Q_{\text{scfh}} = \frac{3.78 \times 10^5 A_2 P_1}{\text{SG}\sqrt{RT}} \left\{ \frac{k}{k-1} \left[\left(\frac{P_2}{P_1}\right)^{2/k} - \left(\frac{P_2}{P_1}\right)^{(k+1/k)} \right] \right\}^{1/2} \tag{10-51}$$

Critical (choked) flow will occur in the nozzle throat when the pressure ratio is

$$r = \frac{P_2}{P_1} = \left(\frac{2}{k+1}\right)^{k/(k-1)} \tag{10-52}$$

Thus, for choked flow, Eq. (10-51) becomes

$$Q_{\text{scfh}} = \frac{3.78 \times 10^5 A_2 P_1}{\text{SG}\sqrt{RT}} \left[\left(\frac{k}{k+1}\right) \left(\frac{2}{k+1}\right)^{2/(k-1)} \right]^{1/2} \tag{10-53}$$

The quantity in square brackets which is a function only of k [fn (k)], represents the dependence of the flow rate on the gas property. Hence it can be used to define a correction factor C_2 that can be used as a multiplier to correct the flow rate for air to that for any other gas:

$$C_2 = \frac{\text{fn}(k)_{\text{gas}}}{\text{fn}(k)_{\text{air}}} = \frac{\left[\left(\frac{k}{k+1}\right) \left(\frac{2}{k+1}\right)^{2/(k-1)} \right]^{1/2}}{0.4839} \tag{10-54}$$

A plot of C_2 vs. k as given by Eq. (10-54) is shown in Fig. 10-22.

D. Viscosity Correction

A correction for fluid viscosity must be applied to the flow coefficient (C_v) for liquids other than water. This viscosity correction factor (F_v) is obtained from Fig. 10-23 by the following procedure, depending upon whether the objective is to find the valve size for a given Q and ΔP, to find Q for a given valve and ΔP, or to find ΔP for a given valve and Q.

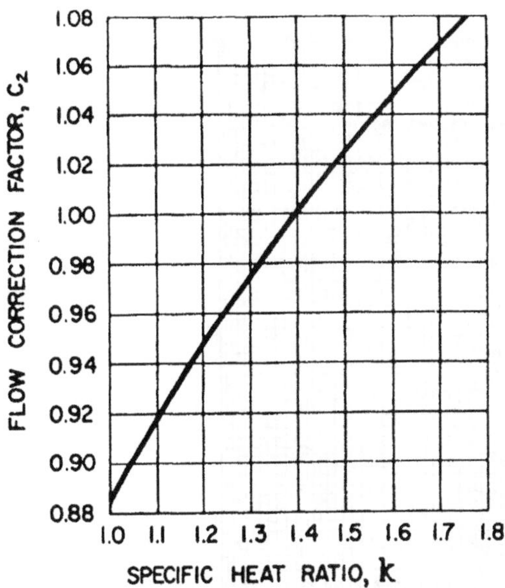

FIGURE **10-22** Correction factor of gas properties.

1. To Find Valve Size

For the given Q and ΔP, calculate the required C_v as follows:

$$C_v = \frac{Q}{\sqrt{\Delta P/\mathrm{SG}}} \tag{10-55}$$

Then determine the Reynolds number from the equation

$$N_{\mathrm{Re}} = 17250 \frac{Q}{\nu_{cs}\sqrt{C_v}} \tag{10-56}$$

where Q is in gpm, ΔP is in psi, and ν_{cs} is the fluid kinematic viscosity (μ/ρ) in centistokes. The viscosity correction factor, F_v, is then read from the middle line on Fig. 10-23 and used to calculate a corrected value of C_v as follows:

$$C_{v_c} = C_v F_v \tag{10-57}$$

The proper valve size and percent opening are then found from the table for the valve flow coefficient (e.g., Table 10-3) at the point where the coefficient is equal to or higher than this corrected value.

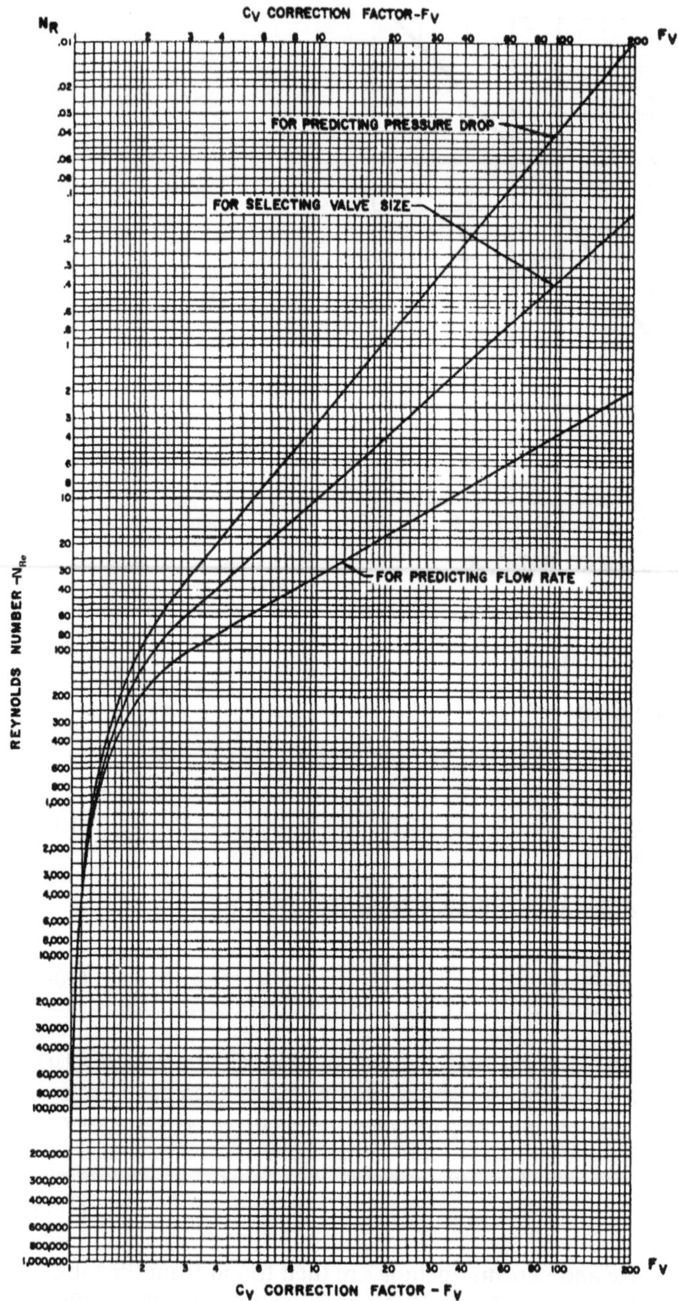

Figure 10-23 Viscosity correction factor for C_V. (From Fisher Controls, 1977.)

2. To Predict Flow Rate

For a given valve (i.e., a given C_v) and given ΔP, the maximum flow rate (Q_{max}) is determined from

$$Q_{max} = C_v \sqrt{\Delta P/\mathrm{SG}} \tag{10-58}$$

The Reynolds number is then calculated from Eq. (10-56), and the viscosity correction factor, F_v, is read from the bottom curve in Fig. 10-23. The corrected flow rate is then

$$Q_c = Q_{max}/F_v \tag{10-59}$$

3. To Predict Pressure Drop

For a given valve (C_v) and given flow rate (Q), calculate the Reynolds number as above and read the viscosity correction factor, F_v, from the top line of Fig. 10-23. The predicted pressure drop across the valve is then

$$\Delta P = \mathrm{SG}\left(\frac{QF_v}{C_v}\right)^2 \tag{10-60}$$

PROBLEMS

Flow Measurement

1. An orifice meter with a hole of 1 in. diameter is inserted into a $1\frac{1}{2}$ in. sch 40 line carrying SAE 10 lube oil at 70°F (SG = 0.93). A manometer using water as the manometer fluid is used to measure the orifice pressure drop and reads 8 in. What is the flow rate of the oil, in gpm?
2. An orifice with a 3 in. diameter hole is mounted in a 4 in. diameter pipeline carrying water. A manometer containing a fluid with an SG of 1.2 connected across the orifice reads 0.25 in. What is the flow rate in the pipe, in gpm?
3. An orifice with a 1 in. diameter hole is installed in a 2 in. sch 40 pipeline carrying SAE 10 lube oil at 100°F. The pipe section where the orifice is installed is vertical, with the flow being upward. Pipe taps are used, which are connected to a manometer containing mercury to measure the pressure drop. If the manometer reading is 3 in., what is the flow rate of the oil, in gpm?
4. The flow rate in a 1.5 in. line can vary from 100 to 1000 bbl/day, so you must install an orifice meter to measure it. If you use a DP cell with a range of 10 in. H_2O to measure the pressure drop across the orifice, what size orifice should you use? After this orifice is installed, you find that the DP cell reads 0.5 in. H_2O. What is the flow rate in bbl/day? The fluid is an oil with an SG = 0.89 and $\mu = 1$ cP.
5. A 4 in. sch 80 pipe carries water from a storage tank on top of a hill to a plant at the bottom of the hill. The pipe is inclined at an angle of 20° to the horizontal.

An orifice meter with a diameter of 1 in. is inserted in the line, and a mercury manometer across the meter reads 2 in. What is the flow rate in gpm?

6. You must size an orifice meter to measure the flow rate of gasoline (SG = 0.72) in a 10 in. ID pipeline at 60°F. The maximum flow rate expected is 1000 gpm, and the maximum pressure differential across the orifice is to be 10 in. of water. What size orifice should you use?
7. A 2 in. sch 40 pipe carries SAE 10 lube oil at 100°F (SG = 0.928). The flow rate may be as high as 55 gpm, and you must select an orifice meter to measure the flow.
   (a) What size orifice should be used if the pressure difference is measured using a DP cell having a full scale range of 100 in.H_2O?
   (b) Using this size orifice, what is the flow rate of oil, in gpm, when the DP cell reads 50 in.H_2O of water?
8. A 2 in. sch 40 pipe carries a 35° API distillate at 50°F (SG = 0.85). The flow rate is measured by an orifice meter which has a diameter of 1.5 in. The pressure drop across the orifice plate is measured by a water manometer connected to flange taps.
   (a) If the manometer reading is 1 in., what is the flow rate of the oil, in gpm?
   (b) What would the diameter of the throat of a venturi meter be that would give the same manometer reading at this flow rate?
   (c) Determine the unrecovered pressure loss for both the orifice and the venturi, in psi.
9. An orifice having a diameter of 1 in. is used to measure the flow rate of SAE 10 lube oil (SG = 0.928) in a 2 in. sch 40 pipe at 70°F. The pressure drop across the orifice is measured by a mercury (SG = 13.6) manometer, which reads 2 cm.
   (a) Calculate the volumetric flow rate of the oil in liters/s.
   (b) What is the temperature rise of the oil as it flows through the orifice, in °F? [$C_v = 0.5$ Btu/(lb$_m$°F).]
   (c) How much power (in horsepower) is required to pump the oil through the orifice? (*Note:* This is the same as the rate of energy dissipated in the flow.)
10. An orifice meter is used to measure the flow rate of CCl_4 in a 2 in. sch 40 pipe. The orifice diameter is 1.25 in., and a mercury manometer attached to the pipe taps across the orifice reads 1/2 in. Calculate the volumetric flow rate of CCl_4 in ft^3/s. (SG of CCl_4 = 1.6.) What is the permanent energy loss in the flow above due to the presence of the orifice in ft lb$_f$/lb$_m$? Express this also as a total overall "unrecovered" pressure loss in psi.
11. An orifice meter is installed in a 6 in. ID pipeline that is inclined upward at an angle of 10° from the horizontal. Benzene is flowing in the pipeline at the flow rate of 10 gpm. The orifice diameter is 3.5 in., and the orifice pressure taps are 9 in. apart.
    (a) What is the pressure drop between the pressure taps, in psi?
    (b) What would be the reading of a water manometer connected to the pressure taps?
12. You are to specify an orifice meter for measuring the flow rate of a 35° API distillate (SG = 0.85) flowing in a 2 in. sch 160 pipe at 70°F. The maximum flow

rate expected is 2000 gph, and the available instrumentation for a differential pressure measurement has a limit of 2 psi. What size hole should the orifice have?

13. You must select an orifice meter for measuring the flow rate of an organic liquid ($SG = 0.8$, $\mu = 15\,cP$) in a 4 in. sch 40 pipe. The maximum flow rate anticipated is 200 gpm, and the orifice pressure difference is to be measured with a mercury manometer having a maximum reading range of 10 in. What size should the orifice be?
14. An oil with an SG of 0.9 and viscosity of 30 cP is transported in a 12 in. sch. 20 pipeline at a maximum flow rate of 1000 gpm. What size orifice should be used to measure the oil flow rate if a DP cell with a full-scale range of 10 in. H_2O is used to measure the pressure drop across the orifice? What size venturi would you use in place of the orifice in the pipeline, everything else being the same?
15. You want to use a venturi meter to measure the flow rate of water, up to 1000 gpm, through an 8 in. sch 40 pipeline. To measure the pressure drop in the venturi, you have a DP cell with a maximum range of 15 in.H_2O pressure difference. What size venturi (i.e., throat diameter) should you specify?
16. Gasoline is pumped through a 2 in. sch 40 pipeline upward into an elevated storage tank at 60°F. An orifice meter is mounted in a vertical section of the line, which uses a DP cell with a maximum range of 10 in.H_2O to measure the pressure drop across the orifice at radius taps. If the maximum flow rate expected in the line is 10 gpm, what size orifice should you use? If a water manometer with a maximum reading of 10 in. is used instead of the DP cell, what would the required orifice diameter be?
17. You have been asked by your boss to select a flow meter to measure the flow rate of gasoline ($SG = 0.85$) at 70°F in a 3 in. sch 40 pipeline. The maximum expected flow rate is 200 gpm, and you have a DP cell (which measures differential pressure) with a range of 0–10 in.H_2O available.
    (a) If you use a venturi meter, what should the diameter of the throat be?
    (b) If you use an orifice meter, what diameter orifice should you use?
    (c) For a venturi meter with a throat diameter of 2.5 in., what would the DP cell read (in inches of water) for a flow rate of 150 gpm?
    (d) For an orifice meter with a diameter of 2.5 in., what would the DP cell read (in inches of water) for a flow rate of 150 gpm?
    (e) How much power (in hp) is consumed by friction loss in each of the meters under the conditions of (c) and (d)?
18. A 2 in. sch 40 pipe is carrying water at a flow rate of 8 gpm. The flow rate is measured by means of an orifice with a 1.6 in. diameter hole. The pressure drop across the orifice is measured using a manometer containing an oil of SG 1.3.
    (a) What is the manometer reading in inches?
    (b) What is the power (in hp) consumed as a consequence of the friction loss due to the orifice plate in the fluid?
19. The flow rate of CO_2 in a 6 in. ID pipeline is measured by an orifice meter with a diameter of 5 in. The pressure upstream of the orifice is 10 psig, and the pressure

drop across the orifice is 30 in.H_2O. If the temperature is 80°F, what is the mass flow rate of CO_2?

20. An orifice meter is installed in a vertical section of a piping system, in which SAE 10 lube oil is flowing upward (at 100°F). The pipe is 2 in. sch 40, and the orifice diameter is 1 in. The pressure drop across the orifice is measured by a manometer containing mercury as the manometer fluid. The pressure taps are pipe taps ($2\frac{1}{2}$ in. ID upstream and 8 in. ID downstream), and the manometer reading is 3 in. What is the flow rate of the oil in the pipe, in gpm?
21. You must install an orifice meter in a pipeline to measure the flow rate of 35.6° API crude oil, at 80°F. The pipeline diameter is 18 in, sch 40, and the maximum expected flow rate is 300 gpm. If the pressure drop across the orifice is limited to 30 in.H_2O or less, what size orifice should be installed? What is the maximum permanent pressure loss that would be expected through this orifice, in psi?
22. You are to specify an orifice meter for measuring the flow rate of a 35° API distillate (SG = 0.85) flowing in a 2 in. sch 160 pipe at 70°F. The maximum flow rate expected is 2000 gal/hr and the available instrumentation for the differential pressure measurement has a limit of 2 psi. What size orifice should be installed?
23. A 6 in. sch 40 pipeline is designed to carry SAE 30 lube oil at 80°F (SG = 0.87) at a maximum velocity of 10 ft/s. You must install an orifice meter in the line to measure the oil flow rate. If the maximum pressure drop to be permitted across the orifice is 40 in. H_2O, what size orifice should be used? If a venturi meter is used instead of an orifice, everything else being the same, how large should the throat be?
24. An orifice meter with a diameter of 3 in. is mounted in a 4 in. sch 40 pipeline carrying an oil with a viscosity of 30 cP and an SG of 0.85. A mercury manometer attached to the orifice meter reads 1 in. If the pumping stations along the pipeline operate with a suction (inlet) pressure of 10 psig and a discharge (outlet) pressure of 160 psig, how far apart should the pump stations be, if the pipeline is horizontal?
25. A 35° API oil at 50°F is transported in a 2 in. sch 40 pipeline. The oil flow rate is measured by an orifice meter that is 1.5 in. in diameter, using a water manometer.
    (a) If the manometer reading is 1 in., what is the oil flow rate, in gpm?
    (b) If a venturi meter is used instead of the orifice meter, what should the diameter of the venturi throat be to give the same reading as the orifice meter at the same flow rate?
    (c) Determine the unrecovered pressure loss for both the orifice and venturi meters.
26. A 6 in. sch 40 pipeline carries a petroleum fraction (viscosity 15 cP, SG 0.85) at a velocity of 7.5 ft/s, from a storage tank at 1 atm pressure to a plant site. The line contains 1500 ft of straight pipe, 25 90° flanged elbows, and four open globe valves. The oil level in the storage tank is 15 ft above ground, and the pipeline discharge at a point 10 ft above ground, at a pressure of 10 psig.

(a) What is the required flow capacity (in gpm) and the head (pressure) to be specified for the pump needed to move the oil?
(b) If the pump is 85% efficient, what horsepower motor is required to drive it?
(c) If a 4 in. diameter orifice is inserted in the line to measure the flow rate, what would the pressure drop reading across it be at the specified flow rate?

27. Water drains by gravity out of the bottom of a large tank, through a horizontal 1 cm ID tube, 5 m long, that has a venturi meter mounted in the middle of the tube. The level in the tank is 4 ft above the tube, and a single open vertical tube is attached to the throat of the venturi. What is the smallest diameter of the venturi throat for which no air will be sucked through the tube attached to the throat? What is the flow rate of the water under this condition?
28. Natural gas (CH_4) is flowing in a 6 in. sch 40 pipeline, at 50 psig and 80°F. A 3 in. diameter orifice is installed in the line, which indicates a pressure drop of 20 in.H_2O. What is the gas flow rate, in lb_m/hr and scfm?
29. A solvent (SG = 0.9, $\mu = 0.8$ cP) is transferred from a storage tank to a process unit through a 3 in. sch 40 pipeline, that is 2000 ft long. The line contains 12 elbows, four globe valves, an orifice meter with a diameter of 2.85 in., and a pump having the characteristics shown in Fig. 8-2 with a $7\frac{1}{4}$ in. impeller. The pressures in the storage tank and the process unit are both 1 atm, and the process unit is 60 ft higher than the storage tank. What is the pressure reading across the orifice meter, in in.H_2O?

Control Valves

30. You want to control the flow rate of a liquid in a transfer line at 350 gpm. The pump in the line has the characteristics shown in Fig. 8-2, with an $8\frac{1}{4}$ in. impeller. The line contains 150 ft of 3 in. sch 40 pipe, 10 flanged elbows, four gate valves, and a 3 × 3 control valve. The pressure and elevation at the entrance and exit of the line are the same. The valve has an equal percentage trim with the characteristics given in Table 10-3. What should the valve opening be to achieve the desired flow rate (in terms of percent of total stem travel)? The fluid has a viscosity of 5 cP and a SG of 0.85.
31. A liquid with a viscosity of 25 cP and an SG of 0.87 is pumped from an open tank to another tank in which the pressure is 15 psig. The line is 2 in. sch 40 diameter, 200 ft long, and contains eight flanged elbows, two gate valves, a control valve, and an orifice meter.
    (a) What should the diameter of the orifice in the line be for a flow rate of 100 gpm, if the pressure across the orifice is not to exceed 80 in. of water?
    (b) If the control valve is a 2 × 2 with equal percentage trim (see Table 10-3), what is the percentage opening of the valve at this flow rate? The pump curve can be represented by the equation

$$H_p = 360 - 0.0006Q - 0.0005Q^2$$

where H_p is in ft and Q is in gpm.

32. Water at 60°F is to be transferred at a rate of 250 gpm from the bottom of a storage tank to the bottom of a process vessel. The water level in the storage tank is 5 ft above ground level and the pressure in the tank is 10 psig. In the process vessel the level is 15 ft above ground and the pressure is 20 psig. The transfer line is 150 ft of 3 in. sch 40 pipe, containing eight flanged elbows, three 80% reduced trim gate valves, and a 3 × 3 control valve with the characteristics given in Table 10-3. The pump in the line has the same characteristics as those shown in Fig. 8-2 with an 8 in. impeller, and the control valve has a linear characteristic. If the stem on the control valve is set to provide the desired flow rate under the specified conditions, what should be the valve opening (i.e., the percent of total travel of the valve stem)?
33. A piping system takes water at 60°F from a tank at atmospheric pressure to a plant vessel at 25 psig that is 30 ft higher than the upstream tank. The transfer line contains 300 ft of 3 in. sch 40 pipe, 10 90° els, an orifice meter, a 2 × 3 pump with a $7\frac{3}{4}$ in. impeller (with the characteristic as given in Fig. 8-2) and a 3 × 2 equal percentage control valve with a trim characteristic as given in Table 10-3. A constant flow rate of 200 gpm is required in the system.
    (a) What size orifice should be installed if the DP cell used to measure the pressure drop across the orifice has a maximum range of 25 in.H_2O?
    (b) What is the stem position of the valve (i.e., the percent of total stem travel) that gives the required flow rate?

NOTATION

A	cross sectional area, $[L^2]$
C_o	orifice coefficient, [—]
C_l	C_g/C_v for a given control valve [—]
C_d	discharge coefficient for (any) flow meter, [—]
C_g	control valve discharge coefficient for gas flow, $[L^4tT^{1/2}/M]$
C_v	control valve discharge coefficient for liquid flow, $[L^{7/2}/M^{1/2}]$
D	pipe diameter, [L]
d	orifice, nozzle, or venturi throat diameter, [L]
F	force, $[F = mL/t^2]$
F_v	control valve viscosity correction factor, Fig. 10-23
H_p	pump head, [L]
H_v	head loss across valve, [L]
k	isentropic coefficient, ($k = c_v/c_p$ for ideal gas), [—]
K_f	loss coefficient, [—]
$\dot{m}$	mass flow rate, [M/t]
N_{Re}	Reynolds number, [—]
P	pressure, $[F/L^2 = M/Lt^2]$
P_c	critical pressure, $[F/L^2 = M/Lt^2]$
P_{2c}	pressure at value exit at which flow is chocked, $[F/L^2 = M/Lt^2]$
P_v	vapor pressure, $[F/L^2 = M/Lt^2]$
Q	volumetric flow rate, $[L^3/t]$

R	radius, [L]
r	radial position, [L]
SG	specific gravity, [—]
V	spatial average velocity, [L/t]
v	local velocity, [L/t]
Y	expansion factor, [—]
z	elevation above an arbitrary reference plane, [L]
α	kinetic energy correction factor, [—]
β	d/D, [—]
$\Delta()$	$()_2 - ()_1$
μ	vicosity, [M/Lt]
ρ	density, $[M/L^3]$
ν	kinematic viscosity, $\nu = \mu/\rho$, $[L^2/t]$

Subscripts

1	reference point 1
2	reference point 2
d	at orifice, nozzle, or venturi throat
D	in pipe
o	orifice
scfh	standard cubic feet per hour
v	venturi, viscosity correction, vapor pressure

REFERENCES

Baumann HD. Control Valve Primer. Durham, NC: Instrument Society of America, 1991.

Cheremisinoff NP, PN Cheremisinoff. Instrumentation for Process Flow Engineering. Lancaster, UK: Technomic, 1987.

Crane Company. Flow of fluids through valves, fittings, and pipe. Tech Manual 410. New York: Crane Co, 1991.

Darby R, Control valves: Match the trim to the selection. *Chem Eng,* June 1997, pp 147–152.

Fisher Controls. Control Valve Handbook. 2nd ed., Marshalltown, IA: Fisher Controls, 1977.

Fisher Controls. Catalog 10. Marshalltown, A: Fisher Controls, 1987, Chap 2.

Fisher Controls. Control Valve Source Book. Marshalltown, IA: Fisher Controls, 1990.

Hutchison, JW, ISA Handbook of Control Valves. Durham, NC: Instrument Society of America, 1971.

Miller RW, Flow Measurement Engineering Handbook. New York: McGraw-Hill, 1983.

Olson RM, Essentials of Engineering Fluid Mechanics. 4th ed. New York: Harper & Row, 1980.

White FM, Fluid Mechanics. 3rd ed., New York: McGraw-Hill, 1994.

11

External Flows

I. DRAG COEFFICIENT

When a fluid flows past a solid body or the body moves through the fluid (e.g., Fig. 11-1), the force (F_D) exerted on the body by the fluid is proportional to the relative rate of momentum transported by the fluid ($\rho V^2 A$). This can be expressed in terms of a *drag coefficient* (C_D) that is defined by the equation

$$\frac{F_D}{A} = \frac{C_D}{2} \rho V^2 \tag{11-1}$$

Here, ρ is the density of the fluid, V is the relative velocity between the fluid and the solid body, and A is the cross sectional area of the body normal to the velocity vector V, e.g., $\pi d^2/4$ for a sphere. Note that the definition of the drag coefficient from Eq. (11-1) is analogous to that of the friction factor for flow in a conduit, i.e.,

$$\tau_w = \frac{f}{2} \rho V^2 \tag{11-2}$$

where τ_w is the force exerted by the moving fluid on the wall of the pipe per unit area. In the case of τ_w, however, the area is the total contact area between the fluid and the wall as opposed to the cross-sectional area normal to the flow direction in the case of C_D. One reason for this is that the fluid

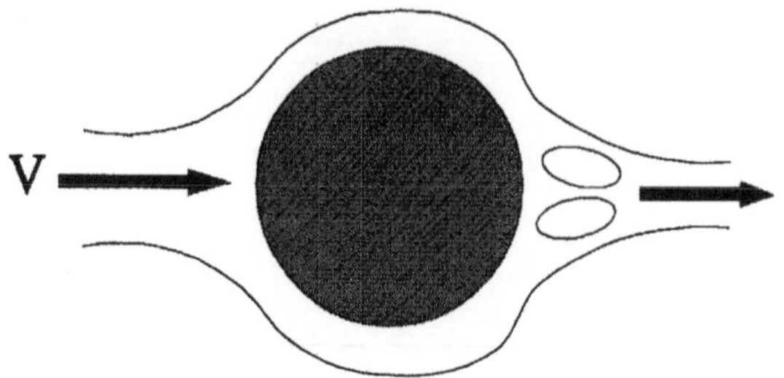

FIGURE 11-1 Drag on a sphere.

interaction with the tube wall is uniform over the entire surface for fully developed flow, whereas for a body immersed in a moving fluid the nature and degree of interaction vary with position around the body.

A. Stokes Flow

If the relative velocity is sufficiently low, the fluid streamlines can follow the contour of the body almost completely all the way around (this is called *creeping flow*). For this case, the microscopic momentum balance equations in spherical coordinates for the two-dimensional flow $[v_r(r, \theta), v_\theta(r, \theta)]$ of a Newtonian fluid were solved by Stokes for the distribution of pressure and the local stress components. These equations can then be integrated over the surface of the sphere to determine the total drag acting on the sphere, two-thirds of which results from viscous drag and one-third from the non-uniform pressure distribution (refered to as *form drag*). The result can be expressed in dimensionless form as a theoretical expression for the drag coefficient:

$$C_D = \frac{24}{N_{Re}} \tag{11-3}$$

where

$$N_{Re} = \frac{dV\rho}{\mu} \tag{11-4}$$

This is known as *Stokes flow*, and Eq. (11-3) has been found be accurate for flow over a sphere for $N_{Re} < 0.1$ and to within about 5% for $N_{Re} < 1$. Note the similarity between Eq. (11-3) and the dimensionless Hagen–Poiseuille equation for laminar tube flow, i.e., $f = 16/N_{Re}$.

B. Form Drag

As the fluid flows over the forward part of the sphere, the velocity increases because the available flow area decreases, and the pressure decreases as a result of the conservation of energy. Conversely, as the fluid flows around the back side of the body, the velocity decreases and the pressure increases. This is not unlike the flow in a diffuser or a converging–diverging duct. The flow behind the sphere into an "adverse pressure gradient" is inherently unstable, so as the velocity (and N_{Re}) increase it becomes more difficult for the streamlines to follow the contour of the body, and they eventually break away from the surface. This condition is called *separation*, although it is the smooth streamline that is separating from the surface, not the fluid itself. When separation occurs eddies or vortices form behind the body as illustrated in Fig. 11-1 and form a "wake" behind the sphere.

As the velocity and N_{Re} increase, the point of streamline separation from the surface moves further upstream and the wake gets larger. The wake region contains circulating eddies of a three-dimensional turbulent nature, so it is a region of relatively high velocity and hence low pressure. Thus the pressure in the wake is lower than that on the front of the sphere, and the product of this pressure difference and the projected area of the wake results in a force acting on the sphere in the direction of the flow, i.e., in the same direction as the drag force. This additional force resulting from the low pressure in the wake increases the *form drag* (the component of the drag due to the pressure distribution, in excess of the viscous drag). The total drag is thus a combination of Stokes drag and wake drag, and the drag coefficient is greater than that given by Eq. (11-3) for $N_{Re} > 0.1$. This is illustrated in Fig. 11-2, which shows C_D vs. N_{Re} for spheres (as well as for cylinders and disks oriented normal to the flow direction). For $\times 10^3 < N_{Re} < 1 \times 10^5$, $C_D = 0.45$ (approximately) for spheres. In this region the wake is maximum, and the streamlines actually separate slightly ahead of the equator of the sphere. The drag at this point is completely dominated by the wake, which is actually larger in diameter than the sphere (see Fig. 11-4(a)).

C. All Reynolds Numbers

For $N_{Re} > 0.1$ (or > 1, within $\sim 5\%$), a variety of expressions for C_D vs. N_{Re} (mostly empirical) have been proposed in the literature. However, a simple and very useful equation, which represents the entire range of C_D vs. N_{Re} reasonably well (within experimental error) up to about $N_{Re} = 2 \times 10^5$ is given by Dallavalle (1948):

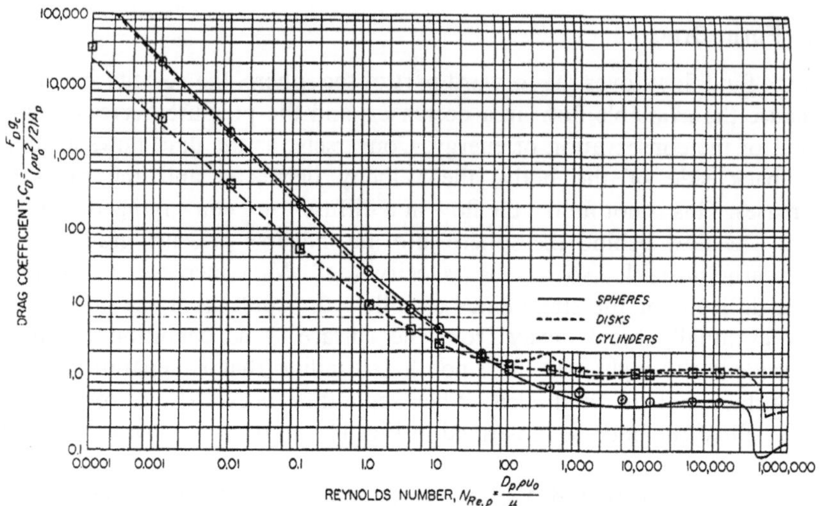

FIGURE 11-2 Drag coefficient for spheres, cylinders, and disks. (From Perry, 1984.) ⊙ Eq. (11-5), spheres. ⊡ Eq. (11-7), cylinders.

$$C_D = \left(0.632 + \frac{4.8}{\sqrt{N_{Re}}}\right)^2 \tag{11-5}$$

[Actually, according to Coulson et al. (1991), this equation was first presented by Wadell (1934).] A comparison of Eq. (11-5) with measured values is shown in Fig. 11-2. A somewhat more accurate equation, although more complex, has been proposed by Khan and Richardson (1987):

$$C_D = \left[\frac{2.25}{N_{Re}^{0.31}} + 0.358 N_{Re}^{0.06}\right]^{3.45} \tag{11-6}$$

Although Eq. (11-6) is more accurate than Eq. (11-5) at intermediate values of N_{Re}, Eq. (11-5) provides a sufficiently accurate prediction for most applications. Also it is simpler to manipulate, so we will prefer it as an analytical expression for the sphere drag coefficient.

D. Cylinder Drag

For flow past a circular cylinder with $L/d \gg 1$ normal to the cylinder axis, the flow is similar to over for a sphere. An equation that adequately represents the cylinder drag coefficient over the entire range of N_{Re} (up to about 2×10^5) that is analogous to the Dallavalle equation is

$$C_D = \left(1.05 + \frac{1.9}{\sqrt{N_{Re}}}\right)^2 \tag{11-7}$$

A comparison of this equation with measured values is also shown in Fig. 11-2.

E. Boundary Layer Effects

As seen in Fig. 11-2, the drag coefficient for the sphere exhibits a sudden drop from 0.45 to about 0.15 (almost 70%) at a Reynolds number of about 2.5×10^5. For the cylinder, the drop is from about 1.1 to about 0.35. This drop is a consequence of the transition of the boundary layer from laminar to turbulent flow and can be explained as follows.

As the fluid encounters the solid boundary and proceeds along the surface, a boundary layer forms as illustrated in Fig. 11-3. The boundary layer is the region of the fluid near a boundary in which viscous forces dominate and the velocity varies with the distance from the wall. Outside the boundary layer the fluid velocity is that of the free stream. Near the wall in the boundary layer the flow is stable, the velocity is low, and the flow is laminar. However, the boundary layer thickness (δ) grows along the plate (in the x direction), in proportion to $N_{Re,x}^{1/2}$ (where $N_{Re,x} = xV\rho/\mu$). As the boundary layer grows, inertial forces increase and it becomes less stable until it reaches a point (at $N_{Re,x} \approx 2 \times 10^5$) where it becomes unstable, i.e., turbulent. Within the turbulent boundary layer, the flow streamlines are no longer parallel to the boundary but break up into a three-dimensional eddy structure.

With regard to the flow over an immersed body (e.g., a sphere), the boundary layer grows from the impact (stagnation) point along the front of

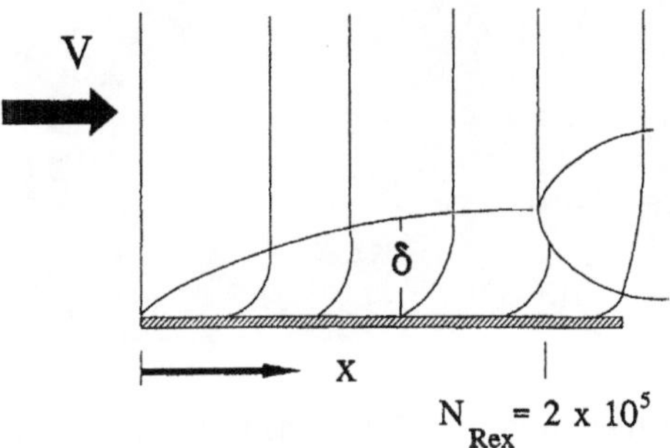

FIGURE 11-3 Boundary layer over a flat plate.

the body and remains laminar until $N_{Re,x} \approx .2 \times 10^5$, where x is the distance traveled along the boundary, at which point it becomes turbulent. If the boundary layer is laminar at the point where streamline separation occurs, the separation point can lie ahead of the equator of the sphere, resulting in a wake diameter that is larger than that of the sphere. However, if the boundary layer becomes turbulent before separation occurs, the three-dimensional eddy structure in the turbulent boundary layer carries momentum components inward toward the surface, which delays the separation of the streamline and tends to stabilize the wake. This delayed separation results in a smaller wake and a corresponding reduction in form drag, which is the cause of the sudden drop in C_D at $N_{Re} \approx 2 \times 10^5$.

This shift in the size of the wake can be rather dramatic, as illustrated in Fig. 11-4, which shows two pictures of a bowling ball falling in water, with the wake clearly visible. The ball on the left shows a large wake because the boundary layer at the separation point is laminar and the separation is ahead of the equator. The ball on the right has a rougher surface, which promotes turbulence, and the boundary layer has become turbulent before separation occurs, resulting in a much smaller wake due to the delayed separation. The primary effect of surface roughness on the flow around immersed objects is to promote transition to the turbulent boundary layer and delay separation of the streamlines and thus to slightly lower the value

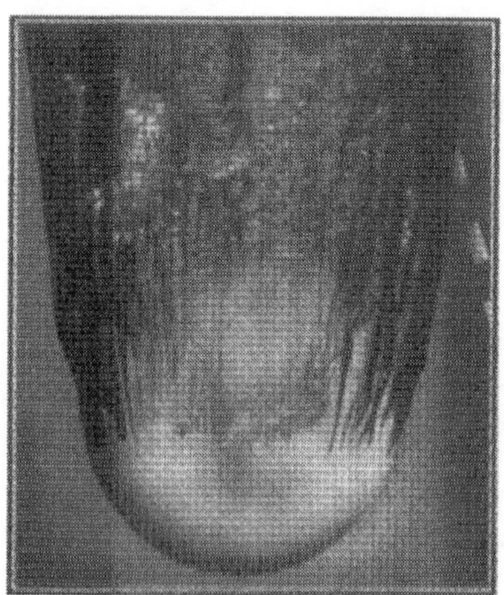

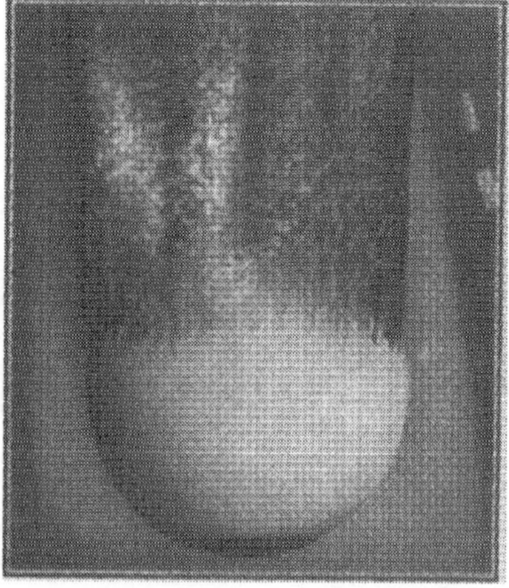

FIGURE 11-4 Two bowling balls falling in still water at 25 ft/s. The ball on the left is smooth, and the one on the right has a patch of sand on the nose. (From Coulson et al., 1991.)

of N_{Re} at which the sudden drop (or "kink") in the $C_D N_{Re}$ curve occurs. This apparent paradox, wherein the promotion of turbulence actually results in lower drag, has been exploited in various ways, such as the dimples on golf balls and the boundary layer "spoilers" on airplane wings and automobiles.

II. FALLING PARTICLES

Many engineering operations involve the separation of solid particles from fluids, in which the motion of the particles is a result of a gravitational (or other potential) force. To illustrate this, consider a spherical solid particle with diameter d and density ρ_s, surrounded by a fluid of density ρ and viscosity μ, which is released and begins to fall (in the $x = -z$ direction) under the influence of gravity. A momentum balance on the particle is simply $\Sigma F_x = ma_x$, where the forces include gravity acting on the solid (F_g), the buoyant force due to the fluid (F_b), and the drag exerted by the fluid (F_D). The inertial term involves the product of the acceleration ($a_x = dV_x/dt$) and the mass (m). The mass that is accelerated includes that of the solid (m_s) as well as the "virtual mass" (m_f) of the fluid that is displaced by the body as it accelerates. It can be shown that the latter is equal to one-half of the total mass of the displaced fluid, i.e., $m_f = \frac{1}{2} m_s(\rho/\rho_s)$. Thus the momentum balance becomes

$$\frac{g(\rho_s - \rho)\pi d^3}{6} - \frac{C_D \rho \pi d^2 V^2}{8} = \frac{\pi d^3(\rho_s + \rho/2)}{6}\frac{dV}{dt} \tag{11-8}$$

At $t = 0$, $V = 0$ and the drag force is zero. As the particle accelerates, the drag force increases, which decreases the acceleration. This process continues until the acceleration drops to zero, at which time the particle falls at a constant velocity because of the balance of forces due to drag and gravity. This steady-state velocity is called the *terminal velocity* of the body and is given by the solution of Eq. (11-8) with the acceleration equal to zero:

$$V_t = \left(\frac{4g\,\Delta\rho\, d}{3\rho C_D}\right)^{1/2} \tag{11-9}$$

where $\Delta\rho = \rho_s - \rho$. It is evident that the velocity cannot be determined until the drag coefficient, which depends on the velocity, is known. If Stokes flow prevails, then $C_D = 24/N_{Re}$, and Eq. (11-9) becomes

$$V_t = \frac{g\,\Delta\rho\, d^2}{18\mu} \tag{11-10}$$

However, the criterion for Stokes flow ($N_{Re} < 1$) cannot be tested until V_t is known, and if it is not valid then Eq. (11-10) will be incorrect. This will be addressed shortly.

There are several types of problems that we may encounter with falling particles, depending upon what is known and what is to be found. All of these problems involve the two primary dimensionless variables C_D and N_{Re}. The former is determined, for gravitation-driven motion, by Eq. (11-9), i.e.,

$$C_D = \frac{4g\,\Delta\rho\, d}{3\rho V_t^2} \tag{11-11}$$

and C_D can be related to N_{Re} by the Dallavalle equation [Eq. (11-5)] over the entire practical range of N_{Re}. The following procedures for the various types of problems apply to Newtonian fluids under all flow conditions.

A. Unknown Velocity

In this case, the unknown velocity (V_t) appears in both the equation for C_D [Eq. (11-11)] and the equation for N_{Re}. Hence, a suitable dimensionless group that does not contain the unknown V can be formulated as follows:

$$C_D N_{Re}^2 = \frac{4d^3\rho g\,\Delta\rho}{3\mu^2} = \frac{4}{3}N_{Ar} \tag{11-12}$$

where N_{Ar} is the Archimedes number (also sometimes called the Galileo number). The most appropriate set of dimensionless variables to use for this problem is thus N_{Ar} and N_{Re}. An equation for N_{Ar} can be obtained by multiplying Eq. (11-5) by N_{Re}^2, and the result can then be rearranged for N_{Re} to give

$$N_{Re} = [(14.42 + 1.827\sqrt{N_{Ar}})^{1/2} - 3.798]^2 \tag{11-13}$$

The procedure for determining the unknown velocity is therefore as follows.

Given : d, ρ, ρ_s, μ *Find* : V_t

1. Calculate the Archimedes number:

$$N_{Ar} = \frac{d^3\rho g\,\Delta\rho}{\mu^2} \tag{11-14}$$

2. Insert this value into Eq. (11-13) and calculate N_{Re}.
3. Determine V_t from N_{Re}, i.e., $V_t = N_{Re}\mu/d\rho$

If $N_{Ar} < 15$, then the system is within the Stokes law range and the terminal velocity is given by Eq. (11-10).

B. Unknown Diameter

It often happens that we know or can measure the particle velocity and wish to know the size of the falling particle. In this case, we can form a dimensionless group that does not contain *d*:

$$\frac{C_D}{N_{Re}} = \frac{4\mu\,\Delta\rho g}{3\rho^2 V_t^3} \tag{11-15}$$

This group can be related to the Reynolds number by dividing Eq. (11-5) by N_{Re} and then solving the resulting equation for $1/N_{Re}^{1/2}$ to give

$$\frac{1}{\sqrt{N_{Re}}} = \left(0.00433 + 0.208\sqrt{\frac{C_D}{N_{Re}}}\right)^{1/2} - 0.0658 \tag{11-16}$$

The two appropriate dimensionless variables are now C_D/N_{Re} and N_{Re}. The procedure is as follows.

Given: V_t, ρ_s, ρ, μ *Find*: d

1. Calculate C_D/N_{Re} from Eq. (11-15).
2. Insert the result into Eq. (11-16) and calculate $1/N_{Re}^{1/2}$ and hence N_{Re}.
3. Calculate $d = \mu N_{Re}/V_t\rho$.

If $C_D/N_{Re} > 30$, the flow is within the Stokes law range, and the diameter can be calculated directly from Eq. (11-10):

$$d = \left(\frac{18\mu V_t}{g\,\Delta\rho}\right)^{1/2} \tag{11-17}$$

C. Unknown Viscosity

The viscosity of a Newtonian fluid can be determined by measuring the terminal velocity of a sphere of known diameter and density if the fluid density is known. If the Reynolds number is low enough for Stokes flow to apply ($N_{Re} < 0.1$), then the viscosity can be determined directly by rearrangement of Eq. (11-10):

$$\mu = \frac{d^2 g\,\Delta\rho}{18\,V_t} \tag{11-18}$$

The Stokes flow criterion is rather stringent. (For example, a 1 mm diameter sphere would have to fall at a rate of 1 mm/s or slower in a fluid with a viscosity of 10 cP and SG = 1 to be in the Stokes range, which means that the density of the solid would have to be within 2% of the density of the fluid!) However, with only a slight loss in accuracy, the Dallavalle equation can be used to extend the useful range of this measurement to a much higher Reynolds number, as follows. From the known quantities, C_D can be calculated from Eq. (11-11). The Dallavalle equation [Eq. (11-5)] can be rearranged to give N_{Re}:

$$N_{Re} = \left(\frac{4.8}{C_D^{1/2} - 0.632} \right)^2 \tag{11-19}$$

The viscosity can then be determined from the known value of N_{Re}:

$$\mu = \frac{dV_t \rho}{N_{Re}} \tag{11-20}$$

Note that when $N_{Re} > 1000$, $C_D \approx 0.45$ (constant). From Eq. (11-19), this gives $\mu = 0$! Although this may seem strange, it is consistent because in this range the drag is dominated by form (wake) drag and viscous forces are negligible. It should be evident that one cannot determine the viscosity from measurements made under conditions that are insensitive to viscosity, which means that the utility of Eq. (11-19) is limited in practice to approximately $N_{Re} < 100$.

III. CORRECTION FACTORS

A. Wall Effects

All expressions so far have assumed that the particles are surrounded by an infinite sea of fluid, i.e. that the boundaries of the fluid container are far enough from the particle that their influence is negligible. For a falling particle, this might seem to be a reasonable assumption if $d/D < 0.01$, say, where D is the container diameter. However, the presence of the wall is felt by the particle over a much greater distance than one might expect. This is because as the particle falls it must displace an equal volume of fluid, which must flow back around the particle to fill the space just vacated by the particle. Thus the relative velocity between the particle and the adjacent fluid is much greater than it would be in an infinite fluid; i.e., the effective "free stream" (relative) velocity is no longer zero, as it would be for an infinite stagnant fluid. A variety of analyses of this problem have been performed, as reviewed by Chhabra (1992). These represent the wall effect by a wall correction factor (K_w) which is a multiplier for the "infinite fluid" terminal velocity. (this is also equivalent to correcting the Stokes' law drag force by a factor of K_w). The following equation due to Francis (see, e.g., Chhabra, 1992) is claimed to be valid for $d/D < 0.97$ and $N_{Re} < 1$:

$$K_{w_o} = \left(\frac{1 - d/D}{1 - 0.475d/D}\right)^4 \tag{11-21}$$

For larger Reynolds numbers, the following expression is claimed to be valid for $d/D < 0.8$ and $N_{Re} > 1000$:

$$K_{w_\infty} = 1 - (d/D)^{1.5} \tag{11-22}$$

Although these wall correction factors appear to be independent of Reynolds number for small (Stokes) and large (> 1000) values of N_{Re}, the value of K_w is a function of both N_{Re} and d/D for intermediate Reynolds numbers (Chhabra, 1992).

B. Drops and Bubbles

Because of surface tension forces, very small drops and bubbles are nearly rigid and behave much like rigid particles. However, larger fluid drops or bubbles may experience a considerably different settling behavior, because the shear stress on the drop surface can be transmitted to the fluid inside the drop, which in turn results in circulation of the internal fluid. This internal circulation dissipates energy, which is extracted from the energy of the bubble motion and is equivalent to an additional drag force. For Stokes flow of spherical drops or bubbles (e.g., $N_{Re} < 1$), it has been shown by Hadamard and Rybcznski (see, e.g., Grace, 1983) that the drag coefficient can be corrected for this effect as follows:

$$C_d = \frac{24}{N_{Re}}\left(\frac{\kappa + 2/3}{\kappa + 1}\right) \tag{11-23}$$

where $\kappa = \mu_i/\mu_o$, μ_i being the viscosity of dispersed ("inside") fluid and μ_o the viscosity of the continuous ("outside") fluid.

For larger Reynolds numbers ($1 < N_{Re} < 500$), Rivkind and Ryskind (see Grace, 1983) proposed the following equation for the drag coefficient for spherical drops and bubbles:

$$C_D = \frac{1}{\kappa + 1}\left[\kappa\left(\frac{24}{N_{Re}} + \frac{4}{N_{Re}^{1/3}}\right) + \frac{14.9}{N_{Re}^{0.78}}\right] \tag{11-24}$$

As the drop or bubble gets larger, however, it will become distorted owing to the unbalanced forces around it. The viscous shear stresses tend to elongate the shape, whereas the pressure distribution tends to flatten it out in the direction normal to the flow. Thus the shape tends to progress from spherical to ellipsoidal to a "spherical cap" form as the size increases. Above a certain size, the deformation is so great that the drag force is approximately proportional to the volume and the terminal velocity becomes nearly independent of size.

IV. NON-NEWTONIAN FLUIDS

The motion of solid particles, drops, or bubbles through non-Newtonian fluid media is encountered frequently and has been the subject of considerable research (see, e.g., Chhabra, 1992). We will present some relations here that are applicable to purely viscous non-Newtonian fluids, although there is also much interest and activity in viscoelastic fluids. Despite the relatively large amount of work that has been done in this area, there is still no general agreement as to the "right," or even the "best," description of the drag on a sphere in non-Newtonian fluids. This is due not only to the complexity of the equations that must be solved for the various models but also to the difficulty in obtaining good, reliable, representative data for fluids with well characterized unambiguous rheological properties.

A. Power Law Fluids

The usual approach for non-Newtonian fluids is to start with known results for Newtonian fluids and modify them to account for the non-Newtonian properties. For example, the definition of the Reynolds number for a power law fluid can be obtained by replacing the viscosity in the Newtonian definition by an appropriate shear rate dependent viscosity function. If the characteristic shear rate for flow over a sphere is taken to be V/d, for example, then the power law viscosity function becomes

$$\mu \to \eta(\dot{\gamma}) \cong m\left(\frac{V}{d}\right)^{n-1} \tag{11-25}$$

and the corresponding expression for the Reynolds number is

$$N_{\mathrm{Re,pl}} = \frac{\rho V^{2-n} d^n}{m} \tag{11-26}$$

The corresponding creeping flow drag coefficient can be characterized by a correction factor (X) to the Stokes law drag coefficient:

$$C_{\mathrm{D}} = X \frac{24}{N_{\mathrm{Re,pl}}} \tag{11-27}$$

A variety of theoretical expressions, as well as experimental values, for the correction factor X as a function of the power law flow index (n) were summarized by Chhabra (1992).

In a series of papers, Chhabra (1995), Tripathi et al. (1994), and Tripathi and Chhabra (1995) presented the results of numerical calculations for the drag on spheroidal particles in a power law fluid in terms of $C_D = \text{fn}(N_{Re}, n)$. Darby (1996) analyzed these results and showed that this function can be expressed in a form equivalent to the Dallavalle equation, which applies over the entire range of n and N_{Re} as given by Chhabra. This equation is

$$C_D = \left(C_1 + \frac{4.8}{\sqrt{N_{Re,pl}/X}} \right)^2 \tag{11-28}$$

where both X and C_1 are functions of the flow index n. These functions were determined by empirically fitting the following equations to the values given by Chhabra (1995):

$$\frac{1}{C_1} = \left[\left(\frac{1.82}{n} \right)^8 + 34 \right]^{1/8} \tag{11-29}$$

$$X = \frac{1.33 + 0.37n}{1 + 0.7n^{3.7}} \tag{11-30}$$

The agreement between these values of C_1 and X and the values given by Tripathi et al. (1994) and Chhabra (1995) is shown in Figs. 11-5 and 11-6. Equations (11-28)–(11-30) are equivalent to the Dallavalle equation for a sphere in a power law fluid. A comparison of the values of C_D predicted by Eq. (11-28) with the values given by Tripathi et al. and Tripathi and Chhabra and Chhabra is shown in Fig. 11-7. The deviation is the greatest for highly dilatant fluids, in the Reynolds number range of about 5–50, although the agreement is quite reasonable above and below this range, and for pseudoplastic fluids over the entire range of Reynolds number. We will illustrate the application of these equations by outlining the procedure for solving the "unknown velocity" and the "unknown diameter" problems.

1. Unknown Velocity

The expressions for C_D and $N_{Re,pl}$ can be combined to give a group that is independent of V:

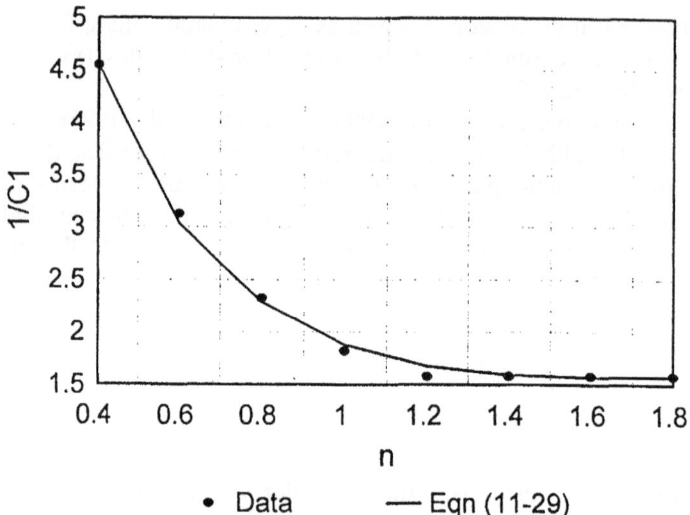

FIGURE 11-5 Plot of $1/C_1$ vs. n for power law fluid. Line is Eq. (11-29). Data points are from Chhabra (1995) and Tripathi et al. (1994).

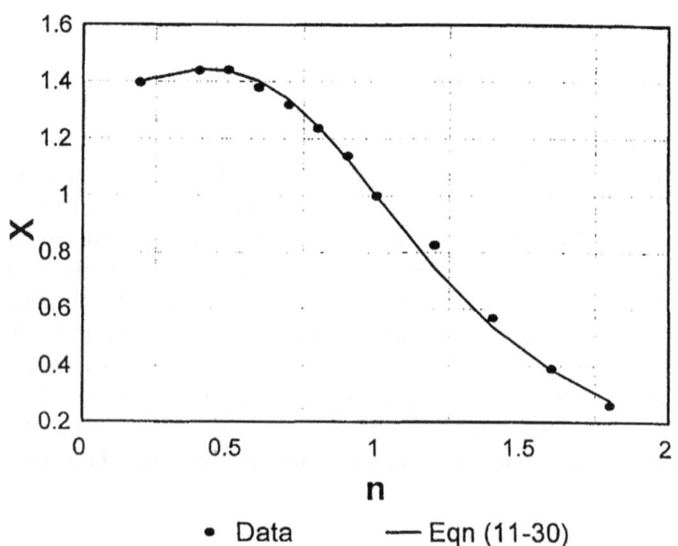

FIGURE 11-6 Plot of X vs. n for power law fluid. Line is Eq. (11-30). Data points are from Tripathi et al. (1994) and Chhabra (1995).

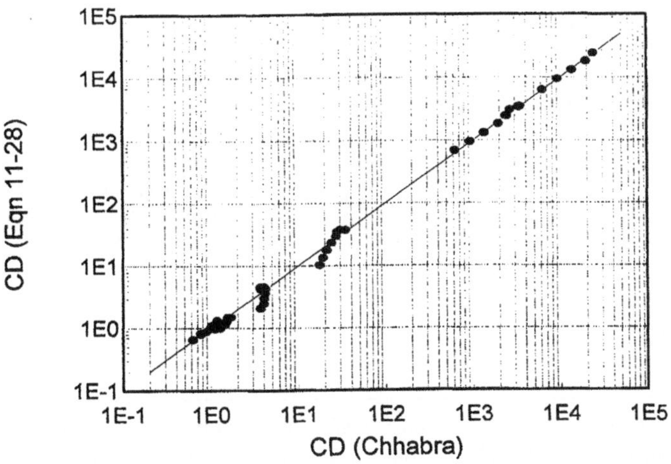

Figure 11-7 Comparison of Eq. (11-28) with result sof Tripathi et al. (1994) and Chhabra (1995).

$$C_D^{2-n}\left(\frac{N_{\text{Re,pl}}}{X}\right)^2 = \left(\frac{\rho}{Xm}\right)^2\left(\frac{4g\,\Delta\rho}{3\rho}\right)^{2-n} d^{n+2} = N_{\text{d}} \tag{11-31}$$

which is similar to Chhabra's D^+ parameter. Using Eq. (11-28) to eliminate C_{D} gives

$$N_{\text{d}} = \left[C_1\left(\frac{N_{\text{Re,pl}}}{X}\right)^{1/(2-n)} + 4.8\left(\frac{N_{\text{Re,pl}}}{X}\right)^{n/2(2-n)}\right]^{2(2-n)} \tag{11-32}$$

Although this equation cannot be solved analytically for $N_{\text{Re,pl}}$, it can be solved by iteration (or by using the "solve" command on a calculator or spreadsheet), because all other parameters are known. The unknown velocity is then given by

$$V = \left[\frac{mN_{\text{Re,pl}}}{\rho d^n}\right]^{1/(2-n)} \tag{11-33}$$

2. Unknown Diameter

The diameter can be eliminated from the expressions for C_{D} and $N_{\text{Re,pl}}$ as follows:

$$\frac{C_{\text{D}}^n X}{N_{\text{Re,pl}}} = \left(\frac{4g\,\Delta\rho}{3\rho}\right)^n\left(\frac{Xm}{\rho}\right)V^{-(n+2)} = N_{\text{v}} \tag{11-34}$$

Again using Eq. (11-28) to eliminate C_D gives

$$N_V = \left[C_1 \left(\frac{X}{N_{Re,pl}} \right)^{1/2n} + 4.8 \left(\frac{X}{N_{Re,pl}} \right)^{(1+n)/2n} \right]^{2n} \tag{11-35}$$

As before, everything in this equation is known except for $N_{Re,pl}$, which can be determined by iteration (or by using the "solve" spreadsheet or calculator command). When this is found, the unknown diameter is given by

$$d = \left(\frac{m N_{Re,pl}}{\rho V^{2-n}} \right)^{1/n} \tag{11-36}$$

Example 11-1: Unknown Velocity and Unknown Diameter of a Sphere Settling in a Power Law Fluid. Table 11-1 summarizes the procedure, and Table 11-2 shows the results of a spreadsheet calculation for an application of this method to the three examples given by Chhabra (1995). Examples 1 and 2 are "unknown velocity" problems, and Example 3 is an "unknown diameter" problem. The line labeled "Equation" refers to Eq. (11-32) for the unknown velocity cases, and Eq. (11-35) for the unknown diameter case. The "Stokes" value is from Eq. (11-9), which only applies for $N_{Re,pl} < 1$ (e.g., Example 1 only). It is seen that the solutions for Examples 1 and 2 are virtually identical to Chhabra's values and the one for Example 3 is within 5% of Chhabra's. The values labeled "Data" were obtained by iteration using the data from Fig. 4 of Tripathi et al. (1994). These values are only approximate, because they were obtained by interpolating from the (very compressed) log scale of the plot.

TABLE 11-1 Procedure for Determining Unknown Velocity or Unknown Diameter for Particles Settling in a Power Law Fluid

Problem	Unknown velocity	Unknown diameter
Given:	Particle diameter (d) and fluid properties (m, n, and ρ)	Particle settling velocity (V) and fluid properties (m, n, and ρ)
Step 1	Using value of n, calculate C_1 and X from Eqs. (11-29) and (11-30)	Using value of n, calculate C_1 and X from and (11-29) and (11-30)
Step 2	Calculate N_d from Eq. (11-31).	Calculate N_v from Eq. (11-34).
Step 3	Solve Eq. (11-32) for $N_{Re,pl}$ by iteration (or using "solve" function).	Solve Eqn. (11-35) for $N_{Re,pl}$ by iteration (or using "solve" function).
Step 4	Get V from Eq. (11-33).	Get d from Eq. (11-36).

TABLE 2 Comparison of Calculated Settling Properties Using Eq. (11-28) with Literature Values[a] Chhabra (1995)

	Example 1	Example 2	Example 3
Given data	$d = 0.002$ m $m = 1.3$ Pa s $n = 0.6$ $\rho = 1.002$ kg/m^3 $\rho_s = 7780$ kg/m^3	$d = 0.002$ m $m = 0.015$ Pa s $n = 0.8$ $\rho = 1.050$ kg/m^3 $\rho_s = 2500$ kg/m^3	$V = 0.2$ m/s $m = 0.08$ Pa s $n = 0.5$ $\rho = 1.005$ kg/m^3 $\rho_s = 8714$ kg/m^3
Calculated values	$X = 1.403$ $C_1 = 0.329$ $N_d = 15.4$ $N_{Re,pl} = 0.082$	$X = 1.244$ $C_1 = 0.437$ $N_d = 2830$ $N_{Re,pl} = 55.7$	$X = 1.438$ $C_1 = 0.275$ $N_d = 0.064$ $N_{Re,pl} = 29.8$
Equation:	$V = 0.0208$ m/s	$V = 0.165$ m/s	$d = 7.05 \times 10^{-4}$ m
Chhabra (1995)	$V = 0.0211 s^{-1}$	$V = 0.167$ m/s	$d = 6.67 \times 10^{-4}$ m
Stokes' law	$V = 0.0206$ m/s	$V = 0.514$ m/s	$d = 5.31 \times 10^{-4}$ m
Tripathi et al. (1994) – Data		$V = 0.167$ m/s $N_{Re,pl} = 57$ $C_d = 1.3$	$d = 7.18 \times 10^{-4}$ $N_{Re,pl} = 30$ $C_d = 1.8$

The method shown here has several advantages over that reported by Chhabra (1995), namely,

1. All expressions are given in equation form, and it is not necessary to read and interpolate any plots to solve the problems (i.e., the empirical data are represented analytically by curve-fitting equations).
2. The method is more general, in that it is a direct extension of the technique of solving similar problems for Newtonian fluids and applies over all values of Reynolds number.
3. Only one calculation procedure is required, regardless of the value of the Reynolds number for the specific problem.
4. The calculation procedure is simple and straightforward and can be done quickly using a spreadsheet.

B. Wall Effects

The wall effect for particles settling in non-Newtonian fluids appears to be significantly smaller than for Newtonian fluids. For power law fluids, the wall correction factor in creeping flow, as well as for very high Reynolds numbers, appears to be independent of Reynolds number. For creeping flow, the wall correction factor given by Chhabra (1992) is

$$K_{W_0} = 1 - 1.6d/D \tag{11-37}$$

whereas for high Reynolds numbers he gives

$$K_{W_\infty} = 1 - 3(d/D)^{3.5} \tag{11-38}$$

For intermediate Reynolds numbers, the wall factor depends upon the Reynolds number as well as d/D. Over a range of $10^{-2} < N_{\mathrm{Re,pl}} < 10^3$, $0 < d/D < 0.5$, and $0.53 < n < 0.95$, the following equation describes the Reynolds number dependence of the wall factor quite well:

$$\frac{1/K_W - 1/K_{W_\infty}}{1/K_{W_0} - 1/K_{W_\infty}} = [1 + 1.3N^2_{\mathrm{Re,pl}}]^{-0.33} \tag{11-39}$$

C. Carreau Fluids

As discussed in Chapter 3, the Carreau viscosity model is one of the most general and useful and reduces to many of the common two-parameter models (power law, Ellis, Sisko, Bingham, etc.) as special cases. This model can be written as

$$\eta = \eta_\infty + (\eta_0 - \eta_\infty)[1 + (\lambda\dot{\gamma})^2]^{(n-1)/2} \tag{11-40}$$

where $n = 1 - 2p$ is the flow index for the power law region [p is the shear thinning parameter in the form of this model given in Eq. (3-26)]. Because the shear conditions surrounding particles virtually never reach the levels corresponding to the high shear viscosity (η_∞), this parameter can be neglected and the parameters reduced to three: η_0, λ, and n. Chhabra and Uhlherr (1988) determined the Stokes flow correction factor for this model, which is a function of the dimensionless parameters n and $N_\lambda = \lambda V/d$. The following equation represents their results for the C_D correction factor over a wide range of data to ±10%, for $0.4 < n < 1$ and $0 < N_\lambda < 400$:

$$X = \frac{1}{[1 + (0.275N_\lambda)^2]^{(1-n)/2}} \tag{11-41}$$

where the Stokes equation uses η_0 for the viscosity in the Reynolds number.

D. Bingham Plastics

A particle will not fall through a fluid with a yield stress unless the weight of the particle is sufficient to overcome the yield stress. Because the stress is not uniform around the particle and the distribution is very difficult to determine, it is not possible to determine the critical "yield" criterion exactly. However, it should be possible to characterize this state by a dimensionless "gravity yield" parameter:

$$Y_G = \frac{\tau_0}{gd\Delta\rho} \tag{11-42}$$

By equating the vertical component of the yield stress over the surface of the sphere to the weight of the particle, a critical value of $Y_G = 0.17$ is obtained (Chhabra, 1992). Experimentally, however, the results appear to fall into groups: one for which $Y_G \approx 0.2$ and one for which $Y_G \approx 0.04$–0.08. There seems to be no consensus as to the correct value, and the difference may well be due to the fact that the yield stress is not an unambiguous empirical parameter, inasmuch as values determined from "static" measurements can differ significantly from the values determined from "dynamic" measurements.

With regard to the drag on a sphere moving in a Bingham plastic medium, the drag coefficient (C_D) must be a function of the Reynolds number as well as of either the Hedstrom number or the Bingham number ($N_{Bi} = N_{He}/N_{Re} = \tau_0 d/\mu_\infty V$). One approach is to reconsider the Reynolds number from the perspective of the ratio of inertial to viscous momentum flux. For a Newtonian fluid in a tube, this is equivalent to

Newtonian:

$$N_{Re} = \frac{DV\rho}{\mu} = \frac{8\rho V^2}{\mu(8V/D)} = \frac{8\rho V^2}{\tau_w} \tag{11-43}$$

which follows from the Hagen–Poiseuille equation, because $\tau_w = \mu(8V/D)$ is the drag per unit area of the wall and the shear rate at the wall in the pipe is $\dot{\gamma}_w = 8V/D$. By analogy, the drag force per unit area on a sphere is $F/A = C_D\rho V^2/2$, which for Stokes flow (i.e., $C_D = 24/N_{Re}$) becomes $F/A = 12\mu V/d$. If F/A for the sphere is considered to be analogous to the "wall stress" (τ) on the sphere, the corresponding "effective wall shear rate" is $12V/d$. Thus the sphere Reynolds number could be written

$$N_{Re} = \frac{dV\rho}{\mu} = \frac{12\rho V^2}{\mu(12V/d)} = \frac{12\rho V^2}{\tau} \tag{11-44}$$

For a Bingham plastic, the corresponding expression would be

Bingham plastic:

$$\frac{12\rho V^2}{\tau} = \frac{12\rho V^2}{\mu_\infty(12V/d) + \tau_0} = \frac{N_{Re}}{1 + N_{Bi}/12} \tag{11-45}$$

Equation (11-45) could be used in place of the traditional Reynolds number for correlating the drag coefficient.

Another approach is to consider the effective shear rate over the sphere to be V/d, as was done in Eq. (11-25) for the power law fluid. If this approach is applied to a sphere in a Bingham plastic, the result is

$$N_{\mathrm{Re,BP}} = \frac{N_{\mathrm{Re}}}{1 + N_{\mathrm{Bi}}} \tag{11-46}$$

This is similar to the analysis obtained by Ainsley and Smith (see Chhabra, 1992) using the slip line theory from soil mechanics, which results in a dimensionless group called the plasticity number:

$$N_{\mathrm{pl}} = \frac{N_{\mathrm{Re}}}{1 + 2\pi N_{\mathrm{Bi}}/24} \tag{11-47}$$

A finite element analysis [as reported by Chhabra and Richardson (1999)] resulted in an equivalent Stokes' law correction factor $X(= C_{\mathrm{D}} N_{\mathrm{Re}}/24)$ that is a function of N_{Bi} for $N_{\mathrm{Bi}} < 1000$:

$$X = 1 + a N_{\mathrm{Bi}}^{b} \tag{11-48}$$

where $a = 2.93$ and $b = 0.83$ for a sphere in an unbounded fluid, and $2.93 > a > 1.63$ and $0.83 < b < 0.95$ for $0 < d/D < 0.5$. Also, based upon available data, Chhabra and Uhlherr (1988) found that the "Stokes flow" relation ($C_{\mathrm{D}} = 24X/N_{\mathrm{Re}}$) applies up to $N_{\mathrm{Re}} \le 100 N_{\mathrm{Bi}}^{0.4}$ for Bingham plastics. Equation (11-48) is equivalent to a Bingham plastic Reynolds number ($N_{\mathrm{Re,BP}}$) or plasticity number (N_{pl}) of

$$N_{\mathrm{Re,BP}} = \frac{N_{\mathrm{Re}}}{1 + 2.93 N_{\mathrm{Bi}}^{0.83}} \tag{11-49}$$

Unfortunately, there are insufficient experimental data reported in the literature to verify or confirm any of these expressions. Thus, for lack of any other information, Eq. (11-49) is recommended, because it is based on the most detailed analysis. This can be extended beyond the Stokes flow region by incorporating Eq. (11-49) into the equivalent Dallavalle equation,

$$C_{\mathrm{D}} = \left(0.632 + \frac{4.8}{\sqrt{N_{\mathrm{Re,BP}}}}\right)^2 \tag{11-50}$$

which can be used to solve the "unknown velocity" and "unknown diameter" problems as previously discussed. However in this case rearrangement of the dimensionless variables C_{D} and $N_{\mathrm{Re,BP}}$ into an alternative set of dimensionless groups in which the unknown is in only one group is not possible owing to the form of $N_{\mathrm{Re,BP}}$. Thus the procedure would be to equate Eqs. (11-50) and (11-11) and solve the resulting equation directly by iteration for the unknown V or d (as the case requires).

PROBLEMS

1. By careful streamlining, it is possible to reduce the drag coefficient of an automobile from 0.4 to 0.25. How much power would this save at (a) 40 mph and a (b) 60 mph, assuming that the effective projected area of the car is 25 ft^2?
2. If your pickup truck has a drag coefficient equivalent to a 5 ft diameter disk and the same projected frontal area, how much horsepower is required to overcome wind drag at 40 mph? What horsepower is required at 70 mph?
3. You take a tumble while water skiing. The handle attached to the tow rope falls beneath the water and remains perpendicular to the direction of the boat's heading. If the handle is 1 in. in diameter and 1 ft long and the boat is moving at 20 mph, how much horsepower is required to pull the handle through the water?
4. Your new car is reported to have a drag coefficient of 0.3. If the cross-sectional area of the car is 20 ft^2, how much horsepower is used to overcome wind resistance at (a) 40 mph? (b) 55 mph? (c) 70 mph? (d) 100 mph? ($T = 70°F$).
5. The supports for a tall chimney must be designed to withstand a 120 mph wind. If the chimney is 10 ft in diameter and 40 ft high, what is the wind force on the chimney at this speed? $T = 50°F$.
6. A speedboat is propelled by a water jet motor that takes water in at the bow through a 10 cm diameter duct and discharges it through an 50 mm diameter nozzle at a rate of 80 kg/s. Neglecting friction in the motor and internal ducts, and assuming that the drag coefficient for the boat hull is the same as for a 1 m diameter sphere, determine:
   (a) The static thrust developed by the motor when it is stationary.
   (b) The maximum velocity attainable by the boat.
   (c) The power (kW) required to drive the motor.
   (Assume seawater density 1030 kg/m^3, viscosity 1.2 cP.)
7. After blowing up a balloon, you release it without tying off the opening, and it flies out of your hand. If the diameter of the balloon is 6 in., the pressure inside it is 1 psig, and the opening is 1/2 in. in diameter, what is the balloon velocity? You may neglect friction in the escaping air and the weight of the balloon and assume that an instantaneous steady state (i.e., a pseudo steady state) applies.
8. A mixture of titanium ($SG = 4.5$) and silica ($SG = 2.65$) particles, with diameters ranging from 50 to 300 μm, is dropped into a tank in which water is flowing upward. What is the velocity of the water if all the silica particles are carried out with the water?
9. A small sample of ground coal is introduced into the top of a column of water 30 cm high, and the time required for the particles to settle out is measured. If it takes 26 s for the first particle to reach the bottom and 18 hr for all particles to settle, what is the range of particle sizes in the sample? ($T = 60°F$, $SG_{coal} = 1.4$.)
10. You want to determine the viscosity of an oil that has an SG of 0.9. To do this, you drop a spherical glass bead (SG = 2.7) with a diameter of 0.5 mm into a large vertical column of the oil and measure its settling velocity. If the measured velocity is 3.5 cm/s, what is the viscosity of the oil?

11. A solid particle with a diameter of 5 mm and SG = 1.5 is released in a liquid with a viscosity of 10 P and SG = 1. How long will it take for the particle to reach 99% of its terminal velocity after it is released?
12. A hot air popcorn popper operates by blowing air through the popping chamber, which carries the popped corn up through a duct and out of the popper leaving the unpopped grains behind. The unpopped grains weigh 0.15 g, half of which is water, and have an equivalent spherical diameter of 4 mm. The popped corn loses half of the water to steam, and has an equivalent diameter of 12 mm. What are the upper and lower limits of the air volumetric flow rate at 200°F over which the popper will operate properly, for a duct diameter of 8 cm?
13. You have a granular solid with SG = 4, which has particle sizes of 300 μm and smaller. You want to separate out all of the particles with a diameter of 20 μm and smaller by pumping water upward through a slurry of the particles in a column with a diameter of 10 cm. What flow rate is required to ensure that all particles less than 20 μm are swept out of the top of the column? If the slurry is pumped upward into the bottom of the column through a vertical tube, what should the diameter of this tube be to ensure that none of the particles settle out in it?
14. You want to perform an experiment that illustrates the wake behind a sphere falling in water at the point where the boundary layer undergoes transition from laminar to turbulent. (See Fig. 11-4.) If the sphere is made of steel with a density of 500 lb_m/ft^3, what should the diameter be?
15. You have a sample of crushed coal containing a range of particle sizes from 1 to 1000 μm in diameter. You wish to separate the particles according to size by entrainment, by dropping them into a vertical column of water that is flowing upward. If the water velocity in the column is 3 cm/s, which particles will be swept out of the top of the column, and which will settle to the bottom? (SG of the solid is 2.5.)
16. A gravity settling chamber consists of a horizontal rectangular duct 6 m long, 3.6 m wide, and 3 m high. The chamber is used to trap sulfuric acid mist droplets entrained in an air stream. The droplets settle out as the air passes horizontally through the duct and can be assumed to behave as rigid spheres. If the air stream has a flow rate of 6.5 m^3/s, what is the diameter of the largest particle that will not be trapped in the duct? ($\rho_{acid} = 1.74$ g/cm^3; $\rho_{air} = 0.01$ g/cm^3; $\mu_{air} = 0.02$ cP; $\mu_{acid} = 2$ cP.)
17. A small sample of a coal slurry containing particles with equivalent spherical diameters from 1 to 500 μm is introduced into the top of a water column 30 cm high. The particles that fall to the bottom are continuously collected and weighed to determine the particle size distribution in the slurry. If the solid SG is 1.4 and the water viscosity is 1 cP, over what time range must the data be obtained in order to collect and weigh all the particles in the sample?
18. Construct a plot of C_D versus $N_{Re,BP}$ for a sphere falling in a Bingham plastic fluid over the range of $1 < N_{Re} < 100$ and $10 < N_{Bi} < 1000$ using Eq. (11-50). Compare the curves for this relation based on Eqs. (11-45), (11-46), (11-47), and (11-49).

19. The viscosity of applesauce at 80° F was measured to be 24.2 poise (P) at a shear rate of 10 s^{-1} and 1.45 P at 500 s^{-1}. The density of the applesauce is 1.5 g/cm^3. Determine the terminal velocity of a solid sphere 1 cm in diameter with a density of 3.0 g/cm^3 falling in the applesauce, if the fluid is described by (a) the power law model; (b) the Bingham plastic model [use Eq. (11-49)].
20. Determine the size of the smallest sphere of SG = 3 that will settle in applesauce with properties given in Problem 19, assuming that it is best described by the Bingham plastic model [Eq. (11-49)]. Find the terminal velocity of the sphere that has a diameter twice this size.

NOTATION

A	cross-sectional area of particle normal to flow direction, [L^2]
C_D	particle drag coefficient, [—]
d	particle diameter, [L]
F_D	drag force on particle, [$F = ML/t^2$]
g	acceleration due to gravity, [L/t^2]
K_{w_0}	low Reynolds number wall correction factor, [—]
K_{w_∞}	high Reynolds number wall correction factor, [—]
m	power law consistency coefficient, [$M/(Lt^{2-n})$]
n	power law flow index, [—]
N_{Ar}	Archimedes number, Eq. (11-14), [—]
$N_{Re,BP}$	Bingham plastic Reynolds number, Eq. (11-47), [—]
N_{Bi}	Bingham number ($= N_{Re}/N_{He} = d\tau_o/\mu_\infty V$), [—]
N_{Re}	Reynolds number, [—]
$N_{Re,pl}$	Power law Reynolds number, [—]
N_λ	dimensionless time constant ($= \lambda V/d$), [—]
V	relative velocity between fluid and particle, [L/t]
X	correction factor to Stokes' law to account for non-Newtonian properties, [—]
κ	μ_i/μ_o, [—]
η_o	low shear limiting viscosity, [M/Lt]
η_∞	high shear limiting viscosity, [M/Lt]
$\Delta\rho$	$\rho_s - \rho$, [M/L^3]
ρ	density, [M/L^3]
λ	Carreau fluid time constant parameter, [t]
μ	viscosity (constant), [M/Lt]
μ_∞	Bingham plastic limiting viscosity, [M/Lt]
τ_0	Bingham plastic yield stress, [$F/L^2 = M/Lt^2$]

Subscripts

i	distributed ("inside") liquid phase
o	continuous ("outside") liquid phase
s	solid
t	terminal velocity condition

REFERENCES

Chhabra RP. Bubbles, Drops, and Particles in Non-Newtonian Fluids. Boca Raton, FL: CRC Press, 1992.

Chhabra RP. Calculating settling velocities of particles. Chem. Eng, September 1995, p 133.

Chhabra RP, D De Kee. Transport Processes in Bubbles, Drops, and Particles. Washington, DCP: Hemisphere, 1992.

Chhabra RP, JF Richardson. Non-Newtonian Flow in the Process Industries. Stoneham, MA: Butterworth-Heinemann, 1999.

Chhabra RP, PHT Uhlherr. Static equilibrium and motion of spheres in viscoplastic liquids. In: NP Cheremisinoff, ed. Encyclopedia of Fluid Mechanics, Vol. 7. Houston, TX: Gulf Pub Co, 1988, Chap 21.

Coulson JM, JF Richardson, JR Blackhurst, JH Harker. Chemical Engineering, Vol. 2, 4th ed. New York: Pergamon Press, 1991.

Dallavalle JM, Micrometrics. 2nd ed. Pitman, 1948.

Darby R. Determine settling rates of particles in non-Newtonian fluids. Chem Eng 103(12): 107-112, 1996.

Grace JR. Hydrodynamics of liquid drops in immiscible liquids. In: Cheremisinoff, NP, R Gupta, eds. Handbook of Fluids in Motion. Ann Arbor Science, 1983, Chap 38.

Khan AR, JF Richardson. Chem Eng Commun **62**:135, 1987.

Perry JH, ed. Chemical Engineers' Handbook. 6th ed., New York: McGraw-Hill, 1984.

Tripathi A, RP Chhabra. Drag on spheroidal particles in dilatant fluids. AIChE J **41**:728, 1995.

Tripathi A, RP Chhabra, T Sundararajan. Power law fluid flow over spheroidal particles. Ind Eng Chem Res **33**:403, 1994.

Wadell H. J Franklin Inst **217**:459, 1934.

12

Fluid–Solid Separations by Free Settling

I. FLUID–SOLID SEPARATIONS

The separation of suspended solids from a carrier fluid is a requirement in many engineering operations. The most appropriate method for achieving this depends upon the specific properties of the system, the most important being the size and density of the solid particles and the solids concentration (the "solids loading") of the feed stream. For example, for relatively dilute systems (~ 10% or less) of relatively large particles (~ 100 μm or more) of fairly dense solids, a gravity settling tank may be appropriate, whereas for more dilute systems of smaller and/or lighter particles, a centrifuge may be more appropriate. For very fine particles, or where a very high separation efficiency is required, a "barrier" system such as a filter or membrane may be needed. For highly concentrated systems, a gravity thickener may be adequate or, for more stringent requirements, a filter may be needed.

In this chapter, we will consider separation processes for relatively dilute systems, in which the effects of particle–particle interaction are relatively unimportant (e.g., gravity and centrifugal separation). Situations in which particle–particle interactions are negligible are referred to as *free settling*, as opposed to *hindered settling*, in which such interactions are important. Figure 12-1 shows the approximate regions of solids concentration and density corresponding to free and

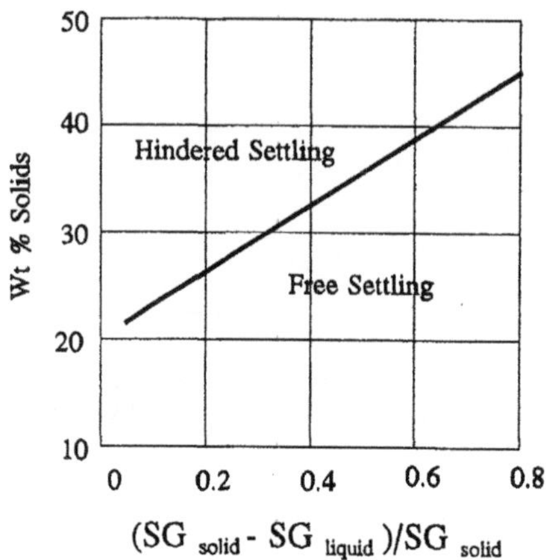

FIGURE 12-1 Regions of hindered and free settling.

hindered settling. In a Chapter 14 we will consider systems that are controlled by hindered settling or interparticle interaction (e.g., filtration and sedimentation processes).

II. GRAVITY SETTLING

Solid particles can be removed from a dilute suspension by passing the suspension through a vessel that is large enough that the vertical component of the fluid velocity is lower than the terminal velocity of the particles and the residence time is sufficiently long to allow the particles to settle out. A typical gravity settler is illustrated in Fig. 12-2. If the upward velocity of the liquid (Q/A) is less than the terminal velocity of the particles (V_t), the particles will settle to the bottom; otherwise, they will be carried out with the overflow. If Stokes flow is applicable (i.e., $N_{Re} < 1$), the diameter of the smallest particle that will settle out is

$$d = \left(\frac{18\mu Q}{g\,\Delta\rho A}\right)^{1/2} \tag{12-1}$$

If Stokes flow is not applicable (or even if it is), the Dallavalle equation in the form of Eq. (11-16) can be used to determine the Reynolds number, and hence the diameter, of the smallest setting particle:

$$\frac{1}{\sqrt{N_{Re}}} = \left(0.00433 + 0.208\sqrt{\frac{C_D}{N_{Re}}}\right)^{1/2} - 0.0658 \tag{12-2}$$

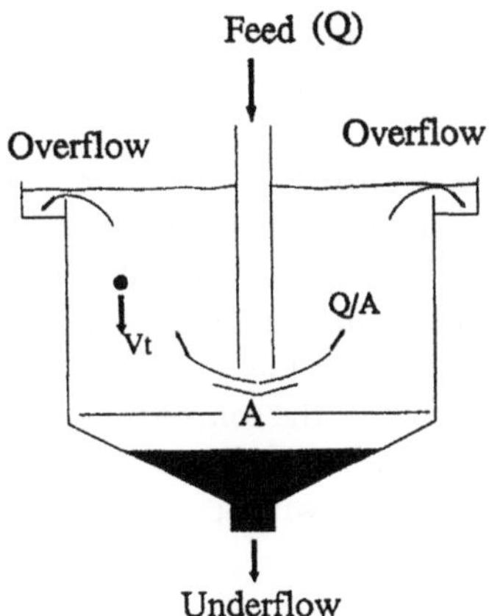

FIGURE 12-2 Gravity settling tank.

where

$$\frac{C_D}{N_{Re}} = \frac{4\mu \Delta \rho g}{3\rho^2 V_t^3}, \qquad d = \frac{N_{Re}\mu A}{Q\rho} \tag{12-3}$$

Alternatively, it may be necessary to determine the maximum capacity (e.g., flow rate, Q) at which particles of a given size, d, will (or will not) settle out. This can also be obtained directly from the Dallavalle equation in the form of Eq. (11-13), by solving for the unknown flow rate:

$$Q = \frac{\mu A}{D\rho}[(14.42 + 1.827\sqrt{N_{Ar}})^{1/2} - 3.798]^2 \tag{12-4}$$

where

$$N_{Ar} = \frac{d^3 \rho g \Delta \rho}{\mu^2} \tag{12-5}$$

III. CENTRIFUGAL SEPARATION

A. Fluid–Solid Separation

For very small particles or low density solids, the terminal velocity may be too low to enable separation by gravity settling in a reasonably sized tank. However, the separation can possibly be carried out in a centrifuge, which operates on the same principle as the gravity settler but employs the (radial) acceleration in a rotating system ($\omega^2 r$) in place of the vertical gravitational acceleration as the driving force. Centrifuges can be designed to operate at very high rotating speeds, which may be equivalent to many g's of acceleration.

A simplified schematic of a particle in a centrifuge is illustrated in Fig. 12-3. It is assumed that any particle that impacts on the wall of the centrifuge (at r_2) before reaching the outlet will be trapped, and all others won't. (It might seem that any particle that impacts the outlet weir barrier would be trapped. However, the fluid circulates around this outlet corner, setting up eddies that could sweep these particles out of the centrifuge.) It is thus necessary to determine how far the particle will travel in the radial direction while in the centrifuge. To do this, we start with a radial force (momentum) balance on the particle:

$$F_{cf} - F_b - F_D = m_e \frac{dV_r}{dt} \tag{12-6}$$

where F_{cf} is the centrifugal force on the particle, F_b is the buoyant force (equal to the centrifugal force acting on the displaced fluid), $F_D$is the drag force, and m_e is the "effective" mass of the particle, which includes the solid particle and the "virtual mass" of the displaced fluid (i.e. half the actual mass of displaced fluid). Equation (12-6) thus becomes

$$(\rho_s - \rho)\left(\frac{\pi d^3}{6}\right)\omega^2 r - \rho V_r^2 C_D\left(\frac{\pi d^2}{8}\right) = \left(\rho_s + \frac{\rho}{2}\right)\left(\frac{\pi d^3}{6}\right)\frac{dV_r}{dt} \tag{12-7}$$

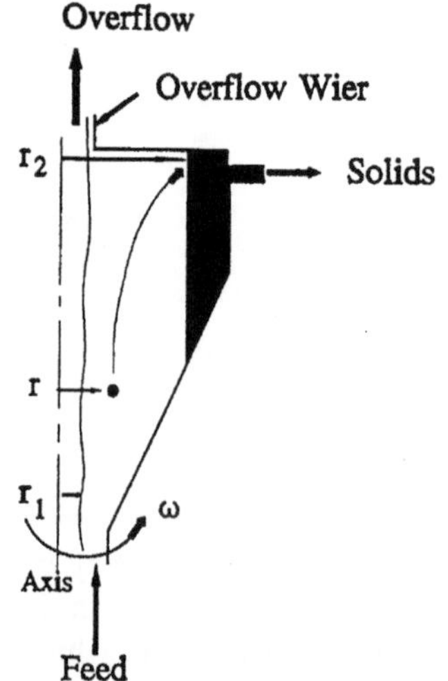

FIGURE 12-3 Particle in a centrifuge.

When the particle reaches its terminal (radial) velocity, $dV_r/dt = 0$, and Eq. (12-7) can be solved for V_{rt}, (the radial terminal velocity):

$$V_{rt} = \left(\frac{4\,\Delta\rho\, d\omega^2 r}{3\rho C_D}\right)^{1/2} \tag{12-8}$$

If $N_{Re} < 1$, Stokes' law holds, and $C_D = 24/N_{Re}$, in which case Eq. (12-8) becomes

$$V_{rt} = \frac{dr}{dt} = \frac{\Delta\rho d^2\omega^2 r}{18\mu} \tag{12-9}$$

This shows that the terminal velocity is not a constant but increases with r, because the (centrifugal) driving force increases with r. Assuming that all of the fluid is rotating at the same speed as the centrifuge, integration of Eq. (12-9) gives

$$\ln\left(\frac{r_2}{r_1}\right) = \frac{\Delta\rho d^2\omega^2}{18\mu} t \tag{12-10}$$

where t is the time required for the particle to travel a radial distance from r_1 to r_2. The time available for this to occur is the residence time of the particle in the centrifuge, $t = \tilde{V}/Q$, where $\tilde{V}$ is the volume of fluid in the centrifuge. If the region occupied by the fluid is cylindrical, then $\tilde{V} = \pi(r_2^2 - r_1^2)L$. The smallest particle that will travel from the surface of the fluid (r_1) to the wall (r_2) in time t is

$$d = \left(\frac{18\mu Q \ln(r_2/r_1)}{\Delta\rho\omega^2\tilde{V}}\right)^{1/2} \tag{12-11}$$

Rearranging Eq. (12-11) to solve for Q gives

$$Q = \frac{\Delta\rho d^2\omega^2\tilde{V}}{18\mu\ln(r_2/r_1)} = \frac{\Delta\rho g d^2}{18\mu}\left(\frac{\tilde{V}\omega^2}{g\ln(r_2/r_1)}\right) \tag{12-12}$$

which can also be written

$$Q = V_t\Sigma, \qquad \Sigma = \left(\frac{\tilde{V}\omega^2}{g\ln(r_2/r_1)}\right) \tag{12-13}$$

Here, V_t is the terminal velocity of the particle in a gravitational field and Σ is the cross-sectional area of the gravity settling tank that would be required to remove the same size particles as the centrifuge. This can be extremely large if the centrifuge operates at a speed corresponding to many g's.

This analysis is based on the assumption that Stokes' law applies, i.e., $N_{Re} < 1$. This is frequently a bad assumption, because many industrial centrifuges operate under conditions where $N_{Re} > 1$. If such is the case, an analytical solution to the problem is still possible by using the Dallavalle equation for C_D, rearranged to solve for N_{Re} as follows:

$$N_{Re} = \frac{d\rho}{\mu}\frac{dr}{dt} = [(14.42 + 1.827\sqrt{N_{Ar}})^{1/2} - 3.797]^2 \tag{12-14}$$

where

$$N_{Ar} = \frac{d^3 \rho \omega^2 r \,\Delta\rho}{\mu^2} \tag{12-15}$$

Equation (12-14) can be integrated from r_1 to r_2 to give

$$t = N_{12} \frac{\mu}{d^2 \omega^2 \,\Delta\rho} \tag{12-16}$$

where

$$N_{12} = 0.599(N_{Re_2} - N_{Re_1}) + 13.65(\sqrt{N_{Re_2}} - \sqrt{N_{Re_1}}) + 17.29 \ln\left(\frac{N_{Re_2}}{N_{Re_1}}\right) + 48.34\left(\frac{1}{\sqrt{N_{Re_2}}} - \frac{1}{\sqrt{N_{Re_1}}}\right) \tag{12-17}$$

The values of N_{Re_2} and N_{Re_1} are computed using Eq. (12-14) and the values of N_{Ar_2} and N_{Ar_1} at r_1 and r_2, respectively. Since $t = \tilde{V}/Q$, Eq. (12-16) can be rearranged to solve for Q:

$$Q = \frac{\Delta\rho\, d^2 \omega^2 \tilde{V}}{\mu N_{12}} = \frac{\Delta\rho\, d^2 \omega^2 \tilde{V}}{18\mu \ln(r_2/r_1)} \left[18 \frac{\ln(r_2/r_1)}{N_{12}}\right] \tag{12-18}$$

where the term in brackets is a "correction factor" that can be applied to the Stokes flow solution to account for non-Stokes conditions.

For separating very fine solids, emulsions, and immiscible liquids, a disk-bowl centrifuge is frequently used in which the settling occurs in the spaces between a stack of conical disks, as illustrated in Fig. 12-4. The advantage of this arrangement is that the particles have a much smaller radial distance to travel before striking a wall and being trapped. The disadvantage is that the carrier fluid circulating between the disks has a higher velocity in the restricted spaces, which can retard the settling motion of the particles. Separation will occur only when $V_{rt} > V_{rf}$, where V_{rt} is the radial terminal velocity of the particle and V_{rf} is the radial velocity component of the carrier fluid in the region where the fluid flow is in the inward radial direction.

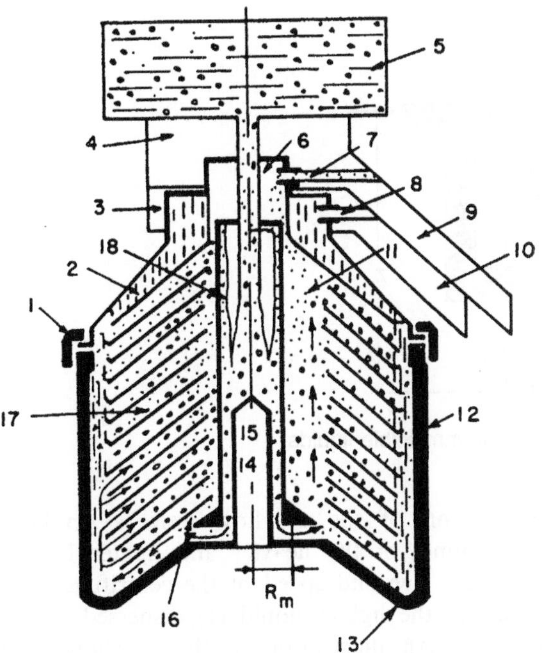

FIGURE 12-4 Schematic of disk-bowl centrifuge: *1*, Ring; *2*, bowl; *3,4*, collectors for products; *5*, feed tank; *6*, tube; *7,8*, discharge nozzles; *9,10*, funnels for collectors; *11*, through channels; *12*, bowl; *13*, bottom; *14*, thick-walled tube; *15* hole for guide; *16*, disk fixator; *17*, disks; *18* central tube (From Azbel and Cheremisinoff, 1983.)

B. Separation of Immiscible Liquids

The problem of separating immiscible liquids in a centrifuge can best be understood by first considering the static gravity separation of immiscible liquids, as illustrated in Fig. 12-5, where the subscript 1 represents the lighter liquid and 2 represents the heavier liquid. In a continuous system, the static head of the heavier liquid in the overflow pipe must be balanced by the combined head of the lighter and heavier liquids in the separator, i.e.,

$$\rho_2 z g = \rho_2 z_2 g + \rho_1 z_1 g \tag{12-19}$$

or

$$z = z_2 + z_1 \frac{\rho_1}{\rho_2} \tag{12-20}$$

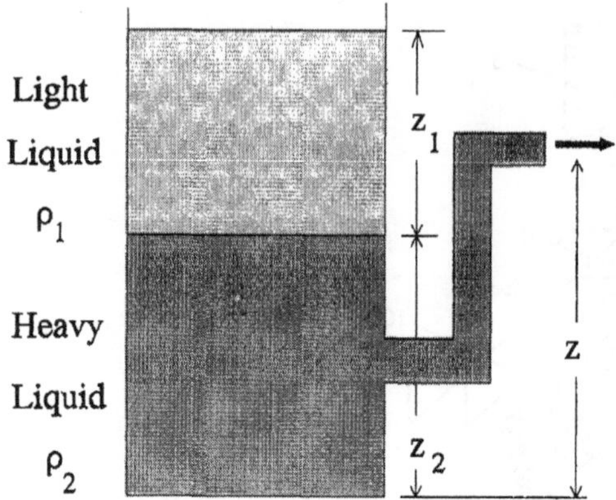

FIGURE 12-5 Gravity separation of immiscible liquids.

In a centrifuge, the position of the overflow weir is similarly determined by the relative amounts of the heavier and lighter liquids and their densities, along with the size and speed of the centrifuge. The feed stream may consist of either the lighter liquid (1) dispersed in the heavier liquid (2) or vice versa. An illustration of the overflow weir positions is shown in Fig. 12-6. Because there is no slip at the interface

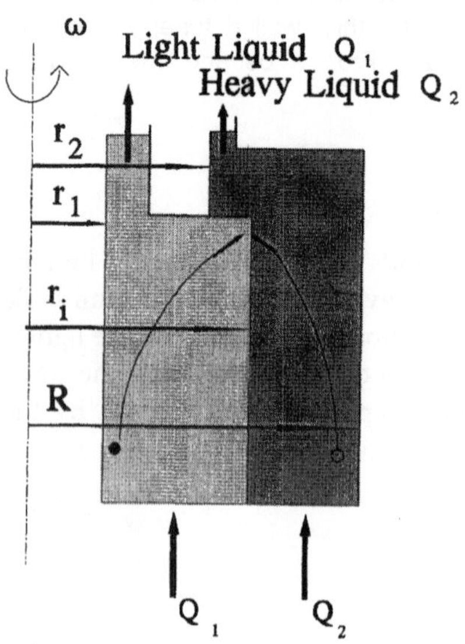

FIGURE 12-6 Centrifugal separation of immiscible liquids.

between the liquids, the axial velocity must be the same at that point for both fluids:

$$V_1 = \frac{Q_1}{A_1} = \frac{Q_1}{\pi(r_i^2 - r_1^2)}$$

$$V_2 = \frac{Q_2}{A_2} = \frac{Q_2}{\pi(R^2 - r_i^2)} \tag{12-21}$$

or

$$\frac{Q_1}{Q_2} = \frac{r_i^2 - r_1^2}{R^2 - r_i^2} \tag{12-22}$$

This provides a relationship between the locations of the interface (r_i) and the inner weir (r) and the relative feed rates of the two liquids. Also, the residence time for each of the two liquids in the centrifuge must be the same, i.e.,

$$t = \frac{\tilde{V}_T}{Q_T} = \frac{\pi L(R^2 - r_1^2)}{Q_2 + Q_1} \tag{12-23}$$

For drops of the lighter liquid (1) dispersed in the heavier liquid (2), assuming that Stokes flow applies, the time required for the drops to travel from the maximum radius (R) to the interface (r_i) is

$$t = \frac{18\mu_2}{\Delta\rho\, d^2\omega^2} \ln\left(\frac{R}{r_i}\right) = \frac{\pi L(R^2 - r_1^2)}{Q_2 + Q_1} \tag{12-24}$$

For the case of drops of the heavier liquid (2) dispersed in the lighter liquid (1), the corresponding time required for the maximum radial travel from the surface (r_1) to the interface (r_i) is

$$t = \frac{18\mu_1}{\Delta\rho\, d^2\omega^2} \ln\left(\frac{r_i}{r_1}\right) = \frac{\pi L(R^2 - r_1^2)}{Q_2 + Q_1} \tag{12-25}$$

Equations (12-22) and (12-24) or (12-25) determine the locations of the light liquid weir (r_1) and the interface (r_i) for given feed rates, centrifuge size, and operating conditions.

The proper location of the heavy liquid weir (r_2) can be determined by a balance of the radial pressure difference through the liquid layers, which is analogous to the gravity head balance in the gravity separator in Fig. 12-5. The radial pressure gradient due to centrifugal force is

$$\frac{dP}{dr} = \rho\omega^2 r \qquad \text{or} \qquad \Delta P = \frac{1}{2}\rho\omega^2\,\Delta r^2 \tag{12-26}$$

Since both the heavy liquid surface at r_2 and the light liquid surface at r_1 are at atmospheric pressure, the sum of the pressure differences from r_1 to R to r_2 must be zero:

$$\frac{1}{2}\rho_2\omega^2(r_2^2 - R^2) + \frac{1}{2}\rho_2\omega^2(R^2 - r_i^2) + \frac{1}{2}\rho_1\omega^2(r_i^2 - r_1^2) = 0 \tag{12-27}$$

which can be rearranged to give

$$\frac{\rho_2}{\rho_1} = \frac{r_i^2 - r_1^2}{r_i^2 - r_2^2} \tag{12-28}$$

Solving for r_2 gives

$$r_2^2 = \frac{\rho_1}{\rho_2}\left[r_i^2\left(\frac{\rho_2}{\rho_1} - 1\right) + r_1^2\right] \tag{12-29}$$

Equations (12-22), (12-24) or (12-25), and (12-29) thus determine the three design parameters r_i, r_1, and r_2. These equations can be arranged in dimensionless form. From Eq. (12-22),

$$\left(\frac{r_1}{R}\right)^2 = \left(\frac{r_i}{R}\right)^2\left(1 + \frac{Q_1}{Q_2}\right) - \frac{Q_1}{Q_2} \tag{12-30}$$

For drops of the light liquid in the heavy liquid, Eq. (12-24) becomes

$$\left(\frac{r_1}{R}\right)^2 = 1 - \frac{18\mu_2(Q_1 + Q_2)}{\pi L R^2\, \Delta\rho\, d^2\omega^2}\ln\left(\frac{R}{r_i}\right) \tag{12-31}$$

For drops of the heavy liquid in the light liquid, Eq. (12-25) becomes

$$\left(\frac{r_1}{R}\right)^2 = 1 - \frac{18\mu_1(Q_1 + Q_2)}{\pi L R^2\, \Delta\rho\, d^2\omega^2}\ln\left[\left(\frac{r_i}{R}\right)\left(\frac{R}{r_1}\right)\right] \tag{12-32}$$

Also, Eq. (12-29) is equivalent to

$$\left(\frac{r_2}{R}\right)^2 = \frac{\rho_1}{\rho_2}\left[\frac{r_i^2}{R^2}\left(\frac{\rho_2}{\rho_1} - 1\right) + \frac{r_1^2}{R^2}\right] \tag{12-33}$$

These three equations can be solved simultaneously (by iteration) for r_1/R, r_2/R, and r_i/R. It is assumed that the size of the suspended drops is known as well as the density and viscosity of the liquids and the overall dimensions and speed of the centrifuge.

IV. CYCLONE SEPARATIONS

A. General Characteristics

Centrifugal force can also be used to separate solid particles from fluids by inducing the fluid to undergo a rotating or spiraling flow pattern in a stationary vessel (e.g., a cyclone) that has no moving parts. Cyclones are widely used to remove small particles from gas streams ("aerocyclones") and suspended solids from liquid streams ("hydrocyclones").

A typical cyclone is illustrated in Fig. 12-7 (this is sometimes referred to as a "reverse flow" cyclone). The suspension enters through a rectangular or circular duct tangential to the cylindrical separator, which usually has a conical bottom. The circulating flow generates a rotating vortex motion that imparts centrifugal force to the particles which are thrown outward to the walls of the vessel, where they fall by gravity to the conical bottom and are removed. The carrier fluid spirals inward and downward to the cylindrical exit duct (also referred to as the "vortex finder"), from which it travels back up and leaves the vessel at the top. The separation is not perfect, and some solid particles leave in the overflow as well as the underflow. The particle size for which 50% leaves in the overflow and 50% leaves in the underflow is called the *cut size*.

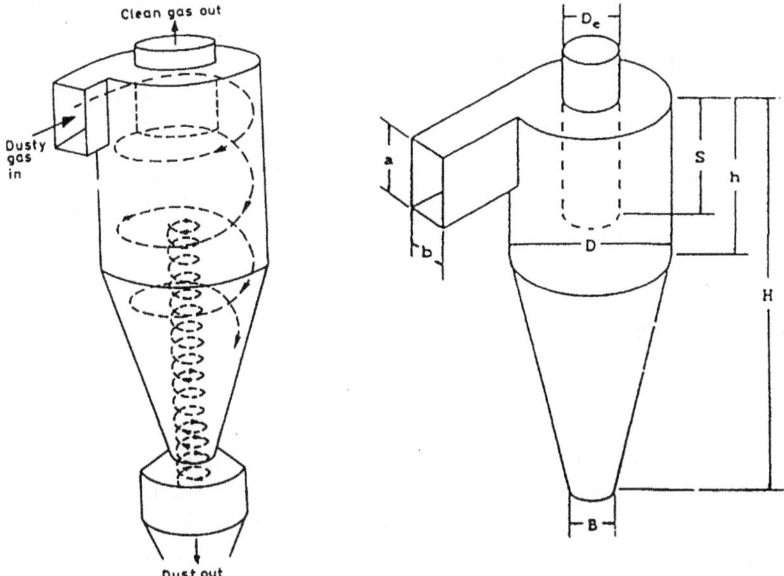

Figure 12-7 Typical reverse flow cyclone.

The diameter of a hydrocyclone can range from 10 mm to 2.5 m, cut sizes from 2 to 250 μm, and flow rate (capacities) from 0.1 to 7200 m^3/hr. Pressure drop can range from 0.3 to 6 atm (Svarovsky, 1984). For aerocyclones, very little fluid leaves with the solids underflow, although for hydrocyclones the underflow solids content is typically 45–50% by volume. Aerocyclones can achieve effective separation for particles as small as 2–5 μm.

Advantages of the cyclone include (Svarosky, 1984)

1. Versatility. Virtually any slurry or suspension can be concentrated, liquids degassed, or the solids classified by size, density, or shape.
2. Simplicity and economy. They have no moving parts and little maintenance.
3. Small size. Low residence times, and relatively fast response.
4. High shear forces, which can break up agglomerates, etc.

The primary disadvantages are:

1. Inflexibility. A given design is not easily adapted to a range of conditions. Performance is strongly dependent upon flow rate and feed composition, and the turndown ratio (range of operation) is small.
2. Limited separation performance in terms of the sharpness of the cut, range of cut size, etc.
3. Susceptibility to erosion.
4. High shear prevents the use of flocculents to aid the separation, as can be done in gravity settlers.

An increase in any one operating parameter generally increases all others as well. For example, increasing the flow rate will increase both separation efficiency and pressure drop, and vice versa.

B. Aerocyclones

1. Velocity Distribution

Although the dominant velocity component in the cyclone is in the angular (tangential) direction, the swirling flow field includes significant velocity components in the radial and axial directions as well, which complicate the motion and make a rigorous analysis impossible. This complex flow field also results in significant particle–particle collisions, which cause some particles of a given size to be carried out in both the overhead and underflow discharge, thus affecting the separation efficiency.

Cyclone analysis and design is not an exact science, and there are a variety of approaches to the analysis of cyclone performance. A critical review of the various methods for analyzing hydrocyclones is given by Svarovsky (1996), and a review of different approaches to aerocyclone analysis is given by Leith and Jones (1997). There are a number different approaches to the analysis of aerocyclones, one of the most comprehensive being that of Bhonet et al. (1997). The presentation here follows that of Leith and Jones (1997), which outlines the basic principles and some of the practical "working relations." The reader is referred to other works, especially those of Bhonet (1983) and Bhonet et al. (1997), for more details on specific cyclone design.

The performance of a cyclone is dependent upon the geometry as described by the values of the various dimensionless "length ratios" (see Fig. 12-7): $a/D, b/D, D_e/D, S/D, h/D, H/D$, and B/D. Typical values of these ratios for various "standard designs" are given in Table 12-1.

The complex three-dimensional flow pattern within the cyclone is dominated by the radial (V_r) and tangential (V_θ) velocity components. The vertical component is also significant but plays only an indirect role in the separation. The tangential velocity in the vortex varies with the distance from the axis in a complex manner, which can be described by the equation

$$V_\theta r^n = \text{constant} \tag{12-34}$$

For a uniform angular velocity ($\omega = \text{constant}$, i.e., a "solid body rotation"), $n = -1$, whereas for a uniform tangential velocity ("plug flow") $n = 0$, and for inviscid free vortex flow $\omega = c/r^2$, i.e., $n = 1$. Empirically, the exponent n has been found to be typically between 0.5 and 0.9. The maximum value of V_θ occurs in the vicinity of the outlet or exit duct (vortex finder) at $r = D_e/2$.

TABLE 12-1 Standard Designs for Reverse Flow Cyclones

Ref.[a]	Duty	D	a/D	b/D	D_e/D	S/D	h/D	H/D	B/D	K_f	Q/D^2 (m/hΩ)
1	High η	1	0.5	0.2	0.5	0.5	1.5	4.0	0.375	6.4	3500
2	High η	1	0.44	0.21	0.4	0.5	1.4	3.9	0.4	9.2	4940
3	Gen	1	0.5	0.25	0.5	0.625	2.0	4.0	0.25	8.0	6860
2	Gen	1	0.5	0.25	0.5	0.6	1.75	3.75	0.4	7.6	6680
1	High Q	1	0.75	0.375	0.75	0.875	1.5	4.0	0.375	7.2	16500
2	High Q	1	0.8	0.35	0.75	0.85	1.7	3.7	0.4	1.0	12500

[a] 1, Stairmand (1951); 2, Swift (1969); 3, Lapple (1951).
Source: Leith and Jones (1997).

For aerocyclones, the exponent n has been correlated with the cyclone diameter by the expression

$$n = 0.67 D_m^{0.14} \tag{12-35}$$

where D_m is the cyclone diameter in meters. The exponent also decreases as the temperature increases according to

$$\frac{1-n}{1-n_1} = \left(\frac{T}{T_1}\right)^{0.3} \tag{12-36}$$

There is a "core" of rotating flow below the gas exit duct (vortex finder), in which the velocity decreaes as the radius decreases and is nearly zero at the axis.

2. Pressure Drop

The pressure drops throughout the cyclone owing to several factors: (1) gas expansion, (2) vortex formation, (3) friction loss, and (4) changes in kinetic energy. The total pressure drop can be expressed in terms of an equivalent loss coefficient, K_f:

$$\Delta P = \frac{K_f}{2} \rho_G V_i^2 \tag{12-37}$$

where V_i is the gas inlet velocity, $V_i = Q/ab$. A variety of expressions have been developed for K_f, but one of the simplest that gives reasonable results is

$$K_f = 16 \frac{ab}{D_e^2} \tag{12-38}$$

This (and other) expressions may be accurate to only about ±50% or so, and more reliable pressure drop informaiton can be obtained only by experimental testing on a specific geometry. Typical values of K_f for the "standard" designs are given in Table 12-1.

3. Separation Efficiency

The efficiency of a cyclone (η) is defined as the fraction of particles of a given size that are separated by the cyclone. The efficiency increases with

1. Increasing particle diameter (d) and density
2. Increasing gas velocity
3. Decreasing cyclone diameter
4. Increasing cyclone length
5. Venting of some of the gas through the bottom solids exit
6. Wetting of the walls

A typical plot of efficiency versus particle diameter is shown in Fig. 12-8. This is called a *grade efficiency* curve. Although the efficiency varies with the particle size, a more easily determined characteristic is the "cut diameter" (d_{50}), the particle size that is collected with 50% efficiency.

The particles are subject to centrifugal, inertial, and drag forces as they are carried in the spriraling flow, and it is assumed that the particles that strike the outer wall before the fluid reaches the vortex finder will be collected. It is assumed that the tangential velocity of the particle is the same as that of the fluid ($V_{\mathrm{p}\theta} = V_\theta$) but that the radial velocity is not ($V_{\mathrm{pr}} \neq V_\mathrm{r}$), because the particles move radially toward the wall relative to the fluid. The centrifugal force acting on the particle is

$$F_\mathrm{c} = m_\mathrm{p}\omega^2 r = \frac{\pi d^3 \rho_\mathrm{s} V_\theta^2}{6r} \tag{12-39}$$

Assuming Stokes flow, the drag force is

$$F_\mathrm{d} = 3\pi\mu d(V_{\mathrm{pr}} - V_\mathrm{r}) \tag{12-40}$$

Equation (12-34) provides a relationship between the tangential velocity at any point and that at the wall:

$$V_\theta r^n = V_{\theta\mathrm{w}} r_\mathrm{w}^n \tag{12-41}$$

Although the velocity right at the wall is zero, the boundary layer at the wall is quite small, so this equation applies up to the boundary layer very near the wall. Setting the sum of the forces equal to the particle acceleration and

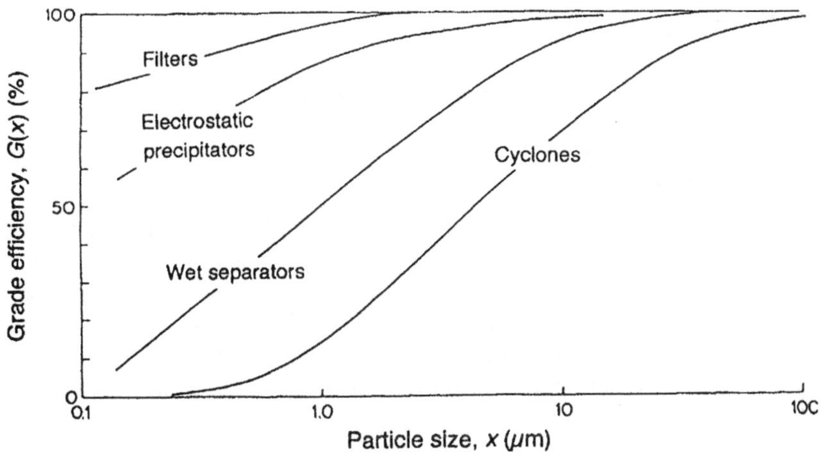

FIGURE 12-8 Typical cyclone grade efficiency curve.

substituting $V_{pr} = dr/dt$ and $V_\theta = V_{\theta r}(r_w/r)^n$ gives the governing equation for the particle radial position:

$$\frac{d^2r}{dt^2} + \frac{18\mu}{d^2\rho_s}\frac{dr}{dt} - \left(\frac{V_{\theta w}^2 r_w^{2n}}{r^{2n+1}} + \frac{18\mu V_r}{d^2\rho_s}\right) = 0 \tag{12-42}$$

There is no general solution to this equation, and various analysis have been based on specific approximations or simplifications of the equation.

One approximation considers the time it takes for the particle to travel from the entrance point, r_i, to the wall, $r_w = D/2$, relative to the residence time of the fluid in the cyclone. By neglecting the acceleration term and the fluid radial velocity and assuming that the velocity of the fluid at the entrance is the same as the tangential velocity at the wall ($V_i = V_{\theta w}$), Eq. (12-42) can be integrated to give the time required for the particle to travel from its initial position (r_i) to the wall ($D/2$). If this time is equal to or less than the residence time of the fluid in the cyclone, that particle will be trapped. The result gives the size of the smallest particle that will be trapped completely (in principle):

$$d_{100} = \left[\frac{9\mu D^2(1 - 2r_i/D)^{2n+2}}{4(n+1)V_i^2\rho_s t}\right]^{1/2} \tag{12-43}$$

The residence time is related to the "number of turns" (N) that the fluid makes in the vortex, which can vary from 0.2 to 10, with an average value of 5. If the 50% "cut diameter" particle is assumed to enter at $(D - b)/2$, with a residence time of

$$t = \frac{\pi DN}{V_i}\left(1 - \frac{D_e}{2D}\right) \tag{12-44}$$

and it is assumed that $n = 0$, the cut diameter is

$$d_{50} = \left(\frac{9\mu b}{2\pi\rho_s V_i N}\right)^{1/2} \tag{12-45}$$

Another approach is to consider the particle for which the drag force of the gas at the edge of the core where the velocity is maximum just balances the centrifugal force. This reduces Eq. (12-42) to a "steady state," with no net acceleration or velocity of this particle. The maximum velocity is given by Eq. (12-41) applied at the edge of the core: $V_{\theta w}^2 r_w^{2n} = V_\theta^2 r_{core}^{2n}$. When this is introduced into Eq. (12-42), the result is

$$d_{100} = \left(\frac{9Q\mu}{\pi(H - S)\rho_s V_{max}^2}\right)^{1/2} \tag{12-46}$$

Although this predicts that all particles larger than d_{100} will be trapped and all smaller particles will escape, the actual grade efficiency depends on particle size because of the variation of the inward radial velocity of the gas.

Leith and Licht (1972) incorporated the effect of turbulent reentrainment of the solids in a solution of Eq. (12-42) to derive the following expression for the grade efficiency:

$$\eta = 1 - \exp[-2(N_G N_{St})^{1/(2n+2)}] \tag{12-47}$$

where

$$N_{St} = \frac{d^2 \rho_s V_i (n+1)}{18\mu D} \tag{12-48}$$

is the Stokes number and N_G is a dimensionless geometric parameter,

$$N_G = \frac{\pi D^2}{ab}\left\{2\left[1 - \left(\frac{D_e}{D}\right)^2\right]\left(\frac{S}{D} - \frac{a}{2D}\right) + \frac{1}{3}\left(\frac{S + z_c - h}{D}\right)\right.$$
$$\left. \times \left[1 + \frac{d_c}{D} + \left(\frac{d_c}{D}\right)^2\right] + \frac{h}{D} - \left(\frac{D_e}{D}\right)^2 \frac{z_c}{D} - \frac{S}{D}\right\} \tag{12-49}$$

where z_c is the core length, given by

$$z_c = 2.3 D_e \left(\frac{D^2}{ab}\right)^{1/3} \tag{12-50}$$

and d_e is the core diameter, given by

$$d_c = D - (D - B)\left(\frac{S + z_c - h}{H - h}\right) \tag{12-51}$$

Equation (12-47) implies that the efficiency increases as N_G and/or N_{St} increases.

These equations can serve as a guide for estimating performance but cannot be expected to provide precise predicted behavior. However, they can be used effectively to scale experimental results for similar designs of different sizes operating under various conditions. For example, two cyclones of a given design should have the same efficiency when the value of N_{St} is the same for both. That is, if a given cyclone has a known efficiency for particles of diameter d_1, a similar cyclone will have the same efficiency for particles of diameter d_2, where

$$d_2 = d_1\left[\frac{Q_1}{Q_2}\left(\frac{\rho_{s1}}{\rho_{s2}}\right)\left(\frac{\mu_2}{\mu_1}\right)\left(\frac{D_2}{D_1}\right)\right]^{1/2} \tag{12-52}$$

Thus, the grade efficiency of the similar cyclone can be constructed from the grade efficiency of the known (tested) cyclone.

4. Other Effects

Increasing the solids loading of the feed increases the collection efficiency and decreases the pressure drop. The effect on pressure drop is given by

$$(\Delta P)_c = \frac{(\Delta P)_o}{1 + 0.0086 C_i^{1/2}} \tag{12-53}$$

where C_i is the inlet solids loading in g/m^3. The effect on overall collection efficiency is given by

$$\frac{100 - \eta}{100 - \eta_1} = \left(\frac{C_{i1}}{C_i}\right)^{0.182} \tag{12-54}$$

If the velocity near the wall is too high, particles will bounce off the wall and become reentrained. The inlet velocity above which this occurs is given by the empirical correlation

$$V_{ic} = 2400 \frac{\mu \rho_s}{\rho_G^2} \left(\frac{D^{0.2}(b/D)^{1.2}}{1 - b/D}\right) \tag{12-55}$$

The cyclone efficiency increases with V_i up to about $1.25V_{ic}$, after which reentrainment results in a decrease in efficiency.

C. Hydrocyclones

A similar approach to the analysis of hydrocyclones was presented by Svarovsky (1984, 1990). He deduced that the system can be described in terms of three dimensionless groups in addition to various dimensionless geometric parameters. These groups are the Stokes number,

$$N_{St50} = \frac{V_{tr}}{V_i} = \frac{\Delta \rho \, d_{50}^2 Q}{4\pi \mu D^3} \tag{12-56}$$

the Euler number, which is equivalent to the loss coefficient, K_f,

$$N_{Eu} = \frac{\Delta P}{\rho V_i^2/2} = \frac{\pi^2 \Delta P D^4}{8\rho Q^2} \tag{12-57}$$

and the Reynolds number,

$$N_{Re} = \frac{D V_i \rho}{\mu} = \frac{4 Q \rho}{\pi D \mu} \tag{12-58}$$

In each of these groups, the characteristic length is the cyclone diameter, D, and the characteristic velocity is $V_i = 4Q/\pi D^2$. Various empirical hydrocyclone models indicate that the relationship between these groups is

$$N_{St50}N_{Eu} = C \tag{12-59}$$

and

$$N_{Eu} = K_p N_{Re}^{n_p} \tag{12-60}$$

The quantities C, K_p, and n_p are empirical constants, with the same values for a given family of geometrically similar cyclones. The value of C ranges from 0.06 to 0.33, the exponent n_p varies from zero to 0.8, and K_p ranges from 2.6 to 6300. A summary of these parameters corresponding to some known hydrocyclone designs is given in Table 12-2. The references in this table are found in Svarovsky (1981), and the notation in this table is as follows. $D_i = 2r_i = (4ab/\pi)^{1/2}$ is the equivalent diameter of the inlet, $D_o = D_e$ is the gas exit diameter, $l = S$ is the length of the vortex finder, and $L = H$ is the total length of the hydrocyclone. These equations can be used to predict the performance of a given cyclone as follows.

Equation (12-60) can be solved for the capacity, Q, to give

$$Q^{2+n_p} = \frac{\pi^2 \Delta P D^4}{8\rho K_p}\left(\frac{\pi D\mu}{4\rho}\right)^{n_p} \tag{12-61}$$

and the cut size obtained from Eq. (12-59):

$$d_{50}^2 = \frac{4\pi N_{St50}N_{Eu}D^3\mu}{K_p Q\,\Delta\rho}\left(\frac{\pi D\mu}{4Q\rho}\right)^{n_p} \tag{12-62}$$

In reality, extensive data by Medronho (Antunes and Medronho, 1992) indicate that the product $N_{St50}N_{Eu}$ is not constant but depends on the ratio of underflow to feed (R) and the feed volumetric concentration (C_v) and N_{Eu} also depends on C_v as well as N_{Re}. The Reitema and Bradley geometries are two common families of geometrically similar designs, as defined by the geometry parameters in Table 12-3 (Antunes and Medronho, 1992).

The Bradley hydrocyclone has a lower capacity than the Reitema geometry but is more efficient. For the Rietema cyclone geometry the correlatins are (Antunes and Medronho, 1992)

$$N_{St50}N_{Eu} = 0.0474[\ln(1/R)]^{0.742}\exp(8.96C_v) \tag{12-63}$$

and

$$N_{Eu} = 371.5N_{Re}^{0.116}\exp(-2.12C_v) \tag{12-64}$$

Table 12-2 Operational Parameters for Some Known Hydrocyclone Designs[a]

Cyclone type and size of hydrocyclone	Geometrical proportions					Scale-up constants			Running cost criterion
	D_i/D	D_o/D	l/D	L/D	Angle θ (deg)	$N_{St50}N_{Eu}$	K_p	n_p	$N_{St50}N_{Eu}$
Rietema's design (optimum separation), $D = 0.075$ m	0.28	0.34	0.4	5	20	0.0611	316	0.134	2.12
Bradley's desigh, $D = 0.038$ m	0.133 (1/7.5)	0.20 (1/5)	0.33 (1/3)	6.85	9	0.1111	446.5	0.323	2.17
Mozley cyclone, $D = 0.022$ m	0.154 (1/6.5)	0.214 (3/14)	0.57 (4/7)	7.43	6	0.1203	6381	0	3.20
Mozley cyclone, $D = 0.044$ m	0.160 (1/6.25)	0.25 (1/4)	0.57 (4/7)	7.71	6	0.1508	4451	0	4.88
Mozley cyclone, $D = 0.044$ m	0.197 (1/5)	0.32 (1/3)	0.57 (4/7)	7.71	6	0.2182	3441	0	8.70
Warman 3 in. Model R, $D = 0.076$ m	0.29 (1/3.5)	0.20 (1/5)	0.31	4.0	15	0.1079	2.618	0.8	2.07
RW 2515 (AKW), $D = 0.125$ m	0.20 (1/5)	0.32 (1/3)	0.8	6.24	15	0.1642	2458	0	6.66
Hi-Klone model 2, $D = 0.097$ m	0.175	0.25	0.92 (0.59)*	5.6	10		873.5	0.2	
Hi-Klone model 3, $D = 0.1125$ m	0.15	0.20	0.80 (0.51)*	5.4	10		815.5	0.2	
Demco, $D = 0.051$ m	0.217	0.50	1.0	4.7	25				
Demco, $D = 0.102$ m	0.244	0.303	0.833	3.9	20				

[a] represents a modification; blank spaces in table are where the figures are not known.
Source: Svarovsky (1984).

TABLE 12-3 Families of Geometrically Similar Cyclones

Cyclone	$2r_{in.}/D$	D_e/D	$2S/D$	$2h/D$	$2H/D$	Cone angle
Rietema	0.28	0.34	0.4	—	5.0	10–20°
Bradley	1/7	1/5	1/3	1/2	—	9°

where

$$R = 1218(B/D)^{4.75} N_{Eu}^{0.30} \tag{12-65}$$

For the Bradley geometry, the corresponding correlations are (Antunes and Medronho, 1992)

$$N_{St50} N_{Eu} = 0.055[\ln(1/R)]^{0.66} \exp(12C_v) \tag{12-66}$$

and

$$N_{Eu} = 258 N_{Re}^{0.37} \tag{12-67}$$

where

$$R = 1.21 \times 10^6 (B/D)^{2.63} N_{Eu}^{-1.12} \tag{12-68}$$

These equations can be used to either predict the performance of a given cyclone or size the cyclone for given conditions. For example, if the definitions of N_{Eu} and N_{Re} from Eqs. (12-57) and (12-58) are substituted into Eq. (12-67) and the result rearranged for D, the result is

$$D = 7.0 \frac{0.31 Q^{0.54}}{\mu^{0.085} \Delta P^{0.23}} \tag{12-69}$$

which is dimensionally consistent.

PROBLEMS

Free Settling Fluid–Particle Separations

1. A slurry containing solid particles having a density of 2.4 g/cm^3 and ranging in diameter from 0.001 to 0.1 in. is fed to a settling tank 10 ft in diameter. Water is pumped into the tank at the bottom and overflows the top, carrying some of the particles with it. If it is desired to separate out all particles of diameter 0.02 in. and smaller, what flow rate of the water in gpm, is required?
2. A handful of sand and gravel is dropped into a tank of water 5 ft deep. The time required for the solids to reach the bottom is measured and found to vary from

3 to 20 s. If the solid particles behave as equivalent spheres and have an SG of 2.4, what is the range of equivalent particle diameters?

3. It is desired to determine the size of pulverized coal particles by measuring the time it takes them to fall a given distance in a known fluid. It is found that the coal particles (SG = 1.35) take a time ranging from 5 s to 1,000 min to fall 23 cm through a column of methanol (SG = 0.785, $\mu = 0.88$ cP). What is the size range of the particles in terms of their equivalent spherical diameters? Assume that the particles are falling at their terminal velocities at all times.
4. A water slurry containing coal particles (SG = 1.35) is pumped into the bottom of a large tank (10 ft diameter, 6 ft high), at a rate of 500 gal/hr, and overflows the top. What is the largest coal particle that will be carried out in the overflow? If the flow rate is increased to 5000 gal/hr, what size particles would you expect in the overflow? The slurry properties can be taken to be the same as for water.
5. To determine the settling characteristics of a sediment, you drop a sample of the material into a column of water. You measure the time it takes for the solids to fall a distance of 2 ft and find that it ranges from 1 to 20 s. If the solid SG = 2.5, what is the range of particle sizes in the sediment, in terms of the diameters of equivalent spheres?
6. You want to separate all the coal particles having a diameter of 100 μm or larger from a slurry. To do this, the slurry is pumped into the bottom of the large tank. It flows upward and flows over the top of the tank, where it is collected in a trough. If the solid coal has SG = 1.4 and the total flow rate is 250 gpm, how big should the tank be?
7. A gravity settling chamber consists of a horizontal rectangular duct 6 m long, 3.6 m wide, and 3 m high. The duct is used to trap sulfuric acid mist droplets entrained in an air stream. The droplets settle out as the air passes through the duct and can be assumed to behave as rigid spheres. If the air stream has a flow rate of 6.5 m^3/s, what is the diameter of the largest particle that will not be trapped by the duct? (Acid: $\rho = 1.75$ g/cm^3, $\mu = 3$ cP. Air: $\rho = 0.0075$ g/cm^3, $\mu = 0.02$ cP.)
8. Solid particles of diameter 0.1 mm and density 2 g/cm3 are to be separated from air in a settling chamber. If the air flow rate is 100 ft^3/s and the maximum height of the chamber is 4 ft, what should its minimum length and width be for all the particles to hit the bottom before exiting the chamber? (Air: $\rho = 0.075$ lbm/ft^3, $\mu = 0.018$ cP.)
9. A settling tank contains solid particles that have a wide range of sizes. Water is pumped into the tank from the bottom and overflows the top, at a rate of 10,000 gph. If the tank diameter is 3 ft, what separation of particle size is achieved? (That is, what size particles are carried out the top of the tank, assuming that the particles are spherical?) Solids density = 150 lb_m/ft^3.
10. You want to use a viscous Newtonian fluid to transport small granite particles through a horizontal 1 in. ID pipeline 100 ft long. The granite particles have a diameter of 1.5 mm and SG = 4.0. The SG of the fluid can be assumed to be 0.95. The fluid should be pumped as fast as possible to minimize settling of the

particles in the pipe but must be kept in laminar flow, so you design the system to operate at a pipe Reynolds number of 1000. The flow rate must be fast enough that the particles will not settle a distance greater than half the ID of the pipe, from the entrance to the exit. What should the viscosity of the fluid be, and what should the flow rate be (in gpm) at which it is pumped through the pipe?

11. An aqueous slurry containing particles with the size distribution shown below is fed to a 20 ft diameter settling tank [see, e.g., McCabe (1993) or Perry et al. (1997) for definition of mesh sizes].

Tyler mesh size	% of total solids in feed
8/10	5.0
10/14	12.0
14/20	26.0
20/28	32.0
28/35	21.0
35/48	4.0

The feed enters near the center of the tank, and the liquid flows upward and overflows the top of the tank. The solids loading of the feed is 0.5 lb_m of solids per gallon of slurry, and the feed rate is 50,000 gpm. What is the total solids concentration and the particle size distribution in the overflow? Density of solids is 100 lb_m/ft^3. Assume that (1) the particles are spherical; (2) the particles in the tank are unhindered; and (3) the feed and overflow have the same properties as water.

12. A water stream contacts a bed of particles with diameters ranging from 1 to 1000 μm and SG = 2.5. The water stream flows upward at a rate of 3 cm/s. What size particles will be carried out by the stream, and what size will be left behind?

13. An aqueous slurry containing particles with SG = 4 and a range of sizes up to 300 μm flows upward through a small tube into a larger vertical chamber with a diameter of 10 cm. You want the liquid to carry all of the solids through the small tube, but you want only those particles with diameters less than 20 μm to be carried out the top of the larger chamber. (a) What should the flow rate of the slurry be (in gpm). (b) What size should the smaller tube be?

14. A dilute aqueous $CaCO_3$ slurry is pumped into the bottom of a classifier at a rate of 0.4 m^3/s, and overflows the top. The density of the solids is 2.71 g/cm^3.
    (a) what should the diameter of the classifier be if the overflow is to contain no particles larger than 0.2 mm in diameter?
    (b) The same slurry as in (a) is sent to a centrifuge that operates at 5000 rpm. The centrifuge diameter is 20 cm, its length is 30 cm, and the liquid layer thickness is 20% of the centrifuge radius. What is the maximum flow rate that the centrifuge can handle and achieve the same separation as the classifier?

15. A centrifuge that has a 40 cm ID and in 30 cm long has an overflow weir that is 5 cm wide. The centrifuge operates at a speed of 3600 rpm.
   (a) What is the maximum capacity of the centrifuge (in gpm) for which particles with a diameter of 25 μm and SG = 1.4 can be separated from the suspension?
   (b) What would be the diameter of a settling tank that would do the same job?
   (c) If the centrifuge ID was 30 cm, how fast would it have to rotate to do the same job, everything else being equal?
16. Solid particles with a diameter of 10 μm and $SG = 2.5$ are to be removed from an aqueous suspension in a centrifuge. The centrifuge has an inner radius of 1 ft, an outer radius of 2 ft, and a length of 1 ft. If the required capacity of the centrifuge is 100 gpm, what should the operating speed (in rpm) be?
17. A centrifuge is used to remove solid particles with a diameter of 5 μm and SG = 1.25 from a dilute aqueous stream. The centrifuge rotates at 1200 rpm and is 3 ft high, the radial distance to the liquid surface is 10 in., and the radial distance to the wall is 14 in.
   (a) Assuming that the particles must strike the centrifuge wall to be removed, what is the maximum capacity of this centrifuge, in gpm?
   (b) What is the diameter, in feet, of the gravity settling tank that would be required to do the same job?
18. A dilute aqueous slurry containing solids with a diameter of 20 μm and SG = 1.5 is fed to a centrifuge rotating at 3000 rpm. The radius of the centrifuge is 18 in., its length is 24 in., and the overflow weir is 12 in. from the centerline.
   (a) If all the solids are to be removed in the centrifuge, what is the maximum capacity that it can handle (in gpm)?
   (b) What is the diameter of the gravity settling tank that would be required for this separation at the same flow rate?
19. A centrifuge with a radius of 2 ft and a length of 1 ft has an overflow weir located 1 ft from the centerline. If particles with SG = 2.5 and diameters of 10 μm and less are to be removed from an aqueous suspension at a flow rate of 100 gpm, what should the operating speed of the centrifuge be (in rpm)?
20. A centrifuge with a diameter of 20 in. operates at a speed of 1800 rpm. If there is a water layer 3 in. thick on the centrifuge wall, what is the pressure exerted on the wall?
21. A vertical centrifuge, operating at 100 rpm, contains an aqueous suspension of solid particles with SG = 1.3 and radius of 1 mm. When the particles are 10 cm from the axis of rotation, determine the direction in which they are moving relative to a horizontal plane.
22. You are required to design an aerocyclone to remove as much dust as possible from the exhaust coming from a rotary drier. The gas is air at 100°C and 1 atm and flows at a rate of 40,000 m^3/hr. The effluent from the cyclone will go to a scrubber for final cleanup. The maximum loading to the scrubber should be 10 g/m^3, although 8 g/m^3 or less is preferable. Measurements on the stack gas indicate that the solids loading from the drier is 50 g/m^3 (ρ_s = 2.5 g/cm^3). The

pressure drop in the cyclone must be less than 2 kPa. Use the Stairmand standard design parameters from Table 12-1 as the basis for your design.

NOTATION

a inlet height, [L]
b inlet width, [L]
A area, [L^2]
B cyclone bottom exit diameter, [L]
C_D drag coefficient, [—]
C_i inlet solids loading, [M/L^3]
d particle diameter, [L]
d_{50} diameter of 50% cut particle, [L]
d_{100} diameter of smallest trapped particle, [L]
D cyclone diameter, [L]
D_m cyclone diameter in meters, [L]
D_e cyclone top exit diameter, [L]
F force [$F = ML/t^2$]
F_c centrifugal force, [$F = ML/t^2$]
g acceleration due to gravity, [L/t^2]
H total cyclone height, [L]
h height of cyclone cylindrical section, [L]
K_f loss coefficient, [—]
n exponent in Eq. (12-34), [—]
N number of turns in vortex, [—]
N_G dimensionless geometry number, Eq. (12-49), [—]
N_{Ar} Archimedes number, [—]
N_{Eu} Euler number, Eq. (12-57), [—]
N_{Re} Reynolds number, [—]
N_{St} Stokes number, Eq. (12-48), [—]
N_{12} parameter defined by Eq. (12-17), [—]
Q volumetric flow rate, [L^3/t]
r radial position, [L]
t time, [t]
S cyclone vortex finder height, [L]
T temperature, [T]
$\tilde{V}$ volume, [L^3]
V_t terminal velocity, [L/t]
V_t^* gravity settling velocity, [L/t]
z vertical distance measured upward, [L]
η efficiency, [—]
$\Delta(\ )$ $(\ )_2-(\ )_1$
ρ density, [M/L^3]
μ viscosity, [M/Lt]
Σ equivalent gravity settling area for centrifuge, [L^2]
ω angular velocity, [1/t]

Subscripts

1,2 reference points
c core
e exit
G gas
i inlet
o solids-free
s solid
w wall
θ angular direction

REFERENCES

Antunes M, RA Medronho. Bradley cyclones: Design and performance analysis. In: L Svarovsky, MT Thew, eds. Hydrocylcones: Analysis and Applications. Boston: Kluwer, 1992, p 3–13.

Azbel DS, NP Cheremisinoff. Fluid Mechanics and Unit Operations, Ann Arbor, MI: Ann Arbor Science, 1983.

Bohnet M. Design methods for aerocyclones and hydrocyclones. In; NP Cheremisinoff, R Gupta, eds. Handbook of Fluids in Motion. Ann Arbor, MI: Ann Arbor Science, 1983, Chap 32.

Bhonet M, O Gottschalk, M Morweiser. Modern design of aerocyclones. Adv Particle Tech 8(2): 137–161, 1997.

Coulson JM, JF Richardson, JR Blackhurst, JH Harker. Chemical Engineering, Vol 2. 3rd ed. New York: Pergamon Press, 1980.

Lapple C. Chem. Eng 58:144, 1951.

Leith D, DL Jones. Cyclones. In: ME Fayed, L Otten, eds. Handbook of Powder Science and Technology. 2nd ed. London: Chapman and Hall, 1997, chap 15.

Leith D, Licht W. 1972 AIChE Symposium Sr. 68: 196.

McCabe WL, JC Smith, P Hariott. Unit Operations of Chemical Engineering. New York: McGraw-Hill, 1993.

Perry RH, DW Green, JO Malonay. Perry's Classical Engineers Handbook, McGraw-Hill, 1997.

Stairmand CJ. Trans Inst Chem Eng 29:356, 1951.

Svarovsky L. Hydrocyclones. Lancster, PA: Technomic, 1984.

Svarovsky L. Hydrocyclones. In: L Svarvosky, ed. Solid-Liquid Separation. 3td ed. New York: Butterworths, 1990.

Svarovsky L. A critical review of hydrocyclone models. In: D Claxton, L Svarovsky, M Thew, eds. Hydrocyclyones '96. London: Mech Eng Publi, 1996.

Swift P. Steam Heating Eng 38:453, 1969.

13

Flow in Porous Media

I. DESCRIPTION OF POROUS MEDIA

By a "porous medium" is meant a solid, or a collection of solid particles, with sufficient open space in or around the particles to enable a fluid to pass through or around them. There are various conceptual ways of describing a porous medium.

One concept is a continuous solid body with pores in it, such as a brick or a block of sandstone. Such a medium is referred to as *consolidated*, and the pores may be unconnected ("closed cell," or impermeable) or connected ("open cell," or permeable). Another concept is a collection (or "pile") of solid particles in a packed bed, where the fluid can pass through the voids between the particles. This is referred to as *unconsolidated*. A schematic representation is shown in Fig. 13-1. Either of these concepts may be valid, depending upon the specific medium under consideration, and both have been used as the basis for developing the equations that describe fluid flow behavior within the medium. In practice, porous media may range from a "tight" oil bearing rock formation to a packed column containing relatively large packing elements and large void spaces.

The "pile of solid particles" concept is useful for either consolidated or unconsolidated media as a basis for analyzing the flow process, because many consolidated media are actually made up of individual particles that

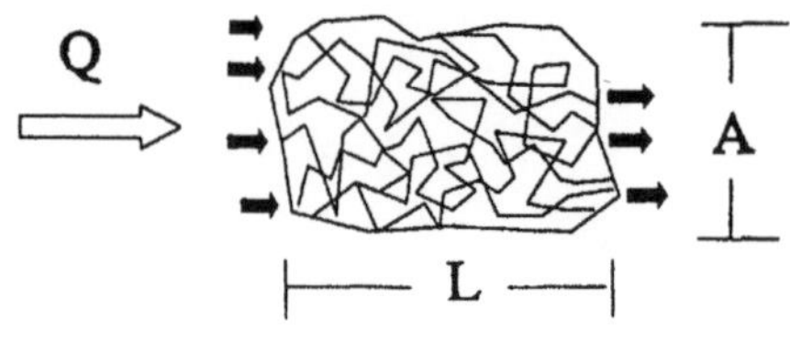

(a)

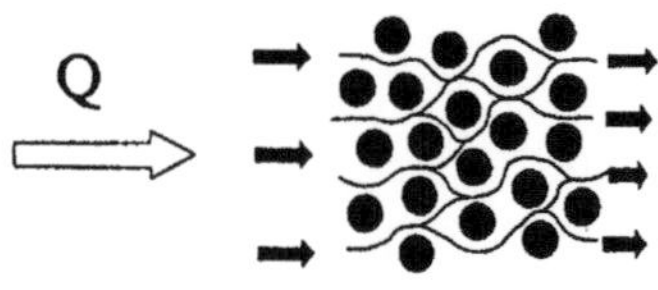

(b)

Figure 13-1 Porous media. (a) Consolidated medium; (b) unconsolidated medium.

are just stuck together (e.g. sandstone). One of the key properties of a porous medium is the porosity ε or void fraction, which is defined by

$$\varepsilon = \frac{\text{Total volume} - \text{Volume of solids}}{\text{Total volume}}$$

$$= 1 - \frac{A_{\text{solid}}}{A} = \frac{A_{\text{voids}}}{A} \tag{13-1}$$

where A_{solid} is the area of the solid phase in a cross section of area A.

We also distinguish between the velocity of approach, or the "superficial" velocity of the fluid,

$$V_s = Q/A \tag{13-2}$$

and the "interstitial" velocity, which is the actual velocity within the pores or voids,

$$V_i = \frac{Q}{\varepsilon A} = \frac{V_s}{\varepsilon} \tag{13-3}$$

A. Hydraulic Diameter

Because the fluid in a porous medium follows a tortuous path through channels of varying size and shape, one method of describing the flow behavior in the pores is to consider the flow path as a "noncircular conduit." This requires an appropriate definition of the hydraulic diameter:

$$D_h = 4\frac{A_i}{W_p} = 4\frac{A_i L}{W_p L} = 4\,\frac{\text{Flow volume}}{\text{Internal wetted surface area}}$$

$$= 4\frac{\varepsilon \times \text{Bed volume}}{(\text{No. of particles})(\text{Surface area/Particle})} \qquad (13\text{-}4)$$

The medium, with overall dimensions AL, is assumed to be made up of a collection of individual particles and may be either consolidated or unconsolidated. The number of particles in the medium can be expressed as

$$\text{No. particles} = \frac{(\text{Bed volume})(\text{Fraction of solids in bed})}{\text{Volume/Particle}}$$

$$= \frac{(\text{Bed volume})(1-\varepsilon)}{\text{Volume/Particle}} \qquad (13\text{-}5)$$

Substitution of this into Eq. (13-4) leads to

$$D_h = 4\,\frac{\varepsilon}{1-\varepsilon}\left(\frac{1}{a_s}\right) \qquad (13\text{-}6)$$

where a_s = (particle surface area)/(particle volume). If the particles are spherical with diameter d, then $a_s = 6/d$. Thus, for a medium composed of uniform spherical particles,

$$D_h = \frac{2d\varepsilon}{3(1-\varepsilon)} \qquad (13\text{-}7)$$

If the particles are not spherical, the parameter d may be replaced by

$$d = \psi d_s = 6/a_s \qquad (13\text{-}8)$$

where ψ is the *sphericity factor*, defined by

$$\psi = \frac{\text{Surface area of a sphere with same volume as the particle}}{\text{Surface area of the particle}} \qquad (13\text{-}9)$$

and d_s is the diameter of a sphere with the same volume as the particle.

B. Porous Medium Friction Factor

The expressions for the hydraulic diameter and the superficial velocity can be incorporated into the definition of the friction factor to give an equivalent expression for the porous medium friction factor:

$$f \equiv \frac{e_f}{(4L/D_h)(V_i^2/2)} = \frac{e_f d\varepsilon}{3L(1-\varepsilon)V_i^2} = \frac{e_f d\varepsilon^3}{3L(1-\varepsilon)V_s^2} \qquad (13\text{-}10)$$

Most references use Eq. (13-10) without the numerical factor of 3 as the definition of the porous medium friction factor, i.e.,

$$f_{\mathrm{PM}} \equiv \frac{e_{\mathrm{f}} d\varepsilon^3}{L(1-\varepsilon)V_{\mathrm{s}}^2} \tag{13-11}$$

C. Porous Medium Reynolds Number

In like fashion, the hydraulic diameter and the superficial velocity can be introduced into the definition of the Reynolds number to give

$$N_{\mathrm{Re}} = \frac{D_{\mathrm{h}} V_{\mathrm{i}} \rho}{\mu} = \frac{2d\varepsilon V_{\mathrm{i}} \rho}{3(1-\varepsilon)\mu} = \frac{2dV_{\mathrm{s}}\rho}{3(1-\varepsilon)\mu} \tag{13-12}$$

Here again, the usual porous medium Reynolds number is defined by Eq. (13-12) without the numerical factor (2/3):

$$N_{\mathrm{Re,PM}} = \frac{dV_{\mathrm{s}}\rho}{(1-\varepsilon)\mu} \tag{13-13}$$

II. FRICTION LOSS IN POROUS MEDIA

A. Laminar Flow

By analogy with laminar flow in a tube, the friction factor in laminar flow would be

$$f = \frac{16}{N_{\mathrm{Re}}} \quad \text{or} \quad f_{\mathrm{PM}} = \frac{72}{N_{\mathrm{Re,PM}}} \tag{13-14}$$

However, this expression assumes that the total resistance to flow is due to the shear deformation of the fluid, as in a uniform pipe. In reality the resistance is a result of both shear and stretching (extensional) deformation as the fluid moves through the nonuniform converging–diverging flow cross section within the pores. The "stretching resistance" is the product of the extension (stretch) rate and the extensional viscosity. The extension rate in porous media is of the same order as the shear rate, and the extensional viscosity for a Newtonian fluid is three times the shear viscosity. Thus, in practice a value of 150–180 instead of 72 is in closer agreement with observations at low Reynolds numbers, i.e.,

$$f_{\mathrm{PM}} = \frac{180}{N_{\mathrm{Re,PM}}} \quad \text{for } N_{\mathrm{Re,PM}} < 10 \tag{13-15}$$

This is known as the *Blake–Kozeny* equation and, as noted, applies for $N_{\mathrm{Re,PM}} < 10$.

B. Turbulent Flow

At high Reynolds numbers (high turbulence levels), the flow is dominated by inertial forces and "wall roughness," as in pipe flow. The porous medium can be considered an "extremely rough" conduit, with $\varepsilon/d \sim 1$. Thus, the flow at a sufficiently high Reynolds number should be fully turbulent and the friction factor should be constant. This has been confirmed by observations, with the value of the constant equal to approximately 1.75:

$$f_{PM} = 1.75 \qquad \text{for } N_{Re,PM} > 1000 \tag{13-16}$$

This is known as the *Burke–Plummer* equation and, as noted, applies for $N_{Re,PM} > 1000$.

C. All Reynolds Numbers

An expression that adequately represents the porous medium friction factor over all values of Reynolds number is

$$f_{PM} = 1.75 + \frac{180}{N_{Re,PM}} \tag{13-17}$$

This equation with a value of 150 instead of 180 is called the *Ergun equation* and is simply the sum of Eqs (13-15) and (13-16). (The more recent references favor the value of 180, which is also more conservative.) Obviously, for $N_{Re,PM} < 10$ the first term is small relative to the second, and the Ergun equation reduces to the Blake–Kozeny equation. Likewise, for $N_{Re,PM} > 1000$ the first term is much larger than the second, and the equation reduces to the Burke–Plummer equation.

If the definitions of f_{PM} and $N_{Re,PM}$ are inserted into the Ergun equation, the resulting expression for the frictional energy loss (dissipation) per unit mass of fluid in the medium is

$$e_f = 1.75 \frac{V_s^2}{d} \left(\frac{1-\varepsilon}{\varepsilon^3} \right) L + 180 \frac{V_s \mu (1-\varepsilon)^2 L}{d^2 \varepsilon^3 \rho} \tag{13-18}$$

III. PERMEABILITY

The "permeability" of a porous medium (K) is defined as the proportionality constant that relates the flow rate through the medium to the pressure drop, the cross-sectional area, the fluid viscosity, and net flow length through the medium:

$$Q = K \frac{-\Delta P A}{\mu L} \tag{13-19}$$

This equation defines the permeability (K) and is known as *Darcy's law*. The most common unit for the permeability is the "darcy," which is defined as the flow rate in cm^3/s that results when a pressure drop of 1 atm is applied to a porous medium that is 1 cm^2 in cross-sectional area and 1 cm long, for a fluid with viscosity of 1 cP. It should be evident that the dimensions of the darcy are L^2, and the conversion factors are (approximately) 10^{-8} cm^2/darcy $\cong 10^{-11}$ ft^2/darcy. The flow properties of tight, crude oil bearing, rock formations are often described in permeability units of millidarcies.

If the Blake–Kozeny equation for laminar flow is used to describe the friction loss, which is then equated to $\Delta P/\rho$ from the Bernoulli equation, the resulting expression for the flow rate is

$$Q = \frac{-\Delta P\,A}{\mu L}\left(\frac{d^2\varepsilon^3}{180(1-\varepsilon)^2}\right) \tag{13-20}$$

By comparison of Eqs. (13-19) and (13-20), it is evident that the permeability is identical to the term in brackets in Eq. (13-20), which shows how the permeability is related to the equivalent particle size and porosity of the medium. Since Eq. (13-20) applies only for laminar flow, it is evident that the permeability has no meaning under turbulent flow conditions.

IV. MULTIDIMENSIONAL FLOW

Flow in a porous medium in two or three dimensions is important in situations such as the production of crude oil from reservoir formations. Thus, it is of interest to consider this situation briefly and to point out some characteristics of the governing equations.

Consider the flow of an incompressible fluid through a two-dimensional porous medium, as illustrated in Fig. 13-2. Assuming that the kinetic energy change is negligible and that the flow is laminar as characterized by Darcy's law, the Bernoulli equation becomes

$$-\left(\frac{\Delta P}{\rho} + g\,\Delta z\right) = e_f = \frac{\mu V_s L}{K\rho} \tag{13-21}$$

or

$$\Delta\left(\frac{\Phi}{\rho}\right) = -\frac{\mu V_s L}{K\rho} \tag{13-22}$$

where the density cancels out if the fluid is incompressible. Equation (13-22) can be applied in both the x and y directions, by taking $L = \Delta x$ for the x direction and $L = \Delta y$ for the y direction:

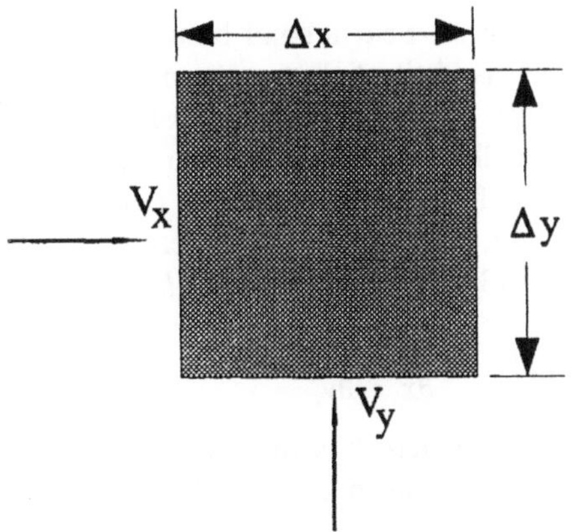

FIGURE 13-2 Two-dimensional flow in a porous medium.

$$\frac{\Delta\Phi}{\Delta x} = -\frac{\mu V_x}{K} = \frac{\partial\Phi}{\partial x} \tag{13-23}$$

and

$$\frac{\Delta\Phi}{\Delta y} = -\frac{\mu V_y}{K} = \frac{\partial\Phi}{\partial y} \tag{13-24}$$

If Eq. (13-23) is differentiated with respect to x and Eq. (13-24) is differentiated with respect to y and the results are added, assuming μ and K to be constant, we get

$$\frac{\partial^2\Phi}{\partial x^2} + \frac{\partial^2\Phi}{\partial y^2} = -\frac{\mu}{K}\left(\frac{\partial V_x}{\partial x} + \frac{\partial V_y}{\partial y}\right) = 0 \tag{13-25}$$

For an incompressible fluid, the term in parentheses is zero as a result of the conservation of mass (e.g., the microscopic continuity equation). Equation (13-25) can be generalized to three dimensions as

$$\nabla^2\Phi = 0 \tag{13-26}$$

which is called the *Laplace equation*. The solution of this equation, along with appropriate boundary conditions, determines the potential (e.g., pressure) distribution within the medium. The derivatives of this potential then determine the velocity distribution in the medium [e.g., Eqs. (13-23) and (13-24)]. The Laplace equation thus governs the three-dimensional (potential) flow of an inviscid fluid. Note that the Laplace equation follows from Eq. (13-25) for either an incompressible viscous fluid, by virtue of the continuity equation, or for any flow with negligible viscosity effects (e.g., compressible flow outside the boundary layer near a solid boundary). It is interesting that the same equation governs both of these extreme cases.

The Laplace equation also applies to the distribution of electrical potential and current flow in an electrically conducting medium as well as the temperature distribution and heat flow in a thermally conducting medium. For example, if $\Phi \Rightarrow E$, $V \Rightarrow i$, and $\mu/K \Rightarrow r_e$, where r_e is the electrical resistivity ($r_e = RA/\Delta x$), Eq. (13-22) becomes *Ohm's law*:

$$\frac{\partial E}{\partial x} = -r_e i_x, \qquad \nabla^2 E = 0, \qquad \text{and} \qquad \frac{\partial i_x}{\partial x} + \frac{\partial i_y}{\partial y} = 0 \tag{13-27}$$

Also, with $\Phi \Rightarrow T$, $V \Rightarrow q$, and $K/\mu \Rightarrow k$, where k is the thermal conductivity, the same equations govern the flow of heat in a thermally conducting medium (e.g., *Fourier's law*):

$$\frac{\partial T}{\partial x} = -\frac{1}{k} q_x, \qquad \nabla^2 T = 0, \qquad \text{and} \qquad \frac{\partial q_x}{\partial x} + \frac{\partial q_y}{\partial y} = 0 \tag{13-28}$$

By making use of these analogies, electrical analog models can be constructed that can be used to determine the pressure and flow distribution in a porous medium from measurements of voltage and current distribution in a conducting medium, for example. The process becomes more complex, however, when the local permeability varies with position within the medium, which is often the case.

V. PACKED COLUMNS

At the other end of the spectrum from a "porous rock" is the unconsolidated medium composed of beds of relatively large scale packing elements. These elements may include a variety of shapes, such as rings, saddles, grids, and meshes, which are generally used to provide a large gas/liquid interface for promoting mass transfer in such operations as distillation or absorption or liquid–liquid extraction. A typical application might be the removal of an impurity from a gas stream by selective absorption by a solvent in an absorption column filled with packing. The gas (or lighter liquid, in the case of liquid–liquid extraction) typically enters the bottom of the column, and the heavier liquid enters the top and drains by gravity, the flow being countercurrent as illustrated in Fig. 13-3.

For single-phase flow through packed beds, the pressure drop can generally be predicted adequately by the Ergun equation. However, because the flow in packed columns is normally countercurrent two-phase flow, this situation is more complex. The effect of increasing the liquid mass flow rate (L) on the pressure drop through the column for a given gas mass flow rate (G), starting with dry packing, is illustrated in Fig. 13-4. The pressure drop

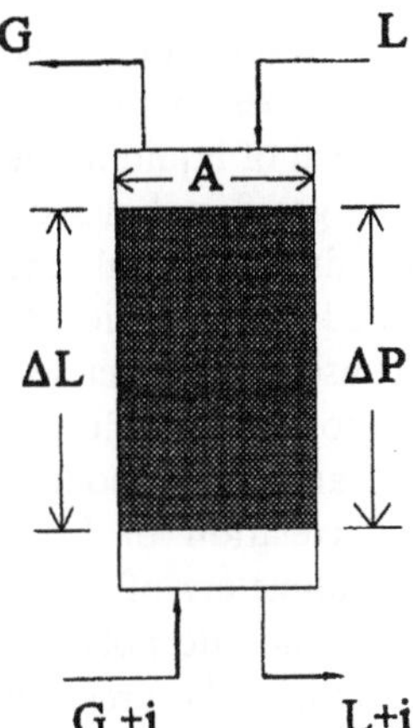

FIGURE 13-3 Schematic of packed column.

for wet drained packing is higher than for dry packing, because the liquid occupies some of the void space between packing elements even in the "drained" condition. As the liquid flow rate increases, the liquid occupies an increasing portion of the void space, so the area available to the gas is reduced and the total pressure drop increases. As the liquid flow rate increases, the curve of ΔP vs. G becomes increasingly nonlinear. The points labeled "l" in Fig. 13-4 are referred to as the "loading" points and indicate points where there is a marked increase in the interaction between the liquid

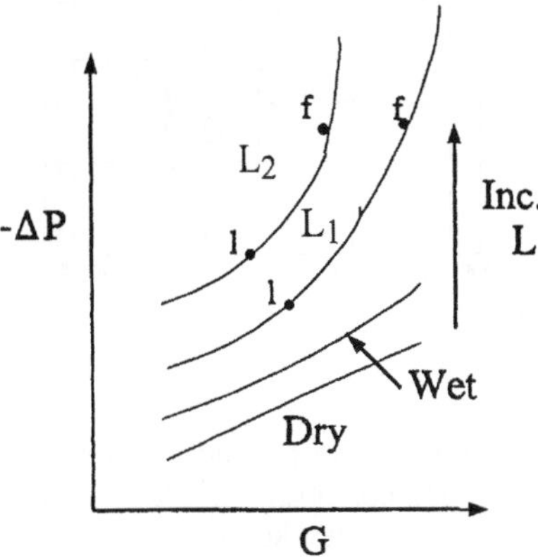

FIGURE 13-4 Effect of liquid rate on ΔP.

and the gas, and this is the desired operation point for the column. The points labeled "f" in Fig. 13-4 are the "flooding" points. At these points, the pressure drop through the column is equal to the static head of liquid in the column. When this occurs, the pressure drop due to the gas flow balances the static head of liquid, so the liquid can no longer drain through the packing by gravity and the column is said to be "flooded." It is obviously undesirable to operate at or near the flooding point, because a slight increase in gas flow at this point will carry the liquid out of the top of the column.

The pressure drop through packed columns, and the flooding conditions, can be estimated from the generalized correlation of Leva (1992), shown in Fig. 13-5. The pressure gradient in millimeters of water per meter of packed height is the parameter on the curves, and interpolation is usually necessary to determine the pressure drop (note that the pressure

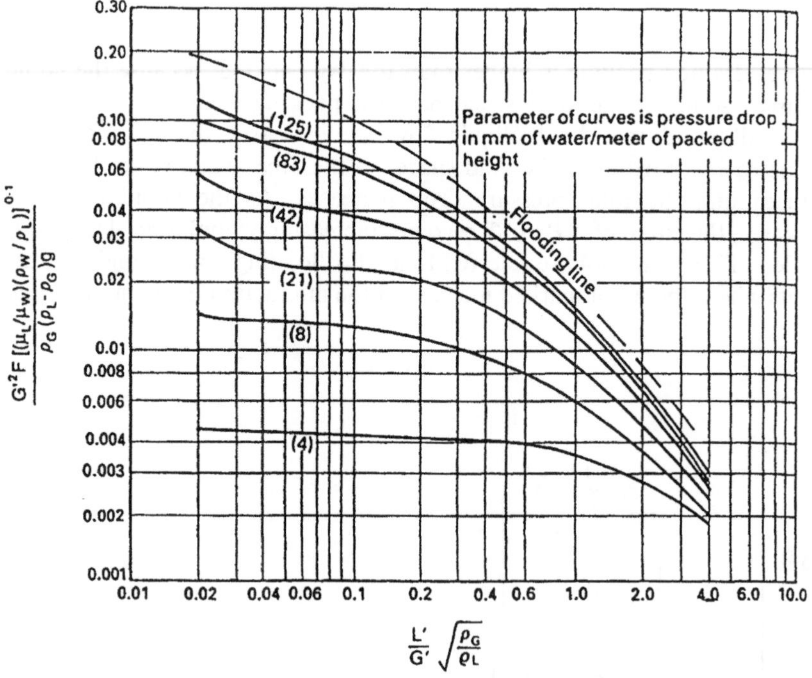

Figure 13-5 Generalized correlation for pressure drop in packed columns L = liquid mass flux [lb$_m$/(s ft^2), kg/(sm^2)]; G = gas mass flux [lb$_m$/(s ft^2), kg/(sm^2)]; ρL = liquid density (lb$_m$/ft^3, kg/m^3), ρ_G = gas density (lb$_m$/ft^3, kg/m^3); F = packing factor (Table 13-1); μ_L = liquid viscosity (mNs/m^2), g = (9.81 m/s^2, 32.2 ft/s^2), w = water at same T and P as column. (From Coulson et al., 1991.)

drop is not linearly proportional to the spacing between the curves). Correction factors for liquid density and viscosity, which are to be applied to the Y axis of this correlation, are also shown. The parameter F in this correlation is called the *packing factor*. Values of F are given in Table 13-1, which shows the dimensions and physical properties of a variety of types of packing. Note that in Table 13-1 the term S_B is equal to $a_s(1 - \varepsilon)$, where a_s is the surface area per unit volume of the packing element. The packing factor F is comparable to the term S_B/ε^3 in the definition of f_{PM}, but is an empirical factor that characterizes the packing somewhat better than S_B/ε^3.

VI. FILTRATION

For fine suspended solids with particles that are too small to be separated from the liquid by gravitational or centrifugal methods, a "barrier" method such as a filter may be used. The liquid is passed through a filter medium (usually a cloth or screen) that provides a support for the solid particles removed from the slurry. In actuality, the pores in the filter medium are frequently larger than the particles, which penetrate some distance into the medium before being trapped. The layer of solids that builds up on the surface of the medium is called the cake, and it is the cake that provides the actual filtration. The pressure–flow characteristics of the porous cake primarily determine the performance of the filter.

A. Governing Equations

A schematic of the flow through the cake and filter medium is shown in Fig. 13-6. The slurry flow rate is Q, and the total volume of filtrate that passes through the filter is $\tilde{V}$. The flow through the cake and filter medium is inevitably laminar, so the resistance can be described by Darcy's law and the permeability of the medium (K):

$$\frac{-\Delta P}{L} = \frac{Q\mu}{KA} \tag{13-29}$$

Applying this relationship across both the cake and the filter medium in series gives

$$P_1 - P_2 = \left(\frac{L}{K}\right)_{\text{cake}} \left(\frac{Q\mu}{A}\right) \tag{13-30}$$

and

$$P_2 - P_3 = \left(\frac{L}{K}\right)_{\text{FM}} \left(\frac{Q\mu}{A}\right) \tag{13-31}$$

TABLE 13-1 Design Data for Various Column Packings

	Size		Wall thickness		Number		Bed density		Contact surface S_B		Free space %	Packing factor	
	in.	mm	in.	mm	ft^{-3}	m^{-3}	lb/ft^3	kg/m^3	ft^3/ft^3	m^3/m^3	(100 ε)	ft^3/ft^3	m^3/m^3
Ceramic Räschig rings	0.25	6.35	0.03	0.76	85,600	2,022,600	60	961	242	794	62	1600	5250
	0.38	9.65	0.05	1.27	24,700	972,175	61	977	157	575	67	1000	3280
	0.50	12.7	0.07	1.78	10,700	377,825	55	881	112	368	64	640	2100
	0.75	19.05	0.09	2.29	3090	109,110	50	801	73	240	72	255	840
	1.0	25.4	0.14	3.56	1350	47,670	42	673	58	190	71	160	525
	1.25	31.75			670	23,660	46	737			71	125	410
	1.5	38.1			387	13,665	43	689			73	95	310
	2.0	50.8	0.25	6.35	164	5790	41	657	29	95	74	65	210
	3.0	76.2			50	1765	35	561			78	36	120
Metal Räschig rings	0.25	6.35	0.03125	0.794	88,000	3,107,345	133	2131			72	700	2300
	0.38	9.65	0.03125	0.794	27,000	953,390	94	1506			81	390	1280
	0.50	12.7	0.03125	0.794	11,400	402,540	75	1201	127	417	85	300	980
	0.75	19.05	0.03125	0.794	3340	117,940	52	833	84	276	89	185	605
(*Note*: Bed densities	0.75	19.05	0.0625	1.59	3140	110,875	94	1506			80	230	750
are for mild steel;	1.0	25.0	0.03125	0.794	1430	50,494	39	625	63	207	92	115	375
multiply by 1.105,	1.0	25.0	0.0625	1.59	1310	46,260	71	1137			86	137	450
1.12, 1.37, 1.115 for	1.25	31.75	0.0625	1.59	725	25,600	62	993			87	110	360
stainless steel,	1.5	38.1	0.0625	1.59	400	14,124	49	785			90	83	270
copper, aluminum,	2.0	50.8	0.0625	1.59	168	5932	37	593	31	102	92	57	190
and monel,	3.0	76.2	0.0625	1.59	51	1800	25	400	22	72	95	32	105
respectively)													
Carbon Räschig rings	0.25	6.35	0.0625	1.59	85,000	3,001,410	46	737	212	696	55	1600	5250
	0.50	12.7	0.0625	1.59	10,600	374,290	27	433	114	374	74	410	1350
	0.75	19.05	0.125	3.175	3140	110,875	34	545	75	246	67	280	920
	1.0	25.0	0.125	3.175	1325	46,787	27	433	57	187	74	160	525
	1.25	31.75			678	23,940	31	496			69	125	410
	1.5	38.1			392	13,842	34	545			67	130	425
	2.0	50.8	0.250	6.35	166	5862	27	433	29	95	74	65	210
	3.0	76.2	0.312	7.92	49	1730	23	368	19	62	78	36	120
Metal Pall rings	0.625	15.9	0.018	0.46	5950	210,098	37	593	104	341	93	70	230
(*Note*: Bed densities	1.0	25.4	0.024	0.61	1400	49,435	30	481	64	210	94	48	160
are for mild steel)	1.25	31.75	0.030	0.76	375	13,240	24	385	39	128	95	28	92
	2.0	50.8	0.036	0.915	170	6003	22	353	31	102	96	20	66

	3.5	76.2	0.048	1.219	33	1165	17	273	20	65.6	97	16	52
Plastic Pall Rings	0.625	15.9	0.03	0.762	6050	213,630	7.0	112	104	341	87	97	320
(*Note*: Bed densities	1.0	25.4	0.04	1.016	1440	50,848	5.5	88	63	207	90	52	170
are for polypropylene)	1.5	38.1	0.04	1.016	390	13,770	4.75	76	39	128	91	40	130
	2.0	50.8	0.06	1.524	180	6356	4.25	68	31	102	92	25	82
	3.5	88.9	0.06	1.524	33	1165	4.0	64	26	85	92	16	52
Ceramic Intalox	0.25	6.35			117,500	4,149,010	54	865			65	725	2400
saddles	0.38	9.65			49,800	1,758,475	50	801			67	330	1080
	0.50	12.7			18,300	646,186	46	737			71	200	660
	0.75	19.05			5640	199,150	44	705			73	145	475
	1.0	25.4			2150	75,918	42	673			73	92	300
	1.5	38.1			675	23,835	39	625	59	194	76	52	170
	2.0	50.8			250	8828	38	609			76	40	130
	3.0	76.2			52	1836	36	577			79	22	72
Plastic Super Intalox	No. 1				1620	57,200	6.0	96	63	207	90	33	108
	No. 2				190	6710	3.75	60	33	108	93	21	69
	No. 3				42	1483	3.25	52	27	88.6	94	16	52
Intalox metal		25		4770		168,425					96.7	41	135
		40		1420		50,140					97.3	25	82
		50		416		14,685					97.8	16	52
		70		131		4625					98.1	13	43
Hy-Pak	No. 1				850	30,014	19	304			96	43	140
(*Note*: Bed densities	No. 1				107	3778	14	224			97	18	59
are for mild steel)	No. 3				31	1095	13	208			97	15	49
Plastic Cascade Mini	No. 1											25	82
Rings	No. 2											15	49
	No. 3											12	39
Metal Cascade Mini	No. 0											55	180
Rings	No. 1											34	110
	No. 2											22	72
	No. 3											14	46
	No. 4											10	33
Ceramic Cascade Mini	No. 2											38	125
Rings	No. 3											24	79
	No. 5											18	59

Note: The packing factor F replaces the term S_B/ε^3. Use of the given value of F in Fig. 13-5 permits more predictable performance of designs incorporating packed beds, because the values quoted are derived from operating characteristics of the packings rather than from their physical dimensions.

Source: Coulson et al. (1991).

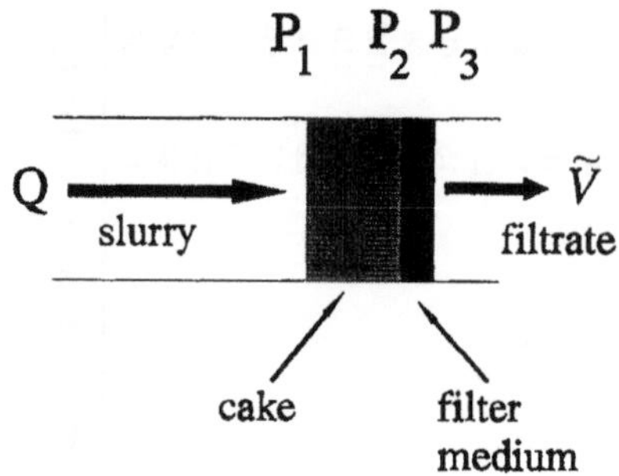

FIGURE 13-6 Schematic of flow through filter cake and medium.

The total pressure drop across the filter is the sum of these:

$$P_1 - P_3 = \frac{Q\mu}{A}\left[\left(\frac{L}{K}\right)_{\text{cake}} + \left(\frac{L}{K}\right)_{\text{FM}}\right] \tag{13-32}$$

The term $(L/K)_{\text{cake}}$ is the resistance of the cake, and $(L/K)_{\text{FM}}$ is the resistance of the filter medium. The latter is higher for a "dirty" filter medium than for a clean one, but once the initial particles become embedded in the medium and the cake starts to build up, it remains constant. The cake resistance, on the other hand, continues to increase with time as the cake thickness increases. The cake thickness is directly proportional to the volume of solids that have been deposited from the slurry and inversely proportional to the area:

$$L_{\text{cake}} = \frac{\tilde{V}_{\text{cake}}}{A} = \frac{\tilde{V}_{\text{solids}}}{A(1-\varepsilon)} = \frac{M_{\text{solids}}}{A\rho_s(1-\varepsilon)} \tag{13-33}$$

Now $M_{\text{solids}}/\tilde{V}$ is the mass of solids per unit volume of liquid in the slurry feed (e.g., the "solids loading" of the slurry), and $\tilde{V}$ is the volume of liquid (filtrate) that has passed to deposit M_{solids} on the cake. Thus, the cake thickness can be expressed as

$$L_{\text{cake}} = \left(\frac{M_{\text{solids}}}{\tilde{V}}\right)\left(\frac{\tilde{V}}{A}\right)\frac{1}{\rho_s(1-\varepsilon)} = W\frac{\tilde{V}}{A} \tag{13-34}$$

where $W = (M_{\text{solids}}/\tilde{V})/\rho_s(1-\varepsilon)$ is a property of the specific slurry or cake. The density of the cake is given by

$$\rho_c = (1-\varepsilon)\rho_s + \varepsilon\rho_{\text{liq}} \tag{13-35}$$

Substituting Eq. (13-34) into Eq. (13-32) and rearranging results in the basic equation governing the filter performance:

$$\frac{Q}{A} = \frac{1}{A}\frac{d\tilde{V}}{dt} = \frac{P_1 - P_3}{\mu(\tilde{V}W/AK + a)} \tag{13-36}$$

where a is the filter medium resistance, i.e., $a = (L/K)_{FM}$.

It should be recognized that the operation of a filter is an unsteady cyclic process. As the cake builds up and its resistance increases with time, either the flow rate (Q) will drop or the pressure drop (ΔP) will increase with time. The specific behavior depends on how the filter is operated, as follows.

B. Constant Pressure Operation

If the slurry is fed to the filter by a centrifugal pump that delivers (approximately) a constant head, or if the filter is operated by a controlled vacuum, the pressure drop will remain essentially constant during operation and the flow rate will drop as the cake thickness (resistance) increases. In this case, Eq. (13-36) can be integrated for constant pressure to give

$$C_1\left(\frac{\tilde{V}}{A}\right)^2 + C_2\left(\frac{\tilde{V}}{A}\right) = (-\Delta P)t \tag{13-37}$$

where $C_1 = \mu W/2K$ and $C_2 = \mu a$, both being assumed to be independent of pressure (we will consider compressible cakes later). In Eq. (13-37), t is the time required to pass volume $\tilde{V}$ of filtrate through the filter.

Since C_1 and C_2 are unique properties of a specific slurry–cake system, it is usually more appropriate to determine their values from laboratory tests using samples of the specific slurry and filter medium that are to be evaluated in the plant. For this purpose, it is more convenient to rearrange Eq. (13-37) in the form

$$\left(\frac{-\Delta P\, t}{\tilde{V}/A}\right) = C_1\left(\frac{\tilde{V}}{A}\right) + C_2 \tag{13-38}$$

If $\tilde{V}$ is measured as a function of t in a lab experiment for given values of ΔP and A, the data can be arranged in the form of Eq. (13-38). When the left hand side is plotted versus $\tilde{V}/A$, the result should be a straight line with slope C_1 and intercept C_2 (which are easily determined by linear regression).

C. Constant Flow Operation

If the slurry is fed to the filter by a positive displacement pump, the flow rate will be constant regardless of the pressure drop, which will increase with time. In this case, noting that $\tilde{V} = Qt$, Eq. (13-36) can be rearranged to give

$$-\Delta P = 2C_1\left(\frac{Q}{A}\right)^2 t + C_2\frac{Q}{A} \tag{13-39}$$

This shows that for given Q and A, the plot of ΔP versus t should be straight and the system constants C_1 and C_2 can be determined from the slope $2C_1 (Q/A)^2$ and intercept $C_2(Q/A)$.

It is evident that the filter performance is governed by the system constants C_1 and C_2 regardless of whether the operation is at constant pressure or constant flow rate and that these constants can be evaluated from laboratory data taken under either type of operation and used to analyze the performance of the plant filter for either type of operation.

D. Cycle Time

As mentioned earlier, the operation of a filter is cyclic. The filtration process proceeds (and the pressure increases or the flow rate drops) until either the cake has built up to fill the space available for it or the pressure drop reaches the operational limit. At that point, the filtration must cease and the cake must be removed. There is often a wash cycle prior to removal of the cake in order to remove the slurry carrier liquid from the pores of the cake using a clean liquid. The pressure–flow behavior during the wash period is a steady-state operation, controlled by the maximum cake and filter medium resistance, because no solids are deposited during this period. The cake can be removed by physically disassembling the filter, removing the cake and the filter medium (as for a plate-and-frame filter), then reassembling the filter and starting the cycle over. Or, in the case of a rotary drum filter, the cake removal is part of the rotating drum cycle, which is continuous although the filtration operation is still cyclic (this will be discussed below).

The variable t in the foregoing equations is the actual time (t_{filter}) that is required to pass a volume $\tilde{V}$ of filtrate through the medium and is only part of the total time of the cycle (t_{cycle}). The rest of the cycle, which may include wash time, disassembly and assembly time, cleaning time, etc., we shall call "dead" time (t_{dead}):

$$t_{\text{cycle}} = t_{\text{filter}} + t_{\text{dead}} \tag{13-40}$$

The net (average) filter capacity is determined by the amount of slurry processed during the total cycle time, not just the "filter" time, and represents the average flow rate ($\bar{Q}$):

$$\bar{Q} = \left(\frac{\tilde{V}_{\text{cycle}}}{t_{\text{cycle}}} \right)_{\text{const}\,\Delta P} = \left(\frac{Q t_{\text{filter}}}{t_{\text{cycle}}} \right)_{\text{const}\,Q} \tag{13-41}$$

E. Plate-and-Frame Filters

A plate-and-frame filter press consists of alternate solid plates and hollow frames in a "sandwich" arrangement. The open frames are covered by the filter medium (e.g., the filter cloth), and the slurry enters through the frames and deposits the cake on the filter medium. The operation is "batch," in that the filter must be disassembled when the cake fills the frame space, then cleaned and reassembled, after which the entire process is repeated. A schematic of a plate-and-frame press is shown in Fig. 13-7. In the arrangement shown, all of the frames are in parallel and the total filter area (which appears in the equations) is

$$A = 2nA_{\text{f}} \tag{13-42}$$

where n is the number of frames and A_{f} is the filter area of (one side) of the frame. The flow rate Q in the equations is the total flow rate, and $Q/A = Q/2nA_{\text{f}}$ is the total flow per unit total filtering area, or the flow rate per filter side per unit area of the filter side.

There are a variety of arrangements that operate in the same manner as the plate- and-frame filter. One is the "leaf filter," which may consist of one or more "frames" that are covered by the filter medium and immersed in the slurry. These filtration devices, are often operated by means of a vacuum that draws the filtrate through the filter, with the cake collecting on the filter medium on the outside of the frame.

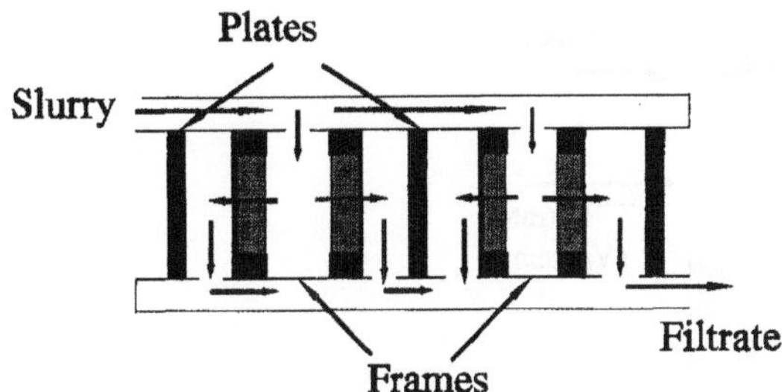

FIGURE 13-7 Plate-and-frame filter.

F. Rotary Drum Filter

The rotary drum filter is a "continuous" filtration device, because it does not have to be shut down during the cycle, although the operation is still cyclic. A schematic is shown in Fig. 13-8. The drum rotates at a rate N (rpm), and the filter area is the total drum surface, i.e., $A = \pi DL$. However, if the fraction of the drum that is in contact with the slurry is f, then the length of time in the cycle during which any one point on the surface is actually filtering is f/N:

$$t_{\text{cycle}} = \frac{1}{N}, \qquad t_{\text{filt}} = \frac{f}{N} \tag{13-43}$$

G. Compressible Cake

The equations presented so far all assume that the cake is incompressible, i.e., that the permeability and density of the cake are constant. For many cakes this is not so, because the cake properties may vary with pressure (flocs, gels, fibers, pulp, etc.). For such cases, the basic filter equation [Eq. (13-36)] can be expressed in the form

$$\frac{Q}{A} = \frac{(-\Delta P)^{1-s}}{\mu[\alpha(\tilde{V}/A)(M_s/\tilde{V}) + a]} = \frac{1}{A}\frac{d\tilde{V}}{dt} \tag{13-44}$$

where the pressure dependence is characterized by the parameter s, and α and a are the pressure-independent properties of the cake. There are several modes of performance of the filter, depending on the value of s:

1. If $s = 0$, then $Q \sim \Delta P$ (the cake is incompressible).
2. If $s < 1$, then Q increases as ΔP increases (slightly compressible).
3. If $s = 1$, then Q is independent of ΔP (compressible).
4. If $s > 1$, then Q decreases as ΔP increases (highly compressible).

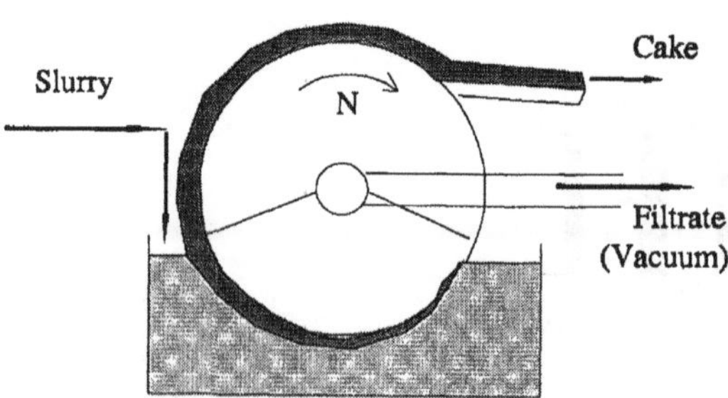

Figure 13-8 Rotary drum filter.

In case 4, the increasing pressure compresses the cake to such as extent that it actually "squeezes off" the flow so that as the pressure increases the flow rate decreases. This situation can be compensated for by adding a "filter aid" to the slurry. This is a rigid dispersed solid that forms an incompressible cake (diatomaceous earth, sand, etc.). This provides "rigidity" to the cake and enhances its permeability, thus increasing the filter capacity (it may seem like a paradox that adding more solids to the slurry feed actually increases the filter performance, but it works!).

The equations that apply for a compressible cake are as follows.

Constant pressure drop:

$$C_1\left(\frac{\tilde{V}}{A}\right)^2 + C_2\left(\frac{\tilde{V}}{A}\right) = t(-\Delta P)^{1-s} \tag{13-45}$$

Constant flow rate:

$$2C_1\left(\frac{Q}{A}\right)^2 t + C_2\left(\frac{Q}{A}\right) = (-\Delta P)^{1-s} \tag{13-46}$$

where $C_1 = (\mu\alpha/2)(M_s/\tilde{V})$ and $C_2 = \mu a$. There are now three parameters that must be determined empirically from laboratory measurements: C_1, C_2, and s. The easiest way to do this would be to use the constant pressure mode in the laboratory (e.g., a Buchner funnel, with a set vacuum pressure difference) and obtain several sets of data for $\tilde{V}$ as a function of t, with each set at a different value of ΔP. For each data set, the plot of $t/\tilde{V}$ vs. $\tilde{V}$ should yield a straight line with a slope of $C_1/A^2(-\Delta P)^{1-s}$ and intercept of $C_2/A(-\Delta P)^{1-s}$. Thus, a log-log plot of either the slope or the intercept versus ΔP should have a slope of $s-1$, which determines s.

PROBLEMS

Porous Media

1. A packed bed is composed of crushed rock with a density of 175 lb_m/ft^3 of such a size and shape that the average ratio of surface area to volume for the particles is 50 $in.^2/in.^3$. The bed is 6 ft deep, has a porosity of 0.3, and is covered by a 2 ft deep layer of water that drains by gravity through the bed. Calculate the flow rate of water through the bed in gpm/ft^2, assuming it exits at 1 atm pressure.
2. An impurity in a water stream at a very small concentration is to be removed in a charcoal trickle bed filter. The filter is in a cylindrical column that is 2 ft in diameter, and the bed is 4 ft deep. The water is kept at a level that is 2 ft above the top of the bed, and it trickles through by gravity flow. If the charcoal particles have a geometric surface area to volume ratio of 48 $in.^{-1}$ and they

pack with a porosity of 0.45, what is the flow rate of water through the column, in gpm?

3. A trickle bed filter is composed of a packed bed of broken rock. The shape of the rock is such that the average ratio of the surface area to volume for the rock particles is 30 in.^{-1} The bed is 2 ft deep, has a porosity of 0.3, and is covered by a layer of water that is 2 ft deep and drains by gravity through the bed.
   (a) Determine the volume flow rate of the water through the bed per unit bed area (in gpm/ft^2).
   (b) If the water is pumped upward through the bed (e.g. to flush it out), calculate the flow rate (in gpm/ft^2 of bed area) that will be required to fluidize the bed.
   (c) Calculate the corresponding flow rate that would sweep the rock particles away with the water. The rock density is 120 $\text{lb}_\text{m}/\text{ft}^3$.

Packed Columns

4. A packed column that is 3 ft in diameter with a packing height of 25 ft is used to absorb an impurity from a methane gas stream using an amine solution absorbent. The gas flow rate is 2000 scfm, and the liquid has a density of 1.2 g/cm3 and a viscosity of 2 cP. If the column operates at 1 atm and 80°F, determine the liquid flow rate at which flooding would occur in the column and the pressure drop at 50% of the flooding liquid rate for the following packings:
   (a) 2 in. ceramic Raschig rings
   (b) 2 in. plastic Pall rings
5. A packed column is used to scrub SO_2 from air by using water. The gas flow rate is 500 scfm/ft^2, and the column operates at 90°F and 1 atm. If the column contains No. 1 plastic Intalox packing, what is the maximum liquid flow rate (per unit cross section of column) that could be used without flooding?
6. A stripping column packed with 2 in. metal Pall rings uses air at 5 psig and 80°C to strip an impurity from an absorber oil ($SG = 0.9$, viscosity $= 5$ cP, $T = 20$°C). If the flow rate of the oil is 500 lb_m/min and that of the air is 20 lb_m/min,
   (a) What is the minimum column diameter that can be used without flooding?
   (b) If the column diameter is 50% greater than the minimum size, what is the pressure drop per ft of column height?
7. A packed column that is 0.6 m in diameter and 4 m high and contains 25 mm Raschig rings is used in a gas absorption process to remove an inpurity from the gas stream by absorbing it in a liquid solvent. The liquid, which has a viscosity of 5 cP and $SG = 1.1$, enters the top of the column at a rate of 2.5 kg/(s m^2), and the gas, which can be assumed to have the same properties as air, enters the bottom of the column at a rate of 0.6 kg/(s m^2). The column operates at atmospheric pressure and 25°C. Determine:
   (a) The pressure drop through the column, in inches of water.
   (b) How high the liquid rate could be increased before the column would flood.
8. A packed column is used to absorb SO_2 from flue gas using an ethanolamine solution. The column is 4 ft in diameter, has a packed height of 20 ft, and is

packed with 2 in. plastic Pall rings. The flue gas is at a temperature of 180°F and has an average molecular weight of 31. The amine solution has a specific gravity of 1.02 and a viscosity at the operating temperature of 1.5 cP. If the gas must leave the column at 25 psig and a flow rate of 10,000 scfm, determine:

(a) The maximum allowable flow rate of the liquid (in gpm) that would result in a pressure drop that is 50% of that at which flooding would occur.
(b) The horsepower that would be required for the blower to move the gas through the column if the blower is 80% efficient.

9. A packed absorption tower is used to remove SO_2 from an air stream by absorption in a solvent. The tower is 5 ft in diameter and 60 ft high and contains 1.5 in. plastic Pall rings. The temperature and pressure in the tower are 90°F and 30 psig. The gas stream flow rate is 6500 scfm. The liquids SG is 1.25, and its viscosity is 25 cP.
   (a) What is the liquid flow rate (in gpm) at which the column will flood?
   (b) If the column operates at a liquid flow rate that is 75% of the flooding value, what is the total pressure drop through the tower in psi?
10. A packed absorption column removes an impurity from a gas stream by contact with a liquid solvent. The column is 3 ft in diameter and contains 25 ft of No. 2 plastic Super Intalox packing. The gas has an MW of 28, enters the column at 120°F, and leaves at 10 psig at a rate of 5000 scfm. The liquid has an SG of 1.15 and a viscosity of 0.8 cP. Determine:
    (a) The flow rate of the liquid in gpm that would be 50% of the flow rate at which the column would flood.
    (b) The pressure drop through the column, in psi.
    (c) The horsepower of the blower required to move the gas through the column if it is 60% efficient.

Filtration

11. A fine aqueous suspension containing 1 lb_m of solids per cubic foot of suspension is to be filtered in a constant pressure filter. It is desired to filter at an average rate of 100 gpm, and the filter cake must be removed when it gets 2 in. thick. What filter area is required? Data: $-\Delta P = 10$ psi, ρ(wet cake) = 85 lb_m/ft^3, K (permeability) = 0.118 Darcy, $a = 2 \times 10^9 ft^{-1}$.
12. An aqueous slurry containing 1.5 lb_m of solid per gallon of liquid is pumped through a filter cloth by a centrifugal pump. If the pump provides a constant pressure drop of 150 psig, how long will it take for the filter cake to build up to a thickness of 2 in.? The density of the filter cake is 30 lb_m/ft^3, and its permeability is 0.01 Darcy.
13. A packed bed that consists of the same medium as that in Problem 3 is to be used to filter solids from an aqueous slurry. To determine the filter properties, you test a small section of the bed, which is 6 in. in diameter and 6 in. deep, in the lab. When the slurry is pumped through this test model at a constant flow rate of 30 gpm, the pressure drop across the bed rises to 2 psia in 10 min. How long will it take to filter 100,000 gal of water from the slurry in a full-sized bed

that is 10 ft in diameter and 2 ft deep, if the slurry is maintained at a depth of 2 ft over the bed and drains by gravity through the bed?

14. A slurry containing 1 lb_m of solids per gallon of water is to be filtered in a plate-and-frame filter with a total filtering area of 60 ft^2. The slurry is fed to the filter by a centrifugal pump that develops a head of 20 psig. How long would it take to build up a layer of filter cake 4 in. thick on the filter medium? Laboratory data were taken on the slurry using a positive displacement pump operating at 5 gpm and 1 ft^2 of filter medium. It was found that the pressure drop increased linearly with time from an initial value of 0.2 psi to a value of 50 psi after 1 min. The density of the dry filter cake was found to be 0.85 g/cm^3.
15. A rotary drum filter 6 ft in diameter and 8 ft long is to be used to filter a slurry. The drum rotates at 0.5 rpm, and one-third of the drum's surface is submerged in the slurry. A vacuum is drawn in the drum so that a constant pressure drop of 10 psi is maintained across the drum and filter cake. You test the slurry in the lab by pumping it at a constant filtrate rate of 20 gpm through 1 ft^2 of the drum filter screen and find that after 1 min the pressure drop is 8 psi and after 3 min the pressure drop is 12 psi. How long will it take to filter 100,000 gal of filtrate from the slurry using the rotary drum?
16. A plate-and-frame filter press contains 16 frames and operates at a constant flow rate of 30 gpm. Each frame has an active filtering area of 4 ft^2, and it takes 15 min to disassemble, clean, and reassemble the press. The press must be shut down for disassembly when the pressure difference builds up to 10 psi. What is the total net filtration rate in gpm for a slurry having properties determined by the following lab test. A sample of the slurry is pumped at a constant pressure differential of 5 psi through 0.25 ft^2 of the filter medium. After 3 min, 1 gal of filtrate has been collected. The resistance of the filter medium may be neglected.
17. A rotary drum filter is used to filter a slurry. The drum rotates at a rate of 3 min/cycle, and 40% of the drum surface is submerged in the slurry. A constant pressure drop at 3 psi is maintained across the filter. If the drum is 5 ft in diameter and 10 ft long, calculate the total net filtration rate in gpm that is possible for a slurry having properties as determined by the following lab test. A sample of the slurry was pumped at a constant flow rate of 1 gpm through 0.25 ft^2 of the filter medium. After 10 min, the pressure difference across the filter had risen to 2.5 psi. The filter medium resistance may be neglected.
18. You must filter 1000 lb_m/min of an aqueous slurry containing 40% solids by weight by using a rotary drum filter, diameter 4 m and length 4 m, that operates at a vacuum of 25 in.Hg with 30% of its surface submerged in the slurry. A lab test is run on a sample of the slurry using 200 cm^2 of the same filter medium and a vacuum of 25 in.Hg. During the first minute of operation, 300 cm^3 of filtrate is collected, and during the second minute an additional 140 cm^3 is collected.
    (a) How fast should the drum be rotated?
    (b) If the drum is rotated at 2 rpm, what would the filter capacity be in pounds of slurry filtered per minute?
19. A rotary drum filter is to be used to filter a lime slurry. The drum rotates at a rate of 0.2 rpm, and 30% of the drum surface is submerged in the slurry. The

filter operates at a constant ΔP of 10 psi. The slurry properties were determined from a lab test at a constant flow rate of 0.5 gpm using 1/2 ft^2 of the filter medium. The test results indicated that the pressure drop rose to 2 psi in 10 s and to 10 psi in 60 s. Calculate the net filtration rate per unit area of the drum under these conditions, in gpm/ft^2.

20. A plate-and-frame filter press operating at a constant ΔP of 150 psi is to be used to filter a sludge containing 2 lb_m of solids per ft^3 of water. The filter must be disassembled and cleaned when the cake thickness builds up to 1 in. The frames have a projected area of 4 ft^2, and the downtime for cleaning is 10 min/frame. The properties of the sludge and cake were determined in a lab test operating at a constant flow rate of 0.2 gpm of filtrate, with a filter area of 1/4 ft^2. The test results show that the pressure drop rises to 3 psi in 20 s and to 8 psi in 60 s. Calculate the overall net filtration rate per frame in the filter, in gpm of filtrate, accounting for the down time. The density of the cake was found to be 150 lb_m/ft^3.
21. A packed bed composed of crushed rock having a density of 175 lb_m/ft^3 is to be used as a filter. The size and shape of the rock particles are such that the average surface area to volume ratio is 50 $in.^2/in.^3$, and the bed porosity is 0.3. A lab test using the slurry to be filtered is run on a bed of the same particles that is 6 in. deep and 6 in. in diameter. The slurry is pumped through this bed at a constant filtrate rate of 10 gpm, and it is found that after 5 min the pressure drop is 5 psi, and after 10 min it is 8 psi. Calculate how long it would take to filter 100,000 gal of filtrate from the slurry in a full-scale bed that is 10 ft in diameter and 2 ft deep, if the slurry is maintained at a depth of 2 ft above the bed and drains through it by gravity. Assume that the slurry density is the same as water.
22. A rotary drum filter has a diameter of 6 ft and a length of 8 ft and rotates at a rate of 30 s/cycle. The filter operates at a vacuum of 500 mmHg, with 30% of its surface submerged. The slurry to be filtered is tested in the lab using 0.5 ft^2 of the drum filter medium in a filter funnel operating at 600 mmHg vacuum. After 5 min of operation, 250 cm^3 of filtrate has collected through the funnel, and after 10 min, a total of 400 cm^3 has collected.
    (a) What would be the net (average) filtration rate of this slurry in the rotary drum filter, in gpm?
    (b) How much could this filtration rate be increased by increasing the speed (i.e., rotation rate) of the drum?
23. A rotary drum filter, 10 ft in diameter and 8 ft long, is to be used to filter a slurry of incompressible solids. The drum rotates at 1.2 rpm, and 40% of its surface is submerged in the slurry at all times. A vacuum in the drum maintains a constant pressure drop of 10 psi across the drum and filter cake. The slurry is tested in the lab by pumping it at a constant rate of 5 gpm through 0.5 ft^2 of the drum filter screen. After 1 min, the pressure drop is 9 psi, and after 3 min it has risen to 15 psi. How long will it take to filter 1 million gal of filtrate from the slurry using the rotary drum? How long would it take if the drum rotated at 3 rpm?

24. A slurry is being filtered at a net rate of 10,000 gal/day by a plate and frame filter with 15 frames, with an active filtering area of 1.5 ft^2 per frame, fed by a positive displacement pump. The pressure drop varies from 2 psi at start-up to 25 psi after 10 min, at which time it is shut down for cleanup. It takes 10 min to disassemble, clean out, and reassemble the filter. Your boss decides that it would be more economical to replace this filter with a rotary drum filter using the same filter medium. The rotary filter operates at a vacuum of 200 mmHg with 30% of its surface submerged and rotates at a rate of 5 min/rev. If the drum length is equal to its diameter, how big should it be?
25. You want to select a rotary drum filter to filter a coal slurry at a rate of 100,000 gal of filtrate per day. The filter operates at a differential pressure of 12 psi, and 30% of the surface is submerged in the slurry at all times. A sample of the slurry is filtered in the lab through a 6 in diameter sample of the filter medium at a constant rate of 1 gpm. After 1 min the pressure drop across this filter is 3 psi, and after 5 min it is 10 psi. If the drum rotates at a rate of 3 rpm, what total filter area is required?
26. A slurry containing 40% solids by volume is delivered to a rotary drum filter that is 4 ft in diameter and 6 ft long and operates at a vacuum of 25 in.Hg. A lab test is run with a 50 cm^2 sample of the filter medium and the slurry, at a constant flow rate of 200 cm^3/min. After 1 min the pressure across the lab filter is 6 psi, and after 3 min it is 16 psi. If 40% of the rotary drum is submerged in the slurry, how fast should it be rotated (rpm) in order to filter the slurry at an average rate of 250 gpm?
27. A slurry is to be filtered with a rotary drum filter that is 5 ft in diameter and 8 ft long, rotates once every 10 s, and has 20% of its surface immersed in the slurry. The drum operates with a vacuum of 20 in.Hg. A lab test was run on a sample of the slurry using 1/4 ft^2 of the filter medium at a constant flow rate of 40 cm^3/s. After 20 s the pressure drop was 30 psi across the lab filter, and after 40 s it was 35 psi. How many gallons of filtrate can be filtered per day in the rotary drum?
28. A rotary drum filter is to be installed in your plant. You run a lab test on the slurry to be filtered using a 0.1 ft^2 sample of the filter medium at a constant pressure drop of 10 psi After 1 min you find that 500 cm^3 of filtrate has passed through the filter, and after 2 min the filtrate volume is 715 cm^3. If the rotary drum filter operates under a vacuum of 25 in.Hg with 25% of its surface submerged, determine:
    (a) The capacity of the rotary drum filter, in gallons of filtrate per square foot of surface area, if it operates at (1) 2 rpm; (2) 5 rpm.
    (b) If the drum has a diameter of 4 ft and a length of 6 ft, what is the total filter capacity in gal/day for each of the operating speeds of 2 and 5 rpm?
29. A slurry of $CaCO_3$ in water at 25°C containing 20% solids by weight is to be filtered in a plate-and-frame filter. The slurry and filter medium are tested in a constant pressure lab filter that has an area of 0.0439 m^2, at a pressure drop of 338 kPa. It is found that 10^{-3} m^3 of filtrate is collected after 9.5 s, and 5×10^{-3} m^3 is collected after 107.3 s. The plate and frame filter has 20 frames, with 0.873 m^2 of filter medium per frame, and operates at a constant flow rate of 0.00462 m^3 of slurry per second. The filter is operated until the pressure drop

reaches 500 kPa, at which time it is shut down for cleaning. The downtime is 15 min per cycle. Determine how much filtrate passes through the filter in each 24 hr period of operation (SG of $CaCO_3$ is 1.6).

30. An algal sludge is to be clarified by filtering. A lab test is run on the sludge using an area A of the filter medium. At a constant pressure drop of 40 kN/m^2, a plot of the time required to collect a volume $\tilde{V}$ of the filtrate times $\Delta P/(\tilde{V}/A)$ vs. $\tilde{V}/A$ gives a straight line with a slope of 1.2×10^6 kN s/m^4 and an intercept of 6.0×10^4 kN s/m^3. A repeat of the data at a pressure drop of 200 kN/m^2 also gave a straight line on the same type of plot, with the same intercept but with a slope of 2.1×10^6 kN s/m^4. When a filter aid was added to the sludge in an amount equal to 20% of the algae by weight, the lab test gave a straight line with the same intercept but with a slope of 1.4×10^6 kN s/m^4.
    (a) What does this tell you about the sludge?
    (b) The sludge is to be filtered using a rotary drum filter, with a diameter of 4 ft and a length of 6 ft, operating at a vacuum of 700 mmHg with 35% of the drum submerged. If the drum is rotated at a rate of 2 rpm, how many gallons of filtrate will be collected in a day, (1) with and (2) without the filter aid?
    (c) What would the answer to (b) be if the drum speed was 4 rpm?
31. A slurry containing 0.2 kg solids/kg water is filtered through a rotary drum filter operating at a pressure difference of 65 kN/m^2. The drum is 0.6 m in diameter and 0.6 m long, rotates once every 350 s, and has 20% of its surface submerged in the slurry.
    (a) If the overall average filtrate flow rate is 0.125 kg/s, the cake is incompressible with a porosity of 50%, and the solids SG = 3.0, determine the maximum thickness of the cake on the drum (you may neglect the filter medium resistance).
    (b) The filter breaks down, and you want to replace it with a plate-and-frame filter of the same overall capacity, which operates at a pressure difference of 275 kN/m^2. The frames are 10 cm thick, and the maximum cake thickness at which the filter will operate properly is 4 cm. It will take 100 s to disassemble the filter, 100 s to clean it out, and 100 s to reassemble it. If the frames are 0.3 m square, how many frames should the filter contain?
32. You want to filter an aqueous slurry using a rotary drum filter, at a total rate (of filtrate) of 10,000 gal/day. The drum rotates at a rate of 0.2 rpm, with 25% of the drum surface submerged in the slurry, at a vacuum of 10 psi. The properties of the slurry are determined from a lab test using a Buchner funnel under a vacuum of 500 mmHg, using a 100 cm^2 sample of the filter medium and the slurry, which resulted in the lab data given below. Determine the total filter area of the rotary drum required for this job.

Time (s)	Volume of filtrate (cm^3)
50	10
100	18
200	31
400	51

33. You want to use a plate and frame filter to filter an aqueous slurry at a rate of 1.8 m^3 per 8 hr day. The filter frames are square, with a length on each side of 0.45 m. The "down time" for the filter press is 300 s plus an additional 100 s per frame for cleaning. The filter operates with a positive displacement pump, and the maximum operation pressure differential for the filter is 45 psi, which is reached after 200 s of operation.
   (a) How many frames must be used in this filter to achieve the required capacity?
   (b) At what flow rate (in gpm) should the pump be operated?

   The following lab data were taken with the slurry at a constant ΔP of 10 psi and a 0.05 m^2 sample of the filter medium:

   After 300 s, the total volume of filtrate was 400 cm^3.
   After 900 s, the total volume of filtrate was 800 cm^3.
34. An aqueous slurry is filtered in a plate-and-frame filter that operates at a constant ΔP of 100 psi. The filter contains 20 frames, each of which has a projected area per side of 900 cm^2. A total filtrate volume of 0.7 m^3 is passed through the filter during a filtration time of 1200 s, and the down time for the filter is 900 s. The resistance of the filter medium is negligible relative to that of the cake. You want to replace the plate and frame filter with a rotary drum filter with the same overall average capacity, using the same filter medium. The drum is 2.2 m in diameter and 1.5 m long and operates at 5 psi vacuum with 25% of the drum surface submerged in the slurry. At what speed, in rpm, should the drum be operated?
35. You must transport a sludge product from an open storage tank to a separations unit at 1 atm, through a 4 in. sch 40 steel pipeline that is 2000 ft long, at a rate of 250 gpm. The sludge is 30% solids by weight in water and has a viscosity of 50 cP and Newtonian properties. The solid particles in the sludge have a density of 3.5 g/cm^3. The pipeline contains four gate valves and six elbows.
   (a) Determine the pump head (in ft) required to do this job. You can select any pump with the characteristics given in Appendix H, and you must find the combination of motor speed, motor horsepower, and impeller diameter that should be used.
   (b) You want to install a long radius venturi meter in the line to monitor the flow rate, and you want the maximum pressure drop to be measured to be equal to or less than 40 in. H_2O. What should the diameter of the venturi throat be?
   (c) At the separations unit, the sludge is fed to a settling tank. The solids settle in the tank, and the water overflows the top. What should the diameter of the tank be if it is desired to limit the size of the particles in the overflow to 100 μm or less?
   (d) If the sludge is fed to a centrifuge instead of the settling tank, at what speed (rpm) should the centrifuge operate to achieve the same separation as the settling tank, if the centrifuge dimensions are $L = D = 1$ ft, $R_1 = R_2/2 = 0.25$ ft?

(e) Suppose the sludge is fed instead to a rotary drum filter that removes all of the solids from the stream. The drum operates at a vacuum of 6 psi, has dimensions $L = D = 4$ ft, and operates with 30% of its surface submerged. A lab test is performed on the sludge using 1 ft^2 of the same filter medium as on the drum, operating at a vacuum of 500 mmHg. In this test, it is found that 8 gal passes the filter in 2 min, and a total of 20 gal passes through in 10 min. At what speed (rpm) should the rotary drum filter be operated?

36. Consider a dilute aqueous slurry containing solid particles with diameters of 0.1–1000 μm and a density of 2.7 g/cm^3, flowing at a rate of 500 gpm.
   (a) If the stream is fed to a settling tank in which all particles with diameter greater than 100 μm are to be removed, what should the tank diameter be?
   (b) The overflow from the settling tank contains almost all of the water plus the fines not removed from the tank. This stream is fed to a centrifuge that has a diameter of 20 in., a length of 18 in., and an overflow dam that is 6 in. from the centerline. At what speed, in rpm, should the centrifuge rotate in order to separate all particles with diameters of 1 μm and larger?
   (c) If the centrifuge rotates at 2500 rpm, what size particles will be removed?
   (d) Instead of the tank and centrifuge, the slurry is fed to a rotary drum filter that has a diameter of 5 ft and a length of 10 ft. The drum operates under a vacuum of 10 in.Hg, with 35% of its surface submerged in the slurry. A lab test is run on the slurry at a constant flow rate of 100 cm^3/min, using 50 cm^2 of the filter medium. In the test filter, the pressure drop reaches 10 mmHg in 1 min and 80 mmHg in 10 min. How fast should the drum rotate (in rpm) to handle the slurry stream?

NOTATION

A	area, [L^2]
a	filter medium resistance, [1/L]
a_s	particle surface area/per unit volume, [1/L]
C_1	filter parameter ($= \mu W/2K$), [M/L^3 t]
C_2	filter parameter ($= \mu a$), [M/L^2 t]
D	diameter, [L]
d	particle diameter, [L]
D_h	hydraulic diameter, [L]
e_f	energy dissipated per unit mass of fluid, [FL/M = L^2/t^2]
G	gas mass flux, [M/L^{2}t]
K	permeability, [L^2]
f_{PM}	porous media friction factor, Eq. (13-11), [—]
L	length, [L], liquid mass flux [M/L^{2}t]
M_{solids}	mass of solids [M]
N	rotation rate, rpm, [1/t]
n	number of frames, [—]
$N_{Re,PM}$	porous media Reynolds number, Eq. (13-13), [—]
P	pressure, [F/L^2 = M/Lt^2]

Q	volumetric flow rate, $[L^3/t]$
s	compressibility parameter, Eq. (13-44), [—]
t	time, [t]
V	velocity, [L/t]
$\tilde{V}$	volume of filtrate, $[L^3]$
W	slurry or cake solids loading parameter $[= (M_{solids}/\tilde{V})/\rho_s(1-\varepsilon)]$, [—]
W_p	wetted perimeter, [L]
x, y, z	coordinate directions, [L]
$\Delta(\ )$	$(\ )_2 - (\ )_1$
ε	porosity or void fraction, [—]
Φ	potential $(= P + \rho g z)$, $[F/L^2 = M/Lt^2]$
μ	viscosity, [M/Lt]
ρ	density, $[M/L^3]$
ψ	sphericity factor, [—]

Subscripts

1,2,3	reference points
f	filter frame side
i	interstitial
s	superficial

REFERENCES

Branan CR. Rules of Thumb for Chemical Engineers. Houston, TX: Gulf Pub Co., 1994.

Coulson JM, JF Richardson, JR Blackhurst, JH Harker. Chemical Engineering, Vol. 2. 4th ed. New York: Pergamon Press, 1991.

Leva M. Reconsider packed tower pressure drop correlations. Chem Eng Prog 88: 65–72, 1992.

Cheremisinoff NP. Liquid Filtration. 2nd ed. New York: Butterworth-Heinemann, 1998.

14

Fluidization and Sedimentation

I. FLUIDIZATION

When a fluid is passed upward through a bed of particles, as illustrated in Fig. 14-1, the pressure drop increases as the fluid velocity increases. The product of the pressure drop and the bed cross sectional area represents a net upward force on the bed, and when this force becomes equal to the weight of the bed (solids and fluid) the bed becomes suspended by the fluid. In this state the particles can move freely within the "bed," which thus behaves much like a boiling liquid. Under these conditions the bed is said to be "fluidized." This freely flowing or bubbling behavior results in a high degree of mixing in the bed, which provides a great advantage for heat or mass transfer efficiency compared with a fixed bed. Fluidized bed operations are found in refineries (i.e., fluid catalytic crackers), polymerization reactors, fluidized bed combustors, etc. If the fluid velocity within the bed is greater than the terminal velocity of the particles, however, the fluid will tend to entrain the particles and carry them out of the bed. If the superficial velocity above the bed (which is less than the interstitial velocity within the bed) is less than the terminal velocity of the particles, they will fall back and remain in the bed. Thus there is a specific range of velocity over which the bed remains in a fluidized state.

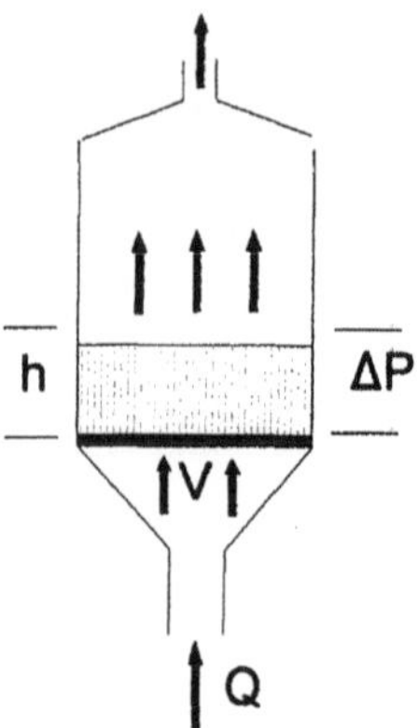

FIGURE 14-1 Fluidized bed.

A. Governing Equations

The Bernoulli equation relates the pressure drop across the bed to the fluid flow rate and the bed properties:

$$\frac{-\Delta P}{\rho_f} - gh = e_f = \frac{f_{PM} h V_s^2}{d}\left(\frac{1-\varepsilon}{\varepsilon^3}\right) \tag{14-1}$$

where the porous medium friction factor is given by the Ergun equation,

$$f_{PM} = 1.75 + \frac{180}{N_{Re,PM}} \tag{14-2}$$

and the porous medium Reynolds number is

$$N_{Re,PM} = \frac{dV_s\rho}{(1-\varepsilon)\mu} \tag{14-3}$$

Now the criterion for incipient fluidization is that the force due to the pressure drop must balance the weight of the bed, i.e.,

$$-\Delta P = \frac{\text{Bed wt.}}{A} = \rho_s(1-\varepsilon)gh + \rho\varepsilon gh \tag{14-4}$$

where the first term on the right is pressure due to the weight of the solids and the second is the weight of the fluid in the bed. When the pressure drop is eliminated from Eqs. (14-4) and (14-1), an equation for the "minimum fluidization velocity" (V_{mf}) results:

$$(\rho_s - \rho)(1-\varepsilon)g = \frac{\rho e_f}{h} = 1.75\,\frac{\rho V_{mf}^2(1-\varepsilon)}{d\varepsilon^3} + 180\,\frac{V_{mf}\mu(1-\varepsilon)^2}{d^2\varepsilon^3} \tag{14-5}$$

which can be written in dimensionless form, as follows:

$$N_{Ar} = 1.75\frac{\hat{N}_{Re}^2}{\varepsilon^3} + 180\left(\frac{1-\varepsilon}{\varepsilon^3}\right)\hat{N}_{Re} \tag{14-6}$$

where

$$N_{Ar} = \frac{\rho g\,\Delta\rho\, d^3}{\mu^2}, \qquad \hat{N}_{Re} = \frac{dV_{mf}\rho}{\mu} \tag{14-7}$$

Equation (14-6) can be solved for the Reynolds number to give

$$\hat{N}_{Re} = (C_1^2 + C_2 N_{Ar})^{1/2} - C_1 \tag{14-8}$$

where

$$C_1 = \frac{180(1-\varepsilon)}{3.5}, \qquad C_2 = \frac{\varepsilon^3}{1.75} \tag{14-9}$$

Equation (14-8) gives the (dimensionless) superficial velocity (V_{mf}) for incipient fluidization.

B. Minimum Bed Voidage

Before the bed can become fluidized, however, the particles must dislodge from their "packed" state, which expands the bed. Thus, the porosity (ε) in Eqs (14-5) and (14-9) is not the initial "packed bed" porosity but the "expanded bed" porosity at the point of minimum fluidization (ε_{mf}), i.e., the "minimum bed voidage" in the bed just prior to fluidization. Actually, the values of C_1 and C_2 in Eq. (14-8) that give the best results for fluidized beds of uniform spherical particles have been found from empirical observations to be:

$$C_1 = 27.2, \qquad C_2 = 0.0408 \tag{14-10}$$

By comparing these empirical values of C_1 and C_2 with Eqs. (14-9), the C_1 value of 27.2 is seen to be equivalent to $\varepsilon_{mf} = 0.471$ and the C_2 value of 0.0408 equivalent to $\varepsilon_{mf} = 0.415$. In actuality, the value of ε_{mf} may vary considerably with the nature of the solid particles, as shown in Fig. 14-2.

C. Nonspherical Particles

Many particles are not spherical and so will not have the same drag properties as spherical particles. The effective diameter for such particles is often characterized by the equivalent *Stokes diameter*, which is the diameter of the sphere that has the same terminal velocity as the particle. This can be determined from a direct measurement of the settling rate of the

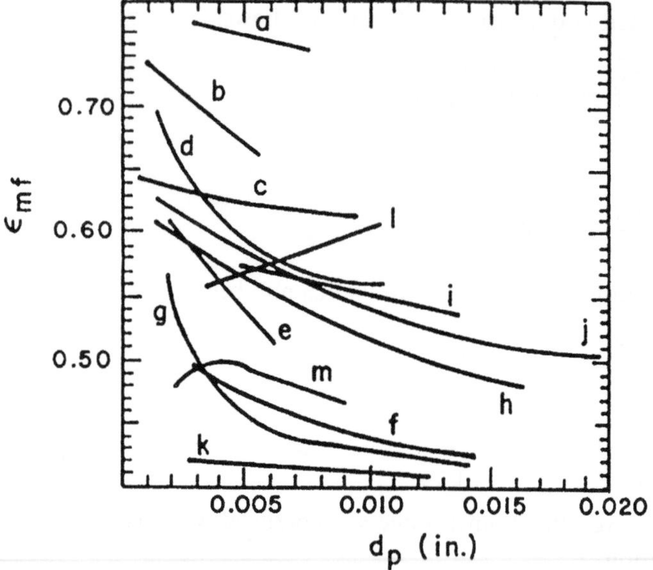

FIGURE 14-2 Values of ε_{mf} for various solids. (From Azbel and Cheremisinoff, 1983.)

particles and provides the best value of equivalent diameter for use in applications involving fluid drag on the particles.

An alternative description of nonspherical particles is often represented by the "sphericity factor" (ψ), which is the number that, when multiplied by the diameter of a sphere with the same volume as the particle (d_s), gives the particle effective diameter (d_p):

$$d_p = \psi d_s \tag{14-11}$$

The sphericity factor is defined as

$$\psi = \frac{\text{Surface area of the sphere with same volume as particle}}{\text{Surface area of the particle}} \tag{14-12}$$

Thus,

$$\psi = \frac{A_s}{A_p} = \frac{A_s/V_s}{A_p/V_p} = \frac{6/d_s}{a_s} \tag{14-13}$$

Equations (14-11) and (14-13) show that $d_p = 6/a_s$, where a_s is the surface-to-volume ratio for the particle (A_p/V_p), as deduced in Chapter 13. Since

$V_p = V_s$ (by definition), equivalent definitions of ψ are

$$\psi = \frac{6}{d_s}\left(\frac{V_p}{A_p}\right) = (6^2\pi)^{1/3}\frac{V_p^{2/3}}{A_p} = \frac{4.84V_p^{2/3}}{A_p} \tag{14-14}$$

The minimum bed porosity at incipient fluidization for nonspherical particles can be estimated from

$$\varepsilon_{mf} \cong (14\psi)^{-1/3} \tag{14-15}$$

For spherical particles ($\psi = 1$) Eq. (14-15) reduces to $\varepsilon_{mf} = 0.415$.

II. SEDIMENTATION

Sedimentation, or thickening, involves increasing the solids content of a slurry or suspension by gravity settling in order to effect separation (or partial separation) of the solids and the fluid. It differs from the gravity settling process that was previously considered in that the solids fraction is relatively high in these systems, so particle settling rates are strongly influenced by the presence of the surrounding particles. This is referred to as *hindered settling*. Fine particles (10 μm or less) tend to behave differently than larger or coarse particles (100 μm or more), because fine particles may exhibit a high degree of flocculation due to the importance of surface forces and high surface area. Figure 12-1 shows a rough illustration of the effect of solids concentration and particle/fluid density ratio on the free and hindered settling regimes.

A. Hindered Settling

A mixture of particles of different sizes can settle in different ways, according to Coulson et al. (1991), as illustrated in Fig. 14-3. Case (a) corresponds to a suspension with a range of particle sizes less than about 6:1. In this case, all the particles settle at about the same velocity in the "constant composition zone" (B), leaving a layer of clear liquid above. As the sediment (D) builds up, however, the liquid that is "squeezed out" of this layer serves to further retard the particles just above it, resulting in a zone of variable composition (C). Case (b) in Fig. 14-3 is less common and corresponds to a broad particle size range, in which the larger particles settle at a rate significantly greater than that of the smaller ones, and consequently there is no constant composition zone.

The settling characteristics of hindered settling systems differ significantly from those of freely settling particles in several ways:

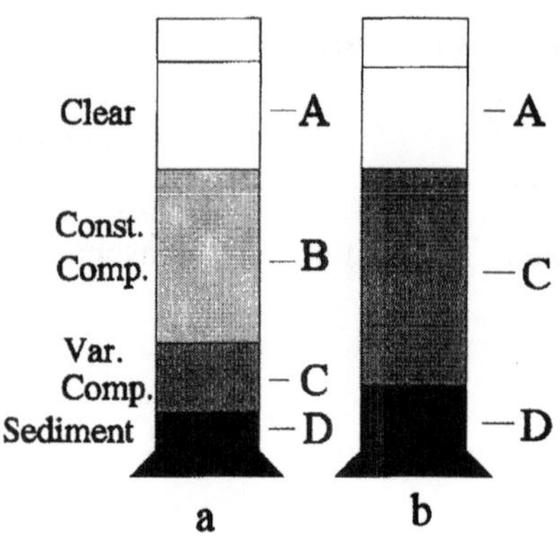

Figure 14-3 Two modes of settling. (a) Narrow particle size range; (b) broad particle size range.

1. The large particles are hindered by the small particles, which increase the effective resistance of the suspending medium for the large particles. At the same time, however, the small particles tend to be "dragged down" by the large particles, so that all particles tend to fall at about the same rate (unless the size range is very large, i.e. greater than 6:1 or so).
2. The upward velocity of the displaced fluid flowing in the interstices between the particles is significant, so the apparent settling velocity (relative to a fixed point) is significantly lower than the particle velocity relative to the fluid.
3. The velocity *gradients* in the suspending fluid flowing upward between the particles are increased, resulting in greater shear forces.
4. Because of the high surface area to volume ratio for small particles, surface forces are important, resulting in flocculation and "clumping" of the smaller particles into larger effective particle groups. This effect is more pronounced in a highly ionic (conducting) fluid, because the electrostatic surface forces that would cause the particles to be repelled are "shorted out" by the conductivity of the surrounding fluid.

There are essentially three different approaches to describing hindered settling. One approach is to define a "correction factor" to the Stokes free settling velocity in an infinite Newtonian fluid (which we will designate V_0), as a function of the solids loading. A second approach is to consider the suspending fluid properties (e.g. viscosity and density) to be modified by the presence of the fine particles. A third approach is to consider the collection or "swarm" of particles equivalent to a moving porous bed, the resistance to flow through the bed being determined by an equivalent of the Kozeny equation. There is insufficient evidence to say that any one of these approaches is any better or worse than the others. For many systems, they may all give comparable results, whereas for others one of these methods may be better or worse than the others.

If all of the solids are relatively fine and/or the slurry is sufficiently concentrated that settling is extremely slow, the slurry can usually be approximated as a uniform continuous medium with properties (viscosity and density) that depend on the solids loading, particle size and density, and interparticle forces (surface charges, conductivity, etc.). Such systems are generally quite non-Newtonian, with properties that can be described by the Bingham plastic or power law models. If the particle size distribution is broad and a significant fraction of the particles are fines (less than about 30 μm or so), the suspending fluid plus fines can be considered to be a continuous medium with a characteristic viscosity and density through which the larger particles must move. Such systems may or may not be non-Newtonian, depending on solids loading, etc., but are most commonly non-Newtonian. If the solids loading is relatively low (below about 10% solids by volume) and/or the particle size and/or density are relatively large, the system will be "heterogeneous" and the larger particles will settle readily. Such systems are usually Newtonian. A summary of the flow behavior of these various systems has been presented by Darby (1986).

B. Fine Particles

For suspensions of fine particles, or systems containing a significant amount of fines, the suspending fluid can be considered to be homogeneous, with the density and viscosity modified by the presence of the fines. These properties depend primarily on the solids loading of the suspension, which may be described in terms of either the porosity or void fraction (ε) or, more commonly, the volume fraction of solids, $\varphi\,(\varphi = 1 - \varepsilon)$. The buoyant force on the particles is due to the difference in density between the solid (ρ_S) and the surrounding suspension (ρ_φ), which is

$$\rho_S - \rho_\varphi = \rho_S - [\rho_S(1-\varepsilon) + \rho\varepsilon] = \varepsilon(\rho_S - \rho) = (1-\varphi)(\rho_S - \rho) \quad (14\text{-}16)$$

where ρ is the density of the suspending fluid.

The viscosity of the suspension (μ_φ) is also modified by the presence of the solids. For uniform spheres at a volumetric fraction of 2% or less,

Einstein (1906) showed that

$$\mu_\varphi = \mu(1 + 2.5\varphi) \tag{14-17}$$

where μ is the viscosity of the suspending fluid. For more concentrated suspensions, a wide variety of expressions have been proposed in the literature (see, e.g., Darby, 1986). For example, Vand (1948) proposed the expression

$$\mu_\varphi = \mu \exp\left(\frac{2.5\varphi}{1 - 0.609\varphi}\right) \tag{14-18}$$

although Mooney (1951) concluded that the constant 0.609 varies between 0.75 and 1.5, depending on the system. Equations (14-16) and (14-18) (or equivalent) may be used to modify the viscosity and density in Stokes' law, i.e.,

$$V_0 = \frac{(\rho_S - \rho)gd^2}{18\mu} \tag{14-19}$$

In this equation, V_0 is the relative velocity between the unhindered particle and the fluid. However, in a hindered suspension this velocity is increased by the velocity of the displaced fluid, which flows back up through the suspension in the void space between the particles. Thus, if V_S is the (superficial) settling velocity of the suspension (e.g., "swarm") and V_L is the velocity of the fluid, the total flux of solids and liquid is $\varphi V_s + (1 - \varphi)V_L$. The relative velocity between the fluid and solids in the swarm is $V_r = V_s - V_L$. If the total net flux is zero (e.g., "batch" settling in a closed-bottom container with no outflow), elimination of V_L gives

$$V_r = \frac{V_s}{1 - \varphi} \tag{14-20}$$

This also shows that $V_L = -\varphi V_s/(1 - \varphi)$, i.e., V_L is negative relative to V_s in batch settling.

From Eqs. (14-16), (14-18), and (14-20), it is seen that the ratio of the settling velocity of the suspension (V_s) to the terminal velocity of a single freely settling sphere (V_0) is

$$\frac{V_s}{V_0} = \frac{(1 - \varphi)^2}{\exp[2.5\varphi/(1 - k_2\varphi)]} \tag{14-21}$$

where the value of the constant k_2 can be from 0.61 to 1.5, depending upon the system. However, Coulson et al. (1991) remark that the use of a modified viscosity for the suspending fluid is more appropriate for the settling of large particles through a suspension of fines than for the uniform settling of a "swarm" of uniform particles with a narrow size distribution. They state that in the latter case the increased resistance is due to the higher velocity gradients in the interstices rather than to an increased viscosity. However, the net effect is essentially the same for either mechanism. This approach, as well as the other two mentioned above, all result in expressions of the general form

$$\frac{V_s}{V_0} = \varepsilon^2 \,\mathrm{fn}(\varepsilon), \qquad \varepsilon = 1 - \varphi \tag{14-22}$$

which is consistent with Eq. (14-21).

A widely quoted empirical expression for the function in Eq. (14-22) is that of Richardson and Zaki (1954):

$$\mathrm{fn}(\varepsilon) = \varepsilon^{\mathrm{n}} \tag{14-23}$$

where

$$n = \begin{cases} 4.65 & \text{for} \quad N_{\mathrm{Re_p}} < 0.2 \\ 4.35 N_{\mathrm{Re_p}}^{-0.03} & \text{for} \quad 0.2 < N_{\mathrm{Re}} < 1 \\ 4.45 N_{\mathrm{Re_p}}^{0.1} & \text{for} \quad 1 < N_{\mathrm{Re_p}} < 500 \\ 2.39 & \text{for} \quad N_{\mathrm{Re_p}} > 500 \end{cases}$$

where $N_{\mathrm{Re_p}}$ is the single particle Reynolds number in an "infinite" fluid. An alternative expression due to Davies et al. (1977) is

$$\frac{V_s}{V_0} = \exp(-k_1\varphi) \tag{14-24}$$

which agrees well with Eq. (14-23) for $k_1 = 5.5$. Another expression for fn(ε), deduced by Steinour (1944) from settling data on tapioca in oil, is

$$\mathrm{fn}(\varepsilon) = 10^{-1.82(1-\varepsilon)} \tag{14-25}$$

Barnea and Mizrahi (1973) considered the effects of the modified density and viscosity of the suspending fluid, as represented by Eq. (14-21), as well as a "crowding" or hindrance effect that decreases the effective space around the particles and increases the drag. This additional "crowding factor" is $1 + k_2\varphi^{1/3}$, which, when included in Eq. (14-21), gives

$$\frac{V_s}{V_0} = \frac{(1-\varphi)^2}{(1+\varphi^{1/3})\exp[5\varphi/(3(1-\varphi)]} \tag{14-26}$$

for the modified Stokes velocity, where the constant 2.5 in Eq. (14-21) has been replaced by 5/3 and the constant k_2 set equal to unity, based upon settling observations.

C. Coarse Particles

Coarser particles (e.g., ~100 μm or larger) have a relatively small specific surface, so flocculation is not common. Also, the suspending fluid surrounding the particles is the liquid phase rather than a "pseudocontinuous" phase of fines in suspension, which would modify the fluid viscosity and density properties. Thus, the properties of the continuous phase can be taken to be those of the pure fluid unaltered by the presence of fine particles. In this case, it can be shown by dimensional analysis that the dimensionless settling velocity V_s/V_0 must be a function of the particle drag coefficient, which in turn is a unique function of the particle Reynolds number, N_{Re_p}, the void fraction (porosity), $\varepsilon = 1 - \varphi$, and the ratio of the particle diameter to container diameter, d/D. Because there is a unique relationship between the drag coefficient, the Reynolds number, and the Archimedes number for settling particles, the result can be expressed in functional form as

$$\frac{V_s}{V_0} = \text{fn}\left\{N_{Ar}, \frac{d}{D}, \varepsilon\right\} \tag{14-27}$$

It has been found that this relationship can be represented by the empirical expression (Coulson et al., 1991)

$$\frac{V_s}{V_0} = \varepsilon^n \left[1 + 2.4\frac{d}{D}\right]^{-1} \tag{14-28}$$

where the exponent n is given by

$$n = \frac{4.8 + 2.4X}{X + 1} \tag{14-29}$$

and

$$X = 0.043 N_{Ar}^{0.57}\left[1 - 2.4\left(\frac{d}{D}\right)^{0.27}\right] \tag{14-30}$$

D. All Flow Regimes

The foregoing expressions give the suspension velocity (V_s) relative to the single particle free settling velocity, V_0, i.e., the Stokes velocity. However, it is not necessary that the particle settling conditions correspond to the Stokes regime to use these equations. As shown in Chapter 11, the Dallavalle equation can be used to calculate the single particle terminal velocity V_0

under any flow conditions from a known value of the Archimedes number, as follows:

$$V_0 = \frac{\mu}{\rho d}[(14.42 + 1.827\sqrt{N_{\mathrm{Ar}}})^{1/2} - 3.798]^2 \tag{14-31}$$

where

$$N_{\mathrm{Ar}} = \frac{d^3 \rho g \, \Delta\rho}{\mu^2} \tag{14-32}$$

This result can also be applied directly to coarse particle "swarms." For fine particle systems, the suspending fluid properties are assumed to be modified by the fines in suspension, which necessitates modifying the fluid properties in the definitions of the Reynolds and Archimedes numbers accordingly. Furthermore, because the particle drag is a direct function of the local relative velocity between the fluid and the solid (the interstitial relative velocity, V_{r}), it is this velocity that must be used in the drag equations (e.g., the modified Dallavalle equation). Since $V_{\mathrm{r}} = V_{\mathrm{s}}/(1 - \varphi) = V_{\mathrm{s}}/\varepsilon$, the appropriate definitions for the Reynolds number and drag coefficient for the suspension (e.g., the particle "swarm") are (after Barnea and Mizrahi, 1973):

$$N_{\mathrm{Re}_\varphi} = \frac{dV_{\mathrm{r}}\rho}{\mu_\varphi} = N_{\mathrm{Re}_0} \frac{V_{\mathrm{s}}}{V_0} \left(\frac{1}{(1 - \varphi) \exp[5\varphi/3(1 - \varphi)} \right) \tag{14-33}$$

and

$$C_{\mathrm{D}_\varphi} = C_{\mathrm{D}_0} \left(\frac{V_0}{V_{\mathrm{s}}} \right)^2 \left(\frac{(1 - \varphi)^2}{1 + \varphi^{1/3}} \right) \tag{14-34}$$

where $N_{\mathrm{Re}_0} = dV_0\rho/\mu$ and $C_{\mathrm{D}_0} = 4gd(\rho_{\mathrm{S}} - \rho)/3\rho V_0^2$ are the Reynolds number and drag coefficient for a single particle in an infinite fluid. Data presented by Barnea and Mizrahi (1973) show that the "swarm" dimensionless groups N_{Re_φ} and C_{D_φ} are related by the same expression as the corresponding groups for single particles, e.g., by the Dallavalle equation:

$$C_{\mathrm{D}_\varphi} = \left(0.6324 + \frac{4.8}{N_{\mathrm{Re}_\varphi}^{1/2}} \right)^2 \tag{14-35}$$

Thus, the settling velocity, or the terminal velocity of the "swarm" can be determined from

$$N_{\mathrm{Re}_\varphi} = [(14.42 + 1.827 N_{\mathrm{Ar}_\varphi}^{1/2})^{1/2} - 3.798]^2 \tag{14-36}$$

where

$$N_{\mathrm{Ar}_\varphi} = \frac{3}{4} C_{\mathrm{D}_\varphi} N_{\mathrm{Re}_\varphi}^2 = \frac{d^3 \rho g(\rho_{\mathrm{S}} - \rho)}{\mu^2(1 + \varphi^{1/3})} \exp\left(\frac{-10\varphi}{3(1 - \varphi)}\right) \tag{14-37}$$

III. GENERALIZED SEDIMENTATION/FLUIDIZATION

The foregoing equations all apply to hindered settling of a suspension (or "swarm") of particles in a stagnant suspending medium. Barnea and Mizrahi (1973) showed that these generalized relations may be applied to fluidization as well, since a fluidized bed may be considered a particle "swarm" suspended by the fluid flowing upward at the terminal velocity of the swarm. In this case the above equations apply with V_s replaced by the velocity V_f, i.e., the superficial velocity of the fluidizing medium. Once N_{Re_φ} is found from Eqs. (14-36) and (14-37), the settling velocity (V_s) is determined from Eq. (14-33). Barnea and Mizrahi (1973) presented data for both settling and fluidization that cover a very wide range of the dimensionless parameters, as shown in Fig. 14-4.

IV. THICKENING

The process of thickening involves the concentration of a slurry, suspension, or sludge, usually by gravity settling. Because concentrated suspensions and/or fine particle dispersions are often involved, the result is usually not a complete separation of the solids from the liquid but is instead a separation into a more concentrated (underflow) stream and a diluted (overflow) stream. Thickeners and clarifiers are essentially identical. The only difference is that the clarifier is designed to produce a clean liquid overflow with a specified purity, whereas the thickener is designed to produce a concentrated underflow product with a specified concentration (Christian, 1994; Tiller and Tarng, 1995; McCabe et al., 1993).

A schematic of a thickener/clarifier is shown in Fig. 14-5. As indicated in Fig. 14-3, several settling regions or zones can be identified, depending on the solids concentration and interparticle interaction. For simplicity, we consider three primary zones, as indicated in Fig. 14-5 (with the understanding that there are transition zones in between). The top, or clarifying, zone contains relatively clear liquid from which most of the particles have settled. Any particles remaining in this zone will settle by free settling. The middle zone is a region of varying composition through which the particles move by hindered settling. The size of this region and the settling rate depend on the local solids concentration. The bottom zone is a highly concentrated settled

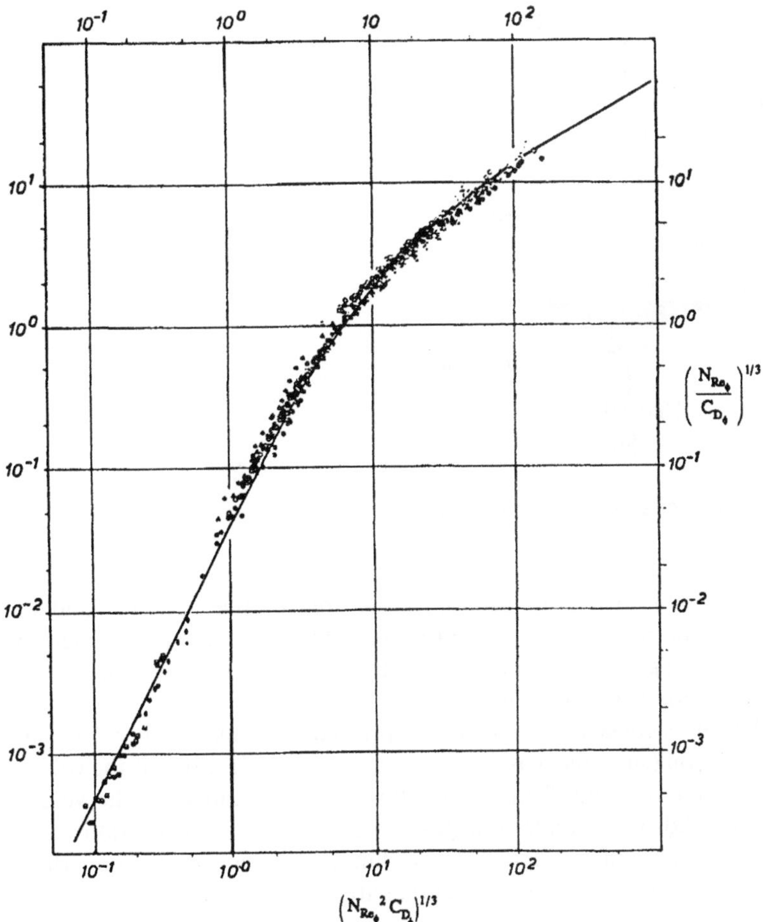

Figure 14-4 Generalized correlation of settling and fluidizing velocities. (From Barnea and Mizrahi, 1973.)

or compressed region containing the settled particles. The particle settling rate in this zone is very slow.

In the top (clarifying) zone the relatively clear liquid moves upward and overflows the top. In the middle zone the solid particles settle as the displaced liquid moves upward, and both the local solids concentration and the settling velocity vary from point to point. In the bottom (compressed) zone, the solids and liquid both move downward at a rate that is determined mainly by the underflow draw-off rate. For a given feed rate and solids

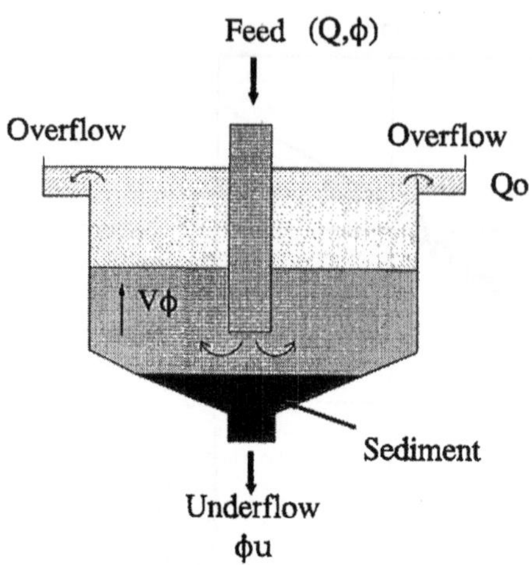

FIGURE 14-5 Schematic of a thickener.

loading, the objective is to determine the area of the thickener and the optimum underflow (draw-off) rate to achieve a specified underflow concentration (φ_u), or the underflow rate and underflow concentration, for stable steady state operation.

The solids concentration can be expressed in terms of either the solids volume fraction (φ) or the mass ratio of solids to fluid (R). If φ_f is the volume fraction of solids in the feed stream (flow rate Q_f) and φ_u is the volume fraction of solids in the underflow (flow rate Q_u), then the solids ratio in the feed, $R_f = [(\text{mass of solids})/(\text{mass of fluid})]_{\text{feed}}$, and in the underflow, $R_u = [(\text{mass of solids})/(\text{mass of liquid})]_u$, are given by

$$R_f = \frac{\varphi_f \rho_S}{(1-\varphi_f)\rho}, \qquad R_u = \frac{\varphi_u \rho_S}{(1-\varphi_u)\rho} \tag{14-38}$$

These relations can be rearranged to give the solids volume fractions in terms of the solids ratio:

$$\varphi_f = \frac{R_f}{R_f + \rho_S/\rho}, \qquad \varphi_u = \frac{R_u}{R_u + \rho_S/\rho} \tag{14-39}$$

Now the total (net) flux of the solids plus liquid moving through the thickener at any point is given by

$$q = \frac{Q}{A} = q_s + q_L = \varphi V_s + (1-\varphi)V_L \tag{14-40}$$

where $q_s = \varphi V_s$ is the local solids flux, defined as the volumetric settling rate of the solids per unit cross-sectional area of the settler, and $q_L = (1 - \varphi)V_L$ is the local liquid flux.

The solids flux depends on the local concentration of solids, the settling velocity of the solids at this concentration *relative to the liquid*, and the net velocity of the liquid. Thus the local solids flux will vary within the thickener because the concentration of solids increases with depth and the amount of liquid that is displaced (upward) by the solids decreases as the solids concentration increases, thus affecting the "upward drag" on the particles. As these two effects act in opposite directions, there will be some point in the thickener at which the actual solids flux is a minimum. This point determines the conditions for stable steady-state operation, as explained below.

The settling behavior of a slurry is normally determined by measuring the velocity of the interface between the top (clear) and middle suspension zones in a *batch settling* test using a closed system (e.g., a graduated cylinder) as illustrated in Fig. 14-3. A typical batch settling curve is shown in Fig. 14-6 (see, e.g., Foust et al., 1980). The initial linear portion of this curve usually corresponds to free (unhindered) settling, and the slope of this region is the free settling velocity, V_0. The nonlinear region of the curve corresponds to hindered settling in which the solids flux in this region depends upon the local solids concentration. This can be determined from the batch settling curve as follows (Kynch, 1952). If the initial height of the suspension with a solids fraction of φ_o is Z_o, at some later time the height of the interface between the clear layer and the hindered settling zone will be $Z(t)$, where the average solids fraction in this zone is $\varphi(t)$. Since the total amount of solids in the system is constant, assuming the amount of solids in the clear layer to be negligible, it follows that

$$Z(t)\varphi(t) = Z_o\varphi_o \quad \text{or} \quad \varphi(t) = \frac{\varphi_o Z_o}{Z(t)} \tag{14-41}$$

Thus, given the initial height and concentration (Z_o, φ_o), the average solids concentration $\varphi(t)$ corresponding to any point on the curve $Z(t)$ can be determined. Furthermore, the hindered settling velocity and batch solids flux at this point can be determined from the slope of the curve at that point, i.e., $V_{sb} = -(dZ/dt)$ and $q_{sb} = \varphi V_{sb}$. Thus, the batch settling curve can be converted to a *batch flux curve*, as shown in Fig. 14-7. The batch flux curve exhibits a maximum and a minimum, because the settling velocity is nearly constant in the free settling region (and the flux is directly proportional to the solids concentration), whereas the settling velocity and the flux drop rapidly with increasing solids concentration in the hindered settling region as explained above. However, the solids flux in the bottom

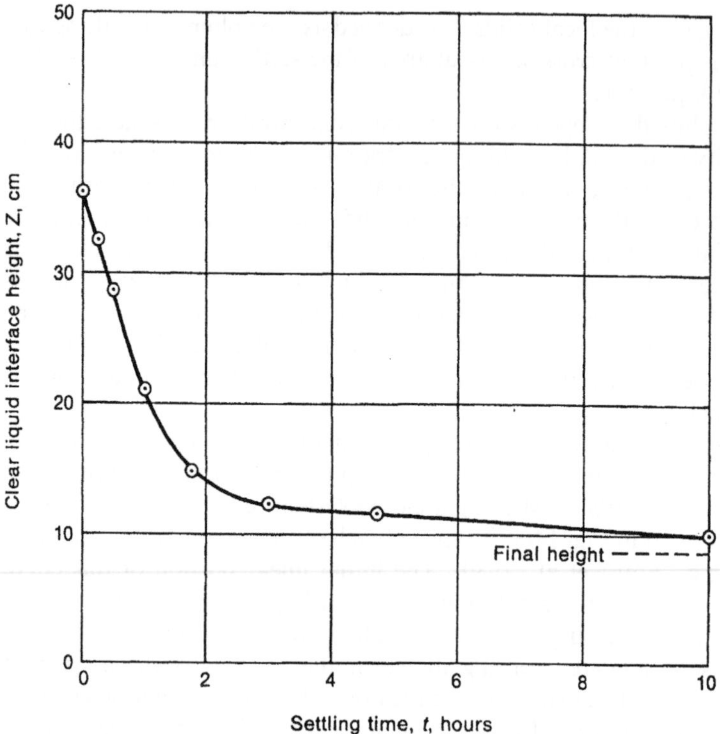

Figure 14-6 Typical batch settling curve for a limestone slurry.

(compressed) zone is much higher because of the high concentration of solids in this zone. The minimum in this curve represents a "pinch" or "critical" condition in the thickener that limits the total solids flux that can be obtained under steady-state (stable) operation.

Because the batch flux data are obtained in a closed system with no outflow, the net solids flux is zero in the batch system and Eq. (14-40) reduces to $V_L = -\varphi V_s/(1-\varphi)$. Note that V_L and V_s are of opposite sign, because the displaced liquid moves upward as the solids settle. The relative velocity between the solids and liquid is $V_r = V_s - V_L$ which, from Eq. (14-20), is $V_r = V_s/(1-\varphi)$. It is this relative velocity that controls the dynamics in the thickener. If the underflow draw-off rate from the thickener is Q_u, the additional solids flux in the thickener due to superimposition of this underflow is $q_u = Q_u/A = V_u$. Thus, the total solids flux at any point in the thickener (q_s) is equal to the settling flux relative to the suspension (i.e., the batch flux q_{sb}) at that point, plus the bulk flux due

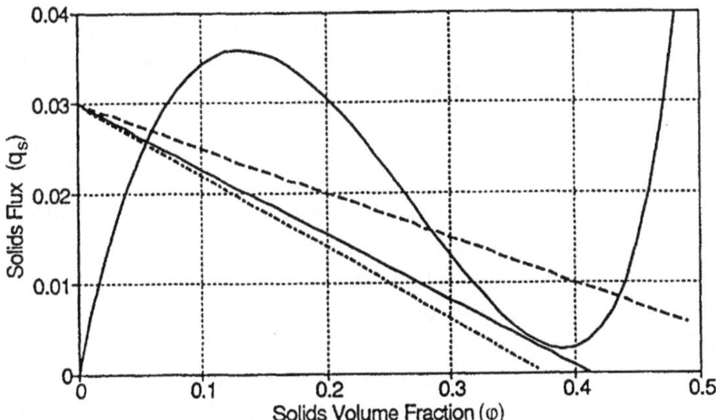

FIGURE 14-7 Typical batch flux curve with operating lines (......) underloaded; (———) properly loaded; (- - - -) overloaded.

to the underflow draw-off rate, φV_u, i.e., $q_s = q_{sb} + \varphi q_u$. Furthermore, at steady state the net local solids flux in the settling zone (q_s) must be equal to that in the underflow, i.e., $q_s = q_u \varphi_u$. Eliminating $q_u$and rearranging leads to

$$q_{sb} = q_s\left(1 - \frac{\varphi}{\varphi_u}\right) \tag{14-42}$$

This equation represents a straight line on the batch flux curve (q_{sb} vs. φ) that passes through the points (q_s, 0) and (0, φ_u). The line intersects the φ axis at φ_u and the q_{sb} axis at q_s, which is the net local solids flux in the thickener at the point where the solids fraction is φ. This line is called the "operating line" for the thickener, and its intersection with the batch flux curve determines the stable operating point for the thickener, as shown in Fig. 14-7. The "properly loaded" operating line is tangent to the batch flux curve. At the tangent point, called the critical (or "pinch") point, the local solids flux corresponds to the steady state value at which the net critical (minimum) settling rate in the thickener equals the total underflow solids rate. The "underloaded" line represents a condition for which the underflow draw-off rate is higher than the critical settling rate, so no sludge layer can build up and excess clear liquid will eventually be drawn out the bottom (i.e. the draw-off rate is too high). The "overloaded" line represents the condition at which the underflow draw-off rate is lower than the critical settling rate, so the bottom solids layer will build up and eventually rise to the overflow (i.e., the underflow rate is too low).

Once the operating line is set, the equations that govern the thickener operation are determined from a solids mass balance as follows. At steady state (stable) operating conditions, the net solids flux is

$$q_s = \frac{Q_S}{A} = \frac{Q_f \varphi_f}{A} = \frac{Q_u \varphi_u}{A} \tag{14-43}$$

This equation relates the thickener area (A) and the feed rate and loading (Q_f, φ_f) to the solids underfow rate (Q_u) and the underflow loading (φ_u), assuming no solids in the overflow. The area of a thickener required for a specified underflow loading can be determined as follows. For a given underflow solids loading (φ_u), the operating line is drawn on the batch flux curve from φ_u on the φ axis tangent to the batch flux curve at the critical point, (q_c, φ_c). The intersection of this line with the vertical axis ($\varphi = 0$) gives the local solids flux (q_s) in the thickener that results in stable or steady-state (properly loaded) conditions. This value is determined from the intersection of the operating line on the q_{sb} axis or from the equation of the operating line that is tangent to the critical point (q_c, φ_c):

$$q_s = \frac{q_c}{1 - \varphi_c/\varphi_u} \tag{14-44}$$

If the feed rate (Q_f) and solids loading (φ_f) are specified, the thickener area A is determined from Eq. (14-43). If it is assumed that none of the solids are carried over with the overflow, the overflow rate Q_o is given by

$$Q_o = Q_f(1 - \varphi_f) - Q_f \varphi_f \frac{1 - \varphi_u}{\varphi_u} \tag{14-45}$$

or

$$\frac{Q_o}{Q_f} = 1 - \frac{\varphi_f}{\varphi_u} \tag{14-46}$$

Likewise, the underflow rate Q_u is given by

$$Q_u = Q_f - Q_o = Q_f - Q_f\left(1 - \frac{\varphi_f}{\varphi_u}\right) \tag{14-47}$$

or

$$\frac{Q_u}{Q_f} = \frac{\varphi_f}{\varphi_u} \tag{14-48}$$

PROBLEMS

1. Calculate the flow rate of air (in scfm) required to fluidize a bed of sand (SG = 2.4), if the air exits the bed at 1 atm, 70°F. The sand grains have an

equivalent diameter of 500 μm and the bed is 2 ft in diameter and 1 ft deep, with a porosity of 0.35. What flow rate of air would be required to blow the sand away?

2. Calculate the flow rate of water (in gpm) required to fluidize a bed of 1/16 in. diameter lead shot (SG = 11.3). The bed is 1 ft in diameter, 1 ft deep, and has a porosity of 0.18. What water flow rate would be required to sweep the bed away?
3. Calculate the range of water velocities that will fluidize a bed of glass spheres (SG = 2.1) if the sphere diameter is: (a) 2 mm, (b) 1 mm, (c) 0.1 mm.
4. A coal gasification reactor operates with particles of 500 μm diameter and density of 1.4 g/cm^3. The gas may be assumed to have properties of air at 1,000°F and 30 atm. Determine the range of superficial gas velocity over which the bed is in a fluidized state.
5. A bed of coal particles, 2 ft in diameter and 6 ft deep, is to be fluidized using a hydrocarbon liquid with a viscosity 15 cP and a density of 0.9 g/cm^3. The coal particles have a density of 1.4 g/cm^3 and an equivalent spherical diameter of 1/8 in. If the bed porosity is 0.4:
   (a) Determine the range of liquid superficial velocities over which the bed is fluidized.
   (b) Repeat the problem using the "particle swarm" (Barnea and Mizrah, 1973) "swarm terminal velocity" approach, assuming (1) $\varphi = 1 - \varepsilon$; (2) $\varphi = 1 - \varepsilon_{mf}$.
6. A catalyst having spherical particles with $d_p = 50\,\mu\text{m}$ and $\rho_s = 1.65$ g/cm^3 is to be used to contact a hydrocarbon vapor in a fluidized reactor at 900°F, 1 atm. At operating conditions, the fluid viscosity is 0.02 cP and its density is 0.21 lb_m/ft^3. Determine the range of fluidized bed operation, i.e., calculate;
   (a) Minimum fluidization velocity for $\varepsilon_{mf} = 0.42$.
   (b) The particle terminal velocity.
7. A fluidized bed reactor contains catalyst particles with a mean diameter of 500 μm and a density of 2.5 g/cm^3. The reactor feed has properties equivalent to 35° API distillate at 400°F. Determine the range of superficial velocities over which the bed will be in a fluidized state.
8. Water is pumped upward through a bed of 1 mm diameter iron oxide particles ($SG = 5.3$). If the bed porosity is 0.45, over what range of superficial water velocity will the bed be fluidized?
9. A fluidized bed combustor is 2 m in diameter and is fed with air at 250°F, 10 psig, at a rate of 2000 scfm. The coal has a density of 1.6 g/cm^3 and a shape factor of 0.85. The flue gas from the combustor has an average MW of 35 and leaves the combustor at a rate of 2100 scfm at 2500°F and 1 atm. What is the size range of the coal particles that can be fluidized in this system?
10. A fluidized bed incinerator, 3 m in diameter and 0.56 m high, operates at 850°C using a sand bed. The sand density is 2.5 g/cm^3, and the average sand grain has a mass of 0.16 mg and a sphericity of 0.85. In the stationary (packed) state, the bed porosity is 35%. Find:
    (a) The range of air velocities that will fluidize the bed.

(b) The compressor power required if the bed is operated at 10 times the minimum fluidizing velocity and the compressor efficiency is 70%. The compressor takes air in from the atmosphere at 20°C, and the gases leave the bed at 1 atm.

11. Determine the range of flow rates (in gpm) that will fluidize a bed of 1 mm cubic silica particles (SG = 2.5) with water. The bed is 10 in. in diameter, 15 in. deep.

12. Determine the range of velocities over which a bed of granite particles ($SG = 4$) would be fluidized using
    (a) Water at 70°F
    (b) Air at 70°F and 20 psig.

13. Calculate the velocity of water that would be required to fluidize spherical particles with SG = 1.6 and diameter of 1.5 mm, in a tube with a diameter of 10 mm. Also, determine the water velocity that would sweep the particles out of the tube. Assume:
    (a) The bed starts as a packed bed and is fluidized when the pressure drop due to friction through the bed balances the weight of the bed.
    (b) The bed is considered to be a "swarm" of particles falling at the terminal velocity of the "swarm."
    (c) Compare the results of (a) and (b). Comment on any uncertainties or limitations in your results.

14. You want to fluidize a bed of solid particles using water. The particles are cubical, with a length on each side of 1/8 in. and on a SG of 1.2.
    (a) What is the sphericity factor for these particles, and what is their equivalent diameter?
    (b) What is the approximate bed porosity at the point of fluidization of the bed?
    (c) What velocity of water would be required to fluidize the bed?
    (d) What velocity of water would sweep the particles out of the bed?

15. Solid particles with a density of 1.4 g/cm^3 and a diameter of 0.01 cm are fed from a hopper into a line where they are mixed with water, which is draining by gravity from an open tank, to form a slurry having 0.4 lb_m of solids per lb_m of water. The slurry is transported by a centrifugal pump through a 6 in. sch 40 pipeline that is 0.5 mi long, at a rate of 1000 gpm. The slurry can be described as a Bingham plastic with a yield stress of 120 dyn/cm^2 and a limiting viscosity of 50 cP.
    (a) If the pipeline is at 60°F, and the pump is 60% efficient with a required NPSH of 15 ft, what horsepower motor would be required to drive the pump?
    (b) If the pump is 6 ft below the bottom of the water storage tank and the water in the line upstream of the pump is at 90°C ($P_v = 526$ mm Hg), what depth of water in the tank would be required to prevent the pump from cavitating?
    (c) A venturi meter is installed in the line to measure the slurry flow rate. If the maximum pressure drop reading for the venturi is to be 29 in. of water, what diameter should the venturi throat be?
    (d) The slurry is discharged from the pipeline to a settling tank, where it is desired to concentrate the slurry to 1 lb_m of solids per lb_m of water (in

the underflow). Determine the required diameter of the settling tank and the volumetric flow rates of the overflow (Q_o) and underflow (Q_u), in gpm.

(e) If the slurry were to be sent to a rotary drum filter instead to remove all of the solids, determine the required size of the drum (assuming that the drum length and diameter are equal). The drum rotates at 3 rpm, with 25% of its surface submerged in the slurry, and operates at a vacuum of 20 in.Hg. Lab test data taken on the slurry with 0.5 ft^2 of the filter medium, at a constant flow rate of 3 gpm, indicated a pressure drop of 1.5 psi after 1 min of filtration and 2.3 psi after 2 min of operation.

16. A sludge is to be clarified in a thickener that is 50 ft in diameter. The sludge contains 35% solids by volume (SG = 1.8) in water, with an average particle size of 25 μm. The sludge is pumped into the center of the tank, where the solids are allowed to settle and the clarified liquid overflows the top. Estimate the maximum flow rate of the sludge (in gpm) that this thickener can handle. Assume that the solids are uniformly distributed across the tank and that all particle motion is vertical.

17. In a batch thickener, an aqueous sludge containing 35% by volume of solids (SG = 1.6), with an average particle size of 50 μm, is allowed to settle. The sludge is fed to the settler at a rate of 1000 gpm, and the clear liquid overflows the top. Estimate the minimum tank diameter required for this separation.

18. Ground coal is slurried with water in a pit, and the slurry is pumped out of the pit at a rate of 500 gpm with a centrifugal pump and into a classifier. The classifier inlet is 50 ft above the slurry level in the pit. The piping system consists of an equivalent length of 350 ft of 5 in. sch 40 pipe and discharges into the classifier at 2 psig. The slurry may be assumed to be a Newtonian fluid with a viscosity of 30 cP, a density of 75 lb_m/ft^3, and a vapor pressure of 30 mmHg. The solid coal has an SG = 1.5.
    (a) How much power would be required to pump the slurry?
    (b) Using the pump characteristic charts in Appendix H, select the best pump for this job. Specify the pump size, motor speed (rpm), and impeller diameter that you would use. Also determine the pump efficiency and NPSH requirement.
    (c) What is the maximum height above the level of the slurry in the pit that the pump could be located without cavitating?
    (d) A Venturi meter is located in a vertical section of the line to monitor the slurry flow rate. The meter has a 4 in. diameter throat, and the pressure taps are 1 ft apart. If a DP cell (transducer) is used to measure the pressure difference between the taps, what would it read (in inches of water)?
    (e) A 90° flanged elbow is located in the line at a point where the pressure (upstream of the elbow) is 10 psig. What are the forces transmitted to the pipe by the elbow from the fluid inside the elbow. (Neglect the weight of the fluid.)

(f) The classifier consists of three collection tanks in series that are full of water. The slurry enters at the top on the side of the first tank, and leaves at the top on the opposite side, which is 5 ft from the entrance. The solids settle into the tank as the slurry flows into it and then overflows into the next tank. The space through which the slurry flows above the tank is 2 ft wide and 3 ft high. All particles for which the settling time in the space above the collection tank is less than the residence time of the fluid flowing in the space over the collection tank will be trapped in that tank. Determine the diameter of the largest particle that will not settle into each of the three collection tanks. Assume that the particles are equivalent spheres and that they fall at their terminal velocity.

(g) The suspension leaving the classifier is transferred to a rotary drum filter to remove the remaining solids. The drum operates at a constant pressure difference of 5 psi and rotates at a rate of 2 rpm with 20% of the surface submerged. Lab tests on a sample of the suspension through the same filter medium were conducted at a constant flow rate of 1 gpm through 0.25 ft^2 of the medium. It was found that the pressure drop increased to 2.5 psi after 10 min, and the resistance of the medium was negligible. How much filter area would be required to filter the liquid?

19. You want to concentrate a slurry from 5% (by vol) solids to 30% (by vol) in a thickener. The solids density is 200 lb_m/ft^3, and that of the liquid is 62.4 lb_m/ft^3. A batch settling test was run on the slurry, and the analysis of the test yielded the following information:

Vol. fraction solids φ	Settling rate $[lb_m/hrft^2)]$
0.05	73.6
0.075	82.6
0.5	79.8
0.125	70.7
0.15	66
0.2	78
0.25	120
0.3	200

(a) If the feed flow rate of the slurry is 500 gpm, what should the area of the thickener tank be?

(b) What are the overflow and underflow rates?

20. You must determine the maximum feed rate that a thickener can handle to concentrate a waste suspension from 5% solids by volume to 40% solids by volume. The thickener has a diameter of 40 ft. A batch flux test in the laboratory for the settled height versus time was analyzed to give the data below for the solids flux versus solids volume fraction. Determine:

(a) The proper feed rate of liquid in gpm.

(b) The overflow liquid rate in gpm.
(c) The underflow liquid rate in gpm.

Vol. fraction solids φ	Settling rate $[\mathrm{lb_m/hrft^2})]$
0.03	0.15
0.05	0.38
0.075	0.46
0.10	0.40
0.13	0.33
0.15	0.31
0.20	0.38
0.25	0.60
0.30	0.80

NOTATION

A	area, $[\mathrm{L^2}]$
a_s	particle surface area/volume, [1/L]
C_D	drag coefficient, [—]
C_{D_φ}	swarm drag coefficient, [—]
D	container diameter, [L]
d	particle diameter, [L]
e_f	energy dissipated per unit mass of fluid, $[\mathrm{FL/M = L^2/t^2}]$
f_{PM}	porous media friction factor, [—]
P	pressure, $[\mathrm{F/L^2 = M/Lt^2}]$
g	acceleration due to gravity, $[\mathrm{L/t^2}]$
h	height of bed, [L]
N_{Ar}	Archimedes number, [—]
N_{Ar_φ}	swarm Archimedes number, [—]
$\hat{N}_{Ar}$	Reynolds number defined by Eq. (14-7), [—]
N_{Re_φ}	swarm Reynolds number, Eq. (14-33), [—]
$N_{Re,PM}$	porous media Reynolds number, [—]
t	time, [t]
V	velocity, [L/t]
$\Delta(\,)$	$(\,)_2 - (\,)_1$
ε	porosity or void fraction, [—]
μ	viscosity, [M/Lt]
φ	volume fraction of solids, [—]
ρ	density, $[\mathrm{M/L^3}]$
ψ	sphericity factor, [—]

Subscripts

c	critical point
f	fluid, feed
i	inlet
L	liquid
mf	minimum fluidization condition
o	infinitely dilute condition, overflow
p	particle
S	solid
s	superficial or solid "swarm", or spherical
u	underflow
φ	solid suspension of volume fraction φ

REFERENCES

Azbel DS, NP Cheremisinoff. Fluid Mechanics and Unit Operations. Ann Arbor, MI: Ann Arbor Science, 1983.

Barnea E, J Mizrahi. A generalized approach to the fluid dynamics of particulate systems, Part I. Chem Eng J 5:171–189, 1973.

Christian JB. Chem Eng Prog, July 1994, p 50.

Coulson JM, JF Richardson, JR Blackhurst, JH Harker. Chemical Engineering, Vol 2. 4th ed. New York: Pergamon Press, 1991.

Darby R. Hydrodynamics of slurries and suspensions. In: NP Cheremisinoff, ed. Encyclopedia of Fluid Mechanics, Vol. 5. New York: Marcel Dekker, 1986, pp 49–92.

Davies L, D Dollimore, GB McBride. Powder Technol 16:45, 1977.

Einstein A. Ann Phys 19:289, 1906.

Foust AS, LA Wenzel, CW Clump, L Maus, LB Anderson. Principles of Unit Operations. 2nd ed. New York: Wiley, 1980.

Kynch GJ. Trans Faraday Soc 51:61, 1952.

McCabe WL, JC Smith, P Harriott. Unit Operations of Chemical Engineering. 5th ed. New York: McGraw-Hill, 1993.

Mooney M. J Colloid Sci 6:162, 1951.

Richardson JF, WN Zaki. Chem Eng Sci 3:65, 1954.

Steinour HH. Ind Eng Chem 36:618, 840, 901, 1944.

Svarovsky L. Hydrocyclones. Lancaster, PA: Technomic, 1984.

Tiller FM, D Trang. Chem Eng Prog. March 1995, p. 75.

Vand V. J Phys Colloid Chem 52:217, 1948.

15

Two-Phase Flow

I. SCOPE

The term "two-phase flow" covers an extremely broad range of situations, and it is possible to address only a small portion of this spectrum in one book, let alone one chapter. Two-phase flow includes any combination of two of the three phases solid, liquid, and gas, i.e., solid–liquid, gas–liquid, solid–gas, or liquid–liquid. Also, if both phases are fluids (combinations of liquid and/or gas), either of the phases may be continuous and the other distributed (e.g., gas in liquid or liquid in gas). Furthermore, the mass ratio of the two phases may be fixed or variable throughout the system. Examples of the former are nonvolatile liquids with solids or noncondensable gases, whereas examples of the latter are flashing liquids, soluble solids in liquids, partly miscible liquids in liquids, etc. In addition, in pipe flows the two phases may be uniformly distributed over the cross section (i.e., homogeneous) or they may be separated, and the conditions under which these states prevail are different for horizontal flow than for vertical flow.

For uniformly distributed homogeneous flows, the fluid properties can be described in terms of averages over the flow cross section. Such flows can be described as "one dimensional," as opposed to separated or heterogeneous flows, in which the phase distribution varies over the cross section.

We will focus on two-phase flow in pipes, which includes the transport of solids as slurries and suspensions in a continuous liquid phase, pneumatic transport of solid particles in a continuous gas phase, and mixtures of gas or vapor with liquids in which either phase may be continuous. Although it may appear that only one additional "variable" is added to the single-phase problems previously considered, the complexity of two-phase flow is greater by orders of magnitude. It is emphasized that this exposition is only an introduction to the subject, and literally thousands of articles can be found in the literature on various aspects of two-phase flow. It should also be realized that if these problems were simple or straightforward, the number of papers required to describe them would be orders of magnitude smaller. Useful information can be found in Brodkey (1967), Butterworth and Hewitt (1977), Chisholm (1983), Darby (1986), Fan and Zhu (1998), Govier and Aziz (1972), Hedstroni (1982), Holland and Bragg (1995), Klinzing et al. (1997), Levy (1999), Molerus (1993), Shook and Rocco (1991) and Wallis (1969) among others.

II. DEFINITIONS

Before proceeding further, it is appropriate to define the various flow rates, velocities, and concentrations for two-phase flow. There is a bewildering variety of notation in the literature relative to two-phase flow, and we will attempt to use a notation that is consistent with the definitions below for solid–liquid, solid–gas, and liquid–gas systems.

The subscripts m, L, S, and G will represent the local two-phase mixture, liquid phase, solid phase and gas phase, respectively. The definitions below are given in terms of solid–liquid (S–L) mixtures, where the solid is the more dense distributed phase and the liquid the less dense continuous phase. The same definitions can be applied to gas–liquid (G–L) flows if the subscript S is replaced by L (the more dense phase) and the L by G (the less dense phase). The symbol φ is used for the volume fraction of the more dense phase, and ε is the volume fraction of the less dense phase (obviously $\varphi = 1 - \varepsilon$). An important distinction is made between (φ, ε) and $(\varphi_m, \varepsilon_m)$. The former (φ, ε) refers to the overall flow-average (equilibrium) values entering the pipe, i.e.,

$$\varphi = \frac{Q_S}{Q_S + Q_L} = 1 - \varepsilon \tag{15-1}$$

whereas the latter $(\varphi_m, \varepsilon_m)$ refers to the local values at a given position in the pipe. These are different $(\varphi_m \neq \varphi, \varepsilon_m \neq \varepsilon)$ when the local velocities of the two phases are not the same (i.e., when *slip* is significant), as will be shown below.

Mass flow rate ($\dot{m}$) *and volume flow rate* (Q):

$$\dot{m}_{\mathrm{m}} = \dot{m}_{\mathrm{S}} + \dot{m}_{\mathrm{L}} = \rho_{\mathrm{S}} Q_{\mathrm{S}} + \rho_{\mathrm{L}} Q_{\mathrm{L}} = \rho_{\mathrm{m}} Q_{\mathrm{m}} \tag{15-2}$$

Mass flux (G):

$$G_{\mathrm{m}} = \frac{\dot{m}_{\mathrm{m}}}{A} = G_{\mathrm{S}} + G_{\mathrm{L}} = \frac{\dot{m}_{\mathrm{S}} + \dot{m}_{\mathrm{L}}}{A} \tag{15-3}$$

Volume flux:

$$J_{\mathrm{m}} = J_{\mathrm{S}} + J_{\mathrm{L}} = \frac{G_{\mathrm{m}}}{\rho_{\mathrm{m}}} = \frac{G_{\mathrm{S}}}{\rho_{\mathrm{S}}} + \frac{G_{\mathrm{L}}}{\rho_{\mathrm{L}}} = \frac{Q_{\mathrm{S}} + Q_{\mathrm{L}}}{A} = V_{\mathrm{m}} \tag{15-4}$$

Phase velocity:

$$V_{\mathrm{S}} = \frac{J_{\mathrm{S}}}{\varphi} = \frac{J_{\mathrm{S}}}{1-\varepsilon}, \qquad V_{\mathrm{L}} = \frac{J_{\mathrm{L}}}{\varepsilon} = \frac{J_{\mathrm{L}}}{1-\varphi} \tag{15-5}$$

Relative (slip) velocity and slip ratio:

$$V_{\mathrm{r}} = V_{\mathrm{L}} - V_{\mathrm{S}}, \qquad S = \frac{V_{\mathrm{L}}}{V_{\mathrm{S}}} = 1 + \frac{V_{\mathrm{r}}}{V_{\mathrm{S}}} \tag{15-6}$$

Note that the total *volume flux* (J_{m}) of the mixture is the same as the *superficial velocity* (V_{m}), i.e., the total volumetric flow divided by the total flow area. However, the local velocity of each phase (V_i) is greater than the volume flux of that phase (J_i), because each phase occupies only a fraction of the total flow area. The volume flux of each phase is the total volume flow rate of that phase divided by the total flow area.

The relative slip velocity (or slip ratio) is an extremely important variable. It comes into play primarily when the distributed phase density is greater than that of the continuous phase and the heavier phase tends to lag behind the lighter phase for various reasons (explained below). The resulting relative velocity (slip) between the phases determines the drag exerted by the continuous (lighter) phase on the distributed (heavier) phase. One consequence of slip (as shown below) is that the concentration or "holdup" of the more dense phase within the pipe (φ_{m}) is greater than that entering or leaving the pipe, because its residence time is longer. Consequently, the concentrations and local phase velocities within a pipe under slip conditions depend upon the properties and degree of interaction of the phases and cannot be determined solely from a knowledge of the entering and leaving concentrations and flow rates. Slip can be determined only indirectly by measurement of some local flow property within the pipe such as the holdup, local phase velocity, or local mixture density.

For example, if φ is the solids volume fraction entering the pipe at velocity V, and φ_m is the local volume fraction in the pipe where the solid velocity is V_S, a component balance gives

$$\frac{V_S}{V} = \frac{\varphi}{\varphi_m} \quad \text{and} \quad \frac{V_L}{V} = \frac{1-\varphi}{1-\varphi_m} \tag{15-7}$$

Substituting these expressions into the definition of the slip velocity (and dividing by the entering velocity, V, to make the results dimensionless) gives

$$\bar{V}_r = \frac{V_r}{V} = \frac{V_L - V_S}{V_L + V_S} = \frac{S-1}{S+1} = \frac{\varphi_m - \varphi}{\varphi_m(1-\varphi_m)} \tag{15-8}$$

This can be solved for φ_m in terms of $\bar{V}_r$ and φ:

$$\varphi_m = \frac{1}{2}\left\{1 - \frac{1}{\bar{V}_r} + \left[\left(\frac{1}{\bar{V}_r} - 1\right)^2 + \frac{4\varphi}{\bar{V}_r}\right]^{1/2}\right\} \tag{15-9}$$

For example, if the entering solids fraction φ is 0.4, the corresponding values of the local solids fraction φ_m for relative slip velocities ($\bar{V}_r$) of 0.01, 0.1, and 0.5 are 0.403, 0.424, and 0.525, respectively. There are many "theoretical" expressions for slip, but practical applications depend on experimental observations and correlations (which will be presented later). In gas–liquid or gas–solid flows, φ_m will vary along the pipe, because the gas expands as the pressure drops and speeds up as it expands, which tends to increase the slip, which in turn increases the holdup of the denser phase.

The mass fraction (x) of the less dense phase (which, for gas–liquid flows, is called the *quality*) is $x = \dot{m}_L/(\dot{m}_S + \dot{m}_L)$, so the mass flow ratio can be written

$$\frac{\dot{m}_L}{\dot{m}_S} = \frac{x}{1-x} = \frac{\rho_L V_L A \varepsilon_m}{\rho_S V_S A(1-\varepsilon_m)} = S\left(\frac{\rho_L}{\rho_S}\right)\left(\frac{\varepsilon_m}{1-\varepsilon_m}\right) \tag{15-10}$$

This can be rearranged to give the less dense phase volume fraction in terms of the mass fraction and slip ratio:

$$\varepsilon_m = \frac{x}{x + S(1-x)\rho_L/\rho_S} \tag{15-11}$$

The local density of the mixture is given by

$$\rho_m = \varepsilon_m \rho_L + (1-\varepsilon_m)\rho_S \tag{15-12}$$

which depends on the slip ratio S through Eq. (15-11). The corresponding expression for the local in situ holdup of the more dense phase is

$$\varphi_m = 1 - \varepsilon_m = \frac{S(1-x)(\rho_L/\rho_S)}{x + S(1-x)(\rho_L/\rho_S)} \tag{15-13}$$

Note that both the local mixture density and the holdup increase as the slip ratio (S) increases. The "no slip" ($S = 1$) density or volume fraction is identical to the equilibrium value entering (or leaving) the pipe.

III. FLUID–SOLID TWO-PHASE PIPE FLOWS

The conveying of solids by a fluid in a pipe can involve a wide range of flow conditions and phase distributions, depending on the density, viscosity, and velocity of the fluid and the density, size, shape, and concentration of the solid particles. The flow regime can vary from essentially uniformly distributed solids in a "pseudohomogeneous" (symmetrical) flow regime for sufficiently small and/or light particles above a minimum concentration to an almost completely segregated or stratified (asymmetrical) transport of a bed of particles on the pipe wall. The demarcation between the "homogeneous" and "heterogeneous" flow regimes depends in a complex manner on the size and density of the solids, the fluid density and viscosity, the velocity of the mixture, and the volume fraction of solids. Figure 15-1 illustrates the approximate effect of particle size, density, and solids loading on these regimes.

Either a liquid or a gas can be used as the carrier fluid, depending on the size and properties of the particles, but there are important differences between hydraulic (liquid) and pneumatic (gas) transport. For example, in liquid (hydraulic) transport the fluid–particle and particle–particle interactions dominate over the particle–wall interactions, whereas in gas (pneumatic) transport the particle–particle and particle–wall interactions tend to dominate over the fluid–particle interactions. A typical "practical" approach, which gives reasonable results for a wide variety of flow conditions in both cases, is to determine the "fluid only" pressure drop and then apply a correction to account for the effect of the particles from the fluid–particle, particle–particle, and/or particle–wall interactions. A great number of publications have been devoted to this subject, and summaries of much of this work are given by Darby (1986), Govier and Aziz (1972), Klinzing et al. (1997), Molerus (1993), and Wasp et al. (1977). This approach will be addressed shortly.

A. Pseudohomogeneous Flows

If the solid particles are very small (e.g., typically less than 100 μm) and/or not tremendously denser than the fluid, and/or the flow is highly turbulent, the mixture may behave as a uniform suspension with essentially continuous properties. In this case, the mixture can be described as a "pseudo single-phase" uniform fluid and the effect of the presence of the particles can be

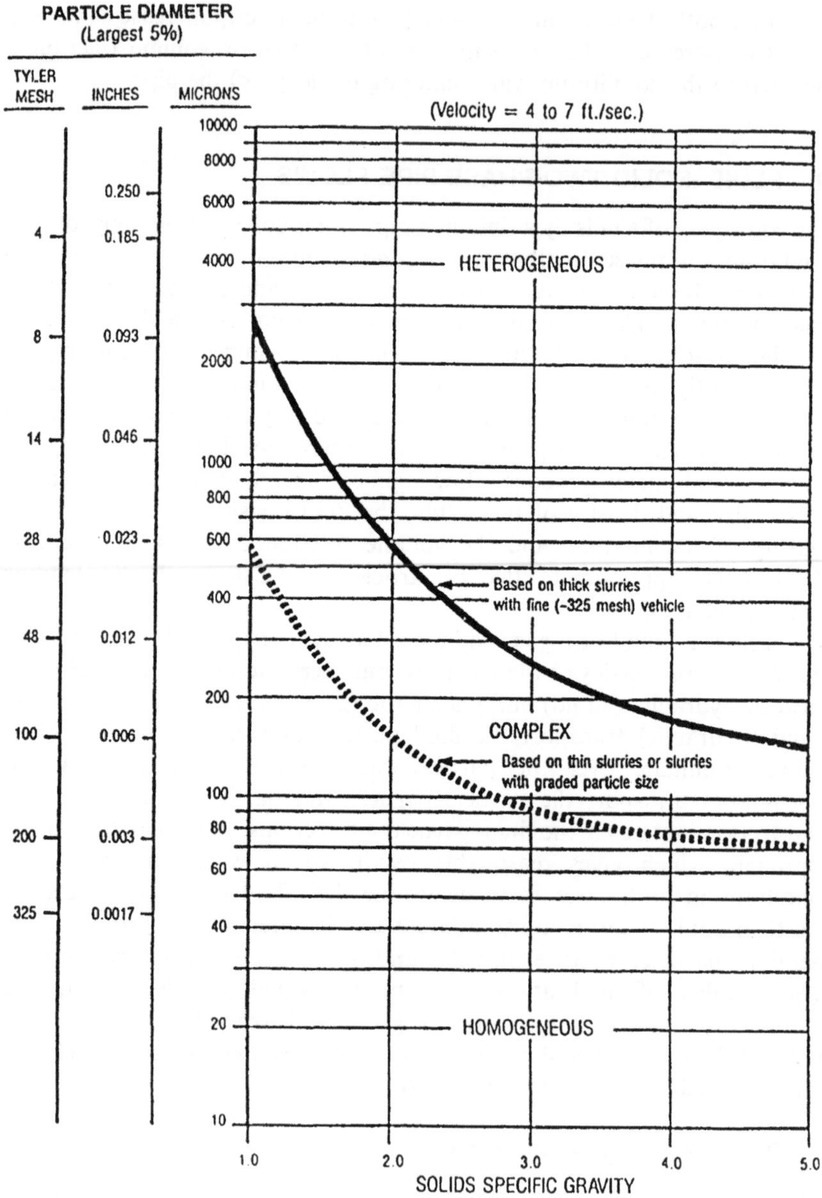

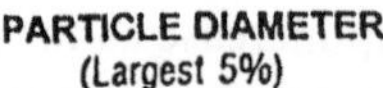

Figure 15-1 Approximate slurry flow regimes (From Aude et al., 1971.)

accounted for by appropriate modification of the fluid properties (density and viscosity). For relatively dilute suspensions (5% by volume or less), the mixture will behave as a Newtonian fluid with a viscosity given by the Einstein equation,

$$\mu = \mu_L(1 + 2.5\varphi) \tag{15-14}$$

where μ_L is the viscosity of the suspending (continuous) Newtonian fluid and $\varphi = 1 - \varepsilon$ is the volume fraction of solids. The density of the mixture is given by

$$\rho_m = \rho_L(1 - \varphi) + \rho_S\varphi \tag{15-15}$$

For greater concentrations of fine particles the suspension is more likely to be non-Newtonian, in which case the viscous properties can probably be adequately described by the power law or Bingham plastic models. The pressure drop–flow relationship for pipe flow under these conditions can be determined by the methods presented in Chapters 6 and 7.

B. Heterogeneous Liquid–Solid Flows

Figure 15-2 shows how the pressure gradient and flow regimes in a horizontal pipe depend on velocity for a typical heterogeneous suspension. It is seen that the pressure gradient exhibits a minimum at the "minimum deposit velocity," the velocity at which a significant amount of solids begins to settle in the pipe. A variety of correlations have been proposed in the literature for the minimum deposit velocity, one of the more useful being (Hanks, 1980)

$$V_{md} = 1.32\varphi^{0.186}[2gD(s - 1)]^{1/2}(d/D)^{1.6} \tag{15-16}$$

where $s = \rho_S/\rho_L$. At velocities below V_{md} the solids settle out in a bed along the bottom of the pipe. This bed can build up and plug the pipe if the velocity is too low, or it can be swept along the pipe wall if the velocity is near the minimum deposit velocity. Above the minimum deposit velocity, the particles are suspended but are not uniformly distributed ("symmetrical") until turbulent mixing is high enough to overcome the settling forces. One criterion for a nonsettling suspension is given by Wasp (1977):

$$\frac{V_t}{V^*} \leq 0.022 \tag{15-17}$$

where V_t is the particle terminal velocity and V^* is the *friction velocity*:

$$V^* = \sqrt{\frac{\tau_w}{\rho}} = \sqrt{\frac{\Delta PD}{4\rho L}} \tag{15-18}$$

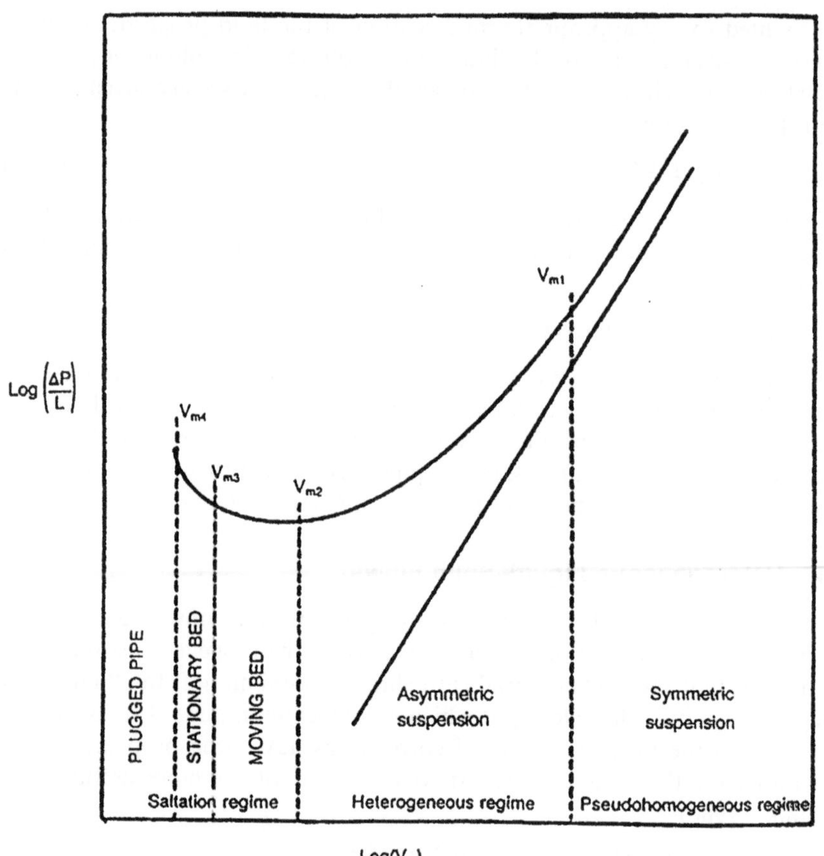

Figure 15-2 Pressure gradient and flow regimes for slurry flow in a horizontal pipe.

For heterogeneous flow, one approach to determining the pressure drop in a pipe is

$$\Delta P_m = \Delta P_L + \Delta P_S \tag{15-19}$$

where $\Delta P_L$is the "fluid only" pressure drop and ΔP_S is an additional pressure drop due to the presence of the solids. For uniformly sized particles in a Newtonian liquid, ΔP_L is determined as for any Newtonian fluid in a pipe. For a broad particle size distribution, the suspension may behave more like a heterogeneous suspension of the larger particles in a carrier vehicle composed of a homogeneous suspension of the finer particles. In this case, the homogeneous carrier will likely be non-Newtonian, and the methods given in Chapter 6 for such fluids should be used to determine ΔP_L.

The procedure for determining ΔP_S that will be presented here is that of Molerus (1993). The basis of the method is a consideration of the extra energy dissipated in the flow as a result of the fluid–particle interaction. This is characterized by the particle terminal settling velocity in an infinite fluid in terms of the drag coefficient, C_d:

$$C_d = \frac{4g(s-1)d}{3V_t^2} \tag{15-20}$$

where $s = \rho_S/\rho_L$. Molerus considered a dimensional analysis of the variables in this system, along with energy dissipation considerations, to arrive at the following dimensionless groups:

$$\frac{V_r}{V}\left(\frac{1}{s}\right)^{1/2} = \frac{\bar{V}_r}{\sqrt{s}} \tag{15-21}$$

$$N_{Fr_p}^2 = \frac{V^2}{(s-1)dg} \tag{15-22}$$

$$N_{Fr_t}^2 = \frac{V_t^2}{(s-1)Dg} \tag{15-23}$$

where V is the overall average velocity in the pipe, $V_r = V_L - V_S$ is the relative ("slip") velocity between the fluid and the solid, V_t is the terminal velocity of the solid particle, d is the particle diameter, D is the tube diameter, N_{Fr_p} is the particle Froude number, and N_{Fr_t} is the tube Froude number. The slip velocity is the key parameter in the mechanism of transport and energy dissipation, because the drag force exerted by the fluid on the particle depends on the *relative* velocity. That is, the fluid must move faster than the particles if it is to carry them along the pipe. The particle terminal velocity is related to the particle drag coefficient and Reynolds number, as discussed in Chapter 11 (e.g., unknown velocity), for either a Newtonian or non-Newtonian carrier medium.

Molerus (1993) developed a "state diagram" that shows a correlation between these dimensionless groups based on an extremely wide range of data covering $25 < D < 315$ mm, $12 < d < 5200\mu$m, and $1270 < \rho_S < 5250$ kg/m^3 for both hydraulic and pneumatic transport. This state diagram is shown in Fig. 15-3 in the form

$$\frac{\bar{V}_{r_0}}{\sqrt{s}} = \text{fn}(\sqrt{s}N_{Fr_p}, N_{Fr_t}^2) \tag{15-24}$$

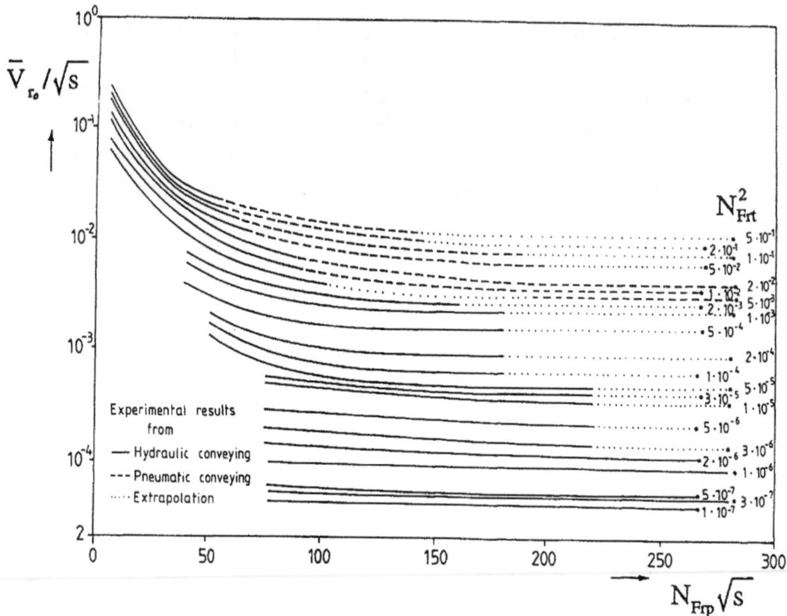

Figure 15-3 State diagram for suspension transport. (From Molerus, 1993.)

where $\bar{V}_{r_0}$ is the dimensionless "single-particle" slip velocity as determined from the diagram, which in turn is used to define the parameter X_0:

$$X_0 = \frac{\bar{V}^2_{r_0}}{1 - \bar{V}_{r_0}} \tag{15-25}$$

Using this value of X_0 and the entering solids volume fraction (φ), a value of X is determined as follows:

$$\text{For } 0 < \varphi < 0.25: \quad X = X_0$$
$$\text{For } \varphi > 0.25: \quad X = X_0 + 0.1 N^2_{\text{Fr}_t}(\varphi - 0.25)$$

The parameter X is the dimensionless solids contribution to the pressure drop:

$$\frac{\Delta P_S}{\varphi \rho_L (s - 1) g L} \left(\frac{V_t}{V}\right)^2 \equiv X \tag{15-26}$$

Knowing X determines ΔP_S, which is added to ΔP_L to get the total pressure drop in the pipe.

The foregoing procedure is straightforward if all the particles are of the same diameter (d). However, if the solid particles cover a broad range of sizes, the procedure must be applied for each particle size (diameter d_i, concentration φ_i) to determine the corresponding contribution of that particle size to the pressure drop ΔP_{Si}, The total solids contribution to the pressure drop is then $\Sigma \Delta P_{\mathrm{S}_i}$. If the carrier vehicle exhibits non-Newtonian properties with a yield stress, particles for which $d \leq \tau_o/(0.2g\,\Delta\rho)$ (approximately) will not fall at all.

For vertical transport, the major difference is that no "bed" can form on the pipe wall but, instead, the pressure gradient must overcome the weight of the solids as well as the fluid/particle drag. Thus the solids holdup and hence the fluid velocity are significantly higher for vertical transport conditions than for horizontal transport. However, vertical flow of slurries and suspensions is generally avoided where possible owing to the much greater possibility of plugging if the velocity drops.

Example 15-1: Determine the pressure gradient (in psi/ft) required to transport a slurry at 300 gpm through a 4 in. sch 40 pipeline. The slurry contains 50% (by weight) solids (SG = 2.5) in water. The slurry contains a bimodal particle size distribution, with half the particles below 100 μm and the other half about 2000 μm. The suspension of fines is stable and constitutes a pseudohomogeneous non-Newtonian vehicle in which the larger particles are suspended. The vehicle can be described as a Bingham plastic with a limiting viscosity of 30 cP and a yield stress of 55 dyn/cm^2.

Solution. First convert the mass fraction of solids to a volume fraction:

$$\varphi = \frac{x}{s - (s-1)x} = 0.286$$

where $s = \rho_S/\rho_L$. Half of the solids is in the non-Newtonian "vehicle," and half will be "settling," with a volume fraction of 0.143. Thus the density of the "vehicle" is

$$\rho_m = \rho_S\varphi + \rho_L(1-\varphi) = 1.215\,\mathrm{g/cm^3}$$

Now calculate the contribution to the pressure gradient due to the continuous Bingham plastic vehicle as well as the contribution from the "nonhomogeneous" solids. For the first part, we use the method presented in Section 6 V.C of Chapter 6 for Bingham plastics. From the given data, we can calculate $N_{\mathrm{Re,BP}} = 9540$ and $N_{\mathrm{He}} = 77,600$. From Eq. (6-62) this gives a friction factor of $f = 0.0629$ and a corresponding pressure gradient of $(\Delta P/L)_f = 2f\rho V^2/D = 1.105$ psi/ft.

The pressure gradient due to the heterogeneous component is determined by the Molerus method. This first requires the determination of the terminal velocity of the settling particles, using the method given in Chapter 11, Section IV. D, for the larger particles settling in a Bingham plastic. This requires determining $N_{Re,BP}$, N_{Bi} and C_d for the particle, all of which depend on V_t. This can be done using an iterative procedure to find V_t, such as the "solve" function on a calculator or spreadsheet. The result is $V_t = 19.5$ cm/s. This is used to calculate the particle and tube Froude numbers, $N^2_{Fr_p} = 25.6$ and $N^2_{Fr_t} = 0.0358$. These values are used with Fig. 15-3 to find $(\bar{V}_r/\sqrt{s}) = 0.05$, which corresponds to a value of $X = 0.00279$. From the definition of X, this gives $(\Delta P/L)_s = 0.0312$ psi/ft and thus a total pressure gradient of $(\Delta P/L)_t = 1.14$ psi/ft. In this case, the pressure drop due to the Bingham plastic "vehicle" is much greater than that due to the heterogeneous particle contribution.

C. Pneumatic Solids Transport

The transport of solid particles by a gaseous medium presents a considerable challenge, because the solid is typically three orders of magnitude more dense that the fluid (compared with hydraulic transport, in which the solid and liquid densities normally differ by less than an order of magnitude). Hence problems that might be associated with instability in hydraulic conveying are greatly magnified in the case of pneumatic conveying. The complete design of a pneumatic conveying system requires proper attention to the prime mover (fan, blower, or compressor), the feeding, mixing, and accelerating conditions and equipment; and the downstream separation equipment as well as the conveying system. A complete description of such a system is beyond the scope of this book, and the interested reader should consult the more specialized literature in the field, such as the extensive treatise of Klinzing et al. (1997).

One major difference between pneumatic transport and hydraulic transport is that the gas–solid interaction for pneumatic transport is generally much smaller than the particle–particle and particle–wall interaction. There are two primary modes of pneumatic transport: *dense phase* and *dilute phase*. In the former, the transport occurs below the *saltation* velocity (which is roughly equivalent to the minimum deposit velocity) in plug flow, dune flow, or sliding bed flow. Dilute phase transport occurs above the saltation velocity in suspended flow. The saltation velocity is not the same as the entrainment or "pickup" velocity, however, which is approximately 50% greater than the saltation velocity. The pressure gradient–velocity relationship is similar to the one for hydraulic transport, as shown in

Fig.15-4, except that transport is possible in the dense phase in which the pressure gradient, though quite large, is still usually not as large as for hydraulic transport. The entire curve shifts up and to the right as the solids mass flux increases. A comparison of typical operating conditions for dilute and dense phase pneumatic transport is shown in Table 15-1.

Although lots of information is available on dilute phase transport that is useful for designing such systems, transport in the dense phase is much more difficult and more sensitive to detailed properties of the specific solids. Thus, because operating experimental data on the particular materials of interest are usually needed for dense phase transport, we will limit our treatment here to the dilute phase.

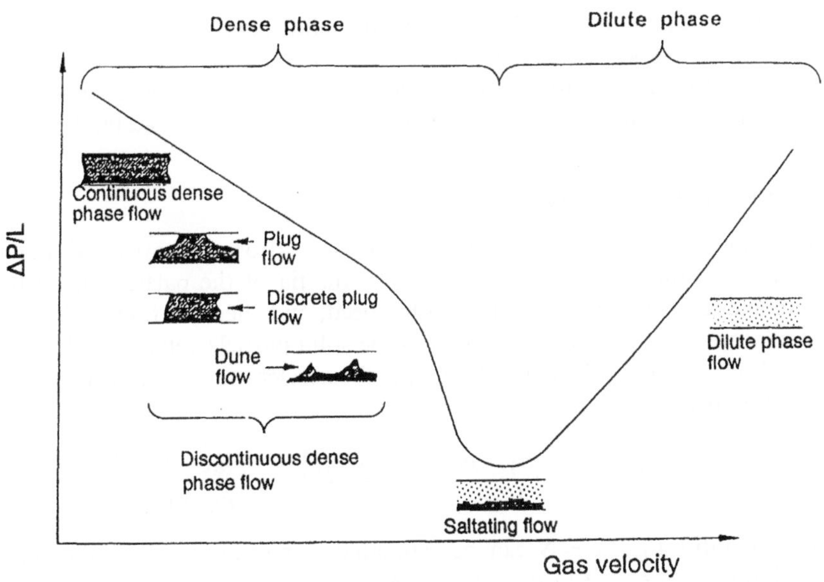

FIGURE 15-4 Pressure gradient–velocity relation for horizontal pneumatic flow.

TABLE 15-1 Dilute vs. Dense Phase Pneumatic Transport

Conveying mode	Solid loading (e.g. kg_S/kg_G	Conveying velocity [ft/s (m/s)]	ΔP [psi (kPa)]	Solids volume fraction
Dilute phase	< 15	$> 35(10)$	$< 15(100)$	$< 1\%$
Dense phase	> 15	$< 35(10)$	$> 15(100)$	$> 30\%$

There are a variety of correlations for the saltation velocity, one of the most popular being that of Rizk (1973):

$$\mu_s = \frac{\dot{m}_S}{\dot{m}_G} = 10^{-\delta} N_{Fr_s}^{\chi} \tag{15-27}$$

where

$$\delta = 1.44d + 1.96, \qquad \chi = 1.1d + 2.5$$

and

$$N_{FR_s} = \frac{V_{g_s}}{\sqrt{gd}} \tag{15-28}$$

Here μ_s is the "solids loading" (mass of solids/mass of gas), V_{g_s} is the saltation gas velocity, and d is the particle diameter in mm. (It should be pointed out that correlations such as this are based, of necessity, on a finite range of conditions and have a relatively broad range of uncertainty, e.g., ±50–60% is not unusual.)

1. Horizontal Transport

Two major effects contribute to the pressure drop in horizontal flow: acceleration and friction loss. Initially the inertia of the particles must be overcome as they are accelerated up to speed, and then the friction loss in the mixture must be overcome. If V_S is the solid particle velocity and $\dot{m}_S = \rho_S V_S (1 - \varepsilon) A$ is the solids mass flow rate, the acceleration component of the pressure drop is

$$\Delta P_{ac_S} + \Delta P_{ac_G} = V_S \frac{\dot{m}_S}{A} + \frac{\rho_G V_G^2}{2} = \frac{\rho_G V_G^2}{2}\left[1 + 2\frac{\dot{m}_S}{\dot{m}_G}\left(\frac{V_S}{V_G}\right)\right] \tag{15-29}$$

The slip ratio $V_G/V_S = S$ can be estimated, for example, from the IGT correlation (see, e.g., Klinzing et al., 1997):

$$\frac{1}{S} = \frac{V_S}{V_G} = 1 - \frac{0.68 d^{0.92} \rho_S^{0.5}}{\rho_G^{0.2} D^{0.54}} \tag{15-30}$$

in which d and D are in meters and ρ_S and ρ_G are in kg/m^3. For vertical transport, the major differences are that no "bed" on the pipe wall is possible, instead, the pressure gradient must overcome the weight of the solids as well as the fluid/particle drag, so that the solids holdup and hence the fluid velocity must be significantly higher under transport conditions.

The steady flow pressure drop in the pipe can be deduced from a momentum balance on a differential slice of the fluid–particle mixture in a constant diameter pipe, as was done in Chapter 5 for single-phase flow (see Fig. 5-6). For steady uniform flow through area A_x,

$$\sum F_x = 0 = dF_{x_p} + dF_{x_g} + dF_{x_w}$$
$$= -A_x dP - [\rho_S(1-\varepsilon_m) + \rho_G \varepsilon_m] g A_x\, dz - [\tau_{w_S} + \tau_{w_G}] W_p\, dX \tag{15-31}$$

where τ_{w_S} and τ_{w_G} are the effective wall stresses resulting from energy dissipation due to the particle–particle as well as particle–wall and gas–wall interaction, and W_p is the wetted perimeter. Dividing by A_x, integrating, and solving for the pressure drop, $-\Delta P = P_1 - P_2$,

$$-\Delta P = [\rho_S(1-\varepsilon_m) + \rho_G \varepsilon_m] g\, \Delta z + (\tau_{w_S} + \tau_{w_G}) 4L/D_h \tag{15-32}$$

where $D_h = 4A_x/W_p$ is the hydraulic diameter. The void fraction ε_m is the volume fraction of gas in the pipe, i.e.,

$$\varepsilon_m = 1 - \frac{\dot{m}_S}{\rho_S V_S A} = \frac{x}{x + S(1-x)\rho_G/\rho_S} \tag{15-33}$$

The wall stresses are related to corresponding friction factors by

$$\tau_{w_S} = \frac{f_S}{2} \rho_S (1-\varepsilon_m) V_S^2 = \frac{\Delta P_{f_S}}{4L/D_h} \tag{15-34a}$$

$$\tau_{w_G} = \frac{f_G}{2} \varepsilon_m P_G V_G^2 = \frac{\Delta P_{f_G}}{4L/D_h} \tag{15-34b}$$

Here ΔP_{f_G} is the pressure drop due to "gas only" flow (i.e., the gas flowing alone in the full pipe cross section). Note that if the pressure drop is less than about 30% of P_1, the incompressible flow equations can be used to determine ΔP_{f_G} by using the average gas density. Otherwise, the compressibility must be considered and the methods in Chapter 9 used to determine ΔP_{f_G}. The pressure drop is related to the pressure ratio P_1/P_2 by

$$P_1 - P_2 = \left(1 - \frac{P_2}{P_1}\right) P_1 \tag{15-35}$$

The solids contribution to the pressure drop, ΔP_{f_S}, is a consequence of both the particle–wall and particle–particle interactions. The latter is reflected in the dependence of the friction factor f_S on the particle diameter, drag coefficient, density, and relative (slip) velocity by (Hinkel, 1953):

$$f_S = \frac{3}{8}\left(\frac{\rho_S}{\rho_G}\right)\left(\frac{D}{d}\right) C_d \left(\frac{V_G - V_S}{V_S}\right)^2 \tag{15-36}$$

A variety of other expressions for f_S have been proposed by various authors (see, e.g., Klinzing et al., 1997), such as that of Yang (1983) for horizontal flow,

$$f_S = 0.117\left(\frac{1-\varepsilon}{\varepsilon^3}\right)\left[(1-\varepsilon)\frac{N_{Re_t}}{N_{Re_p}}\left(\frac{V_G/\varepsilon}{\sqrt{gD}}\right)\right]^{-1.15} \tag{15-37}$$

and for vertical flow,

$$f_S = 0.0206\left(\frac{1-\varepsilon}{\varepsilon^3}\right)\left[(1-\varepsilon)\frac{N_{Re_t}}{N_{Re_p}}\right]^{-0.869} \tag{15-38}$$

where

$$N_{Re_t} = \frac{dV_t\rho_G}{\mu_G}, \qquad N_{Re_p} = \frac{d(V_G/\varepsilon - V_S)\rho_G}{\mu_G} \tag{15-39}$$

and V_t is the particle terminal velocity.

2. Vertical Transport

The principles governing vertical pneumatic transport are the same as those just given, and the method for determining the pressure drop is identical (with an appropriate expression for f_P). However, there is one major distinction in vertical transport, which occurs as the gas velocity is decreased. As the velocity drops, the frictional pressure drop decreases but the slip increases, because the drag force exerted by the gas entraining the particles also decreases. The result is an increase in the solids holdup, with a corresponding increase in the static head opposing the flow, which in turn causes an increase in the pressure drop. A point will be reached at which the gas can no longer entrain all the solids and a slugging, fluidized bed results with large pressure fluctuations. This condition is known as *choking* (not to be confused with the choking that occurs when the gas velocity reaches the speed of sound) and represents the lowest gas velocity at which vertical pneumatic transport can be attained at a specified solids mass flow rate. The choking velocity, V_C, and the corresponding void fraction, ε_C, are related by the two equations (Yang, 1983)

$$\frac{V_C}{V_t} = 1 + \frac{V_S}{V_t(1-\varepsilon_C)} \tag{15-40}$$

and

$$\frac{2gD(\varepsilon_C^{-4.7} - 1)}{(V_C - V_t)^2} = 6.81 \times 10^5\left(\frac{\rho_o}{\rho_S}\right)^{2.2} \tag{15-41}$$

These two equations must be solved simultaneously for V_C and ε_C.

IV. GAS–LIQUID TWO-PHASE PIPE FLOW

The two-phase flow of gases and liquids has been the subject of literally thousands of publications in the literature, and it is clear that we can provide only a brief introduction to the subject here. Although the single phase flow of liquids and gases is relatively straightforward, the two-phase combined flow is orders of magnitude more complex. Two-phase gas–liquid flows are also more complex than fluid–solid flows because of the wider variety of possible flow regimes and the possibility that the liquid may be volatile and/or the gas a condensable vapor, with the result that the mass ratio of the two phases may change throughout the system.

A. Flow Regimes

The configuration or distribution of the two phases in a pipe depends on the phase ratio and the relative velocities of the phases. These regimes can be described qualitatively as illustrated in Fig. 15-5a for horizontal flow and in Fig. 15-5b for vertical flow. The patterns for horizontal flow are seen to be more complex than those for vertical flow because of the asymmetrical effect of gravity. The boundaries or transitions between these regimes have been mapped by various investigators on the basis of observations in terms of various flow and property parameters. A number of these maps have been compared by Rouhani and Sohal (1983). Typical flow regime maps for horizontal and vertical flow are shown in Figs. 15-6a and 15-6b. In Figures 15-5 and 15-6, $G_G = \dot{m}_G/A$ is the mass flux of the gas, $G_L = \dot{m}_L/A$ is the mass flux of the liquid, and λ and Φ are fluid property correction factors:

$$\lambda = \left(\frac{\rho_G}{\rho_A}\right)\left(\frac{\rho_L}{\rho_W}\right)^{1/2} \tag{15-42}$$

$$\Phi = \frac{\sigma_W}{\sigma_L}\left[\frac{\mu_L}{\mu_W}\left(\frac{\rho_W}{\rho_L}\right)^2\right]^{1/2} \tag{15-43}$$

where σ is the surface tension and the subscripts W and A refer to water and air, respectively, at 20°C. A quantitative model for predicting the flow regime map for horizontal flow in terms of five dimensionless variables was developed by Taitel and Duckler (1976).

The momentum equation written for a differential length of pipe containing the two-phase mixture is similar to Eq. (15-31), except that the rate of momentum changes along the tube due to the change in

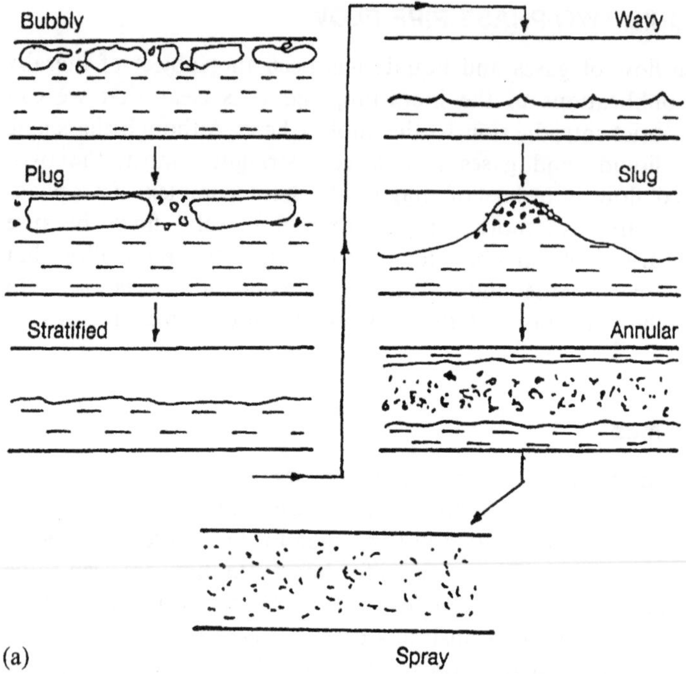

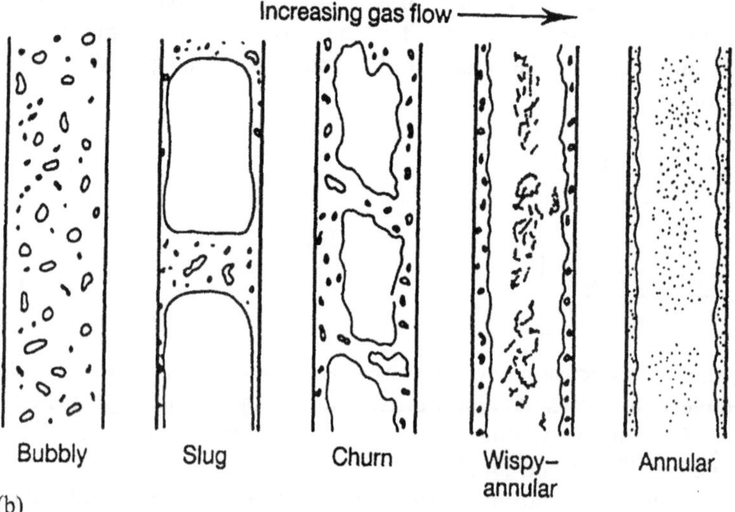

Figure 15-5 Flow regimes in (a) horizontal and (b) vertical gas–liquid flow.

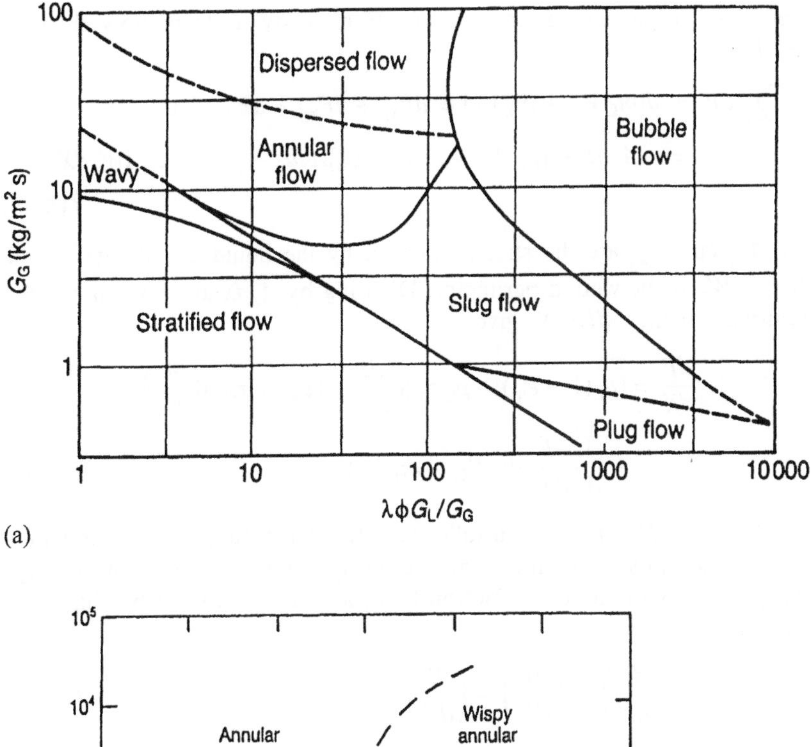

(a)

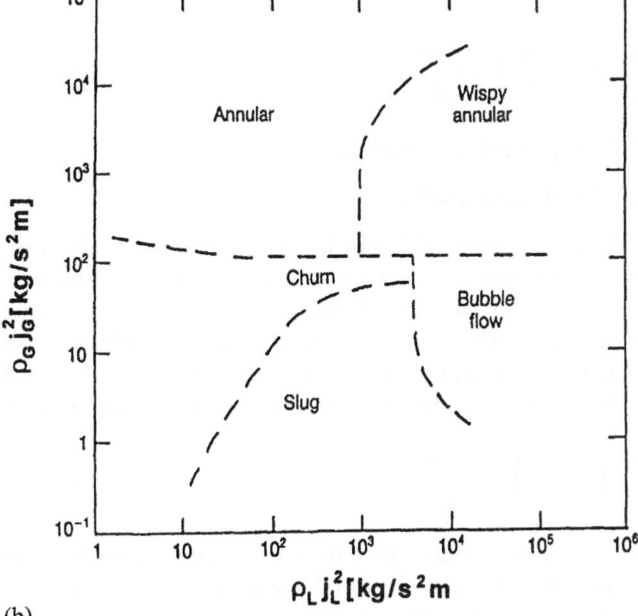

(b)

FIGURE 15-6 Flow regime maps for (a) horizontal and (b) vertical gas–liquid flow. (a, From Baker, 1954; b, from Hewitt and Roberts, 1969.)

velocity as the gas or vapor expands. For steady uniform flow through area A_x,

$$\sum dF_x = d[\dot{m}_G V_G + \dot{m}_L V_L] = dF_{x_P} + dF_{x_G} + dF_{x_W} = 0$$
$$= -A_x dP - [\rho_L(1-\varepsilon_m) + \rho_G \varepsilon_m] A_x dz - [\tau_{W_L} + \tau_{W_G}] W_p dX \tag{15-44}$$

where τ_{W_L} and τ_{W_G} are the stresses exerted by the liquid and the gas on the wall and W_p is the wetted perimeter. Dividing by $A_x dx$ and solving for the pressure gradient, $-dP/dX$, gives

$$-\frac{dP}{dX} = [\rho_L(1-\varepsilon_m) + \rho_G \varepsilon_m] g \frac{dz}{dX} + (\tau_{W_L} + \tau_{W_G})\left(\frac{4}{D_h}\right) + \frac{1}{A_x}\frac{d}{dX}(\dot{m}_G V_G + \dot{m}_L V_L) \tag{15-45}$$

where $D_h = 4A_x/W_p$ is the hydraulic diameter. The total pressure gradient is seen to be composed of three terms resulting from the static head change (gravity), energy dissipation (friction loss), and acceleration (the change in kinetic energy):

$$-\frac{dP}{dX} = -\left(\frac{dP}{dX}\right)_g - \left(\frac{dP}{dX}\right)_f - \left(\frac{dP}{dX}\right)_{acc} \tag{15-46}$$

This is comparable to Eq. (9-14) for pure gas flow.

1. Homogeneous Gas–Liquid Models

In principle, the energy dissipation (friction loss) associated with the gas–liquid, gas–wall, and liquid–wall interactions can be evaluated and summed separately. However, even for distributed (nonhomogeneous) flows it is common practice to evaluate the friction loss as a single term, which, however, depends in a complex manner on the nature of the flow and fluid properties in both phases. This is referred to as the "homogeneous" model:

$$-\left(\frac{dP}{dX}\right)_f = \frac{4f_m}{D_h}\left(\frac{\rho_m V_m^2}{2}\right) = \frac{2f_m G_m^2}{\rho_m D_h} \tag{15-47}$$

The homogeneous model also assumes that both phases are moving at the same velocity, i.e., no slip. Because the total mass flux is constant, the acceleration (or kinetic energy change) term can be written

$$-\left(\frac{dP}{dX}\right)_{acc} = \rho_m V_m \frac{dV_m}{dX} = G_m^2 \frac{dv_m}{dX} \tag{15-48}$$

where $\nu_{\mathrm{m}} = 1/\rho_{\mathrm{m}}$ is the average specific volume of the homogeneous two-phase mixture:

$$\nu_{\mathrm{m}} = \frac{1}{\rho_{\mathrm{m}}} = \frac{x}{\rho_{\mathrm{G}}} + \frac{1-x}{\rho_{\mathrm{L}}} = \nu_{\mathrm{G}} x + (1-x)\nu_{\mathrm{L}} \tag{15-49}$$

and x is the quality (i.e., the mass fraction of gas). For "frozen" flows in which there is no phase change (e.g., air and cold water), the acceleration term is often negligible in steady pipe flow (although it can be appreciable in entrance flows and in nonuniform channels). However, if a phase change occurs (e.g., flashing of hot water or other volatile liquid), this term can be very significant. Evaluating the derivative of ν_{m} from Eq. (15-49) gives

$$\frac{d\nu_{\mathrm{m}}}{dX} = x\frac{d\nu_{\mathrm{G}}}{dX} + (\nu_{\mathrm{G}} - \nu_{\mathrm{L}})\frac{dx}{dX} = x\frac{d\nu_{\mathrm{G}}}{dP}\frac{dP}{dx} + \nu_{\mathrm{GL}}\frac{dx}{dX} \tag{15-50}$$

where $\nu_{\mathrm{GL}} = \nu_{\mathrm{G}} - \nu_{\mathrm{L}}$. The first term on the right describes the effect of the gas expansion on the acceleration for constant mass fraction, and the last term represents the additional acceleration resulting from a phase change from liquid to gas (e.g., a flashing liquid).

Substituting the expressions for the acceleration and friction loss pressure gradients into Eq. (15-45) and rearranging gives

$$-\frac{dP}{dX} = \frac{\dfrac{2f_{\mathrm{m}}G_{\mathrm{m}}^2}{\rho_{\mathrm{m}}D} + G_{\mathrm{m}}^2\nu_{\mathrm{GL}}\dfrac{dx}{dX} + \rho_{\mathrm{m}}g\dfrac{dz}{dX}}{1 + G_{\mathrm{m}}^2 x\dfrac{d\nu_{\mathrm{G}}}{dP}} \tag{15-51}$$

Finding the pressure drop corresponding to a total mass flux G_{m} from this equation requires a stepwise procedure using physical property data from which the densities of both the gas phase and the mixture can be determined as a function of pressure. For example, if the upstream pressure P_1 and the mass flux G_{m} are known, the equation is used to evaluate the pressure gradient at point 1 and hence the change in pressure ΔP over a finite length ΔL, and hence the pressure $P_{1+i} = P_1 - \Delta P$. The densities are then determined at pressure P_{1+i}. And the process is repeated at successive increments until the end of the pipe is reached.

A number of special cases permit simplification of the equation. For example, if the pressure is high and the pressure gradient moderate, the term in the denominator that represents the acceleration due to gas expansion can be neglected. Likewise for "frozen" flow, for which there is no phase change (e.g., air and cold water), the quality x is constant and the second term in the numerator is zero. For flashing flows, the change in quality with length (dx/dX) must be determined from a total energy balance from the pipe inlet (or stagnation) conditions, along with the appropriate vapor–liquid equilibrium data for the flashing liquid. If the Clausius–Clapeyron equation is used, this becomes

$$\left(\frac{\partial \nu_G}{\partial P}\right)_T = -\frac{\nu_{GL}^2 C_p T}{\lambda_{GL}^2} \tag{15-52}$$

where λ_{GL} is the heat of vaporization and ν_{GL} is the change in specific volume at vaporization. For an ideal gas,

$$\left(\frac{\partial \nu_G}{\partial P}\right)_T = -\frac{1}{\rho P}, \qquad \left(\frac{\partial \nu_G}{\partial P}\right)_s = -\frac{P_1^{1/k}}{\rho_1 k P^{(1+k)/k}} \tag{15-53}$$

It should be noted that the derivative is negative, so that at certain conditions the denominator of Eq. (15-51) can be zero, resulting in an infinite pressure gradient. This condition corresponds to the speed of sound, i.e., choked flow. For a nonflashing liquid and an ideal gas mixture, the corresponding maximum (choked) mass flux G_m^* follows directly from the definition of the speed of sound:

$$G_m^* = c_m \rho_m = \rho_m \left[k \left(\frac{\partial P}{\partial \rho_m}\right)_T \right]^{1/2} = \sqrt{\frac{\rho_m k P}{\varepsilon}} \tag{15-54}$$

The ratio of the sonic velocity in a homogeneous two-phase mixture to that in a gas alone is $c_m/c = \sqrt{\rho_G/\varepsilon\rho_m} = \sqrt{\rho_L/\rho_L\varepsilon(1-\varepsilon)}$. This ratio can be much smaller than unity, so choking can occur in a two-phase mixture at a significantly higher downstream pressure than for single phase gas flow (i.e., at a lower pressure drop and a correspondingly lower mass flux).

Evaluation of each term in Eq. (15-51) is straightforward, except for the friction factor. One approach is to treat the two-phase mixture as a "pseudo-single phase" fluid, with appropriate properties. The friction factor is then found from the usual Newtonian methods (Moody diagram, Churchill equation, etc.) using an appropriate Reynolds number:

$$N_{Re,TP} = \frac{DG_m}{\mu_m} \tag{15-55}$$

where μ_m is an appropriate viscosity for the two-phase mixture. A wide variety of methods have been proposed for estimating this viscosity, but one that seems logical is the local volume-weighted average (Duckler et al., 1964b):

$$\mu_m = \varepsilon\mu_G + (1-\varepsilon)\mu_L \tag{15-56}$$

The corresponding density ρ is the "no-slip" or equilibrium density of the mixture:

$$\rho = \varepsilon\rho_G + (1-\varepsilon)\rho_L = \frac{1}{x/\rho_G + (1-x)/\rho_L} \tag{15-57}$$

Note that the frictional pressure gradient is inversely proportional to the fluid density:

$$\left(-\frac{\partial P}{\partial X}\right)_f = \frac{2f_m G_m^2}{\rho D} \tag{15-58}$$

The corresponding pressure gradient for purely liquid flow is

$$\left(-\frac{\partial P}{\partial X}\right)_{fL} = \frac{2f_L G_L^2}{\rho_L D} \tag{15-59}$$

Taking the reference liquid mass flux to be the same as that for the two-phase flow ($G_L = G_m$) and the friction factors to be the same ($f_L = f_m$), then

$$\left(-\frac{\partial P}{\partial X}\right)_{fm} = \frac{\rho_L}{\rho_G}\left(-\frac{\partial P}{\partial X}\right)_{fL} = \left(x\frac{\rho_L}{\rho_G} + 1 - x\right)\left(-\frac{\partial P}{\partial X}\right)_{fL} \tag{15-60}$$

A similar relationship could be written by taking the single phase gas flow as the reference instead of the liquid, i.e., $G_G = G_m$. This is the basis for the *two-phase multiplier* method:

$$\left(-\frac{\partial P}{\partial X}\right)_{fm} = \Phi_R^2\left(-\frac{\partial P}{\partial X}\right)_{fR} \tag{15-61}$$

where R represents a reference single-phase flow, and Φ_R^2 *is the two-phase multiplier*. There are four possible reference flows:

$R = L$	The total mass flow is liquid ($G_m = G_L$)
$R = G$	The total mass flow is gas ($G_m = G_G$)
$R = L_{Lm}$	The total mass flow is that of the liquid only in the mixture [$G_{Lm} = (1-x)G_m$]
$R = G_{Gm}$	The total mass flow is that of the gas only in the mixture ($G_{Gm} = xG_m$)

The two-phase multiplier method is used primarily for separated flows, which will be discussed later.

2. Omega Method for Homogeneous Equilibrium Flow

For homogeneous equilibrium (no-slip) flow in a uniform pipe, the governing equation is [equivalent to Eq. (15-45)]

$$\frac{dP}{dX} + G_m^2 \frac{dv_m}{dX} + \frac{2f_m v_m G_m^2}{2D} + \frac{g\,\Delta z}{V_m L} = 0 \tag{15-62}$$

where $\nu_m = 1/\rho_m$. By integrating over the pipe length L, assuming the friction factor to be constant, this can be rearranged as follows:

$$\frac{4f_m L}{D} = K_f = \int \frac{-\nu_m (1 + G_m^2\, d\nu_m/dP)\, dP}{[G_m^2(\nu_m^2/2) + (gD/4f_m)(\Delta z/L)]} \tag{15-63}$$

Leung (1996) used a linearized two-phase equation of state to evaluate $\nu_m = \text{fn}(P)$:

$$\frac{\nu_m}{\nu_o} = \omega\left(\frac{P_0}{P} - 1\right) + 1 = \frac{\rho_0}{\rho_m} \tag{15-64}$$

where ρ_0 is the two-phase density at the upstream (stagnation) pressure P_0. The parameter ω represents the compressibility of the fluid and can be determined from property data for $\rho = \text{fn}(P)$ at two pressures or estimated from the physical properties at the upstream (stagnation) state. For flashing systems,

$$\omega = \varepsilon_0\left(1 - 2\frac{P_0 \nu_{GL_0}}{\lambda_{GL_0}}\right) + \frac{C_{pL_0} T_0 P_0}{\nu_0}\left(\frac{\nu_{GL_0}}{\lambda_{GL_0}}\right)^2 \tag{15-65}$$

and for nonflashing (frozen) flows,

$$\omega = \varepsilon_0/k \tag{15-66}$$

Using Eq. (15-64), Eq. (15-63) can be written

$$\frac{4f_m L}{D} = K_f = -\int_{\eta_1}^{\eta_2} \frac{[(1-\omega)\eta^2 + \omega\eta](1 - G^{*2}\omega/\eta^2)\, d\eta}{G^{*2}[(1-\omega)\eta + \omega]^2/2 + \eta^2 N_{fi}} \tag{15-67}$$

where $\eta = P/P_0$, $G^* = G_m/(P_0\rho_0)^{1/2}$ and

$$N_{Fi} = \frac{\rho_0 g\,\Delta z}{P_0(4f_m L/D)} \tag{15-68}$$

is the "flow inclination number." From the definition of the speed of sound, it follows that the exit pressure ratio at which choking occurs is given by

$$\eta_{2c} = G_m^* \sqrt{\omega} \tag{15-69}$$

For horizontal flow, Eq. (15-67) can be evaluated analytically to give

$$\frac{4f_m L}{D} = \frac{2}{G^{*2}}\left[\frac{\eta_1 - \eta_2}{1-\omega} + \frac{\omega}{(1-\omega)^2}\ln\left(\frac{(1-\omega)\eta_2 + \omega}{(1-\omega)\eta_1 + \omega}\right)\right] - 2\ln\left[\frac{(1-\omega)\eta_2 + \omega}{(1-\omega)\eta_1 + \omega}\left(\frac{\eta_1}{\eta_2}\right)\right] \tag{15-70}$$

As $\omega \to 1$ [i.e., setting $\omega = 1.001$ in Eq. (15-70)], this reduces to the solution for ideal isothermal gas flow [Eq. (9-17)], and for $\omega = 0$ it reduces to the incompressible flow solution. For inclined pipes, Leung (1996) gives the solution of Eq. (15-67) in graphical form for various values of N_{Fi}.

3. Numerical Solutions

The Omega method is limited to systems for which the linearized two-phase equation of state [Eq. (15-64)] is a good approximation to the two-phase density (i.e., single-component systems that are not too near the critical temperature or pressure and multicomponent mixtures of similar compounds). For other systems, the governing equations for homogeneous flow can be evaluated numerically using either experimental or thermodynamic data for the two-phase $P-\rho$ relation or from a limited amount of data and a more complex nonlinear model for this relation. As an example, the program for such a solution for both homogeneous pipe and nozzle flow is included on a CD that accompanies a CCPS Guidelines book on pressure relief and effluent handling (CCPS, systems 1998). This program is simple to use but does require input data for the density of the two-phase mixture at either two or three pressures.

4. Separated Flow Models

The separated flow models consider that each phase occupies a specified fraction of the flow cross section and account for possible differences in the phase velocities (i.e., slip). There are a variety of such models in the literature, and many of these have been compared against data for various horizontal flow regimes by Duckler et al. (1964a), and later by Ferguson and Spedding (1995).

The "classic" Lockhart–Martinelli (1949) method is based on the two-phase multiplier defined previously for either liquid-only (L_m) or gas-only (G_m) reference flows, i.e.,

$$\left(-\frac{\partial P}{\partial X}\right)_{fm} = \Phi^2_{Lm}\left(-\frac{\partial P}{\partial X}\right)_{fLm} \tag{15-71}$$

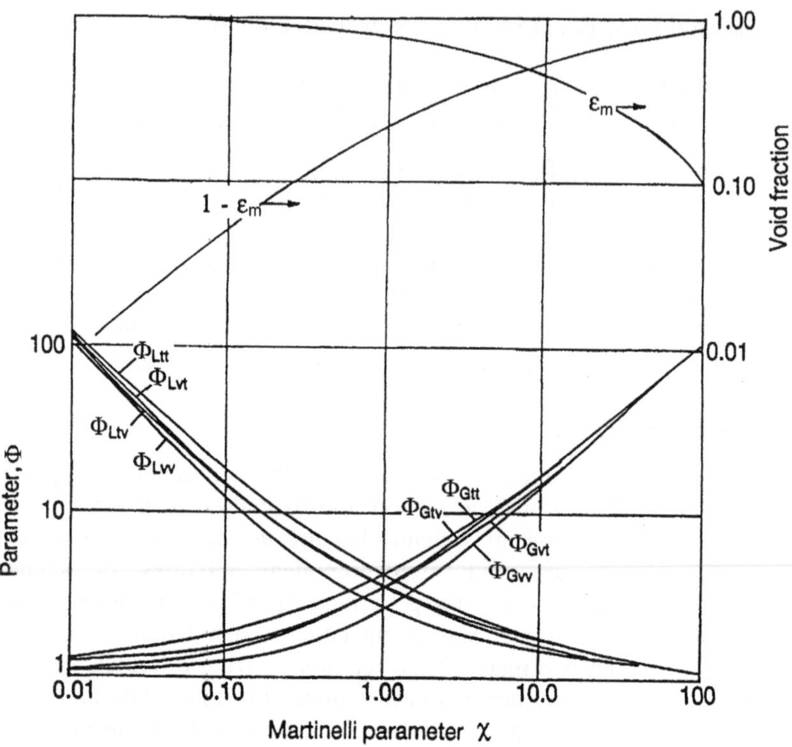

Figure 15-7 Lockhart–Martinelli two-phase multiplier.

or

$$\left(-\frac{\partial P}{\partial X}\right)_{\text{fm}} = \Phi_{\text{Gm}}^2\left(-\frac{\partial P}{\partial X}\right)_{\text{fGm}} \tag{15-72}$$

where the two-phase multiplier Φ is correlated as a function of the parameter χ as shown in Fig. 15-7. There are four curves for each multiplier, depending on the flow regime in each phase, i.e., both turbulent (tt), both laminar (vv), liquid turbulent and gas laminar (tv), or liquid laminar and gas turbulent (vt). The curves can also be represented by the equations

$$\Phi_{\text{Lm}}^2 = 1 + \frac{C}{\chi} + \frac{1}{\chi^2} \tag{15-73}$$

and

$$\Phi_{\text{Gm}}^2 = 1 + C\chi + \chi^2 \tag{15-74}$$

TABLE 15-2 Values of Constant C in Two-Phase Multiplier Equations

Flow state	Liquid	Gas	C
tt	Turbulent	Turbulent	20
vt	Laminar	Turbulent	12
tv	Turbulent	Laminar	10
vv	Laminar	Laminar	5

where the values of C for the various flow combinations are shown in Table 15-2.

The Lockhart–Martinelli correlating parameter χ^2 is defined as

$$\chi^2 = \left(-\frac{\partial P}{\partial X}\right)_{\text{fLm}} \left(-\frac{\partial P}{\partial X}\right)_{\text{fGm}}^{-1} \tag{15-75}$$

where

$$-\left(\frac{\partial P}{\partial X}\right)_{\text{fLm}} = \frac{2f_{\text{Lm}}(1-x)^2 G_{\text{m}}^2}{\rho_{\text{L}} D} \tag{15-76}$$

and

$$\left(-\frac{\partial P}{\partial X}\right)_{\text{fGm}} = \frac{2f_{\text{Gm}} x^2 G_{\text{m}}^2}{\rho_{\text{G}} D} \tag{15-77}$$

Here, f_{Lm} is the tube friction factor based on the "liquid-only" Reynolds number $N_{\text{ReLm}} = (1-x)G_{\text{m}}D/\mu_{\text{L}}$ and f_{Gm} is the friction factor based on the "gas-only" Reynolds number $N_{\text{Re}_{\text{Gm}}} = xG_{\text{m}}D/\mu_{\text{G}}$. The curves cross at $\chi = 1$, and it is best to use the "G" reference curves for $\chi < 1$ and the "L" curves for $\chi > 1$.

Using similarity analysis, Duckler et al. (1964b) deduced that

$$-\left(\frac{\partial P}{\partial X}\right)_{\text{fm}} = \frac{2f_{\text{L}} G_{\text{m}}^2}{\rho_{\text{L}} D}\left(\frac{\rho_{\text{L}}}{\rho_{\text{m}}}\right)\alpha(\varphi)\beta \tag{15-78}$$

or

$$-\left(\frac{\partial P}{\partial X}\right)_{\text{fm}} = \frac{2f_{\text{G}} G_{\text{m}}^2}{\rho_{\text{G}} D}\left(\frac{\rho_{\text{G}}}{\rho_{\text{m}}}\right)\alpha(\varphi)\beta \tag{15-79}$$

which is equivalent to the Martinelli parameters

$$\Phi_{\text{Lm}}^2 = \frac{\rho_{\text{L}}}{\rho_{\text{m}}}\alpha(\varphi)\beta \quad \text{and} \quad \Phi_{\text{Gm}}^2 = \frac{\rho_{\text{G}}}{\rho_{\text{m}}}\alpha(\varphi)\beta \tag{15-80}$$

where

$$\alpha(\varphi) = 1.0 + \frac{-\ln\varphi}{1.281 - 0.478(-\ln\varphi) + 0.444(-\ln\varphi)^2 - 0.094(-\ln\varphi)^3 + 0.00843(-\ln\varphi)^4} \tag{15-81}$$

and

$$\beta = \frac{\rho_L}{\rho_m}\left(\frac{\varphi^2}{\varphi_m}\right) + \frac{\rho_G}{\rho_m}\left(\frac{(1-\varphi)^2}{1-\varphi_m}\right) \tag{15-82}$$

and φ and ρ_m are the equilibrium (no-slip) properties. Another major difference is that Duckler et al. deduced that the friction factors f_L and f_G should both be evaluated at the *mixture* Reynolds number,

$$N_{Re_m} = \frac{DG_m}{\mu_m}\beta \tag{15-83}$$

5. Slip and Holdup

A major complication, especially for separated flows, arises from the effect of slip. Slip occurs because the less dense and less viscous phase exhibits a lower resistance to flow, as well as expansion and acceleration of the gas phase as the pressure drops. The result is an increase in the local holdup of the more dense phase within the pipe (φ_m) (or the corresponding two-phase density, ρ_m), as given by Eq. (15-11). A large number of expressions and correlations for the holdup or (equivalent) slip ratio have appeared in the literature, and the one deduced by Lockhart and Martinelli is shown in Fig. 15-7. Many of these slip models can be summarized in terms of a general equation of the form

$$S = a_0\left(\frac{1-x}{x}\right)^{a_1-1}\left(\frac{\rho_L}{\rho_G}\right)^{a_2-1}\left(\frac{\mu_L}{\mu_G}\right)^{a_3} \tag{15-84}$$

for which the values of the parameters are shown in Table 15-3. Although many additional slip models have been proposed in the literature, it is not clear which of these should be used under a given set of circumstances. In some cases, a constant slip ratio (S) may give satisfactory results. For example, in a comparison of calculated and experimental mass flux data for high velocity air–water flows through nozzles, Jamerson and Fisher (1999) found that $S = 1.1$–1.8 accurately represents the data over a range of $x = 0.02$–0.2, with the value of S increasing as the quality (x) increases.

A general correlation of slip is given by Butterworth and Hewitt (1977):

$$S = 1 + a\left(\frac{Y}{1+bY} - bY\right)^{1/2} \tag{15-85}$$

where

$$Y = \frac{x}{1-x}\left(\frac{\rho_L}{\rho_G}\right) \tag{15-86}$$

TABLE 15-3 Parameters for Slip Model Equation

Model	a_0	a_1	a_2	a_3
Homogeneous	1	1	−1	0
$S = (\rho_L/\rho_G)^{1/2}$ (Fauske, 1962)	1	1	−1/2	0
$S = (\rho_L/\rho_G)^{1/3}$ (Moody, 1965)	1	1	−2/3	0
Thom (1964)	1	1	−0.89	0.18
Baroczy (1966)	1	0.74	−0.65	0.13
Lockhart-Martinelli (tt) 1 (1949)	0.75	−0.417	0.083	
Constant $S = S_0$	S_0	1	−1	0

$$a = 1.578 N_{\mathrm{Re_L}}^{-0.19} \left(\frac{\rho_L}{\rho_G}\right)^{0.22} \tag{15-87}$$

$$b = 0.0273 N_{\mathrm{We}} N_{\mathrm{Re_L}}^{-0.51} \left(\frac{\rho_L}{\rho_G}\right)^{0.08} \tag{15-88}$$

$$N_{\mathrm{Re_L}} = \frac{G_m D}{\mu_L}, \qquad N_{\mathrm{We}} = \frac{G_m^2 D}{\sigma \rho_L} \tag{15-89}$$

An empirical correlation of holdup was developed by Mukherjee and Brill (1983) based on over 1500 measurements of air with oil and kerosene in horizontal, inclined, and vertical flow (inclination of ±90°). Their results for the holdup were correlated by an empirical equation of the form

$$\varepsilon_m = \exp\left[\left(c_1 + c_2 \sin\theta + c_3 \sin^2\theta + c_4 N_L \frac{N_{\mathrm{GV}}^{c_5}}{N_{\mathrm{LV}}^{c_6}}\right)\right] \tag{15-90}$$

where

$$N_L = \mu_L \left[\frac{g}{\rho_L \sigma^3}\right]^{0.25}, \qquad N_{\mathrm{LV}} = J_L \left[\frac{\rho_L}{g\sigma}\right]^{0.25}, \qquad N_{\mathrm{GV}} = J_G \left[\frac{\rho_L}{g\sigma}\right]^{0.25} \tag{15-91}$$

and σ is the liquid surface tension. The constants in Eq. (15-90) are given in Table 15-4 for the various flow inclinations.

A correlation for holdup by Hughmark (1962) was found to represent data quite well for both horizontal and vertical gas-liquid flow over a wide range of conditions. This was found by Duckler et al. (1964a) to be superior

TABLE 15-4 Coefficients for Eq. (15-90)

Flow direction	Flow pattern	C_1	C_2	C_3	C_4	C_5	C_6
Uphill and horizontal	All	−0.3801	0.12988	−0.1198	2.3432	0.47569	0.28866
Downhill	Stratified	−1.3303	4.8081	4.17584	56.262	0.07995	0.50489
	Other	−0.5166	0.78981	0.55163	15.519	0.37177	0.39395

to a number of other relations that were checked against a variety of data. The Hughmark correlation is equivalent to the following expression for slip:

$$S = \frac{1-K+[(1-x)/x]\,\rho_G/\rho_L}{K\,[(1-x)/x]\,\rho_G/\rho_L} \tag{15-92}$$

where the parameter K was found to correlate well with the diensionless parameter Z:

$$Z = N_{Re}^{1/6} N_{Fr}^{1/8}/(1-\varepsilon)^{1/4} \tag{15-93}$$

where ε is the "no-slip" volume fraction of gas. The volume average viscosity of the two phases is used in the Reynolds number, and $N_{Fr} = V^2/gD$ where V is the average velocity of the two-phase mixture. Hughmark presented the correlation between K and Z in graphical form, which can be represented quite well by the expression

$$K = \left(\frac{1}{1+0.12/Z^{0.95}}\right)^{19} \tag{15-94}$$

The presence of slip also means that the acceleration term in the general governing equation [Eq. (15-45)] cannot be evaluated in the same manner as the one for homogeneous flow conditions. When the acceleration term is expanded to account for the difference in phase velocities, the momentum equation, when solved for the total pressure gradient, becomes

$$-\frac{dP}{dX} = \frac{\left[\left(-\dfrac{\partial P}{\partial X}\right)_{fm} + G_m^2 \dfrac{dx}{dX} A(\varphi_m, x) + \rho_m g \dfrac{dz}{dX}\right]}{1 + G_m^2\left[\dfrac{x^2}{\varphi_m}\dfrac{dv}{dP} + \left(\dfrac{\partial \varphi_m}{\partial P}\right)_x \left(\dfrac{(1-x)^2}{\rho_L(1-x)^2} - \dfrac{x^2}{x^2\varphi_m^2}\right)\right]} \tag{15-95}$$

where

$$A(\varphi_m, x) = \left[\frac{2x}{\rho_G \varphi_m} - \frac{2(1-x)}{\rho_L(1-\varphi_m)}\right] + \left(\frac{\partial \varphi_m}{\partial x}\right)_P \left[\frac{(1-x)^2}{\rho_L(1-\varphi_m)^2} - \frac{x^2}{\rho_G \varphi_m^2}\right] \tag{15-96}$$

The pressure drop over a given length of pipe must be determined by a stepwise procedure, as described for homogeneous flow. The major additional complication in this case is evaluation of the holdup (φ_m) or the equivalent slip ratio (S) using one of the above correlations.

In some special cases simplifications are possible that make the process easier. For example,

1. If the denominator of Eq. (15-87) is close to unity
2. If f_m, ρ_L, and ρ_G are nearly constant over the length of pipe.

Example 15-2: Estimate the pressure gradient (in psi/ft) for a two-phase mixture of air and water entering a horizontal 6 in. sch 40 pipe at a total mass flow rate of 6500 lb_m/min at 150 psia, 60°F, with a quality (x) of 0.1 (mass friction air) lb_m air/lb_m water. Compare your answers using the (a) omega (b) Lockhart–Martinelli, and (c) Duckler methods.

Solution. At the entering temperature and pressure, the density of air is 0.7799 lb_m/ft^3, its viscosity is 0.02 cP, the density of water is 62.4 lb_m/ft^3, and its viscosity is 1 cP. The no-slip volume fraction corresponding to the given quality is [by Eq. (15-11)] 0.899, and the corresponding density of the mixture [by Eq. (15-12)] is 7.01 lb_m/ft^3. The viscosity of the mixture, by Eq. (15-56), is 0.119 cP. The slip ratio can be estimated from Eq. (15-84), using the Lockhart–Martinelli constants from Table 15-3, to be $S = 10.28$. Using this value in Eq. (15-13) gives the in situ holdup $\varphi_m = 0.6027$. From the given mass flow rate and diameter, the total mass flux $G_m = 540\,lb_m/(ft^2 s)$.

(a) *Omega method.* Since this is a "frozen" flow (no phase changes), the value of ω is given by Eq. (15-66), with $k = 1.4$ for air, which gives $\omega = 0.642$. From given data, $N_{Re_m} = DG_m/\mu_m = 3.41 \times 10^6$ which, assuming a pipe roughness of 0.0018 in., gives $f = 0.00412$ and a value of $4fL/D = 0.0326$. The pressure gradient is determined from Eq. (15-70), with $G* = G_m/(P_0\rho_0)^{1/2} = 0.245$ and $\eta_1 = 1$. The equation is solved by iteration for $\eta_2 = 0.999354$, or $P_2 = 149.903$ psia. The pressure gradient is thus $(P_1 - P_2)/L = 0.0969$ psi/ft. This pressure gradient will apply until the pressure drops to the choke pressure, which from Eq. (15-69) is 7.19 psia.

(b) *Lockhart–Martinelli method.* Using the "liquid-only" basis, the corresponding Reynolds number is $N_{ReLm} = (1 - x)$ $DG_m/\mu_L = 3.66 \times 10^5$, which gives a value of $f_{Lm} = 0.00419$. Likewise, using the "gas-only" basis gives $N_{Re_{Gm}} = x$ $DG_m/\mu_G = 2.27 \times 10^6$, which gives $f_{Gm} = 0.00383$. These values give the corresponding pressure gradients from Eqs. (15-76) and (15-77) as 0.0135 and 0.012 psi/ft, respectively. The square root of the ratio of these values gives the Lockhart–Martinelli parameter $X = 1.0527$, which, from Eq. (15-73), gives $\Phi^2_{Lm} = 20.9$. The pressure gradient is then calculated from Eq. (15-76) to be 0.283 psi/ft.

(c) *Duckler method.* This method requires determining values for β and α from Eq. (15-82) and (15-81), respectively. The in situ holdup determined above, $\varphi_m = 0.6027$, is used in the equation for β to give a value of 0.377, and the no-slip holdup value of $\varphi =$ 0.101 is used in the equation for α to give a value of 2.416. These values are used in Eq. (15-78), with a value of f = 0.00378, to determine the pressure gradient of 0.122 psi/ft.

The acceleration component to the total pressure gradient, from Eqs. (15-48), (15-50), and (15-53) should also be included for cases (b) and (c) but is negligible in these cases.

The flow regime can be determined from Fig.15-16a, using an ordinate of 1 and an abscissa of 2635 kg/(m^2 s) to be well in the dispersed flow regime, so each of these methods should be applicable.

PROBLEMS

1. An aqueous slurry is composed of 45% solids (by volume). The solids have an SG of 4 and the particle size distribution shown below.

US screen mesh	Mesh opening (μm)	Fraction passing
400	37	0.02
325	44	0.06
200	74	0.08
140	105	0.10
100	149	0.15
60	250	0.18
35	500	0.20
18	1000	0.12
10	2000	0.08
5	4000	0.01

The slurry behaves as a non-Newtonian fluid, which can be described as a Bingham plastic with a yield stress of 40 dyn/cm^2 and a limiting viscosity of 100 cP. Calculate the pressure gradient (in psi/ft) for this slurry flowing at a velocity of 8 ft/s in a 10 in. ID pipe.

2. Repeat Problem 1 but with the slurry described by the power law model, with a consistency of 60 poise and a flow index of 0.18.
3. Spherical polymer pellets with a diameter of 1/8 in. and an SG of 0.96 are to be transported pneumatically using air at 80°F. The pipeline is horizontal, 6 in. ID and 100 ft long, and discharges at atmospheric pressure. It is desired to transport 15% by volume of solids, at a velocity that is 1 ft/s above the minimum deposit velocity.
   (a) What is the pressure of the air that is required at the entrance to the pipe to overcome the friction loss in the pipe? (*Note*: An additional pressure gradient required to accelerate the particles after contacting with the air, but your answer should address only the friction loss.)
   (b) If a section of this pipe is vertical, (1) what would the choking velocity be in this line and (2) what would the pressure gradient (in psi/ft) be at a velocity of 1 ft/s above the choking velocity?
4. Saturated ethylene enters a 4 in. sch 40 pipe at 400 psia. The ethylene flashes as the pressure drops through the pipe, and the quality at any pressure can be estimated by applying a constant enthalpy criterion along the pipe. If the pipe is 80 ft long and discharges at a pressure of 100 psia, what is the mass flow rate through the pipe? Use 50 psi pressure increments in the stepwise calculation procedure.
5. Natural gas (methane) and 40° API crude oil are being pumped through a 6 in. sch 40 pipeline at 80°F. The mixture enters the pipe at 500 psia, a total rate of 6000 lb_m/min, and 6% quality. What is the total pressure gradient in the pipe at this point (in psi/ft)?

NOTATION

A_x	x component of area, [L^2]
c	speed of sound, [L/t]
C_d	particle drag coefficient, [—]
C_p	specific heat, [FL/MT = L^2/t^{2}T]
D	pipe diameter, [L]
D_h	hydraulic diameter, [L]
d	particle diameter, [L]
F	force, [F = ML/t^2]
f	Fanning friction factor, [—]
G	mass flux, [M/L^{2}t]
G^*	dimensionless mass flux, [—]
g	acceleration due to gravity, [L/t^2]
J	volume flux (superficial velocity), [L/t]
k	isentropic exponent (specific heat ratio for ideal gas), [—]

L	length, [L]
$\dot{m}$	mass flow rate, [M/t]
N_{Fi}	flow inclination number, [—]
N_{Fr_p}	particle Froude number, Eq. (15-22), [—]
N_{Fr_s}	solids Froude number, Eq. (15-28), [—]
N_{Fr_t}	pipe Froude number, Eq. (15-23), [—]
N_{Re_t}	particle terminal velocity Reynolds number, Eq. (15-39), [—]
N_{Re_p}	particle relative velocity Reynolds number, Eq. (15-39), [—]
$N_{Re,TP}$	two-phase Reynolds number, [—]
N_{We}	Weber number, Eq. (15-89), [—]
P	pressure, $[F/L^2 = m/Lt^2]$
V	velocity, [L/t]
V^*	friction velocity, Eq. (15-18), [L/t]
V_{md}	minimum deposit velocity, [L/t]
V_r	relative (slip) velocity [L/t]
$\bar{V}_r$	dimensionless slip velocity (V_r/V_m) , [—]
V_t	particle terminal velocity, [L/t]
Q	volumetric flow rate, $[L^3/t]$
S	velocity slip ratio, [—]
s	ρ_S/ρ_L, [—]
X	dimensionless solids contribution to pressure drop, Eq. (15-26), [—]
X	horizontal coordinate direction, [L]
x	mass fraction of less dense phase (quality, for gas–liquid flow), [—]
z	vertical direction measured upward, [L]
Z	dimensionless parameter defined by Eq. (15-93)
ε	volume fraction of the less dense phase, [—]
Φ	property correction factor, Eq. (15-43), [—]
Φ_R	two-phase multiplier with reference to single phase R, [—]
φ	volume fraction of the more dense phase, [—]
λ	latent heat, $[FL/M = L^2/t^2]$
λ	density correction factor, Eq. (15-42), [—]
η	pressure ratio, [—]
μ	viscosity, [M/(Lt)]
μ_s	ratio of mass of solids to mass of gas, [—]
υ	specific volume, $[L^3/M]$
ρ	density, $[M/L^3]$
σ	surface tension, $[F/L = \mu/t^2]$
τ	shear stress, $[F/L^2 = M/(Lt^2)]$
χ	Lockhart–Martinelli correlating parameter, Eq. (15-75), [—]
ω	two-phase equation of state parameter, Eq. (15-65), [—]

Subscripts

1, 2	reference points
A	air

C	choking condition
f	friction loss, fluid
G, g	gas
L	liquid
m	mixture
o	stagnation state
R	reference phase
S	solid
w	wall
W	water

REFERENCES

Aude TC, NT Cowper, and CT Crowe et al. Slurry piping systems: Trends, design methods, guidelines. Chem Eng, June 28, 1971, p 74.

Baker O. Simultaneous flow of oil and gas. Oil Gas J, 53:185–195, 1954.

Baroczy CJ. A systematic correction for two-phase pressure drop. CEP Symp Ser 62(44): 232–249, 1966.

Brodkey RS. the Phenomena of Fluid Motions, Reading, MA: Addison-Wesley, 1967.

Butterworth D, GF Hewitt, eds. Two-Phase Flow and Heat Transfer. Oxford, UK: Oxford Universits Press, 1977.

Chisholm D. Two-Phase Flow in Pipelines and Heat Exchangers. London: The Institution of Chemical Engineers. George Godwin, 1983.

CCPS (Center for Chemical Process Safety). Guidelines for Pressure Relief and Effluent Handling Systems. New York: AIChE, 1998.

Darby R. Hydrodynamics of slurries and suspensions. In: NP Cheremisinoff, ed. Encyclopedia of Fluid Mechanics, Vol. 5. Houston, TX: Gulf Pub 1985, pp 49–92.

Duckler AE, M Wicks III, RG Cleveland. Frictional pressure drop in two-phase flow: A. A comparison of existing correlations for pressure loss and holdup. AIChE J 10:38–43, 1964a.

Duckler AE, M Wicks III, RG Cleveland. Frictional pressure drop in two-phase flow: B. An approach through similarity analysis. AIChE J. 10:44–51, 1964b.

Fan L-S, C Zhu. Principles of Gas–Solid Flows. London: Cambridge Univ Press,1998.

Fauske HK. Contribution to the Theory of Two-Phase, One-Component Critical Flow, Argonne Natl Lab Rep SNL-6673, October 1962.

Ferguson MEG, PL Spedding. Measurement and prediction of pressure drop in two-phase flow. J Chem Technol Biotechnol 62:262–278, 1995.

Govier GW, K Aziz. The Flow of Complex Mixtures in Pipes. New York: Van Nostrand Reinhold, 1972.

Hanks RW. ASME Paper No. 80-PET-45. ASME Energy Sources Technol Conf New Orleans, LA, Feb 3–7, 1980.

Hetsroni G, ed. Handbook of Multiphase Systems. Washington, DC: Hemisphere, 1982.
Hewitt GF, DN Roberts. Studies of Two-Phase Flow Patterns by X-Ray and Flash Photography. Report AERE-M 2159. London: HMSO, 1969.
Hinkel BL. PhD Thesis, Georgia Institute of Technology, 1953 Atlanta, Ga.
Holland FA, DR Bragg. Fluid Flow for Chemical Engineers. 2nd ed. Edward Arnold, 1995.
Hughmark, GA. Holdup in gas-liquid flow. CEP, 58(4):62–65, 1962.
Jamerson SC, HG Fisher. Using constant slip ratios to model non-flashing (frozen) two-phase flow through nozzles. Process Safety Prog 18(2): 89–98, 1999.
Klinzing GE, RD Marcus, F Rizk, LS Leung. Pneumatic Conveying of Solids. 2nd ed. New York: Chapman and Hall, 1997.
Leung JC. Easily size relief devices and piping for two-phase flow. CEP, December 1996, pp 28–50.
Levy S. Two-Phase Flow in Complex Systems. New York: Wiley, 1999.
Lockhart RW, RC Martinelli. Proposed correlation of data for isothermal two-phase, two-component flow in pipes. CEP, 45 (1): 39–48, 1949.
Molerus O. Principles of Flow in Disperse Systems. New York: Chapman and Hall, 1993.
Moody FJ. Maximum flow rate of a single component two-phase mixture. Trans ASME, J Heat Transfer. February, 1965, pp 134–142.
Mukherjee H, JP Brill. Liquid holdup correlations for inclined two-phase flow. J Petrol Tech May 1983, pp 1003–1008.
Rizk F. Dissertation, University of Karlsruhe, 1973.
Rouhani SZ, MS Sohal. Two-phase flow patterns: A review of research results. Prog Nucl Energy 11(3): 219–259, 1983.
Shook CA, MC Roco. Slurry Flow: Principles and Practice. Stoneham, MA: Butterworth-Heinemann, 1991.
Taitel Y, AE Duckler. A model for predicting flow regime transitions in horizontal and near horizontal gas–liquid flow. AIChE J 22(1): 47–55, 1976.
Thom JRS. Prediction of pressure drop during forced circulation boiling of water. Int J Heat Mass Transfer 7:709–724, 1964.
Wallis GB. One-Dimensional Two-Phase Flow. New York: McGraw-Hill, 1969.
Wasp EJ, JP Kenny, RL Gandhi. Solid–Liquid Flow in Slurry Pipeline Transportation. Clausthal, Germany: Trans-Tech, 1977.
Yang WC. Powder Technol 35:143–150, 1983.

Appendix A

Viscosities and Other Properties of Gases and Liquids

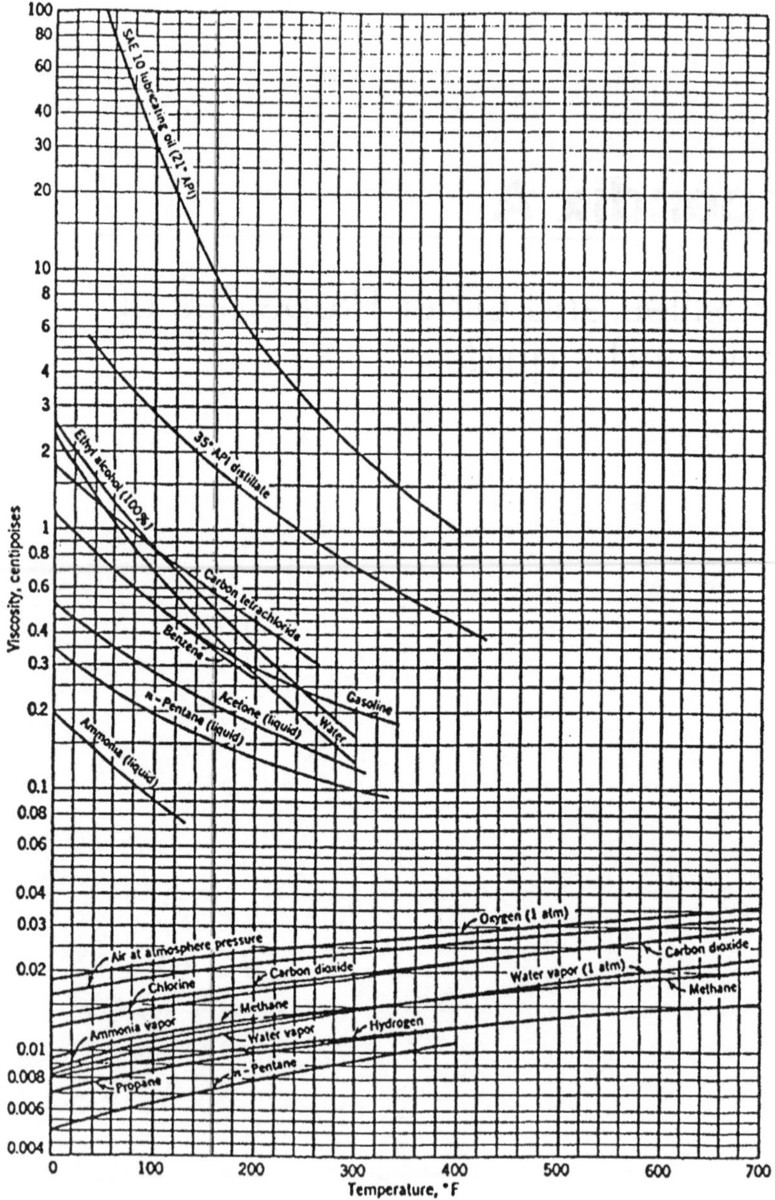

Figure A-1 Viscosities of various fluids at 1 atm pressure. 1 cp = 0.01 g/(cm s) = 6.72×10^{-4} lb_m/(ft s) = 2.42 lb_m/(ft hr) = 2.09×10^{-5} lb_f s/ft^2. (From GG Brown et al., *Unit Operations*, Wiley, New York, 1951, p 586. Reproduced by permission of the publisher.)

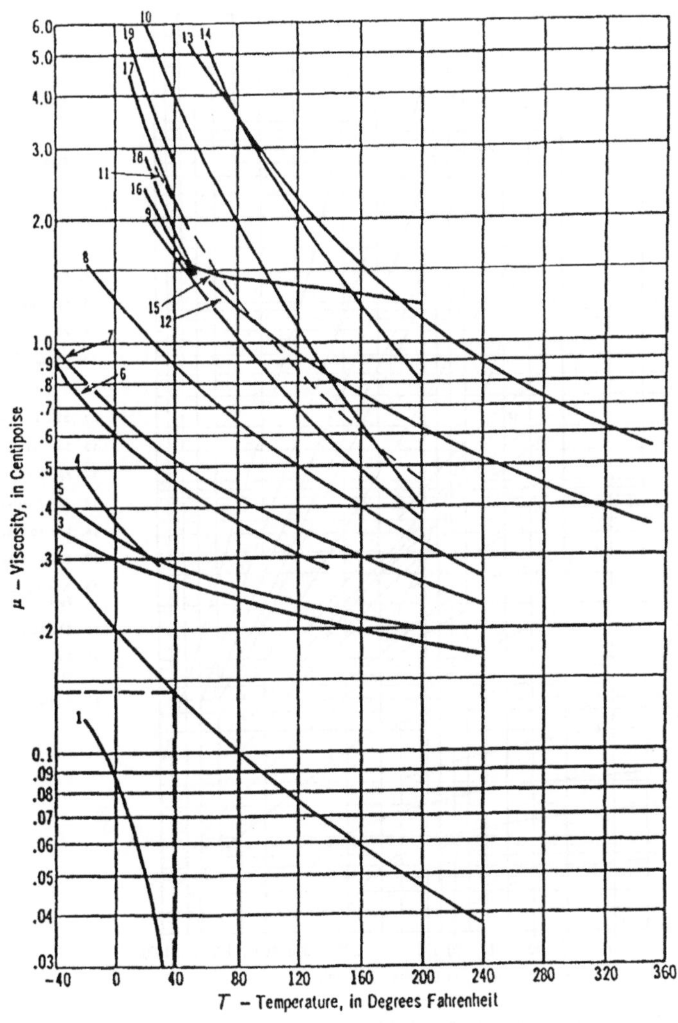

1. Carbon Dioxide . . CO_2
2. Ammonia . . . . NH_3
3. Methyl Chloride . . CH_3Cl
4. Sulphur Dioxide . . SO_2
5. Freon 12 . . . . F-12
6. Freon 114 . . . . F-114
7. Freon 11 . . . . F-11
8. Freon 113 . . . . F-113
9. Ethyl Alcohol
10. Isopropyl Alcohol
11. 20% Sulphuric Acid . . . . . . 20% H_2SO_4
12. Dowtherm E
13. Dowtherm A
14. 20% Sodium Hydroxide . . 20% NaOH
15. Mercury
16. 10% Sodium Chloride Brine . . . 10% NaCl
17. 20% Sodium Chloride Brine . . . 20% NaCl
18. 10% Calcium Chloride Brine . . 10% $CaCl_2$
19. 20% Calcium Chloride Brine . . 20% $CaCl_2$

Example: The viscosity of ammonia at 40 F is 0.14 centipoise.

FIGURE A-2 Viscosity of various liquids. From Crane Technical Paper 4-10, Crane Co. Chicago 1991.

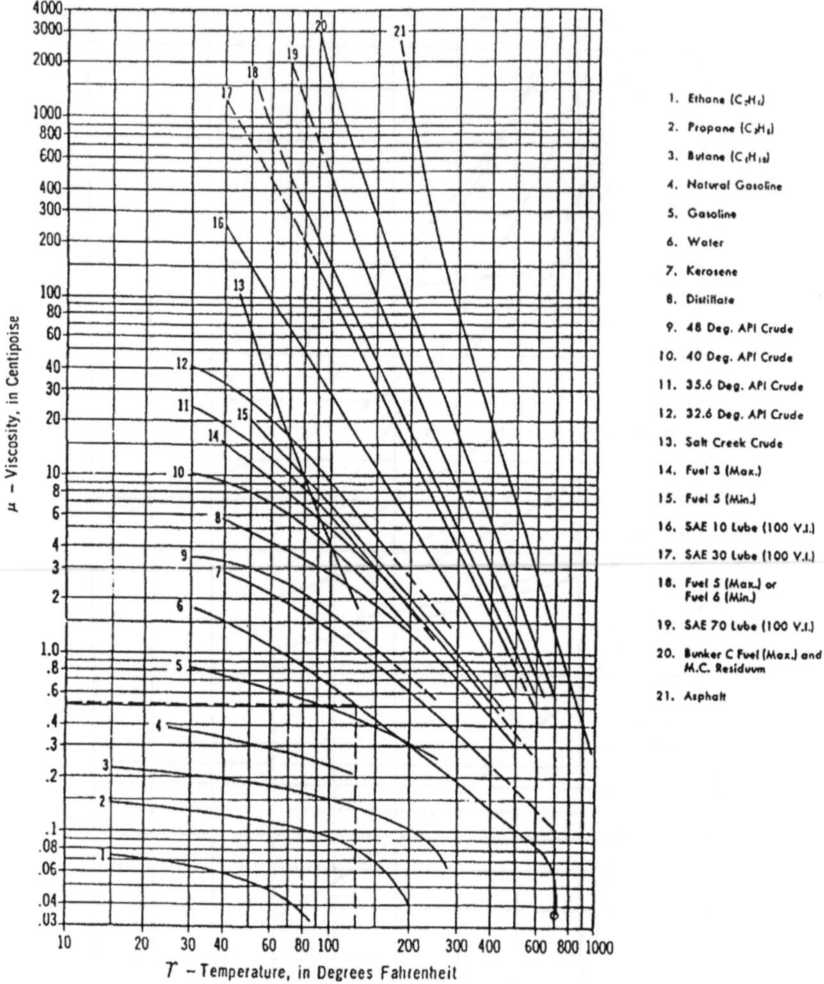

FIGURE A-3 Viscosity of water and liquid petroleum products. From Crane.

TABLE A-1 Viscosities of Liquids (Coordinates apply to Fig. A-4)

Liquid	*X*	*Y*
Acetaldehyde	15.2	4.8
Acetic acid, 100%	12.1	14.2
Acetic acid, 70%	9.5	17.0
Acetic anhydride	12.7	12.8
Acetone, 100%	14.5	7.2
Acetone, 35%	7.9	15.0
Acetonitrile	14.4	7.4
Acrylic acid	12.3	13.9
Allyl alcohol	10.2	14.3
Allyl bromide	14.4	9.6
Allyl iodide	14.0	11.7
Ammonia, 100%	12.6	2.0
Ammonia, 26%	10.1	13.9
Amyl acetate	11.8	12.5
Amyl alcohol	7.5	18.4
Aniline	8.1	18.7
Anisole	12.3	13.5
Arsenic trichloride	13.9	14.5
Benzene	12.5	10.9
Brine, CaCl(S)_2(S), 25%	6.6	15.9
Brine, NaCl, 25%	10.2	16.6
Bromine	14.2	13.2
Bromotoluene	20.0	15.9
Butyl acetate	12.3	11.0
Butyl acrylate	11.5	12.6
Butyl alcohol	8.6	17.2
Butyric acid	12.1	15.3
Carbon dioxide	11.6	0.3
Carbon disuofide	16.1	7.5
Carbon tetrachloride	12.7	13.1
Chlorobenzene	12.3	12.4
Chloroform	14.4	10.2
Chlorosulfonic acid	11.2	18.1
Chlorotoluene, ortho	13.0	13.3
Chlorotoluene, meta	13.3	12.5
Chlorotoluene, para	13.3	12.5
Cresol, meta	2.5	20.8
Cyclohexanol	2.9	24.3
Cyclohexane	9.8	12.9
Dibromomethane	12.7	15.8
Dichloroethane	13.2	12.2
Dichloromethane	14.6	8.9

Table A-1 (*Continued*)

Liquid	X	Y
Diethyl ketone	13.5	9.2
Diethyl oxalate	11.0	16.4
Diethylene glycol	5.0	24.7
Diphenyl	12.0	18.3
Dipropyl ether	13.2	8.6
Dipropyl oxalate	10.3	17.7
Ethyl acetate	13.7	9.1
Ethyl acrylate	12.7	10.4
Ethyl alcohol, 100%	10.5	13.8
Ethyl alcohol, 95%	9.8	14.3
Ethyl alcohol, 40%	6.5	16.6
Ethyl benzene	13.2	11.5
Ethyl bromide	14.5	8.1
2-Ethyl butyl acrylate	11.2	14.0
Ethyl chloride	14.8	6.0
Ethyl ether	14.5	5.3
Ethyl formate	14.2	8.4
2-Ethyl hexyl acrylate	9.0	15.0
Ethyl iodide	14.7	10.3
Ethyl propionate	13.2	9.9
Ethyl propyl ether	14.0	7.0
Ethyl sulfide	13.8	8.9
Ethylene bromide	11.9	15.7
Ethylene chloride	12.7	12.2
Ethylene glycol	6.0	23.6
Ethylidebe chloride	14.1	8.7
Fluorobenzene	13.7	10.4
Formic acid	10.7	15.8
Freon-11	14.4	9.0
Freon-12	16.8	5.6
Freon-21	15.7	7.5
Freon-22	17.2	4.7
Freon-113	12.5	11.4
Glycerol, 100%	2.0	30.0
Glycerol, 50%	6.9	19.6
Heptane	14.1	8.4
Hexane	14.7	7.0
Hydrochloric acid, 31.5%	13.0	16.6
Iodobenzene	12.8	15.9
Isobutyl alcohol	7.1	18.0
Isobutyric acid	12.2	14.4
Isopropyl alcohol	8.2	16.0

TABLE A-1 *(Continued)*

Liquid	*X*	*Y*
Isopropyl bromide	14.1	9.2
Isopropyl chloride	13.9	7.1
Isopropyl iodide	13.7	11.2
Kerosene	10.2	16.9
Linseed oil, raw	7.5	27.2
Mercury	18.4	16.4
Methnol, 100%	12.4	10.5
Methanol, 90%	12.3	11.8
Methanol, 40%	7.8	15.5
Methyl acetate	14.2	8.2
Methyl acrylate	13.0	9.5
Methyl *t*-butyrate	12.3	9.7
Methyl *n*-butyrate	13.2	10.3
Methyl chloride	15.0	3.8
Methyl ethyl ketone	13.9	8.6
Methyl formae	14.2	7.5
Methyl iodide	14.3	9.3
Methyl propionate	13.5	9.0
Mehyl propyl ketone	14.3	9.5
Methyl sulfide	15.3	6.4
Naphthalene	7.9	18.1
Nitric acid	12.8	13.8
Nitric acid, 60%	10.8	17.0
Nitrobenzene	10.6	16.2
Nitrogen dioxide	12.9	8.6
Nitrotoluene	11.0	17.0
Octane	13.7	10.0
Octyl alcohol	6.6	21.1
Pentachloroethane	10.9	17.3
Pentane	14.9	5.2
Phenol	6.9	20.8
Phosphorus tribromide	13.8	16.7
Phosphorus trichloride	16.2	10.9
Propionic acid	12.8	13.8
Propyl acetate	13.1	10.3
Propyl alcohol	9.1	16.5
Propyl bromide	14.5	7.5
Propyl chloride	14.4	7.5
Propyl formate	13.1	9.7
Propyl iodide	14.1	11.6
Sodium	16.4	13.9
Sodium hydroxide, 50%	3.2	25.8

Table A-1 (*Continued*)

Liquid	X	Y
Stannic chloride	13.5	12.8
Succinonitrile	10.1	20.8
Sulfur dioxide	15.2	7.1
Sulfuric acid, 110%	7.2	27.4
Sulfuric acid, 100%	8.0	25.1
Sulfuric acid, 98%	7.0	24.8
Sulfuric acid, 60%	10.2	21.3
Sulfuryl chloride	15.2	12.4
Tetrachloroethane	11.9	15.7
Thiophene	13.2	11.0
Titanium tetrachloride	14.4	12.3
Toluene	13.7	10.4
Trichloroethylene	14.8	10.5
Triethylene glycol	4.7	24.8
Turpentine	11.5	14.9
Vinyl acetate	14.0	8.8
Vinyl toluene	13.4	12.0
Water	10.2	13.0
Xylene, ortho	13.5	12.1
Xylene, meta	13.9	10.6
Xylene, para	13.9	10.9

Source: RH Perry, DW Green, eds. *Perry's Chemical Engineers' Handbook. 7th ed.* New York: McGraw-Hill, 1997. By permission.

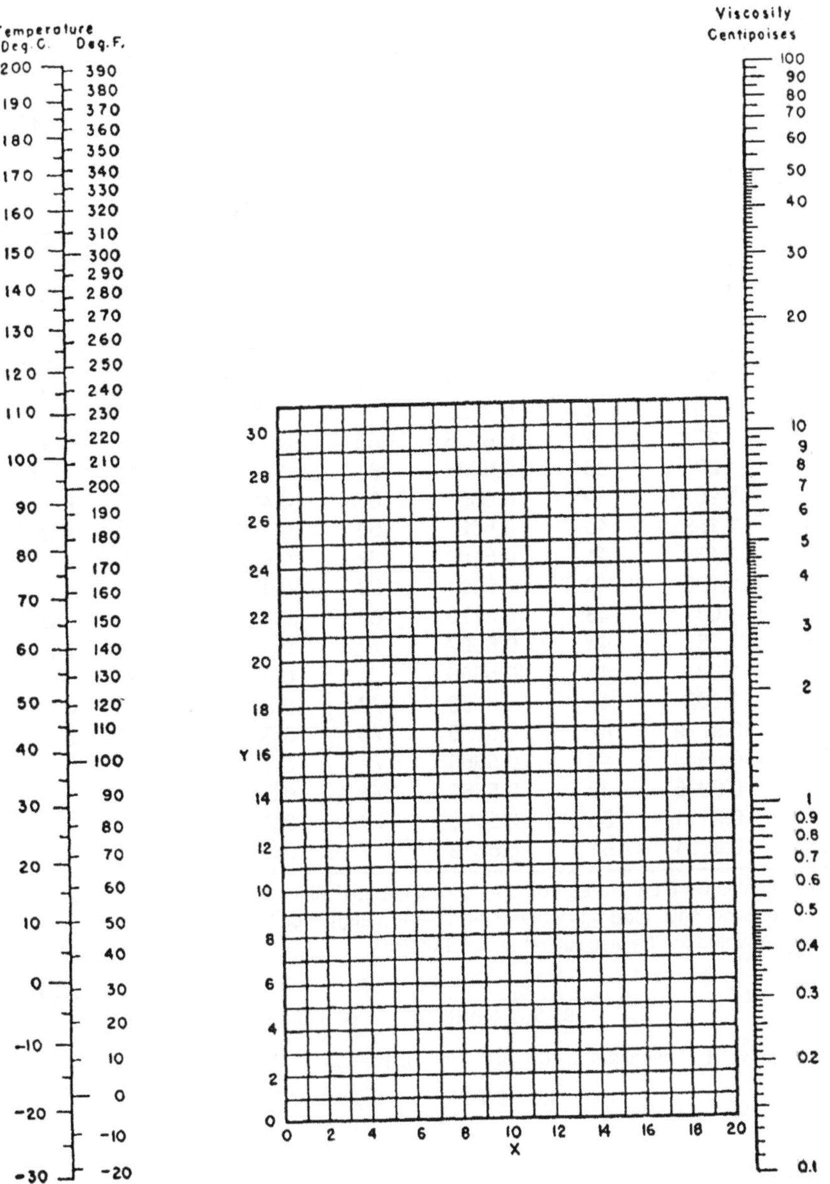

FIGURE A-4 Nomograph for viscosities of liquids at 1 atm. See Table A-1 for coordinates. (To convert centipoise to pascal-seconds, multiply by 0.001.)

Table A-2 The Viscosity of Water at 0–100°C

°C	μ (cP)	°C	μ (cP)	°C	μ (cP)	°C	μ (cP)
0	1.787	26	0.8705	52	0.5290	78	0.3638
1	1.728	27	0.8513	53	0.5204	79	0.3592
2	1.671	28	0.8327	54	0.5121	80	0.3547
3	1.618	29	0.8148	55	0.5040	81	0.3503
4	1.567	30	0.7975	56	0.4961	82	0.2460
5	1.519	31	0.7808	57	0.4884	83	0.3418
6	1.472	32	0.7647	58	0.4809	84	0.3377
7	1.428	33	0.7491	59	0.4736	85	0.3337
8	1.386	34	0.7340	60	0.4665	86	0.3297
9	1.346	35	0.7194	61	0.4596	87	0.3259
10	1.307	36	0.7052	62	0.4528	88	0.3221
11	1.271	37	0.6915	63	0.4462	89	0.3184
12	1.235	38	0.6783	64	0.4398	90	0.3147
13	1.202	39	0.6654	65	0.4335	91	0.3111
14	1.169	40	0.6529	66	0.4273	92	0.3076
15	1.139	41	0.6408	67	0.4213	93	0.3042
16	1.109	42	0.6391	68	0.4155	94	0.3008
17	1.081	43	0.6178	69	0.4098	95	0.2975
18	1.053	44	0.6067	70	0.4042	96	0.2942
19	1.027	45	0.5960	71	0.3987	97	0.2911
20	1.002	46	0.5856	72	0.3934	98	0.2879
21	0.9779	47	0.5755	73	0.3882	99	0.2848
22	0.9548	48	0.5656	74	0.3831	100	0.2818
23	0.9325	49	0.5561	75	0.3781		
24	0.9111	50	0.5468	76	0.3732		
25	0.8904	51	0.5378	77	0.3684		

Table entries were calculated from the following empirical relationships from measurements in viscometers calibrated with water at 20°C (and 1 atm), modified to agree with the currently accepted value for the viscosity at 20° of 1.002 cP:

$$\text{0–20°C}: \quad \log_{10} \eta_T = \frac{1301}{998.333 + 8.1855(T - 20) + 0.00585(T - 20)^2} - 3.30233$$

(RC Hardy, RL Cottingham, J Res NBS *42*:573, 1949.)

$$\text{20–100°C}: \quad \log_{10} \frac{\eta_T}{\eta_{20}} = \frac{1.3272(20 - T) - 0.001053(T - 20)^2}{T + 105}$$

(JF Swindells, NBS, unpublished results.)

TABLE A-3 Physical Properties of Ordinary Water and Common Liquids (SI Units)

Liquid	Temp T (°C)	Density ρ(kg/m^3)	Specific gravity S	Absolute viscosity μ(N s/m^2)	Kinematic viscosity ν(m^2/s)	Surface tension σ(N/m)	Isothermal bulk modulus of elasticity E_p(N/m^2)	Coefficient of thermal expansion α_T (K^{-1})
Water	0	1000	1.000	1.79 E-3	1.79 E-6	7.56 E-2	1.99 E9	6.80 E-5
	3.98	1000	1.000	1.57	1.57	—	—	—
	10	1000	1.000	1.31	1.31	7.42	2.12	8.80
	20	998	0.998	1.00	1.00	7.28	2.21	2.07 E-4
	30	996	0.996	7.98 E-4	7.12	2.26	2.94	
	40	992	0.992	6.53	6.58	6.96	2.29	3.85
	50	988	0.988	5.47	5.48	6.79	2.29	4.58
	60	983	0.983	4.67	4.75	6.62	2.28	5.23
	70	978	0.978	4.04	4.13	6.64	2.24	5.84
	80	972	0.972	3.55	3.65	6.26	2.20	6.41
	90	965	0.965	3.15	3.26	—	2.14	6.96
	100	958	0.958	2.82	2.94	5.89	2.07	7.50
Mercury	0	13600	13.60	1.68 E-3	1.24 E-7	—	2.50 E10	—
	4	13590	13.59	—	—	—	—	—
	20	13550	13.55	1.55	1.14	37.5	2.50 E10	1.82 E-4
	40	13500	13.50	1.45	1.07	—	—	1.82
	60	13450	13.45	1.37	1.02	—	—	1.82
	80	13400	13.40	1.30	9.70 E-8	—	—	1.82
	100	13350	13.35	1.24	9.29	—	—	—
Ethylene glycol	0	—	—	5.70 E-2	—	—	—	—
	20	1110	1.11	1.99	1.79 E-5	—	—	—
	40	1110	1.10	9.13 E-3	8.30 E-6	—	—	—
	60	1090	1.09	4.95	4.54	—	—	—
	80	1070	1.07	3.02	2.82	—	—	—
	100	1060	1.06	1.99	1.88	—	—	—
Methyl alcohol (methanol)	0	810	0.810	8.17 E-4	1.01 E-6	2.45 E-2	9.35 E8	—
	10	801	0.801	—	—	2.26	8.78	—
	20	792	0.792	5.84	7.37 E-7	—	8.23	—
	30	783	0.783	5.10	6.51	—	7.72	—
	40	774	0.774	4.50	5.81	—	7.23	—
	50	765	0.765	3.96	5.18	—	6.78	—
Ethyl alcohol (ethanol)	0	806	0.806	1.77 E-3	2.20 E-6	2.41 E-2	1.02 E9	—
	20	789	0.789	1.20	1.52	—	9.02 E8	—
	40	772	0.772	8.34 E-4	1.08	—	7.89	—
	60	754	0.754	5.92	7.85 E-7	—	6.78	—
Normal octane	0	718	0.718	7.06 E-7	9.83 E-7	—	1.00 E9	—
	16	—	—	5.74	—	—	—	—
	20	702	0.702	5.42	7.72	—	—	—
	25	—	—	—	—	—	8.35 E8	
	40	686	0.686	4.33	6.31	—	7.48	—
Benzene	0	900	0.900	9.12 E-4	1.01 E-6	3.02 E-2	1.23 E9	—
	20	879	0.879	6.52	7.42 E-7	2.76	1.06	—
	40	858	0.857	5.03	5.86	—	9.10 E8	—
	60	836	0.836	3.92	4.69	—	7.78	—
	80	815	0.815	3.29	4.04	—	6.48	—
Kerosene	−18	841	0.841	7.06 E-3	8.40 E-6	—	—	—
	20	814	0.814	1.9	2.37	2.9 E-2	—	—
Lubricating oil	20	871	0.871	1.31 E-6	1.50 E-9	—	—	—
	40	858	0.858	6.81 E-5	7.94 E-8	—	—	—
	60	845	0.845	4.18	4.95	—	—	—
	80	832	0.832	2.83	3.40	—	—	—
	100	820	0.820	2.00	2.44	—	—	—
	120	809	0.809	1.54	1.90	—	—	—

TABLE A-4 Physical Properties of Ordinary Water and Common Liquids (EE units[a])

Liquid	Temp T (°F)	Density $\rho(\mathrm{lb_m/ft^3})$	Specific gravity S	Absolute viscosity $\mu(\mathrm{lb_f s/ft^2})$	Kinematic viscosity $\nu(\mathrm{ft^2/s})$	Surface tension $\sigma(\mathrm{lb_f/ft})$	Isothermal bulk modulus of elasticity $E_p(\mathrm{lb_f/in.^2})$	Coefficient of thermal expansion α_T (°R^{-1})
Water	32	62.4	1.00	3.75 E-5	1.93 E-5	5.18 E-3	2.93 E-5	2.03 E-3
	40	62.4	1.00	3.23	1.66	5.14	2.94	—
	60	62.4	0.999	2.36	1.22	5.04	3.11	—
	80	62.2	0.997	1.80	9.30 E-6	4.92	3.22	—
	100	62.0	0.993	1.42	7.39	4.80	3.27	1.7
	120	61.7	0.988	1.17	6.09	4.65	3.33	—
	140	61.4	0.983	9.81 E-6	5.14	4.54	3.30	—
	160	61.0	0.977	8.38	4.42	4.41	3.26	—
	180	60.6	0.970	7.26	3.85	4.26	3.13	—
	200	60.1	0.963	6.37	3.41	4.12	3.08	1.52
	212	59.8	0.958	5.93	3.19	4.04	3.00	—
Mercury	50	847	13.6	1.07 E-3	1.2 E-6	—	—	1.0 E-4
	200	834	13.4	8.4 E-3	1.0	—	—	1.0 E-4
	300	826	13.2	7.4	9.0 E-7	—	—	—
	400	817	13.1	6.7	8.0	—	—	—
	600	802	12.8	5.8	7.0	—	—	—
Ethylene glycol	68	69.3	1.11	4.16 E-4	1.93 E-4	—	—	—
	104	68.7	1.10	1.91	8.93 E-5	—	—	—
	140	68.0	1.09	1.03	4.89	—	—	—
	176	66.8	1.07	6.31 E-5	3.04	—	—	—
	212	66.2	1.06	4.12	2.02	—	—	—
Methyl alcohol (methanol)	32	50.6	0.810	1.71 E-5	1.09 E-5	1.68 E-3	1.36 E-5	—
	68	50.0	0.801	—	—	1.55	1.9	—
	104	49.4	0.792	1.22	7.93 E-6	—	1.05	—
	140	48.9	0.783	1.07	7.01	—	—	—
	176	48.3	0.774	9.40 E-6	6.25	—	—	—
	212	47.8	0.765	8.27	5.58	—	—	—
Ethyl alcohol (ethanol)	32	50.3	0.806	3.70 E-5	2.37 E-5	1.65 E-3	1.48 E-5	—
	68	49.8	0.789	3.03	1.96	—	1.31	—
	104	49.3	0.789	2.51	1.64	—	1.14	—
	140	48.2	0.772	1.74	1.16	—	9.83 E-4	—
	176	47.7	0.754	1.24	8.45 E-6	—	—	—
	212	47.1	0.745	—	—	—	—	—
Normal octane	32	44.8	0.718	1.47 E-5	1.06 E-5	—	1.45 E-5	—
	68	43.8	0.702	1.13	8.31 E-6	—	—	—
	104	42.8	0.686	9.04 E-6	6.79	—	1.08	—
Benzene	32	56.2	0.900	1.90 E-5	1.09 E-5	2.07 E-3	1.78 E-5	—
	68	54.9	0.879	1.36	7.99 E-6	1.89	1.53	—
	104	53.6	0.858	1.05	6.31	—	1.32	—
	140	52.2	0.836	8.19 E-6	5.05	—	1.13	—
	176	50.9	0.815	6.87	4.35	—	9.40 E-4	—
Kerosene	0	52.5	0.841	1.48 E-4	9.05 E-5	—	—	—
	77	50.8	0.814	3.97 E-5	2.55 E-5	—	—	—
Lubricating oil	68	54.5	0.871	2.74 E-8	1.61 E-8	—	—	—
	104	53.6	0.858	1.42 E-7	8.55 E-7	—	—	—
	140	52.6	0.845	8.73	5.33	—	—	—
	176	51.9	0.832	5.91	3.66	—	—	—
	212	51.2	0.820	4.18	2.63	—	—	—
	248	50.5	0.809	3.22	2.05	—	—	—

[a] EE = English engineering

TABLE A-5 Physical Properties of SAE Oils and Lubricants

		SI units				EE units[a]		
			Kinematic viscosity $\nu(m^2/s)$				Kinematic viscosity $\nu(ft^2/s)$	
Fluid	Temp (°C)	Specific gravity	Minimum	Maximum	Temp. (°F)	Specific gravity	Minimum	Maximum
Oil								
SAE 50	99	—	1.68 E-5	2.27 E-5	210	—	1.81 E-4	2.44 E-4
	99	—	1.29	1.68	210	—	1.08	1.81
	99	—	9.6 E-4	1.29	210	—	1.03 E-2	1.08
	99	—	—	5.7 E-4	210	—	—	6.14 E-3
	−18	0.92	2.60 E-3	1.05 E-2	0	0.92	2.80 E-2	1.13 E-1
	−18	0.92	1.30	2.60 E-2	0	0.92	1.40	2.80 E-2
	−18	0.92	—	1.30	0	0.92	—	1.40
Lubricants								
SAE 250	99	—	4.3 E-5	—	210	—	4.6 E-4	—
140	99	—	2.5	4.3 E-5	210	—	2.7	4.6 E-4
90	99	—	1.4	2.5	210	—	1.5	2.7
85W	99	—	1.1	—	210	—	1.2	—
80W	99	—	7.0 E-6	—	210	—	7.5 E-5	—
75W	99	—	4.2	—	210	—	4.5 E-5	—

[a] EE = English engineering

TABLE A-6 Viscosity of Steam and Water[a]

Temp (°F)	Viscosity of steam and water, μ (cP)														
	1 psia	2 psia	5 psia	10 psia	20 psia	50 psia	100 psia	200 psia	500 psia	1000 psia	2000 psia	5000 psia	7500 psia	10000 psia	12000 psia
Sat. water	0.667	0.524	0.388	0.313	0.255	0.197	0.164	0.138	0.111	0.094	0.078	—	—	—	—
Sat. steam	0.010	0.010	0.011	0.012	0.012	0.013	0.014	0.015	0.017	0.019	0.023	—	—	—	—
1500	0.041	0.041	0.041	0.041	0.041	0.041	0.041	0.041	0.042	0.042	0.042	0.044	0.046	0.048	0.050
1450	0.040	0.040	0.040	0.040	0.040	0.040	0.040	0.040	0.040	0.041	0.041	0.043	0.045	0.047	0.049
1400	0.039	0.039	0.039	0.039	0.039	0.039	0.039	0.039	0.039	0.040	0.040	0.042	0.044	0.047	0.049
1350	0.038	0.038	0.038	0.038	0.038	0.038	0.038	0.038	0.038	0.038	0.039	0.041	0.044	0.046	0.049
1300	0.037	0.037	0.037	0.037	0.037	0.037	0.037	0.037	0.037	0.037	0.038	0.040	0.043	0.045	0.048
1250	0.035	0.035	0.035	0.035	0.035	0.035	0.035	0.036	0.036	0.036	0.037	0.039	0.042	0.045	0.048
1200	0.034	0.034	0.034	0.034	0.034	0.034	0.034	0.034	0.035	0.035	0.036	0.038	0.041	0.045	0.048
1150	0.034	0.034	0.034	0.034	0.034	0.034	0.034	0.034	0.034	0.034	0.034	0.037	0.041	0.045	0.049
1100	0.032	0.032	0.032	0.032	0.032	0.032	0.032	0.032	0.033	0.033	0.034	0.037	0.040	0.045	0.050
1050	0.031	0.031	0.031	0.031	0.031	0.031	0.031	0.031	0.032	0.032	0.033	0.036	0.040	0.047	0.052
1000	0.030	0.030	0.030	0.030	0.030	0.030	0.030	0.030	0.030	0.031	0.032	0.035	0.041	0.049	0.055
950	0.029	0.029	0.029	0.029	0.029	0.029	0.029	0.029	0.029	0.030	0.031	0.035	0.042	0.052	0.059
900	0.028	0.028	0.028	0.028	0.028	0.028	0.028	0.028	0.028	0.028	0.029	0.035	0.045	0.057	0.064
850	0.026	0.026	0.026	0.026	0.026	0.026	0.027	0.027	0.027	0.027	0.028	0.035	0.052	0.064	0.070
800	0.025	0.025	0.025	0.025	0.025	0.025	0.025	0.025	0.026	0.026	0.027	0.040	0.062	0.071	0.075
750	0.024	0.024	0.024	0.024	0.024	0.024	0.024	0.024	0.025	0.025	0.026	0.057	0.071	0.078	0.081
700	0.023	0.023	0.023	0.023	0.023	0.023	0.023	0.023	0.023	0.024	0.026*	0.071	0.079	0.085	0.086
650	0.022	0.022	0.022	0.022	0.022	0.022	0.022	0.022	0.023	0.023	0.023	0.082	0.088	0.092	0.096
600	0.021	0.021	0.021	0.021	0.021	0.021	0.021	0.021	0.021	0.021	0.087	0.091	0.096	0.101	0.104
550	0.020	0.020	0.020	0.020	0.020	0.020	0.020	0.020	0.020	0.019	0.095	0.101	0.105	0.109	0.113
500	0.019	0.019	0.019	0.019	0.019	0.019	0.019	0.018	0.018	0.103	0.105	0.111	0.114	0.119	0.122
450	0.018	0.018	0.018	0.018	0.017	0.017	0.017	0.017	0.115	0.116	0.118	0.123	0.127	0.131	0.135
450	0.106	0.106	0.106	0.106	0.106	0.106	0.106	0.106	0.131	0.132	0.134	0.138	0.143	0.147	0.150
350	0.015	0.015	0.015	0.015	0.015	0.015	0.015	0.152	0.153	0.154	0.155	0.160	0.164	0.168	0.171
300	0.014	0.014	0.014	0.014	0.014	0.014	0.182	0.183	0.183	0.184	0.185	0.190	0.194	0.198	0.201
250	0.013	0.013	0.013	0.013	0.013	0.228	0.228	0.228	0.228	0.229	0.231	0.235	0.238	0.242	0.245
200	0.012	0.012	0.012	0.012	0.300	0.300	0.300	0.300	0.301	0.301	0.303	0.306	0.310	0.313	0.316
150	0.011	0.011	0.427	0.427	0.427	0.427	0.427	0.427	0.427	0.428	0.429	0.431	0.434	0.437	0.439
100	0.680	0.680	0.680	0.680	0.680	0.680	0.680	0.680	0.680	0.680	0.680	0.681	0.682	0.683	0.683
50	1.299	1.299	1.299	1.299	1.299	1.299	1.299	1.299	1.299	1.298	1.296	1.289	1.284	1.279	1.275
32	1.753	1.753	1.753	1.753	1.753	1.753	1.753	1.752	1.751	1.749	1.745	1.733	1.723	1.713	1.705

[a] Values directly below underscored viscosities are for water.

* Critical point.

Table A-7 Viscosities of Gases[a] (Coordinates Apply to Fig. A-5)

Gas	X	Y	$\mu \times 10^7$ P
Acetic acid	7.0	14.6	825 (50°C)
Acetone	8.4	13.2	735
Acetylene	9.3	15.5	1017
Air	10.4	20.4	1812
Ammonia	8.4	16.0	1000
Amylene (β)	8.6	12.2	676
Argon	9.7	22.6	2215
Arsine	8.6	20.0	1575
Benzene	8.7	13.2	746
Bromine	8.8	19.4	1495
Butane (η)	8.6	13.2	735
Butane (iso)	8.6	13.2	744
Butyl acetate (iso)	5.7	16.3	778
Butylene (α)	8.4	13.5	761
Butylene (β)	8.7	13.1	746
Butylene (iso)	8.3	13.9	786
Butyl formate (iso)	6.6	16.0	840
Cadmium	7.8	22.5	5690 (500)
Carbon dioxide	8.9	19.1	1463
Carbon disulfide	8.5	15.8	990
Carbon monoxide	10.5	20.0	1749
Carbon oxysulfide	8.2	17.9	1220
Carbon tetrtachloride	8.0	15.3	966
Chlorine	8.8	18.3	1335
Chloroform	8.8	15.7	1000
Cyanogen	8.2	16.2	1002
Cyclohexane	9.0	12.2	701
Cyclopropane	8.3	14.7	870
Deuterium	11.0	16.2	1240
Diethyl ether	8.8	12.7	730
Dimethyl ether	9.0	15.0	925
Diphenyl ether	8.6	10.4	610 (50)
Diphenyl methane	8.0	10.3	605 (50)
Ethane	9.0	14.5	915
Ethanol	8.2	14.5	835
Ethyl acetate	8.4	13.4	743
Ethyl chloride	8.5	15.6	987
Ethylene	9.5	15.2	1010
Ethyl propionate	12.0	12.4	890
Fluoride	7.3	23.8	2250
Freon-11	8.6	16.2	1298 (93)
Freon-12	9.0	17.4	1496 (93)

Table A-7 (*Continued*)

Gas	X	Y	$\mu \times 10^7$ P
Freon-14	9.5	20.4	1716
Freon-21	9.0	16.7	1389 (93)
Freon-22	9.0	17.7	1554 (93)
Freon-113	11.0	14.0	1166 (93)
Freon-114	9.4	16.4	1364 (93)
Helium	11.3	20.8	1946
Heptane (*n*)	9.6	10.6	618 (50)
Hexane (*n*)	8.4	12.0	644
Hydrogen	11.3	12.4	880
Hydrogen–helium			
10% H_2, 90% He	11.0	20.5	1780 (0)
25% H_2, 75% He	11.0	19.4	1603 (0)
40% H_2, 60% He	10.7	18.4	1431 (0)
60% H_2, 40% He	10.8	16.7	1227 (0)
81% H_2, 19% He	10.5	15.0	1016 (0)
Hydrogen–sulfur dioxide	8.7	18.1	1259 (17)
10% H_2, 90% SO_2	8.7	18.1	1259 (17)
20% H_2, 80% SO_2	8.6	18.2	1277 (17)
50% H_2, 50% SO_2	8.9	18.3	1332 (17)
80% H_2, 20% SO_2	9.7	17.7	1306 (17)
Hydrogen bromide	8.4	21.6	1843
Hydrogen chloride	8.5	19.2	1425
Hydrogen cyanide	7.1	14.5	737
Hydrogen iodide	8.5	21.5	1830
Hydrogen sulfide	8.4	18.0	1265
Iodine	8.7	18.7	1730 (100)
Krypton	9.4	24.0	2480
Mercury	7.4	24.9	4500 (200)
Mercury bromide	8.5	19.0	2253
Mercuric chloride	7.7	18.7	2200 (200)
Mercuric iodide	8.4	18.0	2045 (200)
Mesitylene	9.5	10.2	660 (50)
Methane	9.5	15.8	1092
Methane (deuterated)	9.5	17.6	1290
Methanol	8.3	15.6	935
Methyl acetate	8.4	14.0	870 (50)
Methyl acetylene	8.9	14.3	867
3-Methyl-1-butene	8.0	13.3	716
Methyl butyrate (iso)	6.6	15.8	824
Methyl bromide	8.1	18.7	1327
Methyl bromide	8.1	18.7	1327
Methyl chloride	8.5	16.5	1062

TABLE A-7 (*Continued*)

Gas	X	Y	$\mu \times 10^7$ P
3-Methylene-1-butene	8.0	13.3	716
Methylene chloride	8.5	15.8	989
Methyl formate	5.1	18.0	923
Neon	11.1	25.8	3113
Nitric oxide	10.4	20.8	1899
Nitrogen	10.6	20.0	1766
Nitrous oxide	9.0	19.0	1460
Nonane (*n*)	9.2	8.9	554 (50)
Octane (*n*)	8.8	9.8	586 (50)
Oxygen	10.2	21.6	2026
Pentene (*n*)	8.5	12.3	668
Pentane (iso)	8.9	12.1	685
Phosophene	8.8	17.0	1150
Propane	8.9	13.5	800
Propanol (*n*)	8.4	13.5	770
Propanol (iso)	8.4	13.6	774
Propyl acetate	8.0	14.3	797
Propylene	8.5	14.4	840
Pyridine	8.6	13.3	830 (50)
Silane	9.0	16.8	1148
Stannic chloride	9.1	16.0	1330 (100)
Stannic bromide	9.0	16.7	142 (100)
Sulfur dioxide	8.4	18.2	1250
Thiazole	10.0	14.4	958
Thiophene	8.3	14.2	901 (50)
Toluene	8.6	12.5	686
2,2,3-Trimethylbutane	10.0	10.4	691 (50)
Trimethylethane	8.0	13.0	686
Water	8.0	16.0	1250 (100)
Xenon	9.3	23.0	2255
Zinc	8.0	22.0	5250 (500)

[a] Viscosity at 20°C unless otherwise indicated.

Source: RH Perry, DW Green, eds. *Perry's Chemical Engineers' Handbook. 7th ed.* New York: McGraw-Hill, 1997. By permission.

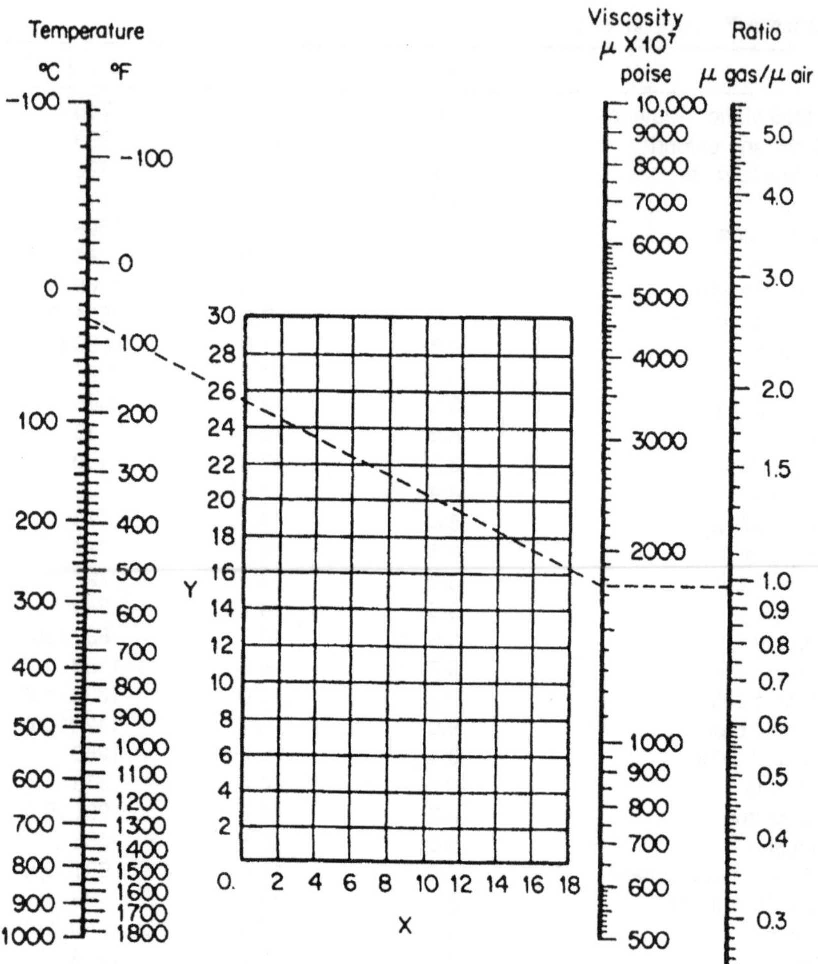

FIGURE A-5 Nomograph for determining absolute viscosity of a gas near ambient pressure and relative viscosity of a gas compared with air. (Coordinates from Table A-7.) To convert from poise to pascal-seconds, multiply by 0.1. (From Beerman, Meas Control, June 1982, pp 154–157.)

The curves for hydrocarbon vapors and natural gases in the chart at the upper right are taken from Maxwell, the curves for all other gases (except helium) in the chart are based upon Sutherland's formula, as folloows:

$$\mu = \mu_0\left(\frac{0.555T_0C}{0.555T + C}\right)\left(\frac{T}{T_0}\right)^{3/2}$$

where

μ = viscosity in cP at temperature T
μ_0 = viscosity, in CP at temperatue T_0
T = absolute temperature, in °R (460 + °F) for which viscosity is desired.
T_0 = absolute temperature, in °R, for which viscosity is known.
C = Sutherland's constant

Note: The variation of viscosity with presssure is small for most gases. For gases given on this page, the correction of viscosity for pressure is less than 10% for pressures up to 500 psi.

Fluid	Approximate values of °C
O_2	127
Air	120
N_2	111
CO_2	240
CO	118
SO_2	416
NH_2	370
H_2	72

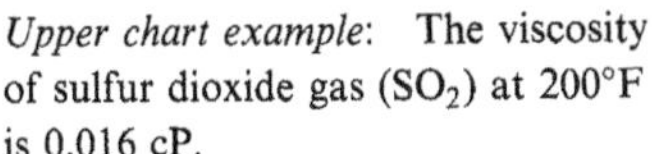

Upper chart example: The viscosity of sulfur dioxide gas (SO_2) at 200°F is 0.016 cP.

Lower chart example: The viscosity of carbon dioxide gas (CO_2) at about 80°F is 0.015 cP.

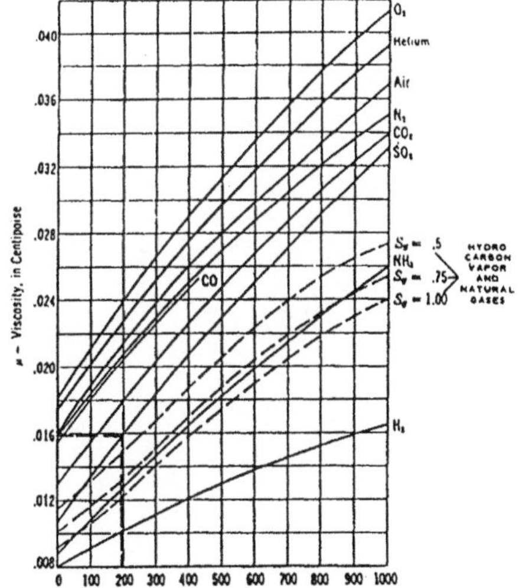

(a)

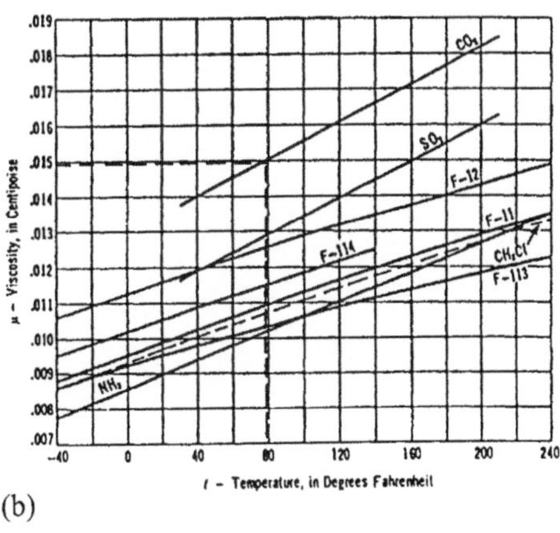

(b)

FIGURE A-6 Viscosity of (a) gases and (b) refrigerant vapors. (From Crane Technical Paper 410, Crane Co., Chicago, 1991.)

Appendix B

Generalized Viscosity Plot

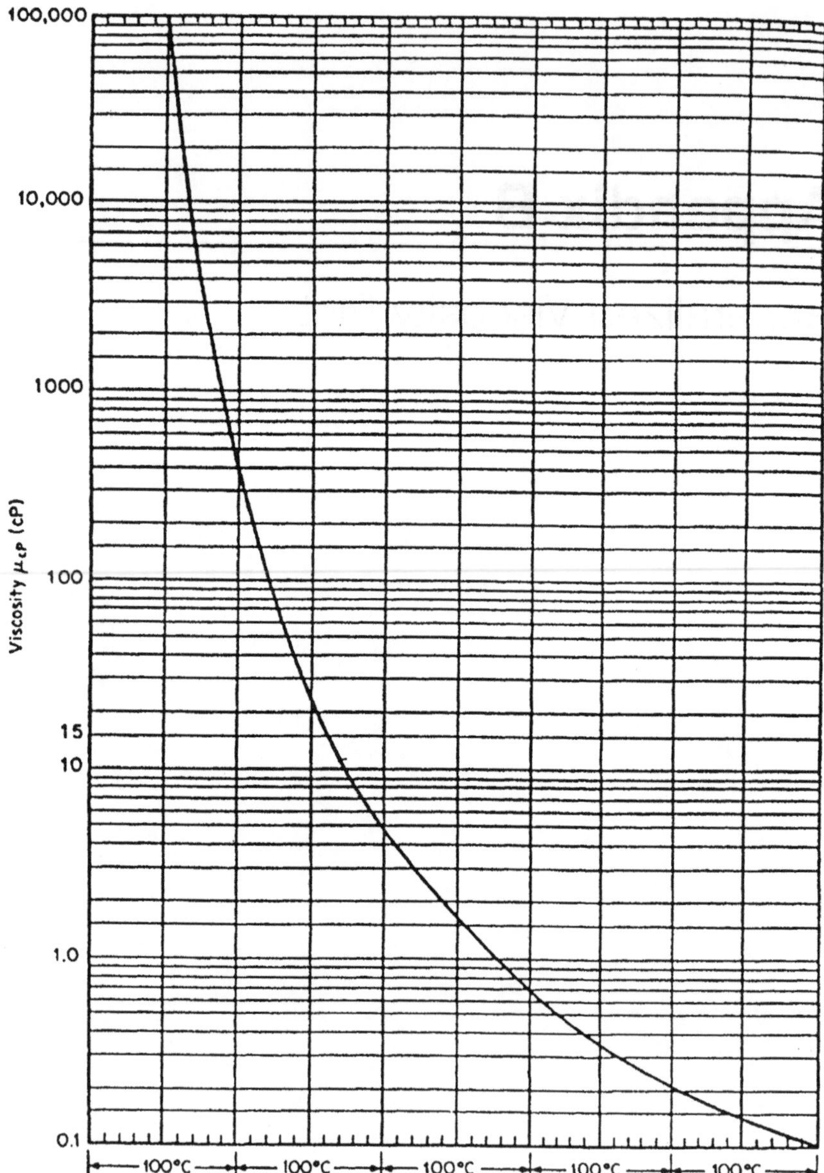

Figure B-1 Curves for estimating viscosity from a single measurement value. (From Gambill, 1959.)

Appendix C

Properties of Gases

TABLE C-1 Physical Properties of Gases (Approximate Values at 68°F and 14.8 psia)

Name of gas	Chemical formula or symbol	Approx. molecular weight, M	Weight density ρ (lb/ft^3)	Specific gravity relative to air, S_0	Individual gas constant, R	Specific heat at room temperature [Btu/(lb °F)] c_p[a]	c_v[b]	Heat capacity per cubic foot c_p	c_v	k equal to c_p/c_v
Acetylene (ethylene)	C_2H_2	26.0	0.0682	0.907	59.4	0.350	0.269	0.0239	0.0184	1.30
Air	—	29.0	0.0752	1.000	53.3	0.241	0.172	0.0181	0.0129	1.40
Ammonia	NH_2	17.0	0.0448	0.596	91.0	0.523	0.396	0.0234	0.0178	1.32
Argon	A	39.9	0.1037	1.379	38.7	0.124	0.074	0.0129	0.0077	1.67
Butane	C_4H_{10}	58.1	0.1554	2.067	26.5	0.395	0.356	0.0614	0.0553	1.11
Carbon	CO_2	44.0	0.1150	1.529	35.1	0.205	0.158	0.0236	0.0181	1.30
Carbon monoxide	CO	28.0	0.0727	0.967	55.2	0.243	0.173	0.0177	0.0126	1.40
Chlorine	Cl_2	70.9	0.1869	2.486	21.8	0.115	0.086	0.0215	0.0162	1.33
Ethane	C_2H_6	30.0	0.0789	1.049	51.5	0.386	0.316	0.0305	0.0250	1.22
Ethylene	C_2H_4	28.0	0.0733	0.975	55.1	0.400	0.329	0.0293	0.0240	1.22
Helium	He	4.0	0.01039	0.1381	386.3	1.250	0.754	0.130	0.0078	1.66
Hydrogen chloride	HCl	36.5	0.0954	1.268	42.4	0.191	0.135	0.0182	0.0129	1.41
Hydrogen	H_2	2.0	0.00523	0.0695	766.8	3.420	2.426	0.0179	0.0127	1.41
Hydrogen sulfide	H_2S	34.1	0.0895	1.190	45.2	0.243	0.187	0.0217	0.0167	1.30
Methane	CH_4	16.0	0.0417	0.554	96.4	0.593	0.449	0.0247	0.0187	1.32
Methyl chloride	CH_2Cl	50.5	0.1342	1.785	30.6	0.240	0.200	0.0322	0.0268	1.20
Natural gas	—	19.5	0.0502	0.667	79.1	0.560	0.441	0.0281	0.0221	1.27
Nitric oxide	NO	30.0	0.0780	1.037	51.5	0.231	0.165	0.0180	0.0129	1.40
Nitrogen	N_2	28.0	0.0727	0.967	55.2	0.247	0.176	0.0180	0.0127	1.41
Nitrous oxide	N_2O	44.0	0.1151	1.530	35.1	0.221	0.169	0.0254	0.0194	1.31
Oxygen	O_2	32.0	0.0831	1.105	48.3	0.217	0.155	0.0180	0.0129	1.40
Propane	C_3H_8	44.1	0.1175	1.562	35.0	0.393	0.342	0.0462	0.0402	1.15
Propene (propylene)	C_3H_6	42.1	0.1091	1.451	36.8	0.358	0.314	0.0391	0.0343	1.14
Sulfur dioxide	SO_2	64.1	0.1703	2.264	24.0	0.154	0.122	0.0262	0.0208	1.26

[a] = Specific heat at constant pressure.

[b] = Specific heat at constant volume.

Weight density values were obtained by multiplying density of air by specific gravity of gas. For values of 60°F, multiply by 1.0154.

Natural gas values were representative only. Exact characteristics require knowledge of specific constituents.

Source: Molecular weight, specific gravity, individual gas constant, and specific heat values were abstracted from, or based on, data in Table 24 of Mark's *Standard Handbook for Mechanical Engineers* (7th ed.).

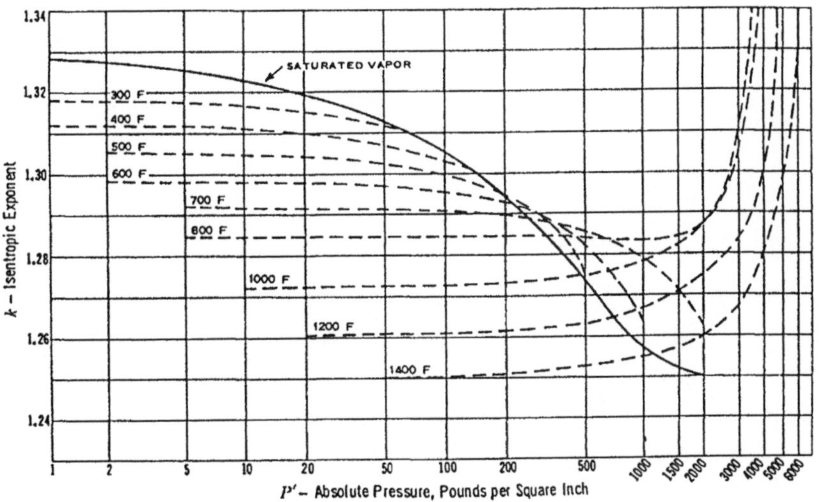

Figure C-1 Steam values of isentropic exponent, k (for small changes in pressure (or volume) along an isentrope, pV^k = constant).

Appendix D

Pressure–Enthalpy Diagrams for Various Compounds

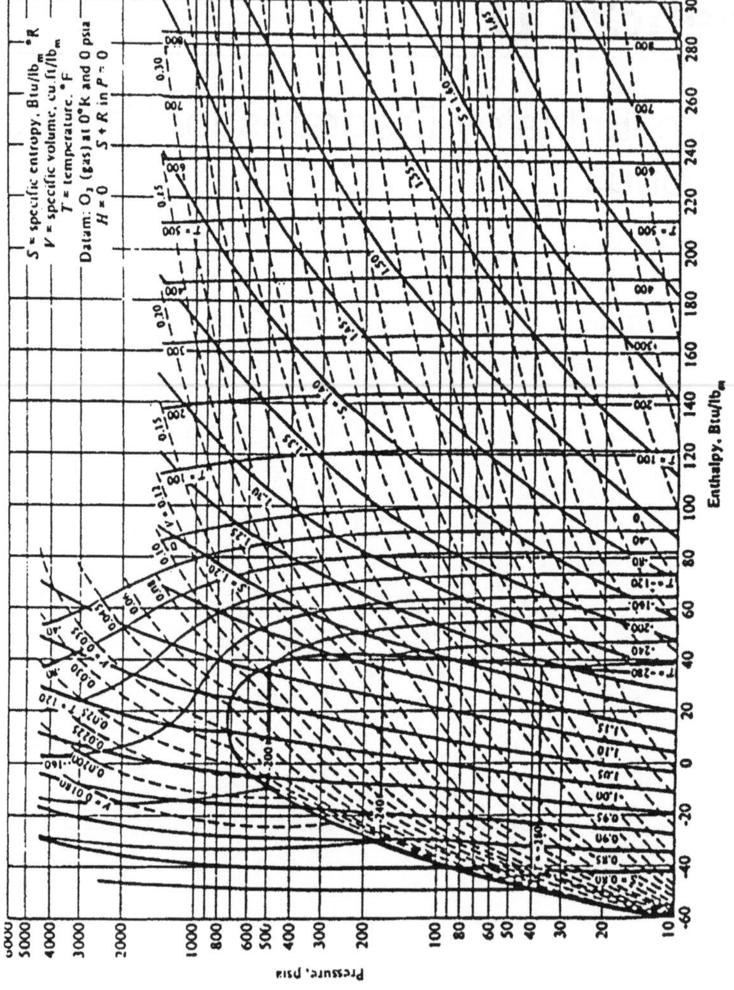

FIGURE D-1 Oxygen pressure–enthalpy diagram. (From LN Canjar, FS Manning. *Thermodynamic Properties and Reduced Correlations for Gases*. Houston, TX: Gulf Pub, 1967. Reproduced by permission.)

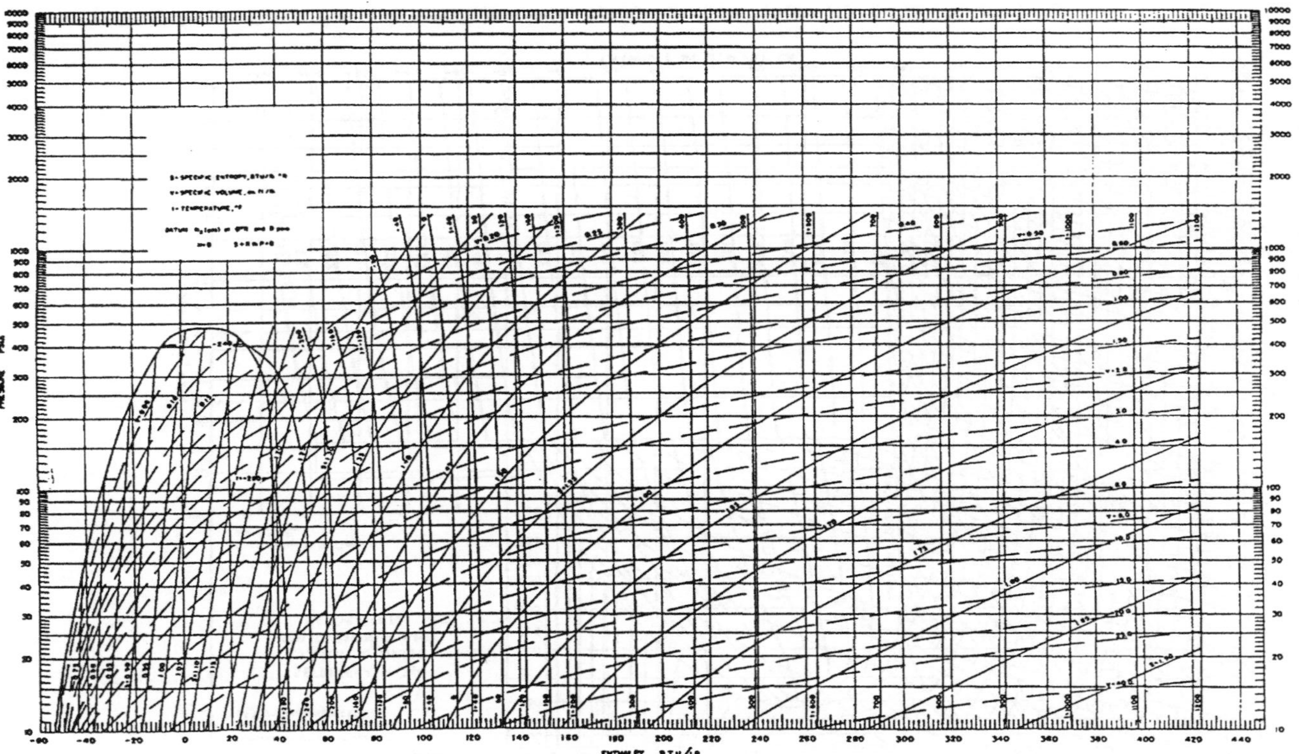

FIGURE D-2 Nitrogen pressure–enthalpy diagram. (From VM Tejada et al. Thermo properties of non-hydrocarbons. *Hydrocarbon Proc Petrol Refiner*, March 1966. Reprinted by permission.)

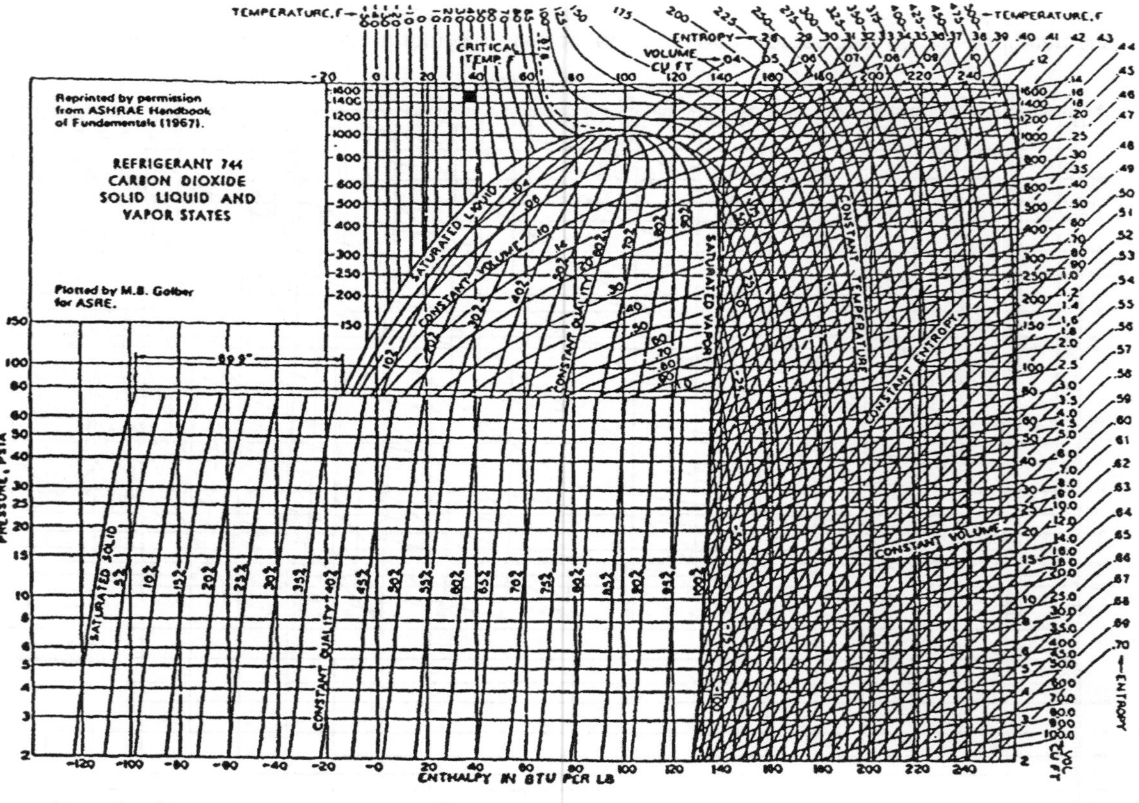

FIGURE D-3 Pressure–enthalpy chart, carbon dioxide. (From *ASHRAE Handbook of Fundamentals*, 1967.)

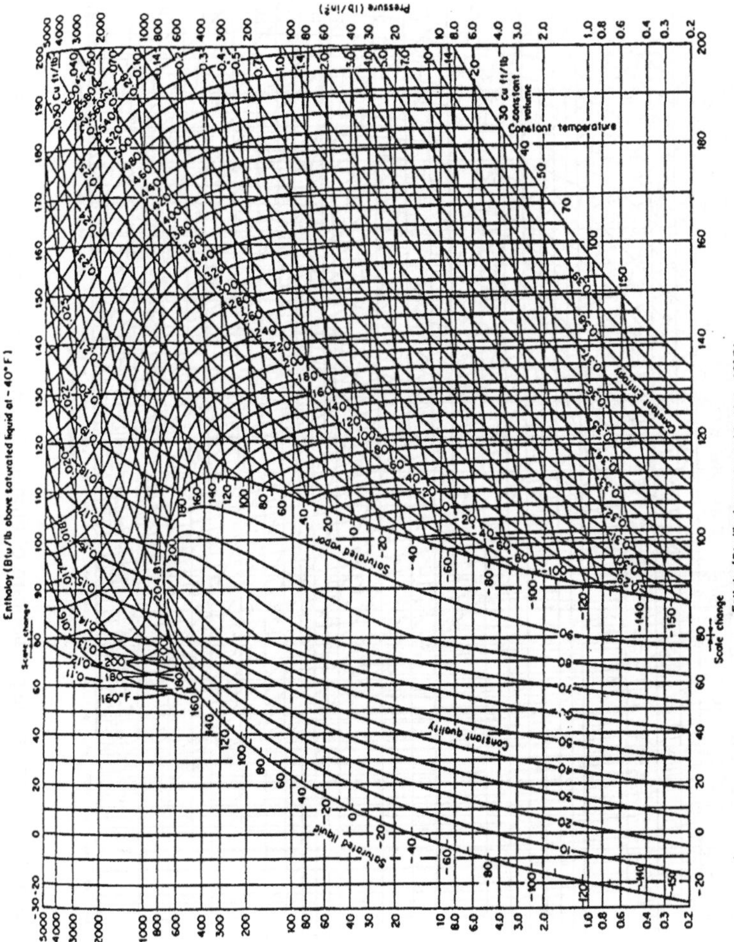

FIGURE D-4 Enthalpy–log pressure diagram for Refrigerant 22. Temperature in °F, volume in ft^3/lb, entropy in Btu/(lb °R), quality in weight percent. (Reprinted by permission of El du Pont de Nemours & Company, 1967.)

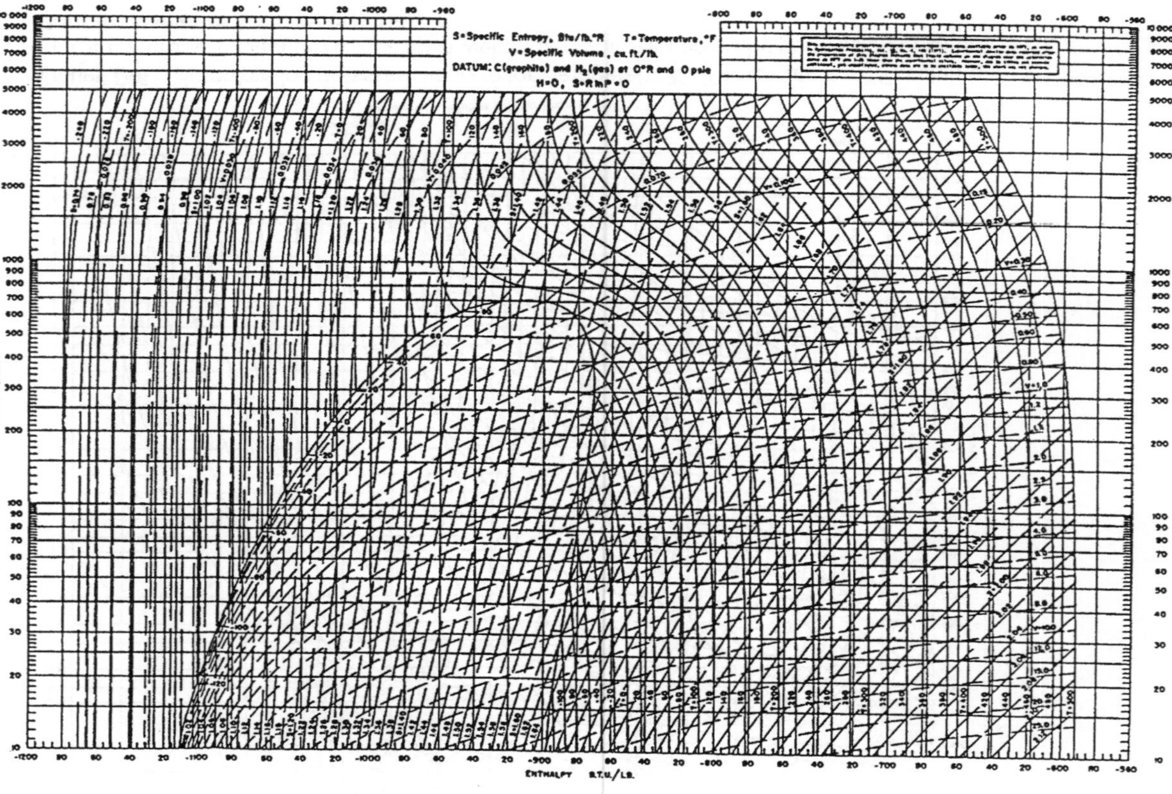

FIGURE D-5 Pressure–enthalpy diagram of ethane. (From *Hydrocarbon Processing* 50(4): 140, 1971.)

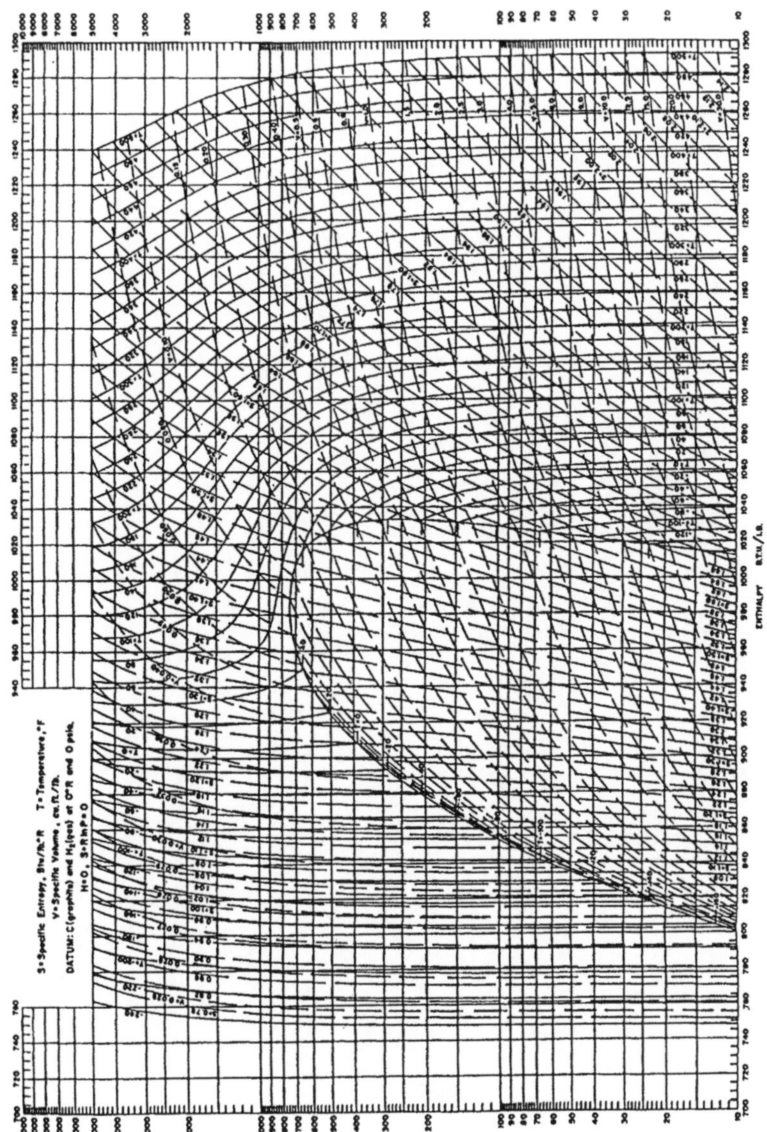

FIGURE D-6 Pressure–enthalpy diagram for ethylene. (From RE Sterling, *Fluid Thermal Properties for Petroleum Systems*. Houston, TX: Gulf Pub. Reprinted by permission.)

Appendix E

Microscopic Conservation Equations in Rectangular, Cylindrical, and Spherical Coordinates

CONTINUITY EQUATION

Rectangular coordinates (x, y, z):

$$\frac{\partial \rho}{\partial t} + \frac{\partial}{\partial x}(\rho \mathrm{v}_x) + \frac{\partial}{\partial y}(\rho \mathrm{v}_y) + \frac{\partial}{\partial z}(\rho \mathrm{v}_z) = 0$$

Cylindrical coordinates (r, θ, z):

$$\frac{\partial \rho}{\partial t} + \frac{1}{r}\frac{\partial}{\partial r}(\rho r \mathrm{v}_r) + \frac{1}{r}\frac{\partial}{\partial \theta}(\rho \mathrm{v}_\theta) + \frac{\partial}{\partial z}(\rho \mathrm{v}_z) = 0$$

Spherical coordinates (r, θ, ϕ):

$$\frac{\partial \rho}{\partial t} + \frac{1}{r^2}\frac{\partial}{\partial r}(\rho r^2 \mathrm{v}_r) + \frac{1}{r \sin\theta}\frac{\partial}{\partial \theta}(\rho \mathrm{v}_\theta \sin\theta) + \frac{1}{r \sin\theta}\frac{\partial}{\partial \phi} = 0$$

MOMENTUM EQUATION IN RECTANGULAR COORDINATES

x component:

$$\rho\left(\frac{\partial v_x}{\partial t} + v_x\frac{\partial v_x}{\partial x} + v_y\frac{\partial v_x}{\partial y} + v_z\frac{\partial v_x}{\partial z}\right)$$

$$= -\frac{\partial P}{\partial x} + \left(\frac{\partial \tau_{xx}}{\partial x} + \frac{\partial \tau_{yx}}{\partial y} + \frac{\partial \tau_{zx}}{\partial z}\right) + \rho g_x$$

y component:

$$\rho\left(\frac{\partial v_y}{\partial t} + v_x\frac{\partial v_y}{\partial x} + v_y\frac{\partial v_y}{\partial y} + v_z\frac{\partial v_y}{\partial z}\right)$$

$$= -\frac{\partial P}{\partial y} + \left(\frac{\partial \tau_{xy}}{\partial x} + \frac{\partial \tau_{yy}}{\partial y} + \frac{\partial \tau_{zy}}{\partial z}\right) + \rho g_y$$

z component:

$$\rho\left(\frac{\partial v_z}{\partial t} + v_x\frac{\partial v_z}{\partial x} + v_y\frac{\partial v_z}{\partial y} + v_z\frac{\partial v_z}{\partial z}\right)$$

$$= -\frac{\partial P}{\partial z} + \left(\frac{\partial \tau_{xz}}{\partial x} + \frac{\partial \tau_{yz}}{\partial y} + \frac{\partial \tau_{zz}}{\partial z}\right) + \rho g_z$$

MOMENTUM EQUATION IN CYLINDRICAL COORDINATES

r component:

$$\rho\left(\frac{\partial v_r}{\partial t} + v_r\frac{\partial v_r}{\partial r} + \frac{v_\theta}{r}\frac{\partial v_r}{\partial \theta} - \frac{v_\theta^2}{r} + v_z\frac{\partial v_r}{\partial z}\right)$$

$$= -\frac{\partial P}{\partial r} + \left(\frac{1}{r}\frac{\partial}{\partial r}(r\tau_{rr}) + \frac{1}{r}\frac{\partial \tau_{r\theta}}{\partial \theta} - \frac{\tau_{\theta\theta}}{r} + \frac{\partial \tau_{rz}}{\partial z}\right) + \rho g_r$$

θ component:

$$\rho\left(\frac{\partial v_\theta}{\partial t} + v_r\frac{\partial v_\theta}{\partial r} + \frac{v_\theta}{r}\frac{\partial v_\theta}{\partial \theta} - \frac{v_r v_\theta}{r} + v_z\frac{\partial v_\theta}{\partial z}\right)$$

$$= -\frac{1}{r}\frac{\partial P}{\partial \theta} + \left(\frac{1}{r^2}\frac{\partial}{\partial r}(r^2\tau_{r\theta}) + \frac{1}{r}\frac{\partial \tau_{\theta\theta}}{\partial \theta} + \frac{\partial \tau_{\theta z}}{\partial z}\right) + \rho g_\theta$$

z component:

$$\rho\left(\frac{\partial \mathrm{v}_z}{\partial t} + \mathrm{v}_r \frac{\partial \mathrm{v}_z}{\partial r} + \frac{\mathrm{v}_\theta}{r}\frac{\partial \mathrm{v}_z}{\partial \theta} + \mathrm{v}_z \frac{\partial \mathrm{v}_z}{\partial z}\right)$$

$$= -\frac{\partial P}{\partial z} + \left(\frac{1}{r}\frac{\partial}{\partial r}(r\tau_{rz}) + \frac{1}{r}\frac{\partial \tau_{\theta z}}{\partial \theta} + \frac{\partial \tau_{zz}}{\partial z}\right) + \rho g_z$$

MOMENTUM EQUATION IN SPHERICAL COORDINATES

r component:

$$\rho\left(\frac{\partial \mathrm{v}_r}{\partial t} + \mathrm{v}_r \frac{\partial \mathrm{v}_r}{\partial r} + \frac{\mathrm{v}_\theta}{r}\frac{\partial \mathrm{v}_r}{\partial \theta} + \frac{\mathrm{v}_\phi}{r\sin\theta}\frac{\partial \mathrm{v}_r}{\partial \phi} - \frac{\mathrm{v}_\theta^2 + \mathrm{v}_\phi^2}{r}\right)$$

$$= -\frac{\partial P}{\partial r} + \left(\frac{1}{r^2}\frac{\partial}{\partial r}(r^2\tau_{rr}) + \frac{1}{r\sin\theta}\frac{\partial}{\partial \theta}(\tau_{r\theta}\sin\theta)\right.$$

$$\left. + \frac{1}{r\sin\theta}\frac{\partial \tau_{r\phi}}{\partial \phi} - \frac{\tau_{\theta\theta} + \tau_{\phi\phi}}{r}\right) + \rho g_r$$

θ component:

$$\rho\left(\frac{\partial \mathrm{v}_\theta}{\partial t} + \mathrm{v}_r \frac{\partial \mathrm{v}_\theta}{\partial r} + \frac{\mathrm{v}_\theta}{r}\frac{\partial \mathrm{v}_\theta}{\partial \theta} + \frac{\mathrm{v}_\phi}{r\sin\theta}\frac{\partial \mathrm{v}_\theta}{\partial \phi} + \frac{\mathrm{v}_r\mathrm{v}_\theta}{r} - \frac{\mathrm{v}_\phi^2\cot\theta}{r}\right)$$

$$= -\frac{1}{r}\frac{\partial P}{\partial \theta} + \left(\frac{1}{r^2}\frac{\partial}{\partial r}(r^2\tau_{r\theta}) + \frac{1}{r\sin\theta}\frac{\partial}{\partial \theta}(\tau_{\theta\theta}\sin\theta) + \frac{1}{r\sin\theta}\frac{\partial \tau_{\theta\phi}}{\partial \phi}\right.$$

$$\left. + \frac{\tau_{r\theta}}{r} - \frac{\cot\theta}{r}\tau_{\phi\phi}\right) + \rho g_\theta$$

φ component:

$$\rho\left(\frac{\partial \mathrm{v}_\phi}{\partial t} + \mathrm{v}_r \frac{\partial \mathrm{v}_\phi}{\partial r} + \frac{\mathrm{v}_\theta}{r}\frac{\partial \mathrm{v}_\phi}{\partial \theta} + \frac{\mathrm{v}_\phi}{r\sin\theta}\frac{\partial \mathrm{v}_\phi}{\partial \phi} + \frac{\mathrm{v}_\phi\mathrm{v}_r}{r} + \frac{\mathrm{v}_\theta\mathrm{v}_\phi}{r}\cot\theta\right)$$

$$= -\frac{1}{r\sin\theta}\frac{\partial P}{\partial \phi} + \left(\frac{1}{r^2}\frac{\partial}{\partial r}(r^2\tau_{r\phi}) + \frac{1}{r}\frac{\partial \tau_{\theta\phi}}{\partial \theta} + \frac{1}{r\sin\theta}\frac{\partial \tau_{\phi\phi}}{\partial \phi}\right.$$

$$\left. + \frac{\tau_{r\phi}}{r} + \frac{2\cot\theta}{r}\tau_{\theta\phi}\right) + \rho g_\phi$$

COMPONENTS OF THE STRESS TENSOR τ

Rectangular coordinates:

$$\tau_{xx} = +\mu\left[2\frac{\partial v_x}{\partial x} - \frac{2}{3}(\nabla \cdot v)\right] + \kappa(\nabla \cdot v)$$

$$\tau_{yy} = +\mu\left[2\frac{\partial v_y}{\partial y} - \frac{2}{3}(\nabla \cdot v)\right] + \kappa(\nabla \cdot v)$$

$$\tau_{zz} = +\mu\left[2\frac{\partial v_z}{\partial z} - \frac{2}{3}(\nabla \cdot v)\right] + \kappa(\nabla \cdot v)$$

$$\tau_{xy} = \tau_{yx} = +\mu\left[\frac{\partial v_x}{\partial y} + \frac{\partial v_y}{\partial x}\right]$$

$$\tau_{yz} = \tau_{zy} = +\mu\left[\frac{\partial v_y}{\partial z} + \frac{\partial v_z}{\partial y}\right]$$

$$\tau_{zx} = \tau_{xz} = +\mu\left[\frac{\partial v_z}{\partial x} + \frac{\partial v_x}{\partial z}\right]$$

$$(\nabla \cdot v) = \frac{\partial v_x}{\partial x} + \frac{\partial v_y}{\partial y} + \frac{\partial v_z}{\partial z}$$

Cylindrical coordinates:

$$\tau_{rr} = +\mu\left[2\frac{\partial v_r}{\partial r} - \frac{2}{3}(\nabla \cdot v)\right] + \kappa(\nabla \cdot v)$$

$$\tau_{\theta\theta} = +\mu\left[2\left(\frac{1}{r}\frac{\partial v_\theta}{\partial \theta} + \frac{v_r}{r}\right) - \frac{2}{3}(\nabla \cdot v)\right] + \kappa(\nabla \cdot v)$$

$$\tau_{zz} = +\mu\left[2\frac{\partial v_z}{\partial z} - \frac{2}{3}(\nabla \cdot v)\right] + \kappa(\nabla \cdot v)$$

$$\tau_{r\theta} = \tau_{\theta r} = +\mu\left[r\frac{\partial}{\partial r}\left(\frac{v_\theta}{r}\right) + \frac{1}{r}\frac{\partial v_r}{\partial \theta}\right]$$

$$\tau_{\theta z} = \tau_{z\theta} = +\mu\left[\frac{\partial v_\theta}{\partial z} + \frac{1}{r}\frac{\partial v_z}{\partial \theta}\right]$$

$$\tau_{zr} = \tau_{rz} = +\mu\left[\frac{\partial v_z}{\partial r} + \frac{\partial v_r}{\partial z}\right]$$

$$(\nabla \cdot v) = \frac{1}{r}\frac{\partial}{\partial r}(rv_r) + \frac{1}{r}\frac{\partial v_\theta}{\partial \theta} + \frac{\partial v_z}{\partial z}$$

Spherical coordinates:

$$\tau_{rr} = +\mu\left[2\frac{\partial \mathrm{v}_r}{\partial r} - \frac{2}{3}(\nabla \cdot \mathrm{v})\right] + \kappa(\nabla \cdot \mathrm{v})$$

$$\tau_{\theta\theta} = +\mu\left[2\left(\frac{1}{r}\frac{\partial \mathrm{v}_\theta}{\partial \theta} + \frac{\mathrm{v}_r}{r}\right) - \frac{2}{3}(\nabla \cdot \mathrm{v})\right] + \kappa(\nabla \cdot \mathrm{v})$$

$$\tau_{\phi\phi} = +\mu\left[2\left(\frac{1}{r\sin\theta}\frac{\partial \mathrm{v}_\phi}{\partial \phi} + \frac{\mathrm{v}_r}{r} + \frac{\mathrm{v}_\theta \cot\theta}{r}\right) - \frac{2}{3}(\nabla \cdot \mathrm{v})\right] + \kappa(\nabla \cdot \mathrm{v})$$

$$\tau_{r\theta} = \tau_{\theta r} = +\mu\left[r\frac{\partial}{\partial r}\left(\frac{\mathrm{v}_\theta}{r}\right) + \frac{1}{r}\frac{\partial \mathrm{v}_r}{\partial \theta}\right]$$

$$\tau_{\theta\phi} = \tau_{\phi\theta} = +\mu\left[\frac{\sin\theta}{r}\frac{\partial}{\partial \theta}\left(\frac{\mathrm{v}_\phi}{\sin\theta}\right) + \frac{1}{r\sin\theta}\frac{\partial \mathrm{v}_\phi}{\partial \phi}\right]$$

$$\tau_{\phi r} = \tau_{r\phi} = +\mu\left[\frac{1}{r\sin\theta}\frac{\partial \mathrm{v}_r}{\partial \phi} + r\frac{\partial}{\partial r}\left(\frac{\mathrm{v}_\phi}{r}\right)\right]$$

$$(\nabla \cdot \mathrm{v}) = \frac{1}{r^2}\frac{\partial}{\partial r}(r^2 \mathrm{v}_r) + \frac{1}{\mathrm{r}\sin\theta}\frac{\partial}{\partial \theta}(\mathrm{v}_\theta \sin\theta) + \frac{1}{r\sin\theta}\frac{\partial \mathrm{v}_\phi}{\partial \phi}$$

Appendix F

Standard Steel Pipe Dimensions and Capacities

TABLE F-1 Standard Steel Pipe Dimensions and Capacities[a]

Nominal pipe size (in.)	Outside diameter (in.)	Schedule no.	Wall thickness (in.)	Inside diameter (in)	Cross-sectional area		Circumference (ft) or surface (ft/ft of length)		Capacity at 1 ft/s velocity	
					Metal ($in.^2$)	Flow (ft^2)	Outside	Inside	U.S. min	lb/h water
$\frac{1}{8}$	0.405	10S	0.949	0.307	0.055	0.00051	0.106	0.0804	0.231	115.5
		40St, 40S	1.068	0.269	0.072	0.00040	0.106	0.0705	0.179	89.5
		80X, 80S	0.095	0.215	0.093	0.00025	0.106	0.0563	0.113	56.5
$\frac{1}{4}$	0.540	10S	0.065	0.410	0.097	0.00092	0.141	0.107	0.412	206.5
		40St, 40S	0.088	0.364	0.125	0.00072	0.141	0.095	0.323	161.5
		80XS, 80S	0.119	0.302	0.157	0.00050	0.141	0.079	0.224	112.0
$\frac{3}{8}$	0.675	10S	0.065	0.545	0.125	0.00162	0.177	0.143	0.727	363.5
		40ST, 40S	0.091	0.493	0.167	0.00133	0.177	0.129	0.596	298.0
		80XS, 80S	0.126	0.423	0.217	0.00098	0.177	0.111	0.440	220.0
$\frac{1}{2}$	0.840	5S	0.065	0.710	0.158	0.00275	0.220	0.186	1.234	617.0
		10S	0.083	0.674	0.197	0.00248	0.220	0.176	1.112	556.0
		40ST, 40S	0.109	0.622	0.250	0.00211	0.220	0.163	0.945	472.0
		80XS, 80S	0.147	0.546	0.320	0.00163	0.220	0.143	0.730	365.0
		160	0.188	0.464	0.385	0.00117	0.220	0.122	0.527	263.5
		XX	0.294	0.252	0.504	0.00035	0.220	0.066	0.155	77.5
$\frac{3}{4}$	1.050	5S	0.065	0.920	0.201	0.00461	0.275	0.241	2.072	1036.0
		10S	0.083	0.884	0.252	0.00426	0.275	0.231	1.903	951.5
		40ST, 40S	0.113	0.824	0.333	0.00371	0.275	0.216	1.665	832.5
		80XS, 80S	0.154	0.742	0.433	0.00300	0.275	0.194	1.345	672.5
		160	0.219	0.612	0.572	0.00204	0.275	0.160	0.917	458.5
		XX	0.308	0.434	0.718	0.00103	0.275	0.114	0.461	230.5
1	1.315	5S	0.065	1.185	0.255	0.00768	0.344	0.310	3.449	1725
		10S	0.109	1.097	0.413	0.00656	0.344	0.287	2.946	1473
		40ST, 40S	0.133	1.049	0.494	0.00600	0.344	0.275	2.690	1345
		80XS, 80S	0.179	0.957	0.639	0.00499	0.344	0.250	2.240	1120
		160	0.250	0.815	0.836	0.00362	0.344	0.213	1.625	812.5
		XX	0.358	0.599	1.076	0.00196	0.344	0.157	0.878	439.0
$1\frac{1}{4}$	1.660	5S	0.065	1.530	0.326	0.01277	0.435	0.401	5.73	2865
		10S	0.109	1.442	0.531	0.01134	0.435	0.378	5.09	2545
		40ST, 40S	0.140	1.380	0.668	0.01040	0.435	0.361	4.57	2285
		80XS, 80S	0.191	1.278	0.881	0.0891	0.435	0.335	3.99	1995
		160	0.250	1.160	1.107	0.00734	0.435	0.304	3.29	1645
		XX	0.382	0.896	1.534	0.00438	0.435	0.235	1.97	985
$1\frac{1}{2}$	1.900	5S	0.065	1.770	0.375	0.01709	0.497	0.463	7.67	3835
		10S	0.109	1.682	0.614	0.01543	0.497	0.440	6.94	3465
		40ST, 40S	0.145	1.610	0.800	0.01414	0.497	0.421	6.34	3170
		80XS, 80S	0.200	1.500	1.069	0.01225	0.497	0.393	5.49	2745
		160	0.281	1.338	1.439	0.00976	0.497	0.350	4.38	2190
		XX	0.400	1.100	1.885	0.00660	0.497	0.288	2.96	1480
2	2.375	5S	0.065	2.245	0.472	0.02749	0.622	0.588	12.34	6170
		10S	0.109	2.157	0.776	0.02538	0.622	0.565	11.39	5695
		40ST, 40S	0.154	2.067	1.075	0.02330	0.622	0.541	10.45	5225
		80ST, 80S	0.218	1.939	1.477	0.02050	0.622	0.508	9.20	4600

TABLE F-1 *(Continued)*

Nominal pipe size (in.)	Outside diameter (in.)	Schedule no.	Wall thickness (in.)	Inside diameter (in)	Cross-sectional area		Circumference (ft) or surface (ft/ft of length)		Capacity at 1 ft/s velocity	
					Metal ($in.^2$)	Flow (ft^2)	Outside	Inside	U.S. gal/min	lb/h water
		160	0.344	1.687	2.195	0.01552	0.622	0.436	6.97	3485
		XX	0.436	1.503	2.656	0.01232	0.622	0.393	5.53	2765
$2\frac{1}{2}$	2.875	5S	0.083	2.709	0.728	0.04003	0.753	0.709	17.97	8985
		10S	0.120	2.635	1.039	0.03787	0.753	0.690	17.00	8500
		40ST, 40S	0.203	2.469	1.704	0.03322	0.753	0.647	14.92	7460
		80XS, 80S	0.276	2.323	2.254	0.2942	0.753	0.608	13.20	6600
		160	0.375	2.125	2.945	0.2463	0.753	0.556	11.07	5535
		XX	0.552	1.771	4.028	0.01711	0.753	0.464	7.68	3840
3	3.500	5S	0.083	3.334	0.891	0.06063	0.916	0.873	27.21	13,605
		10S	0.120	3.260	1.274	0.05796	0.916	0.853	26.02	13,010
		40ST, 40S	0.216	3.068	2.228	0.05130	0.916	0.803	23.00	11,500
		80XS, 80S	0.300	2.900	3.016	0.04587	0.916	0.759	20.55	10,275
		160	0.438	2.624	4.213	0.03755	0.916	0.687	16.86	8430
		XX	0.600	2.300	5.466	0.02885	0.916	0.602	12.95	6475
$3\frac{1}{2}$	4.0	5S	0.083	3.834	1.021	0.08017	1.047	1.004	35.98	17,990
		10S	0.120	3.760	1.463	0.07711	1.047	0.984	34.61	17,305
		40ST, 40S	0.226	3.548	2.680	0.06870	1.047	0.929	30.80	15,400
		80XS, 80S	0.318	3.364	3.678	0.06170	1.047	0.881	27.70	13,850
4	4.5	5S	0.083	4.334	1.152	0.10245	1.178	1.135	46.0	23,000
		10S	0.120	4.260	1.651	0.09898	1.178	1.115	44.4	22,200
		40ST, 40S	0.237	4.026	3.17	0.08840	1.178	1.054	39.6	19,800
		80XS, 80S	0.337	3.826	4.41	0.07986	1.178	1.002	35.8	17,900
4		120	0.438	3.624	5.58	0.07170	1.178	0.949	32.2	16,100
		160	0.531	3.438	6.62	0.06647	1.178	0.900	28.9	14,450
		XX	0.674	3.152	8.10	0.05419	1.178	0.825	24.3	12,150
5	5.563	5S	0.109	5.345	1.87	0.1558	1.456	1.399	69.9	34.950
		10S	0.134	5.295	2.29	0.1529	1.456	1.386	68.6	34,300
		40ST, 40S	0.258	5.047	4.30	0.1390	1.456	1.321	62.3	31,150
		80XS, 80S	0.375	4.813	6.11	0.1263	1.456	1.260	57.7	28,850
		120	0.500	4.563	7.95	0.1136	1.456	1.195	51.0	25,500
		160	0.625	4.313	9.70	0.1015	1.456	1.129	45.5	22,750
		XX	0.750	4.063	11.34	0.0900	1.456	1.064	40.4	20,200
6	6.625	5S	0.109	6.407	2.23	0.2239	1.734	1.677	100.5	50,250
		10S	0.134	6.357	2.73	0.2204	1.734	1.664	98.9	49,450
		40ST, 40S	0.280	6.065	5.58	0.2006	1.734	1.588	90.0	45,000
		80XS, 80S	0.432	5.761	8.40	0.1810	1.734	1.508	81.1	40,550
		120	0.562	5.501	10.70	0.1650	1.734	1.440	73.9	36,950
		160	0.719	5.187	13.34	0.1467	1.734	1.358	65.9	32,950
		XX	0.864	4.897	15.64	0.1308	1.734	1.282	58.7	29,350
8	8.625	5S	0.109	8.407	2.915	0.3855	2.258	2.201	173.0	86,500
		10S	0.148	8.329	3,941	0.3784	2.258	2.180	169.8	84,900
		20	0.250	8.125	6.578	0.3601	2.258	2.127	161.5	80,750

Table F-1 (*Continued*)

Nominal pipe size (in.)	Outside diameter (in.)	Schedule no.	Wall thickness (in.)	Inside diameter (in)	Cross-sectional area: Metal (in.^{2})	Cross-sectional area: Flow (ft^2)	Circumference (ft) or surface (ft/ft of length): Outside	Circumference (ft) or surface (ft/ft of length): Inside	Capacity at 1 ft/s velocity: U.S. gal/min	Capacity at 1 ft/s velocity: lb/h water
		30	0.277	8.071	7.265	0.3553	2.258	2.113	159.4	79,700
		40ST, 40S	0.322	7.981	8.399	0.3474	2.258	2.089	155.7	77,850
		60	0.406	7.813	10.48	0.3329	2.258	2.045	149.4	74,700
		80XS, 80S	0.500	7.625	12.76	0.3171	2.258	1.996	142.3	71,150
		100	0.594	7.437	14.99	0.3017	2.258	1.947	135.4	67,700
		120	0.719	7.187	17.86	0.2817	2.258	1.882	126.4	63,200
		140	0.812	7.001	19.93	0.2673	2.258	1.833	120.0	60,000
		XX	0.875	6.875	21.30	0.2578	2.258	1.800	115.7	57,850
		160	0.906	6.813	21.97	0.2532	2.258	1.784	113.5	56,750
10	10.75	5S	0.134	10.482	4.47	0.5993	2.814	2.744	269.0	134,500
		10S	0.165	10.420	5.49	0.5922	2.814	2.728	265.8	132,900
		20	0.250	10.250	8.25	0.5731	2.814	2.685	257.0	128,500
		30	0.307	10.136	10.07	0.5603	2.814	2.655	252.0	126,000
		40ST, 40S	0.365	10.020	11.91	0.5745	2.814	2.620	246.0	123,000
		80S, 60XS	0.500	9.750	16.10	0.5185	2.814	2.550	233.0	116,500
		80	0.594	9.562	18.95	0.4987	2.814	2.503	223.4	111,700
		100	0.719	9.312	22.66	0.4728	2.814	2.438	212.3	106,150
		120	0.844	9.062	26.27	0.4479	2.814	2.372	201.0	100,500
		140, XX	1.000	8.750	30.63	0.4176	2.814	2.291	188.0	94,000
		160	1.125	8.500	34.02	0.3941	2.814	2.225	177.0	88,500
12	12.75	5S	0.156	12.428	6.17	0.8438	3.338	3.26	378.7	189,350
		10S	0.180	12.390	7.11	0.8373	3.338	3.24	375.8	187,900
		20	0.250	12.250	9.82	0.8185	3.338	3.21	367.0	183,500
		30	0.330	12.090	12.88	0.7972	3.338	3.17	358.0	179,000
		ST, 40S	0.375	12.000	14.58	0.7854	3.338	3.14	352.5	176,250
		40	0.406	11.938	15.74	0.7773	3.338	3.13	349.0	174,500
		XS, 80S	0.500	11.750	19.24	0.7530	3.338	3.08	338.0	169,000
		60	0.562	11.626	21.52	0.7372	3.338	3.04	331.0	165,500
		80	0.688	11.374	26.07	0.7056	3.338	2.98	316.7	158,350
		100	0.844	11.062	31.57	0.6674	3.338	2.90	299.6	149,800
		120, XX	1.000	10.750	36.91	0.6303	3.338	2.81	283.0	141,500
		140	1.125	10.500	41.09	0.6013	3.338	2.75	270.0	135,000
		160	1.312	10.136	47.14	0.5592	3.338	2.65	251.0	125,500
14	14	5S	0.156	13.688	6.78	1.0219	3.665	3.58	459	229,500
		10S	0.188	13.624	8.16	1.0125	3.665	3.57	454	227,000
		10	0.250	13.500	10.80	0.9940	3.665	3.53	446	223,000
		20	0.312	13.376	13.42	0.9750	3.665	3.50	438	219,000
		30, ST	0.375	13.250	16.05	0.9575	3.665	3.47	430	215,000
		40	0.438	13.124	18.66	0.9397	3.665	3.44	422	211,000
		XS	0.500	13.000	21.21	0.9218	3.665	3.40	414	207,000
		60	0.594	12.812	25.02	0.8957	3.665	3.35	402	201,000
		80	0.750	12.500	31.22	0.8522	3.665	3.27	382	191,000
		100	0.938	12.124	38.49	0.8017	3.665	3.17	360	180,000
		120	1.094	11.812	44.36	0.7610	3.665	3.09	342	171,000
		140	1.250	11.500	50.07	0.7213	3.665	3.01	324	162,000
		160	1.406	11.188	55.63	0.6827	3.665	2.93	306	153,000

TABLE F-1 (*Continued*)

Nominal pipe size (in.)	Outside diameter (in.)	Schedule no.	Wall thickness (in.)	Inside diameter (in)	Cross-sectional area		Circumference (ft) or surface (ft/ft of length)		Capacity at 1 ft/s velocity	
					Metal (in.^{2})	Flow (ft^2)	Outside	Inside	U.S. gal/min	lb/h water
16	16	5S	0.165	15.670	8.21	1.3393	4.189	4.10	601	300,500
		10S	0.188	15.624	9.34	1.3314	4.189	4.09	598	299,000
		10	0.250	15.500	12.37	1.3104	4.189	4.06	587	293,500
		20	0.312	15.376	15.38	1.2985	4.89	4.03	578	289,000
		30, ST	0.375	15.250	18.41	1.2680	4.189	3.99	568	284,000
		40, XS	0.500	15.000	24.35	1.2272	4.189	3.93	550	275,000
		60	0.656	14.688	31.62	1.766	4.189	3.85	528	264,000
		80	0.844	14.312	40.19	1.1171	4.189	3.75	501	250,500
		100	1.031	13.939	48.48	1.0596	4.189	3.65	474	237,000
		120	1.219	13.562	56.61	1.0032	4.189	3.55	450	225,000
		140	1.438	13.124	65.79	0.9394	4.189	3.44	422	211,000
		160	1.594	12.812	72.14	0.8953	4.189	3.35	402	201,000
18	18	5S	0.165	17.760	9.25	1.8029	4.712	4.63	764	382,000
		10S	0.188	17.624	10.52	1.6941	4.712	4.61	760	379,400
		10	0.250	17.500	13.94	1.6703	4.712	4.58	750	375,000
		20	0.312	17.376	17.34	1.6468	4.712	4.55	739	369.500
		ST	0.375	17.250	20.76	1.6230	4.712	4.52	728	364,000
		30	0.438	17.124	24.16	1.5993	4.712	4.48	718	359,000
		XS	0.500	17.000	27.49	1.5763	4.712	4.45	707	353,500
		40	0.562	16.876	30.79	1.5533	4.712	4.42	697	348,500
		60	0.750	16.500	40.54	1.4849	4.712	4.32	666	333,000
		80	0.938	16.124	50.28	1.4180	4.712	4.2	636	318,000
		100	1.156	15.688	61.17	1.3423	4.712	4.11	602	301,000
		120	1.375	15.250	71.82	1.2684	4.712	3.99	569	284,500
		140	1.562	14.876	80.66	1.2070	4.712	3.89	540	270,000
		160	1.781	14.438	90.75	1.1370	4.712	3.78	510	255,000
20	20	5S	0.188	19.624	11.70	2.1004	5.236	5.14	943	471,500
		10S	0.218	19.564	13.55	2.0878	5.236	5.12	937	467,500
		10	0.250	19.500	5.51	2.0740	5.236	5.11	930	465,000
		20, ST	0.375	19.250	23.12	2.0211	5.236	5.04	902	451,000
		30, XS	0.500	19.000	30.63	1.9689	5.236	4.97	883	441,500
		40	0.594	18.812	36.21	1.9302	5.236	4.92	866	433,000
		60	0.812	18.376	48.95	1.8417	5.236	4.81	826	413,000
		80	1.031	17.938	61.44	1.7550	5.236	4.70	787	393,500
		100	1.281	17.438	75.33	1.6585	5.236	4.57	744	372,000
		120	1.500	17.000	87.18	1.5763	5.236	4.45	707	353,500
		140	1.750	16.500	100.3	1.4849	5.236	4.32	665	332.500
		160	1.969	16.062	111.5	1.4071	5.236	4.21	632	316,000
24	24	5S	0.218	23.564	16.29	3.0285	6.283	6.17	1359	579,500
		20, 10S	0.250	23.500	18.65	3.012	6.283	6.15	1350	675,000
		20, ST	0.375	23.250	27.83	2.948	6.283	6.09	1325	662,500
		XS	0.500	23.000	36.90	2.885	6.283	6.02	1295	642,500
		30	0.562	22.876	41.39	2.854	6.283	5.99	1281	640,500
		40	0.688	22.624	50.39	2.792	6.283	5.92	1253	626,500
		60	0.969	22.062	70.11	2.655	6.283	5.78	1192	596,000
		80	1.219	21.562	87.24	2.536	6.283	5.64	1138	569,000
		100	1.531	20.938	108.1	2.391	6.283	5.48	1073	536,500

TABLE F-1 (*Continued*)

Nominal pipe size (in.)	Outside diameter (in.)	Schedule no.	Wall thickness (in.)	Inside diameter (in)	Cross-sectional area: Metal (in.^{2})	Cross-sectional area: Flow (ft^2)	Circumference (ft) or surface (ft/ft of length): Outside	Circumference (ft) or surface (ft/ft of length): Inside	Capacity at 1 ft/s velocity: U.S. gal/min	Capacity at 1 ft/s velocity: lb/h water
		120	1.812	20.376	126.3	2.264	6.283	5.33	1016	508,000
		140	2.062	19.876	142.1	2.155	6.283	5.20	965	482,500
		160	2.344	19.312	159.5	2.034	6.283	5.06	913	456,500
30	30	5S	0.250	29.500	23.37	4.746	7.854	7.72	2130	1,065,000
		10, 10S	0.312	29.376	29.10	4.707	7.854	7.69	2110	1,055,000
		ST	0.375	29.250	34.90	4.666	7.854	7.66	2094	1,048,000
		20, XS	0.500	29.000	46.34	4.587	7.854	7.59	2055	1,027,500
		30	0.625	28.750	57.68	4.508	7.854	7.53	2020	1,010,000

[a] 5S, 10S, and 40S are extracted from Stainless Steel Pipe, ANSI B36.19–1976, with permission of the publisher, the American Society of Mechanical Engineers, New York. ST = standard wall, XS = extra strong wall, XX = double extra strong wall, and Schedules 10–160 are extracted from Wrought-Steel and Wrought-Iron Pipe, ANSI B36.10–1975, with permission of the same publisher. Decimal thicknesses for respective pipe sizes represent their nominal or average wall dimensions. Mill tolerances as high as $\pm 12\frac{1}{2}\%$ are permitted.

Plain-end pipe is produced by a square cut. Pipe is also shipped from the mills threaded, with a threaded coupling on one end, or with the ends beveled for welding, or grooved or sized for patented couplings. Weights per foot for threaded and coupled pipe are slightly greater because of the weight of the coupling, but it is not available larger than 12 in or lighter than Schedule 30 sizes 8 through 12 in, or Schedule 406 in and smaller.

To convert inches to millimeters, multiply by 25.4; to convert square inches to square millimeters, multiply by 645; to convert feet to meters, multiply by 0.3048; to convert square feet to square meters, multiply by 0.0929; to convert pounds per foot to kilograms per meter, multiply by 1.49; to convert gallons to cubic meters, multiply by 3.7854×10^{-3}; and to convert pounds to kilograms, multiply by 0.4536.

Appendix G

Flow of Water/Air Through Schedule 40 Pipe

TABLE G-1 Flow of Water Through Schedule 40 Steel Pipe

Discharge		Pressure Drop per 100 feet and Velocity in Schedule 40 Pipe for Water at 60 F.															
Gallons per Minute	Cubic Ft per Second	Velocity Feet per Second	Press Drop Lbs per Sq. In.	Velocity Feet per Second	Press Drop Lbs per Sq. In.	Velocity Feet per Second	Press Drop Lbs per Sq. In.	Velocity Feet per Second	Press Drop Lbs per Sq. In.	Velocity Feet per Second	Press Drop Lbs per Sq. In.	Velocity Feet per Second	Press Drop Lbs per Sq. In.	Velocity Feet per Second	Press Drop Lbs per Sq. In.	Velocity Feet per Second	Press Drop Lbs per Sq. In.
		$\frac{1}{8}''$		$\frac{1}{4}''$		$\frac{3}{8}''$		$\frac{1}{2}''$									
0.2	0.000446	1.13	1.86	0.616	0.359												
0.3	0.000668	1.69	4.22	0.924	0.903	0.504	0.159	0.317	0.061		$\frac{3}{4}''$						
0.4	0.000891	2.26	6.98	1.23	1.61	0.672	0.345	0.422	0.086								
0.5	0.00111	2.82	10.5	1.54	2.39	0.840	0.539	0.528	0.167	0.301	0.033						
0.6	0.00134	3.39	14.7	1.85	3.29	1.01	0.751	0.633	0.240	0.361	0.041						
0.8	0.00178	4.52	25.0	2.46	5.44	1.34	1.25	0.844	0.408	0.481	0.102		1″		$1\frac{1}{4}''$		
1	0.00223	5.65	37.2	3.08	8.28	1.68	1.85	1.06	0.600	0.602	0.155	0.171	0.048			$1\frac{1}{2}''$	
2	0.00446	11.29	134.4	6.16	30.1	3.36	6.58	2.11	2.10	1.20	0.526	0.741	0.164	0.429	0.044		
3	0.00668			9.25	64.1	5.04	13.9	3.17	4.33	1.81	1.09	1.114	0.336	0.644	0.090	0.473	0.043
4	0.00891			12.33	111.2	6.72	23.9	4.22	7.42	2.41	1.83	1.49	0.565	0.858	0.150	0.630	0.071
5	0.01114	2″				8.40	36.7	5.28	11.2	3.01	2.75	1.66	0.835	1.073	0.223	0.788	0.104
6	0.01337	0.574	0.044	$2\frac{1}{2}''$		10.08	51.9	6.3	15.8	3.61	3.84	2.23	1.17	1.29	0.309	0.946	0.145
8	0.01782	0.765	0.073			13.44	91.1	8.45	27.7	4.81	6.60	2.97	1.99	1.72	9.518	1.26	0.241
10	0.02228	0.956	0.108	0.670	0.046			10.56	42.4	6.02	9.99	3.71	2.99	2.15	0.774	1.58	0.361
15	0.03342	1.43	0.224	1.01	0.094		3″			9.03	21.6	5.57	6.36	3.22	1.63	2.37	0.755
20	0.04456	1.91	0.375	1.34	0.158	0.868	0.056		$3\frac{1}{2}''$	12.03	37.8	7.43	10.9	4.29	2.78	3.16	1.28
25	0.05570	2.39	0.561	1.68	0.234	1.09	0.083	0.812	0.041		4″	9.28	16.7	5.37	4.22	3.94	1.93
30	0.06684	2.87	0.786	2.01	0.327	1.30	0.114	0.974	0.056			11.14	23.8	6.44	5.92	4.73	2.72
35	0.07798	3.35	1.05	2.35	0.436	1.52	0.151	1.14	0.071	0.882	0.041	12.99	32.2	7.51	7.90	5.52	3.64
40	0.08912	3.83	1.35	2.68	0.556	1.74	0.192	1.30	0.095	1.01	0.052	14.85	41.5	8.59	10.24	6.30	4.65
45	0.1003	4.30	1.67	3.02	0.668	1.95	0.239	1.46	0.117	1.13	0.064			9.67	12.80	7.09	5.85

50	0.1114	4.78	2.03	3.35	0.839	2.17	0.288	1.62	0.142	1.26	0.076	5″		10.74	15.66	7.88	7.15
60	0.1337	5.74	2.87	4.02	1.18	2.60	0.406	1.95	0.204	1.51	0.107			12.89	22.2	9.47	10.21
70	0.1560	6.70	3.84	4.69	1.59	3.04	0.540	2.27	0.261	1.76	0.047	1.12	0.047			11.05	13.71
80	0.1782	7.65	4.97	5.36	2.03	3.47	0.687	2.60	0.334	2.02	0.180	1.28	0.060			12.62	17.59
90	0.2005	8.60	6.20	6.03	2.53	3.91	0.861	2.92	0.416	2.27	0.074	1.44	0.074	6″		14.20	22.0
100	0.2228	9.56	7.59	6.70	3.09	4.34	1.05	3.25	0.509	2.52	0.272	1.60	0.090	1.11	0.036	15.78	26.9
125	0.2785	11.97	11.76	8.38	4.71	5.43	1.61	4.06	0.769	3.15	0.415	2.01	0.135	1.39	0.055	19.72	41.4
150	0.3342	14.36	16.70	10.05	6.69	6.51	2.24	4.87	1.08	3.78	0.580	2.41	0.190	1.67	0.077		
175	0.3899	16.75	22.3	11.73	8.97	7.60	3.00	5.68	1.44	4.41	0.774	2.81	0.253	1.94	0.102		
200	0.4456	19.14	28.8	13.42	11.68	8.68	3.87	6.49	1.85	5.04	0.985	3.21	0.323	2.22	0.130	8″	
225	0.5013	...	...	15.09	14.63	9.77	4.83	7.30	2.32	5.67	1.23	3.61	0.401	2.50	0.162	1.44	0.043
250	0.557	...	...	...	...	10.85	5.93	8.12	2.84	6.30	1.46	4.01	0.495	2.78	0.195	1.60	0.051
275	0.6127	...	...	...	...	11.94	7.14	8.93	3.40	6.93	1.79	4.41	0.583	3.05	0.234	1.76	0.061
300	0.6684	...	...	...	...	13.00	8.36	9.74	4.02	7.56	2.11	4.81	0.683	3.33	0.275	1.92	0.072
325	0.7241	...	...	...	...	14.12	9.89	10.53	4.09	8.19	2.47	5.21	0.797	3.61	0.320	2.08	0.083
350	0.7798			...	...	...	...	11.36	5.41	8.82	2.84	5.62	0.919	3.89	0.367	2.24	0.095
375	0.8355			...	...	...	...	12.17	6.18	9.45	3.25	6.02	1.05	4.16	0.416	2.40	0.108
400	0.8912			...	...	...	...	12.98	7.03	10.08	3.68	6.42	1.19	4.44	0.471	2.56	0.121
425	0.9469			...	...	...	...	13.80	7.89	10.71	4.12	6.82	1.33	4.72	0.529	2.73	0.136
450	1.003	10″		...	...	...	...	14.61	8.80	11.34	4.60	7.22	1.48	5.00	0.590	2.89	0.151
475	1.059	1.93	0.054			...	...	...	...	11.97	5.12	7.62	1.64	5.27	0.653	3.04	0.166
500	1.114	2.03	0.059			...	...	...	...	12.60	5.65	8.02	1.81	5.55	0.720	3.21	0.182
550	1.225	2.24	0.071			...	...	...	...	13.85	6.79	8.82	2.17	6.11	0.861	3.53	0.219
600	1.337	2.44	0.083			...	...	...	...	15.12	8.04	9.63	2.55	6.66	1.02	3.85	0.258
650	1.448	2.64	0.097	12″		...	...	...	...	...	...	10.43	2.98	7.22	1.18	4.17	0.301
700	1.560	2.85	0.112	2.01	0.047			...	...	...	...	11.23	3.43	7.78	1.35	4.49	0.343
750	1.671	3.05	0.127	2.15	0.054	14″		...	...	...	...	12.03	3.92	8.33	1.55	4.81	0.392
900	1.782	3.25	0.143	2.29	0.061			...	...	...	...	12.83	4.43	8.88	1.75	5.13	0.443
850	1.894	3.46	0.160	2.44	0.068	2.02	0.042	...	...	...	...	13.64	5.00	9.44	1.96	5.45	0.497
900	2.005	3.66	0.179	2.58	0.075	2.13	0.047	...	...	...	...	14.44	5.58	9.99	2.18	5.77	0.554
950	2.117	3.86	0.198	2.72	0.083	2.25	0.052			...	...	15.24	6.21	10.55	2.42	6.09	0.613
1 000	2.228	4.07	0.218	2.87	0.091	2.37	0.057	16″		...	...	16.04	6.84	11.10	2.68	6.41	0.675
1 100	2.451	4.48	0.260	3.15	0.110	2.61	0.068			...	...	17.65	8.23	12.22	3.22	7.05	0.807

TABLE G-1 *(Continued)*

Discharge		Pressure Drop per 100 feet and Velocity in Schedule 40 Pipe for Water at 60 F.															
Gallons per Minute	Cubic Ft per Second	Velocity Feet per Second	Press Drop Lbs per Sq. In.	Velocity Feet per Second	Press Drop Lbs per Sq. In.	Velocity Feet per Second	Press Drop Lbs per Sq. In.	Velocity Feet per Second	Press Drop Lbs per Sq. In.	Velocity Feet per Second	Press Drop Lbs per Sq. In.	Velocity Feet per Second	Press Drop Lbs per Sq. In.	Velocity Feet per Second	Press Drop Lbs per Sq. In.	Velocity Feet per Second	Press Drop Lbs per Sq. In.
1 200	2.674	4.88	0.306	3.44	0.128	2.85	0.080	2.18	0.042	...	...	...	...	13.33	3.81	7.70	0.948
1 300	2.896	5.29	0.355	3.73	0.150	3.08	0.093	2.36	0.048	...	...	...	...	14.43	4.45	8.11	1.11
1 400	3.119	5.70	0.409	4.01	0.171	3.32	0.107	2.54	0.055					15.55	5.13	8.98	1.28
1 500	3.342	6.10	0.466	4.30	0.195	3.56	0.122	2.72	0.063	18″				16.66	5.85	9.62	1.46
1 600	3.565	6.51	0.527	4.59	0.219	3.79	0.138	2.90	0.071					17.77	6.61	10.26	1.65
1 800	4.010	7.32	0.663	5.16	0.276	4.27	0.172	3.27	0.088	2.58	0.050			19.99	8.37	11.54	2.08
2 000	4.456	8.14	0.808	5.73	0.339	4.74	0.209	3.63	0.107	2.87	0.060	20″		22.21	10.3	12.82	2.55
2 500	5.570	10.17	1.24	7.17	0.515	5.93	0.321	4.54	0.163	3.59	0.091					16.03	3.94
3 000	6.684	12.20	1.76	8.60	0.731	7.11	0.451	5.45	0.232	4.10	0.129	3.46	0.075	24″		19.24	5.59
3 500	7.798	14.24	2.38	10.01	0.982	8.30	0.607	6.35	0.312	5.02	0.173	5.05	0.101			22.44	7.56
4 000	8.912	16 27	3.08	11.47	1.27	8.48	0.787	7.26	0.401	5.74	0.222	4.62	0.129	3.19	0.052	25.65	9.80
4 500	10.03	18.31	3.87	12.90	1.60	10.67	0.990	8.17	0.503	6.46	0.280	5.20	0.162	3.59	0.065	28.87	12.2
5 000	11.14	20.35	4.71	14.33	1.95	11.85	1.21	9.08	0.617	7.17	0.340	5.77	0.199	3.99	0.079		
6 000	13.37	24 41	6.47	17.20	2.77	14.21	1.71	10.89	0.877	8.61	0.483	6.91	0.280	4.79	0.111		
7 000	15.60	28.49	9.11	20.07	3.74	16.60	2.31	12.71	1.18	10.04	0.652	8.08	0.376	5.59	0.150		
8 000	17.82		...	22.93	4.48	18.96	2.99	14.52	1.51	11.47	0.839	9.23	0.488	6.18	0.192		
9 000	20.05			25.79	6.09	21.14	3.76	16.34	1.90	12.91	1.05	10.39	0.608	7.18	0.242		
10 000	22.28			28.66	7.46	23.71	4.61	18.15	2.34	14.14	1.28	11.54	0.739	7.98	0.294		
12 000	26.74			14.40	10.7	28.45	6.59	21.79	3.33	17.21	1.83	13.85	1.06	9.58	0.416		
14 000	31.19			...	...	11.19	8.89	25.42	4.49	20.08	2.45	16.16	1.43	11.17	0.562		
16 000	35.65			...	...	...	...	29.05	5.83	22.95	3.18	18.47	1.85	12.77	0.723		
18 000	40.10			...	...	...	...	32.68	7.31	25.82	4.03	20.77	2.32	14.36	0.907		
20 000	44.56			...	...	...	...	36.31	9.03	28.69	4.93	23.08	2.86	15.96	1.12		

For pipe lengths other than 100 feet, the pressure drop is proportional to the length. Thus, for 50 feet of pipe, the pressure drop is approximately one-half the value given in the table . . . for 300 feet, three times the given value, etc.

Velocity is a function of the cross sectional flow area; thus, it is constant for a given flow rate and is independent of pipe length.

TABLE G-2 Flow of Air Through Schedule 40 Steel Pipe

For lengths of pipe other than 100 feet, the pressure drop is proportional to the length. Thus, for 50 feet of pipe, the pressure drop is approximately one-half the value given in the table . . . for 300 feet, thre times the given value, etc.

The pressure drop is also inversely proportional to the absolute pressure and directly proportional to the absolute temperature.

Therefore, to determine the pressure drop for inlet or average pressures other than 100 psi and at temperatures other than 60 F, multiply the values given in the table by the ratio:

$$\left(\frac{100 + 14.7}{P + 14.7}\right)\left(\frac{460 + t}{520}\right)$$

where:

"P" is the inlet or average gauge pressure in pounds per square inch, and,

"t" is the temperature in degrees Fahrenheit under consideration.

The cubic feet per minute of compressed air at any pressure is inversely proportional to the absolute pressure and directly proportional to the absolute temperature.

To determine the cubic feet per minute of compressed air at any temperature and pressure other than standard conditions, multiply the value of cubic feet per minute of free air by the ratio:

$$\left(\frac{14.7}{14.7 + P}\right)\left(\frac{460 + t}{520}\right)$$

Calculations for Pipe Other than Schedule 40

To determine the velocity of water, or the pressure drop of water or air, through pipe other than Schedule 40, use the following formulas:

$$v_a = v_{40}\left(\frac{d_{40}}{d_a}\right)^2$$

$$\Delta P_a = \Delta P_{40}\left(\frac{d_{40}}{d_a}\right)^5$$

Subscript "a" refers to the Schedule of pipe through which velocity or pressure drop is desired.

Subscript "40" refers to the velocity or pressure drop through Schedule 40 pipe, as given in the tables on these facing pages.

Free Air q'_m Cubic Feet Per Minute at 60 F and 14.7 psia	Compressed Air Cubic Feet Per Minute at 60 F and 100 psig	Pressure Drop of Air In Pounds per Square Inch Per 100 Feet of Schedule 40 Pipe For Air at 100 Pounds per Square Inch Gauge Pressure and 60 F Temperature								
		1/8″	1/4″	3/8″	1/2″					
1	0.128	0.361	0.083	0.018						
2	0.256	1.31	0.285	0.064	0.020					
3	0.384	3.06	0.605	0.133	0.042	3/4″				
4	0.513	4.83	1.04	0.226	0.071					
5	0.641	7.45	1.58	0.343	0.106	0.027	1″			
6	0.769	10.6	2.23	0.408	0.148	0.037				
8	1.025	18.6	3.89	0.848	0.255	0.062	0.019			
10	1.282	28.7	5.96	1.26	0.356	0.094	0.029	1¼″	1½″	
15	1.922	...	13.0	2.73	0.834	0.201	0.062			
20	2.563	...	22.8	4.76	1.43	0.345	0.102	0.026		
25	3.204	...	35.6	7.34	2.21	0.526	0.156	0.039	0.019	
30	3.845	...	...	10.5	3.15	0.748	0.219	0.055	0.026	
35	4.486	...	...	14.2	4.24	1.00	0.293	0.073	0.035	
40	5.126	...	...	18.4	5.49	1.30	0.379	0.095	0.044	2″
45	5.767	...	...	23.1	6.90	1.62	0.474	0.116	0.055	
50	6.408			28.5	8.49	1.99	0.578	0.149	0.067	0.019
60	7.690	2½″		40.7	12.2	2.85	0.819	0.200	0.094	0.027
70	8.971			...	16.5	3.83	1.10	0.270	0.126	0.036
80	10.25	0.019		...	21.4	4.96	1.43	0.350	0.162	0.046
90	11.53	0.023		...	27.0	6.25	1.80	0.437	0.203	0.058
100	12.82	0.029	3″		33.2	7.69	2.21	0.634	0.247	0.070
125	16.02	0.044			...	11.9	3.39	0.825	0.380	0.107
150	19.22	0.062	0.021		...	17.0	4.87	1.17	0.537	0.151
175	22.43	0.083	0.028		...	23.1	6.60	1.58	0.727	0.205
200	25.63	0.107	0.036	3½″	...	30.0	8.64	2.05	0.937	0.264
225	28.84	0.134	0.045	0.022		37.9	10.8	2.59	1.19	0.331
250	32.04	0.164	0.055	0.027		...	13.3	3.18	1.45	0.404
275	35.24	0.191	0.066	0.032		...	16.0	3.83	1.75	0.484
300	38.45	0.232	0.078	0.037		...	19.0	4.56	2.07	0.573
325	41.65	0.270	0.090	0.043	4″	...	22.3	5.32	2.42	0.673
350	44.87	0.313	0.104	0.050		...	25.8	6.17	2.80	0.776
375	48.06	0.356	0.119	0.057	0.030	...	29.6	7.05	3.20	0.887
400	51.26	0.402	0.134	0.064	0.034	...	33.6	8.02	3.64	1.00
425	54.47	0.452	0.151	0.072	0.038	...	37.9	9.01	4.09	1.13
450	57.67	0.507	0.168	0.081	0.042	...	...	10.2	4.59	1.26
475	60.88	0.562	0.187	0.089	0.047		...	11.3	5.09	1.40
500	64.08	0.623	0.206	0.099	0.052		...	12.5	5.61	1.55
550	70.49	0.749	0.248	0.118	0.062		...	15.1	6.79	1.87
600	76.90	0.887	0.293	0.139	0.073		...	18.0	8.04	2.21
650	83.30	1.04	0.342	0.163	0.086	5″	...	21.1	9.43	2.60
700	89.71	1.19	0.395	0.188	0.099	0.032		24.3	10.9	3.00
750	96.12	1.36	0.451	0.214	0.113	0.036		27.9	12.6	3.44
800	102.5	1.55	0.513	0.244	0.127	0.041		31.8	14.2	3.90
850	108.9	1.74	0.576	0.274	0.144	0.046		35.9	16.0	4.40
900	115.3	1.95	0.642	0.305	0.160	0.051	6″	40.2	18.0	4.91
950	121.8	2.18	0.715	0.340	0.178	0.057	0.023	...	20.0	5.47
1 000	128.2	2.40	0.788	0.375	0.197	0.063	0.025	...	22.1	6.06
1 100	141.0	2.89	0.948	0.451	0.236	0.075	0.030	...	26.7	7.29
1 200	153.8	3.44	1.13	0.533	0.279	0.089	0.035	...	31.8	8.63
1 300	166.6	4.01	1.32	0.626	0.327	0.103	0.041	...	37.3	10.1
1 400	179.4	4.65	1.52	0.718	0.377	0.119	0.047			11.8
1 500	192.2	5.31	1.74	0.824	0.431	0.136	0.054	8″		13.5
1 600	205.1	6.04	1.97	0.932	0.490	0.154	0.061			15.3
1 800	230.7	7.65	2.50	1.18	0.616	0.193	0.075			19.3
2 000	256.3	9.44	3.06	1.45	0.757	0.237	0.094	0.023	10″	23.9
2 500	320.4	14.7	4.76	2.25	1.17	0.366	0.143	0.035		37.3
3 000	384.5	21.1	6.82	3.20	1.67	0.524	0.204	0.051	0.016	
3 500	448.6	28.8	9.23	4.33	2.26	0.709	0.276	0.068	0.022	
4 000	512.6	37.6	12.1	5.66	2.94	0.919	0.358	0.088	0.028	12″
4 500	576.7	47.6	15.3	7.16	3.69	1.16	0.450	0.111	0.035	
5 000	640.8	...	18.8	8.85	4.56	1.42	0.552	0.136	0.043	0.018
6 000	769.0	...	27.1	12.7	6.57	2.03	0.794	0.195	0.061	0.025
7 000	897.1	...	36.9	17.2	8.94	2.76	1.07	0.262	0.082	0.034
8 000	1025	...	...	22.5	11.7	3.59	1.39	0.339	0.107	0.044
9 000	1153	...	...	28.5	14.9	4.54	1.76	0.427	0.134	0.055
10 000	1282	...	...	35.2	18.4	5.60	2.16	0.526	0.164	0.067
11 000	1410	...	...	...	22.2	6.78	2.62	0.633	0.197	0.081
12 000	1538	...	...	...	26.4	8.07	3.09	0.753	0.234	0.096
13 000	1666	...	...	...	31.0	9.47	3.63	0.884	0.273	0.112
14 000	1794	...	...	...	36.0	11.0	4.21	1.02	0.316	0.129
15 000	1922	...	...	...	...	12.6	4.84	1.17	0.364	0.148
16 000	2051	...	...	...	...	14.3	5.50	1.33	0.411	0.167
18 000	2307	...	...	...	...	18.2	6.96	1.68	0.520	0.213
20 000	2563	...	...	...	...	22.4	8.60	2.01	0.642	0.260
22 000	2820	...	...	...	...	27.1	10.4	2.50	0.771	0.314
24 000	3076	...	...	...	...	32.3	12.4	2.97	0.918	0.371
26 000	3332	...	...	...	...	37.9	14.5	3.49	1.12	0.435
28 000	3588	...	...	...	...	...	16.9	4.04	1.25	0.505
30 000	3845	...	...	...	...	...	19.3	4.64	1.42	0.520

Appendix H

Typical Pump Head Capacity Range Charts

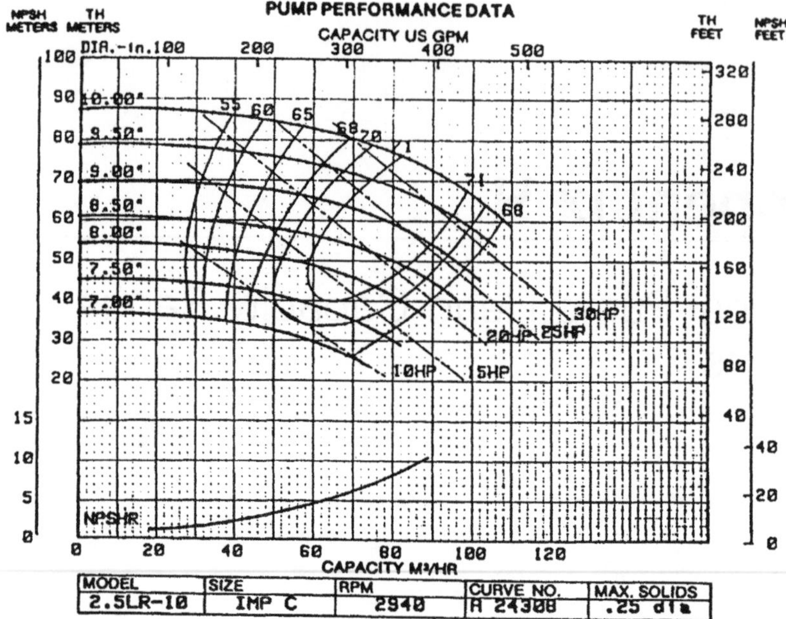

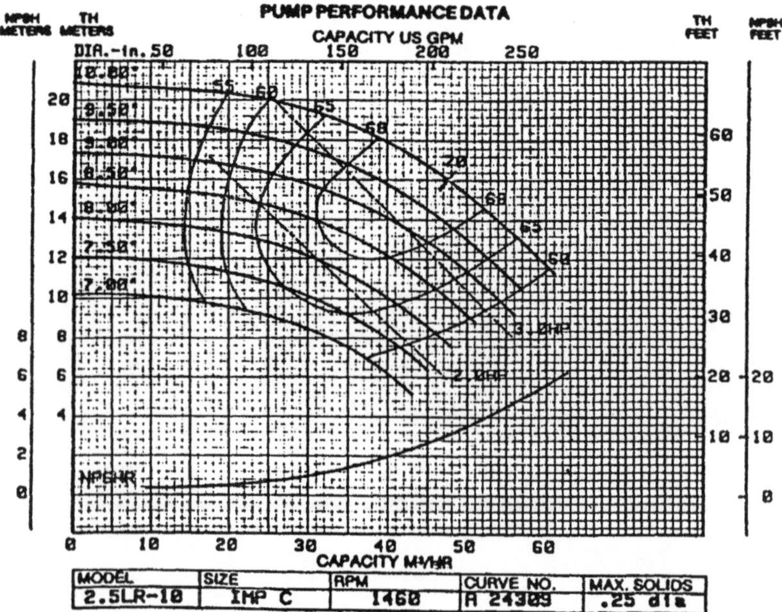

FIGURE H-1 Typical pump characteristic curves.

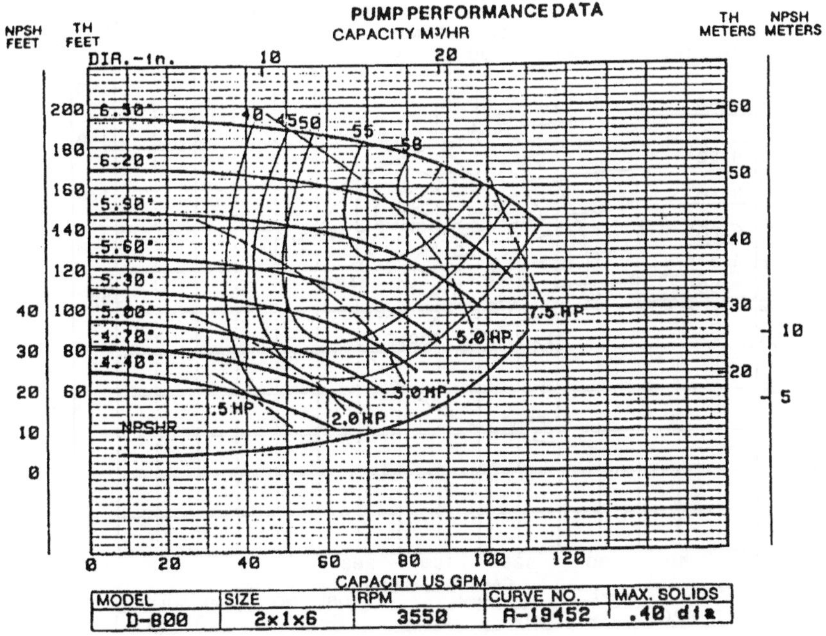
PUMP PERFORMANCE DATA
CAPACITY M³/HR
NPSH FEET
TH FEET
TH METERS
NPSH METERS
DIA.-in.
6.30"
6.20"
5.90"
5.60"
5.30"
5.00"
4.70"
4.40"
40
45
50
55
58
7.5 HP
5.0 HP
3.0 HP
2.0 HP
.5 HP
NPSHR
CAPACITY US GPM
MODEL D-800
SIZE 2x1x6
RPM 3550
CURVE NO. A-19452
MAX. SOLIDS .40 dia

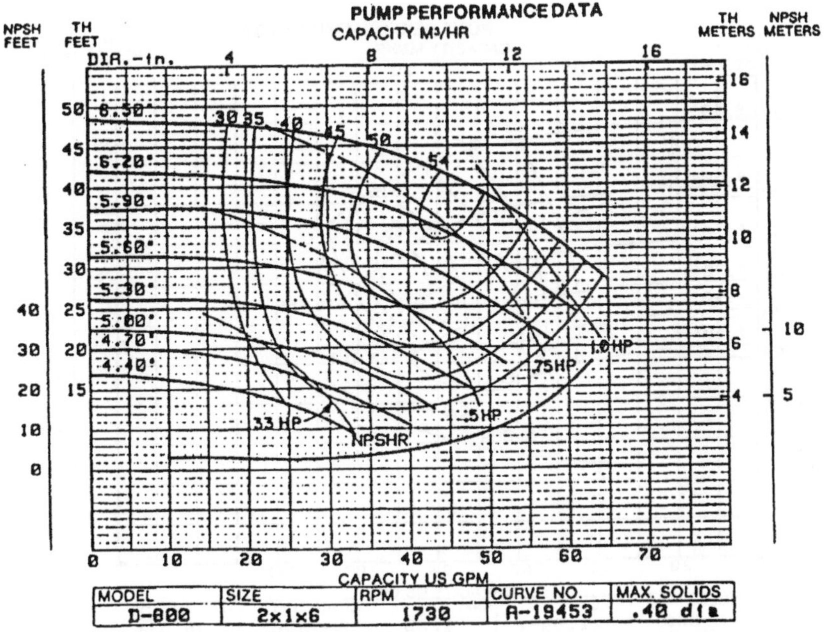
PUMP PERFORMANCE DATA
CAPACITY M³/HR
NPSH FEET
TH FEET
TH METERS
NPSH METERS
DIA.-in.
6.50"
6.20"
5.90"
5.60"
5.30"
5.00"
4.70"
4.40"
30
35
40
45
50
54
1.0 HP
.75 HP
.5 HP
.33 HP
NPSHR
CAPACITY US GPM
MODEL D-800
SIZE 2x1x6
RPM 1730
CURVE NO. A-19453
MAX. SOLIDS .40 dia

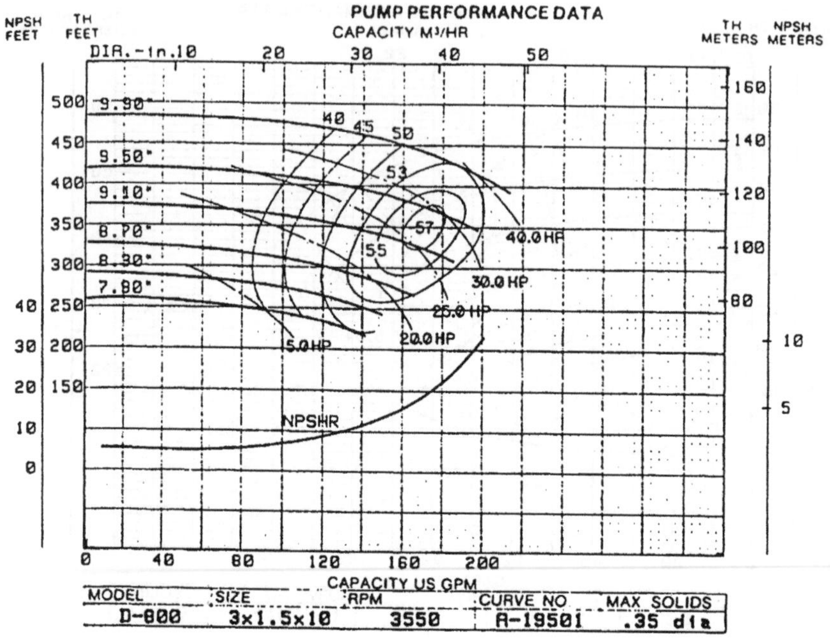

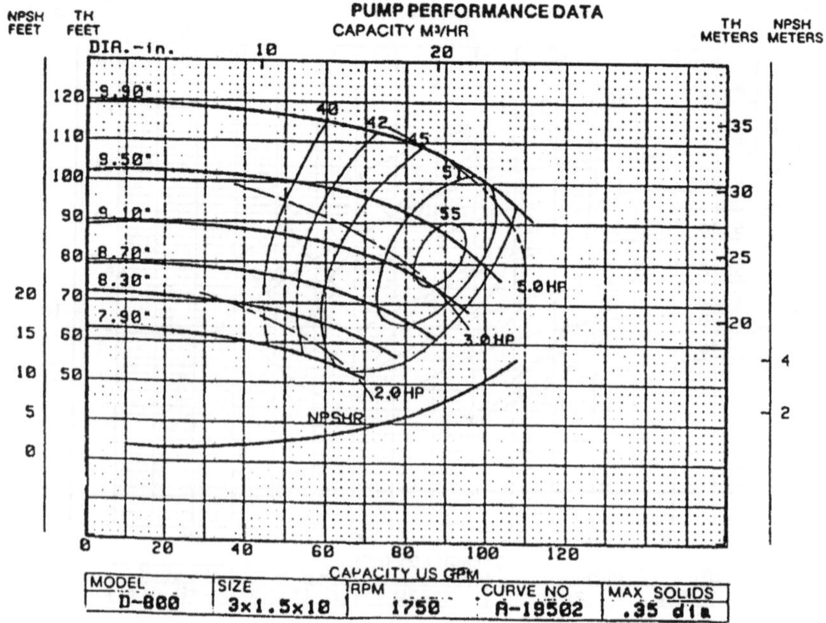

FIGURE H-1 (*Continued*)

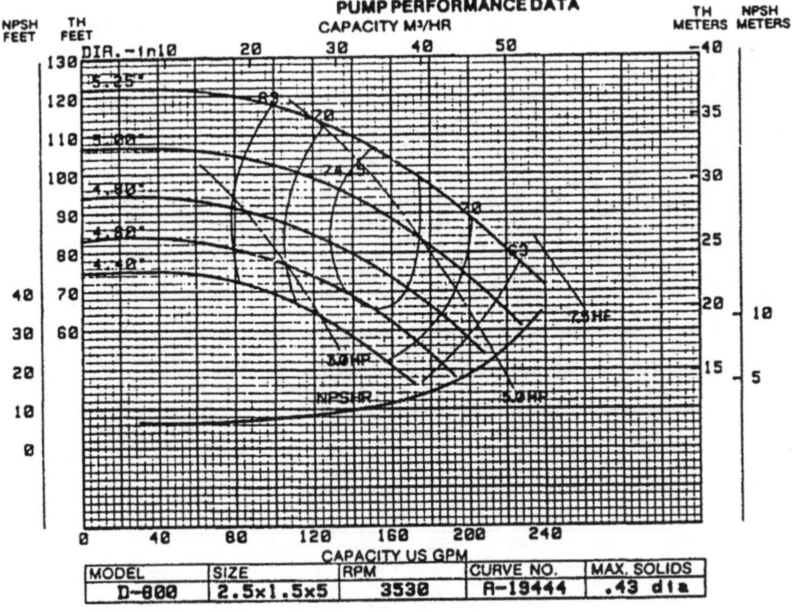

MODEL	SIZE	RPM	CURVE NO.	MAX. SOLIDS
D-800	2.5x1.5x5	3530	A-19444	.43 dia

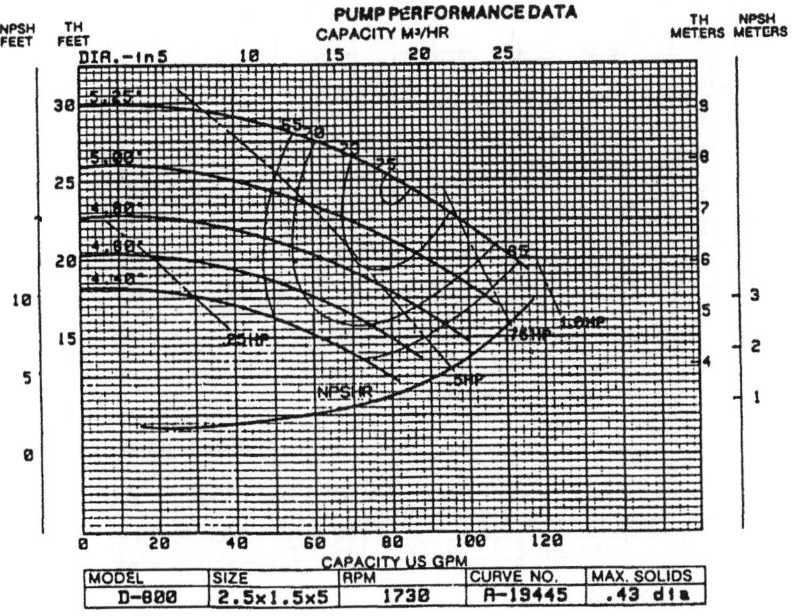

MODEL	SIZE	RPM	CURVE NO.	MAX. SOLIDS
D-800	2.5x1.5x5	1730	A-19445	.43 dia

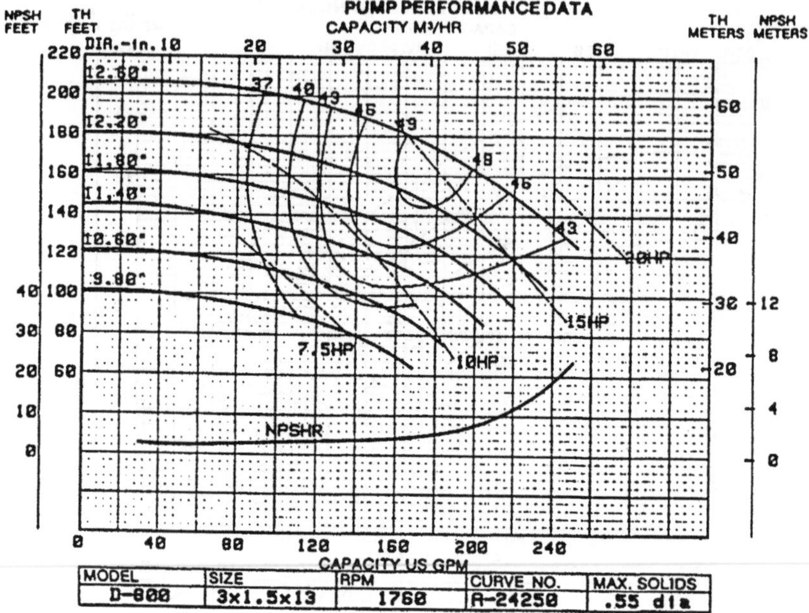

MODEL	SIZE	RPM	CURVE NO.	MAX. SOLIDS
D-800	3x1.5x13	1760	A-24250	.55 dia

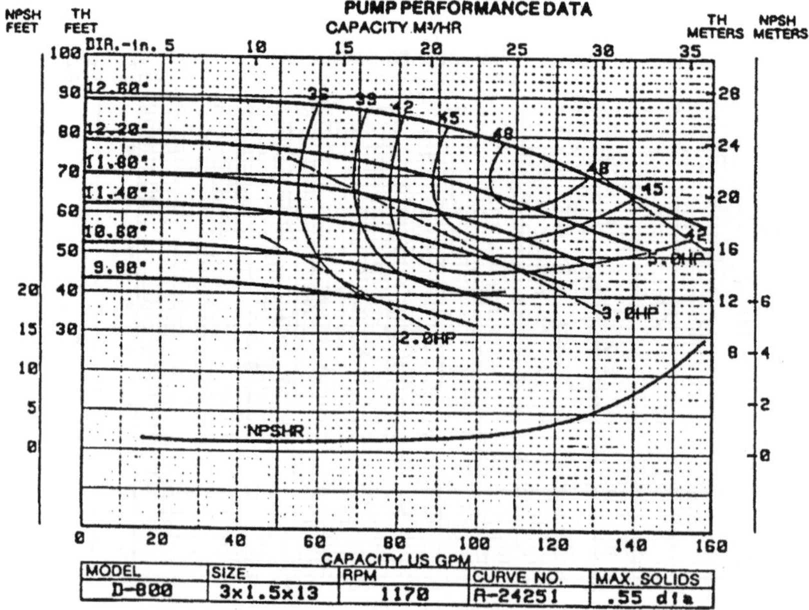

MODEL	SIZE	RPM	CURVE NO.	MAX. SOLIDS
D-800	3x1.5x13	1170	A-24251	.55 dia

FIGURE H-1 (*Continued*)

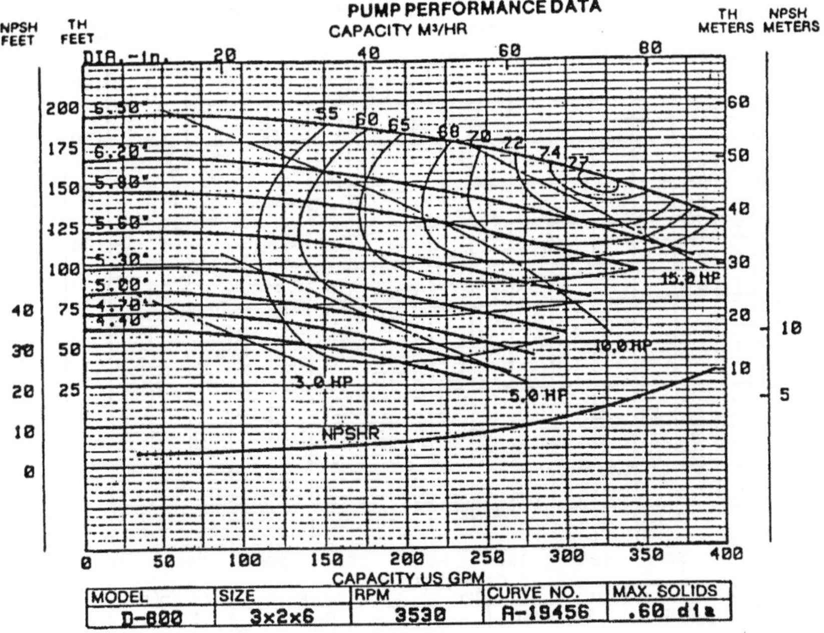
PUMP PERFORMANCE DATA
CAPACITY M³/HR
NPSH FEET
TH FEET
TH METERS
NPSH METERS
DIA.-in.
6.50"
6.20"
5.90"
5.60"
5.30"
5.00"
4.70"
4.40"
55
60
65
68
70
72
74
77
15.0 HP
10.0 HP
5.0 HP
3.0 HP
NPSHR
CAPACITY US GPM
MODEL
D-800
SIZE
3x2x6
RPM
3530
CURVE NO.
A-19456
MAX. SOLIDS
.60 dia

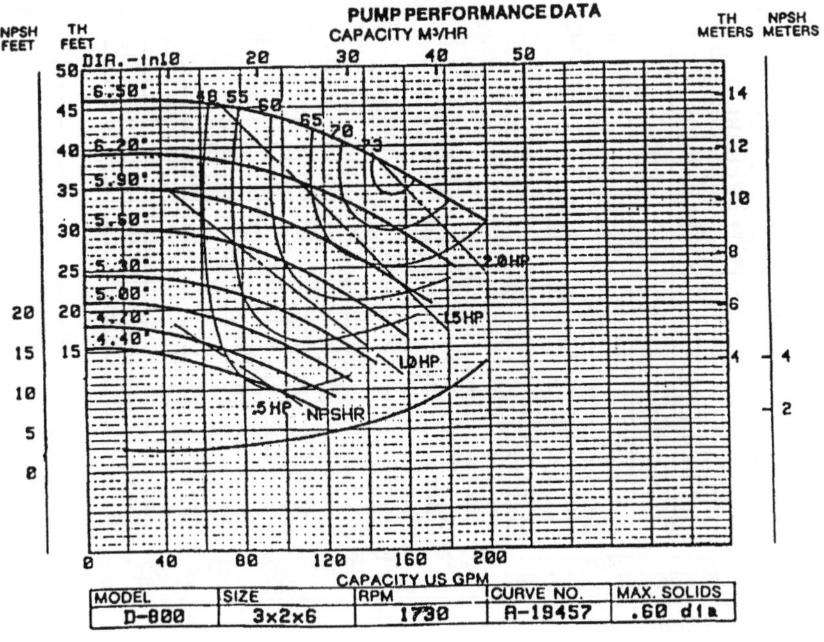
PUMP PERFORMANCE DATA
CAPACITY M³/HR
NPSH FEET
TH FEET
TH METERS
NPSH METERS
DIA.-in
6.50"
6.20"
5.90"
5.60"
5.30"
5.00"
4.70"
4.40"
40
55
60
65
70
73
2.0HP
1.5HP
1.0HP
.5HP
NPSHR
CAPACITY US GPM
MODEL
D-800
SIZE
3x2x6
RPM
1730
CURVE NO.
A-19457
MAX. SOLIDS
.60 dia

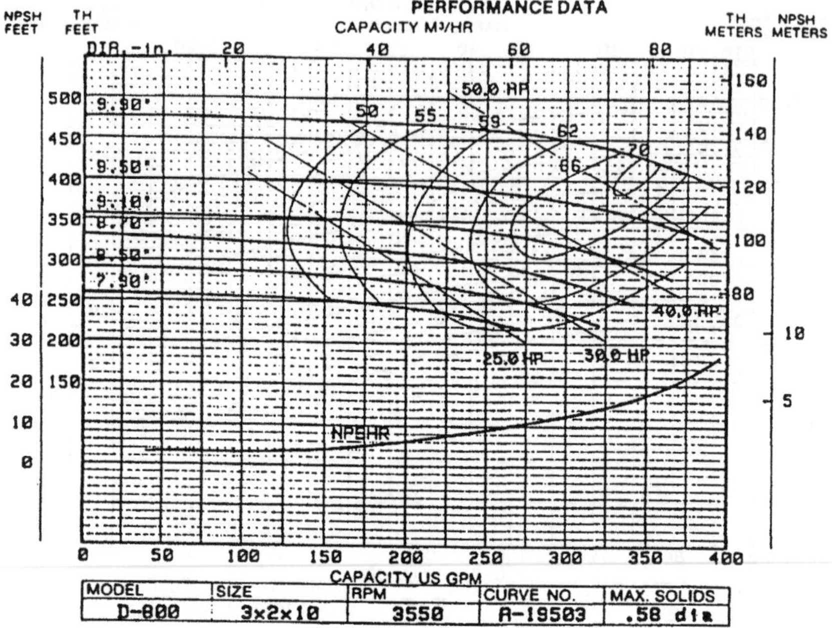

MODEL	SIZE	RPM	CURVE NO.	MAX. SOLIDS
D-800	3x2x10	3550	A-19503	.58 dia

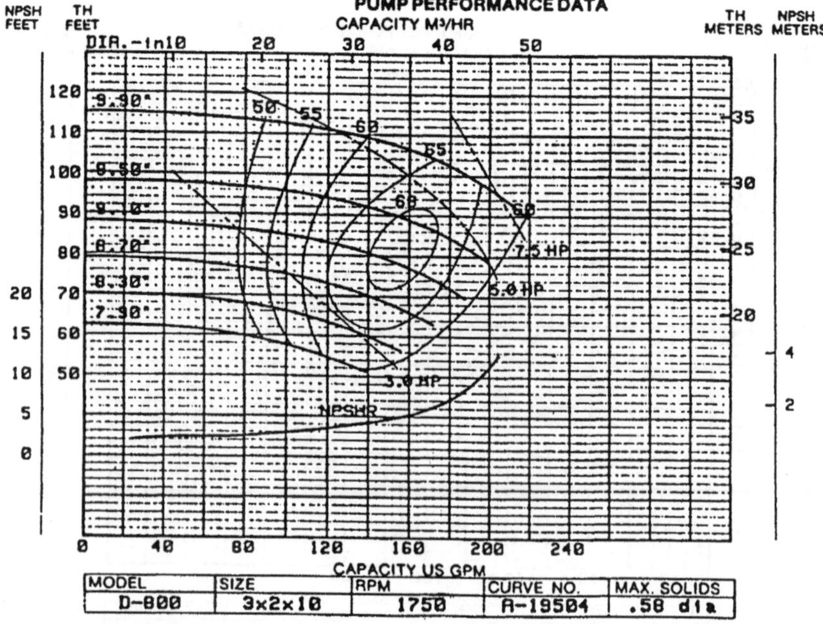

MODEL	SIZE	RPM	CURVE NO.	MAX. SOLIDS
D-800	3x2x10	1750	A-19504	.58 dia

FIGURE H-1 (*Continued*)

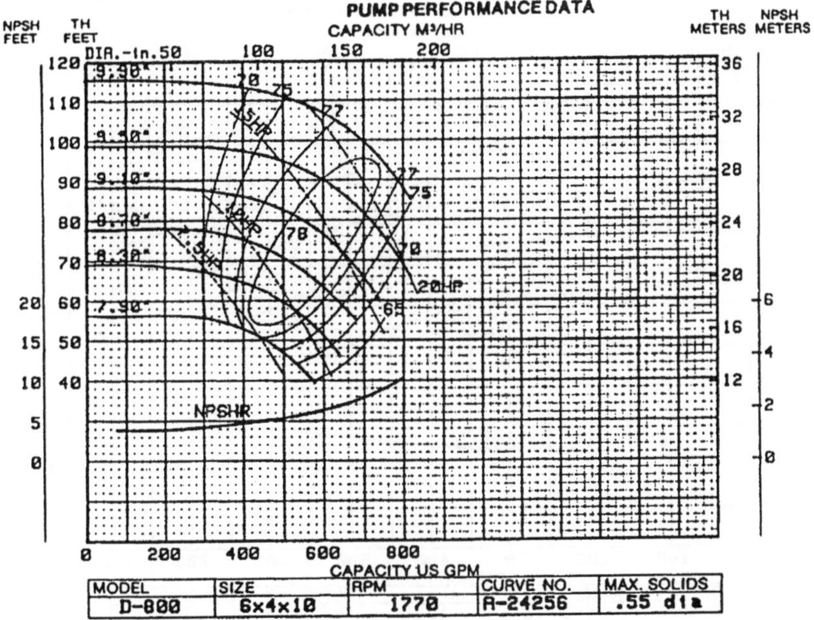
PUMP PERFORMANCE DATA
CAPACITY M³/HR
NPSH FEET
TH FEET
TH METERS
NPSH METERS
DIA.-in.
9.90"
9.50"
9.10"
8.70"
8.30"
7.90"
20HP
NPSHR
CAPACITY US GPM
MODEL D-800
SIZE 6x4x10
RPM 1770
CURVE NO. A-24256
MAX. SOLIDS .55 dia

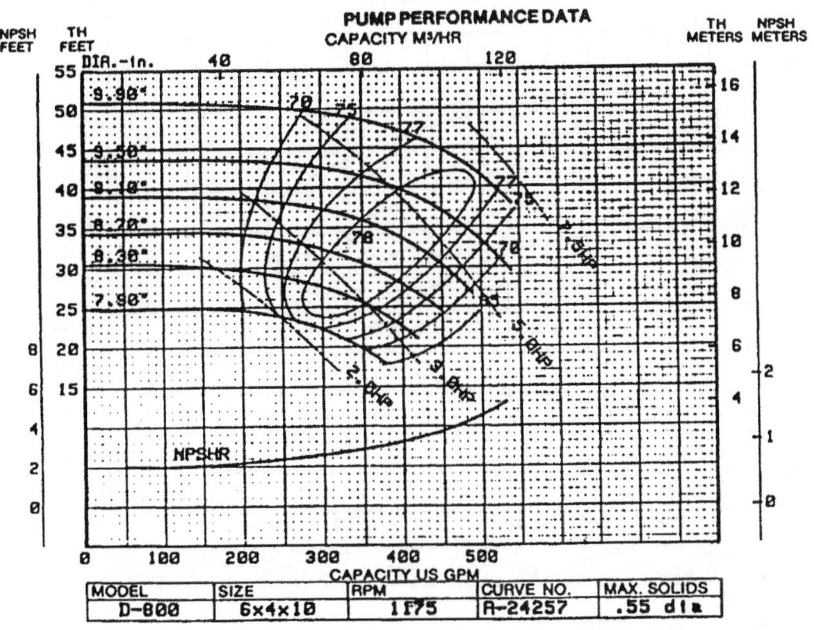
PUMP PERFORMANCE DATA
CAPACITY M³/HR
NPSH FEET
TH FEET
TH METERS
NPSH METERS
DIA.-in.
9.90"
9.50"
9.10"
8.70"
8.30"
7.90"
NPSHR
CAPACITY US GPM
MODEL D-800
SIZE 6x4x10
RPM 1175
CURVE NO. A-24257
MAX. SOLIDS .55 dia

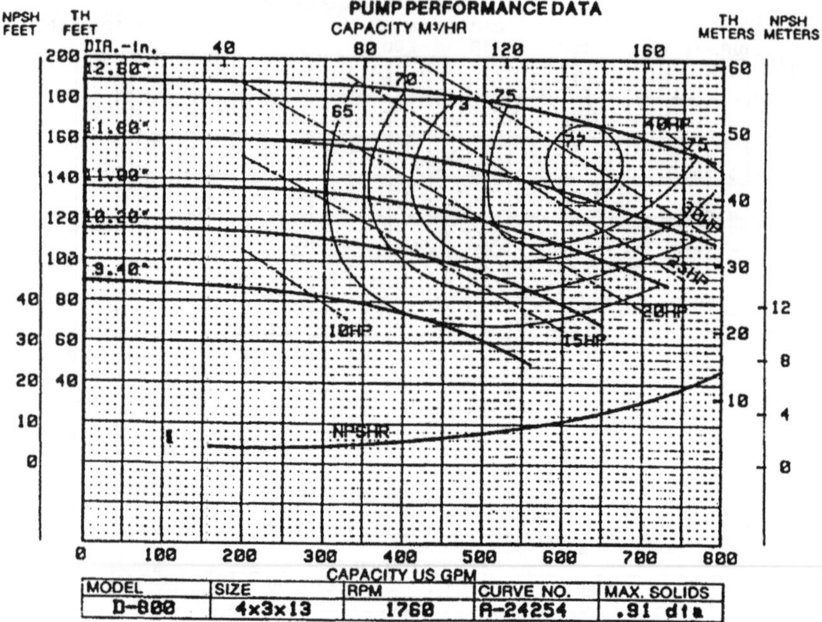

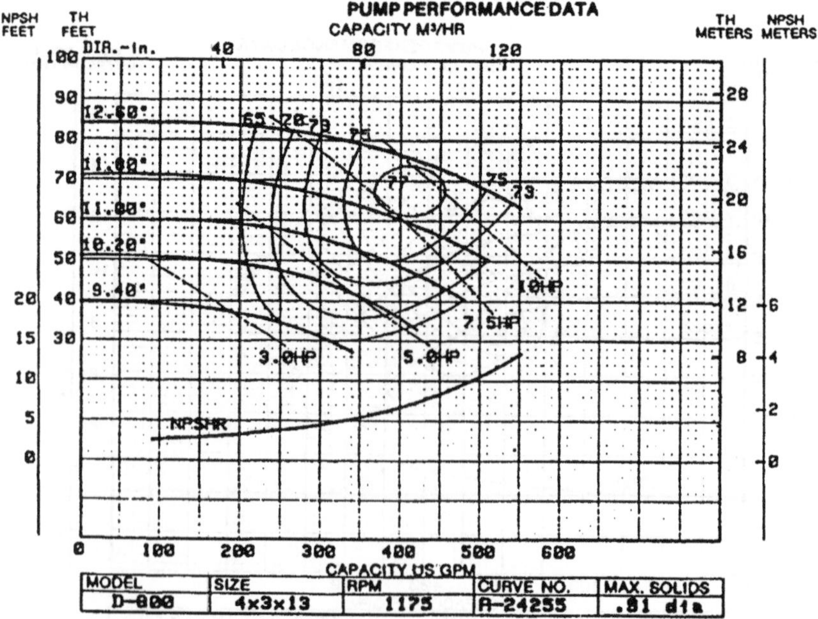

FIGURE H-1 (*Continued*)

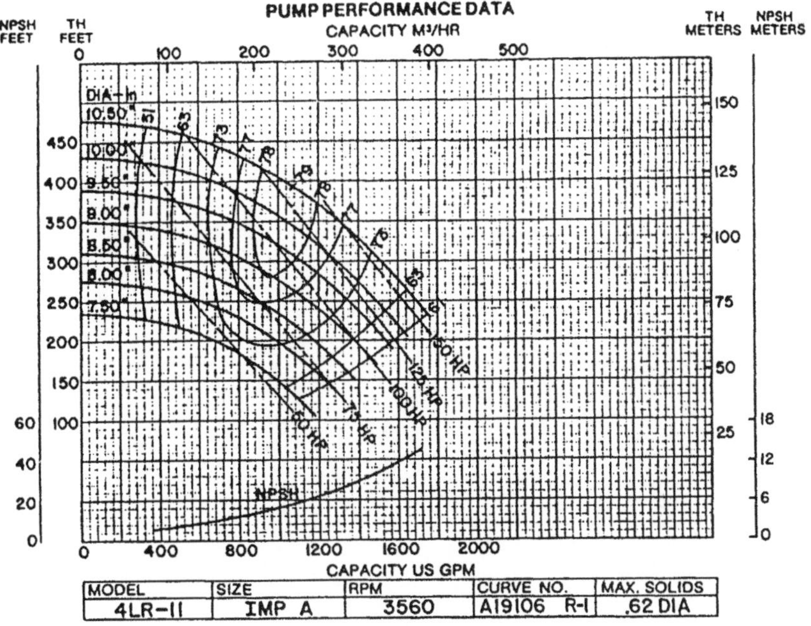

MODEL	SIZE	RPM	CURVE NO.	MAX. SOLIDS
4LR-11	IMP A	3560	A19106 R-1	.62 DIA

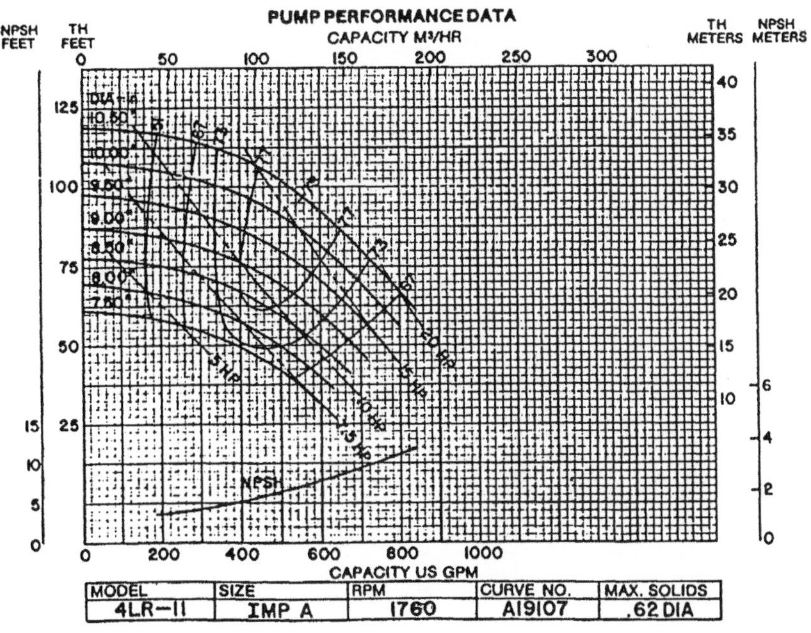

MODEL	SIZE	RPM	CURVE NO.	MAX. SOLIDS
4LR-11	IMP A	1760	A19107	.62 DIA

Appendix I

Fanno Line Tables for Adiabatic Flow of Air in a Constant Area Duct

TABLE I-1 Fanno Line—Adiabatic, Constant Area Flow ($k = 1.400$)

N_{Ma}	$\frac{T}{T*}$	$\frac{P}{P*}$	$\frac{P^0}{P^0*}$	$\frac{\tilde{V}}{\tilde{V}^*}$	$\frac{F}{F*}$	$\frac{4fL}{D}$
0	1.2000	∞	∞	0	∞	∞
0.01	1.2000	10, 9.544	5, 7.874	0.01095	4, 5.650	7, 134.40
0.02	1.1999	5, 4.770	2, 8.942	0.02191	22, 834	1, 778.45
0.03	1.1998	3, 6.511	1, 9.300	0.03286	15, 232	7, 87.08
0.04	1.1996	27, 382	14, 482	0.04381	11, 435	4, 40.35
0.05	1.1994	21, 903	11, 5914	0.05476	9, 1584	2, 80.02
0.06	1.1991	1 8, 251	9, 6659	0.06570	7.6, 428	19, 3.03
0.07	1.1988	15.6, 42	8.2, 915	0.07664	6.5, 620	14, 0.66
0.08	1.1985	13.6, 84	7.2, 616	0.08758	5.7, 529	10, 6.72
0.09	1.1981	12.1, 62	6.4, 614	0.09851	5.1, 249	8, 3.496
0.10	1.1976	10.9, 435	5.8, 218	0.10943	4.6, 236	66, 922
0.11	1.1971	9.9, 465	5.2, 992	0.12035	4.2, 146	54, 688
0.12	1.1966	9.1, 156	4.8, 643	0.13126	3.8, 747	45, 408
0.13	1.1960	8.4, 123	4.4, 968	0.14216	3.58, 80	38, 207
0.14	1.1953	7.8, 093	4.1, 824	0.15306	3.34, 32	32, 51 1
0.15	1.1946	7.2, 866	3.91, 03	0.16395	3.13, 17	27, 932
0.16	1.1939	6.82, 91	3.67, 27	0.17482	2.94, 74	24, 198
0.17	1.1931	6.42, 52	3.46, 35	0.18568	2.78, 55	21, 115
0.18	1.1923	6.06, 62	3.27, 79	0.19654	2.64, 22	18.5, 43
0.19	1.1914	5.74, 48	3.11, 23	0.20739	2.51, 46	16.3, 75
0.20	1.1905	5.45, 55	2.96, 35	0.21822	2.40, 04	14.5, 33
0.21	1.1895	5.19, 36	2.82, 93	0.22904	2.29, 76	12.9, 56
0.22	1.1885	4.95, 54	2.70, 76	0.23984	2.20, 46	11.5, 96
0.23	1.1874	4.73, 78	2.59, 68	0.25063	2.12, 03	10.4, 16
0.24	1.1863	4.53, 83	2.49, 56	0.26141	2.04, 34	9.3, 865
0.25	1.1852	4.35, 46	2.40, 27	0.27217	1.97, 32	8.4, 834
0.26	1.1840	4.18, 50	2.31, 73	0.28291	1.90, 88	7.6, 876
0.27	1.1828	4.02, 80	2.23, 85	0.29364	1.84, 96	6.9, 832
0.28	1.1815	3.88, 20	2.16, 56	0.30435	1.795, 0	6.3, 572
0.29	1.1802	3.74, 60	2.09, 79	0.31504	1.744, 6	5.7, 989
0.30	1.1788	3.61, 90	2.035, 1	0.32572	1.697, 9	5.2, 992
0.31	1.1774	3.50, 02	1.976, 5	0.33637	1.654, 6	4.8, 507
0.32	1.1759	3.38, 88	1.921, 9	0.34700	1.614, 4	4.44, 68
0.33	1.1744	3.28, 40	1.870, 8	0.35762	1.576, 9	4.08, 21
0.34	1.1729	3.18, 53	1.822, 9	0.36822	1.542, 0	3.75, 20

Table I-1 (*Continued*)

N_{Ma}	$\frac{T}{T*}$	$\frac{P}{P*}$	$\frac{P^0}{P^0*}$	$\frac{\tilde{V}}{\tilde{V}^*}$	$\frac{F}{F*}$	$\frac{4fL}{D}$
0.35	1.1713	3.09, 22	1.778, 0	0.37880	1.509, 4	3.45, 25
0.36	1.1697	3.004, 2	1.735, 8	0.38935	1.478, 9	3.18, 01
0.37	1.1680	2.920, 9	1.696, 1	0.39988	1.450, 3	2.93, 20
0.38	1.1663	2.842, 0	1.658, 7	0.41039	1.423, 6	2.70, 55
0.39	1.1646	2.767, 1	1.623, 4	0.42087	1.398, 5	2.49, 83
0.40	1.1628	2.695, 8	1.590, 1	0.43133	1.374, 9	2.30, 85
0.41	1.1610	2.628, 0	1.558, 7	0.44177	1.352, 7	2.13, 44
0.42	1.1591	2.563, 4	1.528, 9	0.45218	1.331, 8	1.97, 44
0.43	11572	2.501, 7	1.500, 7	0.46257	1.312, 2	1.82, 72
0.44	1.1553	2.442, 8	1.473, 9	0.47293	1.293, 7	1.69, 15
0.45	1.1533	2.386, 5	1.448, 6	0.48326	1.276, 3	1.56, 64
0.46	1.1513	2.332, 6	1.424, 6	0.49357	1.259, 8	1.45, 09
0.47	1.1492	2.280, 9	1.401, 8	0.50385	1.244, 3	1.34, 42
0.48	1.1471	2.231, 4	1.380, 1	0.51410	1.229, 6	1.24, 53
0.49	1.1450	2.183, 8	1.359, 5	0.52433	1.215, 8	1.15, 39
0.50	1.1429	2.138, 1	1.339, 9	0.53453	1.202, 7	1.06, 908
0.51	1.1407	2.094, 2	1.321, 2	0.54469	1.190, 3	0.99, 042
0.52	1.1384	2.051, 9	1.303, 4	0.55482	1.178, 6	0.91, 741
0.53	1.1362	2.011, 2	1.286, 4	0.56493	1.167, 5	0.84, 963
0.54	1.1339	1.971, 9	1.270, 2	0.57501	1, 157, 1	0.786, 62
0.55	1.1315	1.934, 1	1.254, 9	0.58506	1.147, 2	0.728, 05
0.56	1.1292	1.897, 6	1.240, 3	0.59507	1.137, 8	0.673, 57
0.57	1.1266	1.862, 3	1.226, 3	0.60505	1.128, 9	0.622, 86
0.58	1.1244	1.828, 2	1.213, 0	0.61500	1.120, 5	0.575, 68
0.59	1.1219	1.795, 2	1.200, 3	0.62492	1.112, 6	0.531, 74
0.60	1.1194	1.763, 4	1.188, 2	0.63481	1.1050, 4	0.490, 81
0.61	1.1169	1.732, 5	1.176, 6	0.64467	1.0979, 3	0.452, 70
0.62	1.1144	1.702, 6	1.165, 6	0.65449	1.0912, 0	0.417, 20
0.63	1.1118	1.673, 7	1.155, 1	0.66427	1.0848, 5	0.384, 11
0.64	1.1091	1.645, 6	1.145, 1	0.67402	1.0788, 3	0.353, 30
0.65	1.10650	1.618, 3	1.135, 6	0.68374	1.0731, 4	0.324, 60
0.66	1.10383	1.591, 9	1.126, 5	0.69342	1.0677, 7	0.297, 85
0.67	1.10114	1.566, 2	1.117, 9	0.70306	1.0627, 1	0.272, 95
0.68	1.09842	1.541, 3	1.109, 7	0.71267	1.0579, 2	0.249, 78
0.69	1.09567	1.517, 0	1.101, 8	0.72225	1.0534, 0	0.228, 21

Table I-1 (*Continued*)

N_{Ma}	$\frac{T}{T*}$	$\frac{P}{P*}$	$\frac{P^0}{P^0*}$	$\frac{\tilde{V}}{\tilde{V}^*}$	$\frac{F}{F*}$	$\frac{4fL}{D}$
0.70	1.09290	1.493, 4	1.0943, 6	0.73179	1.0491, 5	0.208, 14
0.71	1.09010	1.470, 5	1.0872, 9	0.74129	1.0451, 4	0.189, 49
0.72	1.08727	1.448, 2	1.0805, 7	0.75076	1.0413, 7	0.172, 15
0.73	1.08442	1.426, 5	1.0741, 9	0.76019	1.0378, 3	0.156, 06
0.74	1.08155	1.405, 4	1.0681, 5	0.76958	1.0345, 0	0.141, 13
0.75	1.07856	1.384, 8	1.0624, 2	0.77893	1.0313, 7	0.127, 28
0.76	1.07573	1.364, 7	1.0570, 0	0.78825	1.0284, 4	0.114, 46
0.77	1.07279	1.345, 1	1.0518, 8	0.79753	1.0257, 0	0.102, 62
0.78	1.06982	1.326, 0	1.0470, 5	0.80677	1.0231, 4	0.091, 67
0.79	1.06684	1.3074	1.0425, 0	0.81598	1.0207, 5	0.081, 59
0.80	1.06383	1.2892	1.0382, 3	0.82514	1.0185, 3	0.072, 29
0.81	1.06080	1.2715	1.0342, 2	0.83426	1.0164, 6	0.063, 75
0.82	1.05775	1.2542	1.0304, 7	0.84334	1.0145, 5	0.055, 93
0.83	1.05468	1.2373	1.0269, 6	0.85239	1.0127, 8	0.048, 78
0.84	1.05160	1.2208	1.0237, 0	0.86140	1.0111 5	0.042, 26
0.85	1.04849	1.2047	1.0206, 7	0.87037	1.0096, 6	0.036, 32
0.86	1.04537	1.1889	1.0178, 7	0.87929	1.0082, 9	0.030, 97
0.87	1.04223	1.1735	1.0152, 9	0.88818	1.0070, 4	0.026, 13
0.88	1.03907	1.1584	1.0129, 4	0.89703	1.0059, 1	0.021, 80
0.89	1.03589	1.1436	1.0108, 0	0.90583	1.0049, 0	0.017, 93
0.90	1.03270	1.1291, 3	1.0088, 7	0.91459	1.0039, 9	0.0145, 13
0.91	1.02950	1.1150, 0	1.0071, 4	0.92332	1.0031, 8	0.0115, 19
0.92	1.02627	1.1011, 4	1.0056, 0	0.93201	1.0024, 8	0.0089, 16
0.93	1.02304	1.0875, 8	1.0042, 6	0.94065	1.0018, 8	0.0066, 94
0.94	1.01978	1.0743, 0	1.0031, 1	0.94925	1.0013, 6	0.0048, 15
0.95	1.01652	1.0612, 9	1.0021, 5	0.95782	1.0009, 3	0.0032, 80
0.96	1.01324	1.0485, 4	1.0013, 7	0.96634	1.0005, 9	0.0020, 56
0.97	1.00995	1.0360, 5	1.0007, 6	0.97481	1.0003, 3	0.0011, 35
0.98	1.00664	1.0237, 9	1.0003, 3	0.98324	1.0001, 4	0.0004, 93
0.99	1.00333	1.0117, 8	1.0000, 8	0.99164	1.0000, 3	0.0001, 20
1.00	1.00000	1.0000, 0	1.0000, 0	1.00000	1, 0000, 0	0
1.01	0.99666	0.9884, 4	1.0000, 8	1.00831	1.0000, 3	0.0001, 14
1.02	0.99331	0.9771, 1	1.0003, 3	1.01658	1.0001, 3	0.0004, 58
1.03	0.98995	0.9659, 8	1.0007, 3	1.02481	1.0003, 0	0.0010, 13
1.04	0.98658	0.9550, 6	1.0013, 0	1.03300	1.0005, 3	0.0017, 71

TABLE I-1 (*Continued*)

N_{Ma}	$\frac{T}{T*}$	$\frac{P}{P*}$	$\frac{P^0}{P^0*}$	$\frac{\tilde{V}}{\tilde{V}^*}$	$\frac{F}{F*}$	$\frac{4fL}{D}$
1.05	0.98320	0.9443, 5	1.0020, 3	1.04115	1.0008, 2	0.0027, 12
1.06	0.97982	0.9338, 3	1.0029, 1	1.04925	1.0011, 6	0.0038, 37
1.07	0.97642	0.9235, 0	1.0039, 4	1.05731	1.0015, 5	0.0051, 29
1.08	0.97302	0.9133, 5	1.0051, 2	1.06533	1.0020, 0	0.0065, 82
1.09	0.96960	0.9033, 8	1.0064, 5	1.07331	1.0025, 0	0.0081, 85
1.10	0.96618	0.8935, 9	1.0079, 3	1.08124	1.00305	0.0099, 33
1.11	0.96276	0.8839, 7	1.0095, 5	1.08913	1.00365	0.0118, 13
1.12	0.95933	0.8745, 1	1.0113, 1	1.09698	1.00429	0.0138, 24
1.13	0.95589	0.8652, 2	1.0132, 2	1.10479	1.00497	0.0159, 49
1.14	0.95244	0.8560, 8	1.0152, 7	1.11256	1.00569	0.0181, 87
1.15	0.94899	0.8471, 0	1.0174, 6	1.1203	1.00646	0.0205, 3
1.16	0.94554	0.8382, 7	1.0197, 8	1.1280	1.00726	0.0229, 8
1.17	0.94208	0.8295, 8	1.0222, 4	1.1356	1.00810	0.0255, 2
1.18	0.93862	0.8210, 4	1.0248, 4	1.1432	1.00897	0.0281, 4
1.19	0.93515	0.8126, 3	1.0275, 7	1.1508	1.00988	0.0308, 5
1.20	0.93168	0.8043, 6	1.0304, 4	1.1583	1.01082	0.0336, 4
1.21	0.92820	0.7962, 3	1.0334, 4	1.1658	1.01178	0.0365, 0
1.22	0.92473	0.7882, 2	1.0365, 7	1.1732	1.01278	0.394, 2
1.23	0.92125	0.7803, 4	1.0398, 3	1.1806	1.01381	0.0424, 1
1.24	0.91777	0.7725, 8	1.0432, 3	1.1879	1.01486	0.0454, 7
1.25	0.91429	0.7649, 5	1.0467, 6	1.1952	1.01594	0.04858
1.26	0.91080	0.7574, 3	1.0504, 1	1.2025	1.01705	0.05174
1.27	0.90732	0.7500, 3	1.0541, 9	1.2097	1.01818	0.05494
1.28	0.90383	0.7427, 4	1.0580, 9	1.2169	1.01933	0.05820
1.29	0.90035	0.7355, 6	1.0621, 3	1.2240	1.02050	0.06150
1.30	0.89686	0.7284, 8	1.0663, 0	1.2311	1.02169	0.06483
1.31	0.89338	0.7215, 2	1.0706, 0	1.2382	1.02291	0.06820
1.32	0.88989	0.7146, 5	1.0750, 2	1.2452	1.02415	0.07161
1.33	0.88641	0.7078, 9	1.0795, 7	1.2522	1.02540	0.07504
1.34	0.88292	0.7012, 3	1.0842, 4	1.2591	1.02666	0.07850
1.35	0.87944	0.6946, 6	1.0890, 4	1.2660	1.02794	0.08199
1.36	0.87596	0.6881, 8	1.0939, 7	1.2729	1.02924	0.08550
1.37	0.87249	0.6818, 0	1.0990, 2	1.2797	1.03056	0.08904
1.38	0.86901	0.6755, 1	1.1041, 9	1.2864	1.03189	0.09259
1.39	0.06554	0.6693, 1	1.1094, 8	1.2932	1.03323	0.09616

Table I-1 (*Continued*)

N_{Ma}	$\frac{T}{T*}$	$\frac{P}{P*}$	$\frac{P^0}{P^0*}$	$\frac{\tilde{V}}{\tilde{V}^*}$	$\frac{F}{F*}$	$\frac{4fL}{D}$
1.40	0.86207	0.6632, 0	1.1149	1.2999	1.03458	0.09974
1.41	0.85860	0.6571, 7	1.1205	1.3065	1.03595	0.10333
1.42	0.85514	0.6512, 2	1.1262	1.3131	1.03733	0.10694
1.43	0.85168	0.6453, 6	1.1320	1.3197	1.03872	0.11056
1.44	0.84822	0.6395, 8	1.1379	1.3262	1.04012	0.11419
1.45	0.84477	0.6338, 7	1.1440	1.3327	1.04153	0.11782
1.46	0.84133	0.6282, 4	1.1502	1.3392	1.04295	0.12146
1.47	0.83788	0.6226, 9	1.1565	1.3456	1.04438	0.12510
1.48	0.83445	0.6172, 2	1.1629	1.3520	1.04581	0.12875
1.49	0.83101	0.6118, 1	1.1695	1.3583	1.04725	0.13240
1.50	0.82759	0.6064, 8	1.1762	1.3646	1.04870	0.13605
1.51	0.82416	0.6012, 2	1.1830	1.3708	1.05016	0.13970
1.52	0.82075	0.5960, 2	1.1899	1.3770	1.05162	0.14335
1.53	0.81734	0.5908, 9	1.1970	1.3832	1.05309	0.14699
1.54	0.81394	0.5858, 3	1.2043	1.3894	1.05456	0.15063
1.55	0.81054	0.5808, 4	1.2116	1.3955	1.05604	0.15427
1.56	0.80715	0.5759, 1	1.2190	1.4015	1.05752	0.15790
1.57	0.80376	0.5710, 4	1.2266	1.4075	1.05900	0.16152
1.58	0.83038	0.5662, 3	1.2343	1.4135	1.06049	0.16514
1.59	0.79701	0.5614, 8	1.2422	1.4195	1.06198	0.16876
1.60	0.79365	0.5567, 9	1.2502	1.4254	1.06348	0.17236
1.61	0.79030	0.5521, 6	1.2583	1.4313	1.06498	0.17595
1.62	0.78695	0.5475, 9	1.2666	1.4371	1.06648	0.17953
1.63	0.78361	0.5430, 8	1.2750	1.4429	1.06798	0.18311
1.64	0.78028	0.5386, 2	1.2835	1.4487	1.06948	0.18667
1.65	0.77695	0.5342, 1	1.2922	1.4544	1.07098	0.19022
1.66	0.77363	0.5298, 6	1.3010	1.4601	1.07249	0.19376
1.67	0.77033	0.5255, 6	1.3099	1.4657	1.07399	0.19729
1.68	0.76703	0.5213, 1	1.3190	1.4713	1.07550	0.20081
1.69	0.76374	0.5171, 1	1.3282	1.4769	1.07701	0.20431
1.70	0.76046	0.5129, 7	1.3376	1.4825	1.07851	0.20780
1.71	0.75718	0.5088, 7	1.3471	1.4880	1.08002	0.21128
1.72	0.75392	0.5048, 2	1.3567	1.4935	1.08152	0.21474
1.73	0.75067	0.5008, 2	1.3665	1.4989	1.08302	0.21819
1.74	0.74742	0.4968, 6	1.3764	1.5043	1.08453	0.22162

TABLE I-1 (*Continued*)

N_{Ma}	$\frac{T}{T*}$	$\frac{P}{P*}$	$\frac{P^0}{P^0*}$	$\frac{\tilde{V}}{\tilde{V}*}$	$\frac{F}{F*}$	$\frac{4fL}{D}$
1.75	0.74419	0.4929, 5	1.3865	1.5097	1.08603	0.22504
1.76	0.74096	0.4890, 9	1.3967	1.5150	1.08753	0.22844
1.77	0.73774	0.4852, 7	1.4070	1.5203	1.08903	0.23183
1.78	0.73453	0.4814, 9	1.4175	1.5256	1.09053	0.23520
1.79	0.73134	0.4777, 6	1.4282	1.5308	1.09202	0.23855
1.80	0.72816	0.47407	1.4390	1.5360	1.09352	0.24189
1.81	0.72498	0.47042	1.4499	1.5412	1.09500	0.24521
1.82	0.72181	0.46681	1.4610	1.5463	1.09649	0.24851
1.83	0.71865	0.46324	1.4723	1.5514	1.09798	0.25180
1.84	0.71551	0.45972	1.4837	1.5564	1.00946	0.25507
1.85	0.71238	0.45623	1.4952	1.5614	1.1009	0.25832
1.86	0.70925	0.45278	1.5069	1.5664	1.1024	0.26156
1.87	0.70614	0.49937	1.5188	1.5714	1.1039	0.26478
1.88	0.70304	0.44600	1.5308	1.5763	1.1054	0.26798
1.89	0.69995	0.44266	1.5429	1.5812	1.1068	0.27116
1.90	0.69686	0.43936	1.5552	1.5861	1.1083	0.27433
1.91	0.69379	0.43610	1.5677	1.5909	1.1097	0.27748
1.92	0.69074	0.43287	1.5804	1.5957	1.1112	0.28061
1.93	0.68769	0.42967	1.5932	1.6005	1.1126	0.28372
1.94	0.68465	0.42651	1.6062	1.6052	1.1141	0.28681
1.95	0.68162	0.42339	1.6193	1.6099	1.1155	0.28989
1.96	0.67861	0.42030	1.6326	1.6146	1.1170	0.29295
1.97	0.67561	0.41724	1.6461	1.6193	1.1184	0.29599
1.98	0.67262	0.41421	1.6597	1.6239	1.1198	0.29901
1.99	0.66964	0.41121	1.6735	1.6824	1.1213	0.30201
2.00	0.66667	0.40825	1.6875	1.6330	1.1227	0.30499
2.01	0.66371	0.40532	1.7017	1.6375	1.1241	0.30796
2.02	0.66076	0.40241	1.7160	1.6420	1.1255	0.31091
2.03	0.65783	0.39954	1.7305	1.6465	1.1269	0.31384
2.04	0.65491	0.39670	1.7452	1.6509	1.1283	0.31675
2.05	0.65200	0.39389	1.7600	1.6553	1.1297	0.31965
2.06	0.64910	0.39110	1.7750	1.6597	1.1311	0.32253
2.07	0.64621	0.38834	1.7902	1.6640	1.1325	0.32538
2.08	0.64333	0.38562	1.8056	1.6683	1.1339	0.32822
2.09	0.64047	0.38292	1.8212	1.6726	1.1352	0.33104

TABLE I-1 (*Continued*)

N_{Ma}	$\frac{T}{T*}$	$\frac{P}{P*}$	$\frac{P^0}{P^0*}$	$\frac{\tilde{V}}{\tilde{V}^*}$	$\frac{F}{F*}$	$\frac{4fL}{D}$
2.10	0.63762	0.38024	1.8369	1.6769	1.1366	0.33385
2.11	0.63478	0.37760	1.8528	1.6811	1.1380	0.33664
2.12	0.63195	0.37498	1.8690	1.6853	1.1393	0.33940
2.13	0.62914	0.37239	1.8853	1.6895	1.1407	0.34215
2.14	0.62633	0.36982	1.9018	1.6936	1.1420	0.34488
2.15	0.62354	0.36728	1.9185	1.6977	1.1434	0.34760
2.16	0.62076	0.36476	1.9354	1.7018	1.1447	0.35030
2.17	0.61799	0.36227	1.9525	1.7059	1.1460	0.35298
2.18	0.61523	0.35980	1.9698	1.7099	1.1474	0.35564
2.19	0.61249	0.35736	1.9873	1.7139	1.1487	0.35828
2.20	0.60976	0.35494	2.0050	1.7179	1.1500	0.36091
2.21	0.60704	0.35254	2.0228	1.7219	1.1513	0.36352
2.22	0.60433	0.35017	2.0409	1.7258	1.1526	0.36611
2.23	0.60163	0.34782	2.0592	1.7297	1.1539	0.36868
2.24	0.59895	0.34550	2.0777	1.7336	1.1552	0.37124
2.25	0.59627	0.34319	2.0964	1.7374	1.1565	0.37378
2.26	0.59361	0.34091	2.1154	1.7412	1.1578	0.37630
2.27	0.59096	0.33865	2.1345	1.7450	1.1590	0.37881
2.28	0.58833	0.33641	2.1538	1.7488	1.1603	0.38130
2.29	0.58570	0.33420	2.1733	1.7526	1.1616	0.38377
2.30	0.58309	0.33200	2.1931	1.7563	1.1629	0.38623
2.31	0.58049	0.32983	2.2131	1.7600	1.1641	0.38867
2.32	0.57790	0.32767	2.2333	1.7637	1.1653	0.39109
2.33	0.57532	0.32554.	2.2537	1.7673	1.1666	0.39350
2.34	0.57276	0.32342	2.2744	1.7709	1.1678	0.39589
2.35	0.57021	0.32133	2.2953	1.7745	1.1690	0.39826
2.36	0.56767	0.31925	2.3164	1.7781	1.1703	0.40062
2.37	0.56514	0.31720	2.3377	1.7817	1.1715	0.40296
2.38	0.56262	0.31516	2.3593	1.7852	1.1727	0.40528
2.39	0.56011	0.31314	2.3811	1.7887	1.1739	0.40760
2.40	0.55762	0.31114	2.4031	1.7922	1.1751	0.40989
2.41	0.55514	0.30916	2.4254	1.7956	1.1763	0.41216
2.42	0.55267	0.30720	2.4479	1.7991	1.1775	0.41442
2.43	0.55021	0.30525	2.4706	1.8025	1.1786	0.41667
2.44	0.54776	0.30332	2.4936	1.8059	1.1798	0.41891

TABLE I-1 (*Continued*)

N_{Ma}	$\frac{T}{T*}$	$\frac{P}{P*}$	$\frac{P^0}{P^0*}$	$\frac{\tilde{V}}{\tilde{V}^*}$	$\frac{F}{F*}$	$\frac{4fL}{D}$
2.45	0.54533	0.30141	2.5168	1.8092	1.1810	0.42113
2.46	0.54291	0.29952	2.5403	1.8126	1.1821	0.42333
2.47	0.54050	0.29765	2.5640	1.8159	1.1833	0.42551
2.48	0.53810	0.29579	2.5880	1.8192	1.1844	0.42768
2.49	0.53571	0.29395	2.6122	1.8225	1.1856	0.42983
2.50	0.53333	0.29212	2.6367	1.8257	1.1867	0.43197
2.51	0.53097	0.29031	2.6615	1.8290	1.1879	0.43410
2.52	0.52862	0.28852	2.6865	1.8322	1.1890	0.43621
2.53	0.52627	0.28674	2.7117	1.8354	1.1901	0.43831
2.54	0.52394	0.28498	2.7372	1.8386	1.1912	0.44040
2.55	0.52163	0.28323	2.7630	1.8417	1.1923	0.44247
2.56	0.51932	0.28150	2.7891	1.8448	1.1934	0.44452
2.57	0.51702	0.27978	2.8154	1.8479	1.1945	0.44655
2.58	0.51474	0.27808	2.8420	1.8510	1.1956	0.44857
2.59	0.51247	0.27640	2.8689	1.8541	1.1967	0.45059
2.60	0.51020	0.27473	2.8960	1.8571	1.1978	0.45259
2.61	0.50795	0.27307	2.9234	1.8602	1.1989	0.45457
2.62	0.50571	0.27143	2.9511	1.8632	1.2000	0.45654
2.63	0.50349	0.26980	2.9791	1.8662	1.2011	0.45850
2.64	0.50127	0.26818	3.0074	1.8691	1.2021	0.46044
2.65	0.49906	0.26658	3.0359	1.8721	1.2031	0.46237
2.66	0.49687	0.26499	3.0647	1.8750	1.2042	0.46429
2.67	0.49469	0.26342	3.0938	1.8779	1.2052	0.46619
2.68	0.49251	0.26186	3.1234	1.8808	1.2062	0.46807
2.69	0.49035	0.26032	3.1530	1.8837	1.2073	0.46996
2.70	0.48820	0.25878	3.1830	1.8865	1.2083	0.47182
2.71	0.48606	0.25726	3.2133	1.8894	1.2093	0.47367
2.72	0.48393	0.25575	3.2440	1.8922	1.2103	0.47551
2.73	0.48182	0.25426	3.2749	1.8950	1.2113	0.47734
2.74	0.47971	0.25278	3.3061	1.8978	1.2123	0.47915
2.75	0.47761	0.25131	3.3376	1.9005	1.2133	0.48095
2.76	0.47553	0.24985	3.3695	1.9032	1.2143	0.48274
2.77	0.47346	0.24840	3.4017	1.9060	1.2153	0.48452
2.78	0.47139	0.24697	3.4342	1.9087	1.2163	0.48628
2.79	0.46933	0.24555	3.4670	1.9114	1.2173	0.48803

Table I-1 (*Continued*)

N_{Ma}	$\frac{T}{T*}$	$\frac{P}{P*}$	$\frac{P^0}{P^0*}$	$\frac{\tilde{V}}{\tilde{V}*}$	$\frac{F}{F*}$	$\frac{4fL}{D}$
2.80	0.46729	0.24414	3.5001	1.9140	1.2182	0.48976
2.81	0.46526	0.24274	3.5336	1.9167	1.2192	0.49148
2.82	0.46324	0.24135	3.5674	1.9193	1.2202	0.49321
2.83	0.46122	0.23997	3.6015	1.9220	1.2211	0.49491
2.84	0.45922	0.23861	3.6359	1.9246	1.2221	0.49660
2.85	0.45723	0.23726	3.6707	1.9271	1.2230	0.49828
2.86	0.45525	0.23592	3.7058	1.9297	1.2240	0.49995
2.87	0.45328	0.23458	3.7413	1.9322	1.2249	0.50161
2.88	0.45132	0.23326	3.7771	1.9348	1.2258	0.50326
2.89	0.44937	0.23196	3.8133	1.9373	1.2268	0.50489
2.90	0.44743	0.23066	3.8498	1.9398	1.2277	0.50651
2.91	0.44550	0.22937	3.8866	1.9423	1.2286	0.50812
2.92	0.44358	0.22809	3.9238	1.9448	1.2295	0.50973
2.93	0.44167	0.22682	3.9614	1.9472	1.2304	0.51133
2.94	0.43977	0.22556	3.9993	1.9497	1.2313	0.51291
2.95	0.43788	0.22431	4.0376	1.9521	1.2322	0.51447
2.96	0.43600	0.22307	4.0763	1.9545	1.2331	0.51603
2.97	0.43413	0.22185	4.1153	1.9569	1.2340	0.51758
2.98	0.43226	0.22063	4.1547	1.9592	1.2348	0.51912
2.99	0.43041	0.21942	4.1944	1.9616	1.2357	0.52064
3.00	0.42857	0.21822	4.1346	1.9640	1.2366	0.52216
3.50	0.34783	0.16850	6.7896	2.0642	1.2743	0.58643
4.00	0.28571	0.13363	10.719	2.1381	1.3029	0.63306
4.50	0.23762	0.10833	16.562	2.1936	1.3247	0.66764
5.00	0.20000	0.08944	25.000	2.2361	1.3416	0.69381
6.00	0.14634	0.06376	53.180	2.2953	1.3655	0.72987
7.00	0.11111	0.04762	104.14	2.3333	1.3810	0.75281
8.00	0.08696	0.03686	190.11	2.3591	1.3915	0.76820
9.00	0.06977	0.02935	327.19	2.3772	1.3989	0.77898
10.00	0.05714	0.02390	535.94	2.3905	1.4044	0.78683
∞	0	0	∞	2.4495	1.4289	0.82153

Index

An environmentally friendly book printed and bound in England by www.printondemand-worldwide.com

This book is made of chain-of-custody materials; FSC materials for the cover and PEFC materials for the text pages.

#0062 - 160616 - C0 - 229/152/31 [33] - CB - 9780824704445